Analytical Geomicrobiology

A Handbook of Instrumental Techniques

Geomicrobiology is the study of microbes and microbial processes and their role in driving environmental and geological processes ranging from the nano, to the micron, to the meter scale. This growing field has seen major advances in recent years, largely due to the development of new analytical tools and improvements to existing techniques, which allow us to better understand the complex interactions between microbes and their surroundings.

In this comprehensive handbook, expert authors outline the state-of-the-art and emerging analytical techniques used in geomicrobiology. Readers are guided through each technique, including background theory, sample preparation, standard methodology, data collection and analysis, best practices and common pitfalls, and examples of how and where the technique has been applied. The book provides a practical go-to reference for advanced students, researchers, and professional scientists looking to employ techniques commonly used in geomicrobiology.

Janice P. L. Kenney is an assistant professor at MacEwan University, and a member of the Geochemical Society. She is interested in how contaminants, such as radionuclides, are transported in the environment, and she employs techniques in geochemistry and geomicrobiology to better understand the fate of these elements.

Harish Veeramani is a research facilitator at Carleton University, and previously worked as a lecturer in water engineering at the University of Glasgow. He specializes in applied environmental microbiology for the remediation of heavy metal contaminants, including radionuclides.

Daniel S. Alessi is an associate professor and the Encana Chair in Water Resources at the University of Alberta. His research focuses on environmental geomicrobiology and geochemistry, and the role of microbes and minerals in controlling the transport and fate of metals in near-surface environments.

Analytical Geomicrobiology

A Handbook of Instrumental Techniques

Edited by

JANICE P. L. KENNEY
MacEwan University

HARISH VEERAMANI
Carleton University

DANIEL S. ALESSI
University of Alberta

CAMBRIDGE
UNIVERSITY PRESS

University Printing House, Cambridge CB2 8BS, United Kingdom

One Liberty Plaza, 20th Floor, New York, NY 10006, USA

477 Williamstown Road, Port Melbourne, VIC 3207, Australia

314–321, 3rd Floor, Plot 3, Splendor Forum, Jasola District Centre, New Delhi – 110025, India

79 Anson Road, #06–04/06, Singapore 079906

Cambridge University Press is part of the University of Cambridge.

It furthers the University's mission by disseminating knowledge in the pursuit of education, learning, and research at the highest international levels of excellence.

www.cambridge.org
Information on this title: www.cambridge.org/9781107070332
DOI: 10.1017/9781107707399

First published 2019

Printed in the United Kingdom by TJ International Ltd. Padstow Cornwall

A catalogue record for this publication is available from the British Library.

ISBN 978-1-107-07033-2 Hardback

Additional resources for this publication at www.cambridge.org/analyticalgeomicrobiology

Contents

List of Contributors — *page* vii
Foreword — xi
Kurt O. Konhauser

Part I. Standard Techniques in Geomicrobiology — 1

1. **General Geochemistry and Microbiology Techniques** — 3
   Sarrah M. Dunham-Cheatham and Yaqi You

Part II. Advanced Analytical Instrumentation — 61

2. **The Application of Isothermal Titration Calorimetry for Investigating Proton and Metal Interactions on Microbial Surfaces** — 63
   Drew Gorman-Lewis

3. **Potentiometric Titrations to Characterize the Reactivity of Geomicrobial Surfaces** — 79
   Daniel S. Alessi, Shannon L. Flynn, Md. Samrat Alam, Leslie J. Robbins, and Kurt O. Konhauser

4. **Use of Multi-collector ICP-MS for Studying Biogeochemical Metal Cycling** — 93
   Kai Liu, Lingling Wu, and Sherry L. Schiff

Part III. Imaging Techniques — 119

5. **Scanning Probe Microscopy** — 121
   Adam F. Wallace

6. **Applications of Scanning Electron Microscopy in Geomicrobiology** — 148
   Jeremiah Shuster, Gordon Southam, and Frank Reith

7. **Applications of Transmission Electron Microscopy in Geomicrobiology** — 166
   Jeremiah Shuster, Frank Reith, and Gordon Southam

8. **Whole Cell Identification of Microorganisms in Their Natural Environment with Fluorescence *in situ* Hybridization (FISH)** 187
Natuschka M. Lee

Part IV. Spectroscopy 213

9. **X-ray Diffraction Techniques** 215
Daniel K. Unruh and Tori Z. Forbes

10. **Application of Synchrotron X-ray Absorption Spectroscopy and Microscopy Techniques to the Study of Biogeochemical Processes** 238
Maxim I. Boyanov and Kenneth M. Kemner

11. **Bacterial Surfaces in Geochemistry – How Can X-ray Photoelectron Spectroscopy Help?** 262
Madeleine Ramstedt, Laura Leone, and Andrey Shchukarev

12. **Applications of Fourier-transform Infrared Spectroscopy in Geomicrobiology** 288
Janice P. L. Kenney and András Gorzsás

13. **Mössbauer Spectroscopy** 314
James M. Byrne and Andreas Kappler

Part V. Microbiological Techniques 339

14. **Lipid Biomarkers in Geomicrobiology: Analytical Techniques and Applications** 341
Jiasong Fang, Shamik Dasgupta, Li Zhang, and Weiqiang Zhao

15. **Phylogenetic Techniques in Geomicrobiology** 360
Denise M. Akob, Adam C. Mumford, Darren S. Dunlap, and Amisha T. Poret-Peterson

Index 405

Contributors

Denise M. Akob, U.S. Geological Survey, National Research Program, 12201 Sunrise Valley Dr., MS 430, Reston, VA 20192, USA

Md. Samrat Alam, Department of Physical & Environmental Sciences, University of Toronto Scarborough, 1265 Military Trail, Toronto, ON, M1 C 1A4, Canada

Daniel S. Alessi, Department of Earth and Atmospheric Sciences, 1–26 Earth Sciences Building, University of Alberta, Edmonton, AB, T6 G 2E3, Canada

Maxim I. Boyanov, Institute of Chemical Engineering, Bulgarian Academy of Sciences, Sofia, Bulgaria and Biosciences Division, Argonne National Laboratory, Lemont, IL, USA

James M. Byrne, Center for Applied Geoscience (ZAG), Eberhard-Karls-University Tuebingen, Sigwartstrasse 10, 72076 Tuebingen, Germany

Shamik Dasgupta, Deep Sea Science Division, Institute of Deep-Sea Science and Engineering, Chinese Academy of Sciences, Sanya, 572000, Hainan, China

Sarrah M. Dunham-Cheatham, Department of Natural Resources and Environmental Science, University of Nevada, Reno, 89557, USA

Darren S. Dunlap, The Boeing Company, Boeing Research & Technologies, Huntsville, AL 35824, USA

Jiasong Fang, Hadal Science and Technology Research Center, Shanghai Ocean University, 999 Huchenghuan Road, Shanghai 201306, China; Laboratory for Marine Biology and Biotechnology, Qingdao National Laboratory for Marine Science and Technology, Qingdao, China; College of Natural and Computational Sciences, Hawaii Pacific University, Honolulu, HI 96813, USA

Shannon L. Flynn, School of Natural and Environmental Sciences, Newcastle University, Newcastle-upon-Tyne, NE1 7RU, UK

Tori Z. Forbes, Department of Chemistry, University of Iowa, Iowa City, IA 52242, USA

Drew Gorman-Lewis, Department of Earth and Space Sciences, University of Washington, Seattle, WA 98195, USA

András Gorzsás, Department of Chemistry, Umeå University, Umeå, Sweden

Andreas Kappler, Center for Applied Geoscience (ZAG), Eberhard-Karls-University Tuebingen, Sigwartstrasse 10, 72076 Tuebingen, Germany

Kenneth M. Kemner, Biosciences Division, Argonne National Laboratory, Lemont, IL, USA

Janice P. L. Kenney, Department of Physical Sciences, MacEwan University, Edmonton, Alberta, Canada.

Kurt O. Konhauser, Department of Earth and Atmospheric Sciences, 1–26 Earth Sciences Building, University of Alberta, Edmonton, AB, T6 G 2E3, Canada

Natuschka M. Lee, Department of Ecology and Environmental Science, Umeå University, S-901 87 Umeå, Sweden; Biochemical Imaging Centre, at the Department of Medical Biochemistry and Biophysics, Umeå University, S-901 87 Umeå, Sweden

Laura Leone, Department of Chemistry, Umeå University, Sweden

Kai Liu, Department of Earth and Environmental Sciences, University of Waterloo, Waterloo, Ontario, Canada N2 L 3G1; Water Institute, University of Waterloo, Waterloo, Ontario, Canada N2 L 3G1

Adam C. Mumford, U.S. Geological Survey, National Research Program, 12201 Sunrise Valley Dr., MS 430, Reston, VA 20192, USA

Amisha T. Poret-Peterson, U. S. Department of Agriculture, Agricultural Research Service, Davis, CA 95616, USA

Madeleine Ramstedt, Department of Chemistry, Umeå University, Sweden

Frank Reith, School of Biological Sciences, The University of Adelaide, Adelaide, South Australia 5005, Australia; Commonwealth Scientific and Industrial Research Organisation (CSIRO) Land and Water, Contaminant Biogeochemistry and Environmental Toxicology, PMB2 Glen Osmond, South Australia 5064, Australia

Leslie J. Robbins, Department of Earth and Atmospheric Sciences, 1–26 Earth Sciences Building, University of Alberta, Edmonton, AB, T6 G 2E3, Canada

Sherry L. Schiff, Department of Earth and Environmental Sciences, University of Waterloo, Waterloo, Ontario, Canada N2 L 3G1; Water Institute, University of Waterloo, Waterloo, Ontario, Canada N2 L 3G2

Andrey Shchukarev, Department of Chemistry, Umeå University, Sweden

Jeremiah Shuster, School of Biological Sciences, The University of Adelaide, Adelaide, South Australia 5005, Australia; CSIRO Land and Water, Contaminant Biogeochemistry and Environmental Toxicology, PMB2 Glen Osmond, South Australia 5064, Australia

Gordon Southam, School of Earth & Environmental Sciences, The University of Queensland, St. Lucia, Queensland 4072, Australia

Daniel K. Unruh, Department of Chemistry and Biochemistry, Texas Tech University, Lubbock, TX 79409, USA

Adam F. Wallace, Department of Geological Sciences, University of Delaware, Newark, DE 19716, USA

Lingling Wu, Department of Earth and Environmental Sciences, University of Waterloo, Waterloo, Ontario, Canada N2 L 3G1; Water Institute, University of Waterloo, Waterloo, Ontario, Canada N2 L 3G1

Yaqi You, Department of Civil and Environmental Engineering, University of Nevada, Reno, 89557, USA

Li Zhang, Faculty of Earth Sciences, China University of Geosciences, Wuhan, Hubei 740034, China

Weiqiang Zhao, Hadal Science and Technology Research Center, Shanghai Ocean University, 999 Huchenghuan Road, Shanghai 201306, China

Foreword

As a discipline, geomicrobiology's origins can be traced to the microbiologist Lourens Baas-Becking's 1934 book *Geobiology*, in which he argued that although microorganisms exist everywhere, the environment selects which species dominate any given habitat. In the decades since, geomicrobiology has transformed into a scientific discipline that covers all aspects of how the biosphere shapes, and is shaped by, our planet's surface environments and deep-Earth processes alike. Ignited by technological advances in stable isotope and organic geochemistry, electron microscopy, synchrotron radiation, and other surface probing techniques, as well as molecular biology, studies in geomicrobiology are providing new insights into how interacting biological and physical processes have influenced environments, both locally and globally, across our planet's entire 4.5-billion-year history.

A key topic within geomicrobiology includes life's control over elemental cycling, from the weathering and dissolution of rock, to the assimilation of diverse bioessential nutrients necessary for all forms of life, to the diagenetic transformations taking place during sediment burial. These processes cover a vast range of spatial scales, from micron-sized niches to reservoirs as immense as the oceans, and temporal scales from seconds to billions of years. The central theme running through all this research is the recognition that life shapes the environment to the same degree that environmental change drives the spatial and temporal distribution of life. This co-evolution of life and its environment, specifically investigations of the cause-and-effect relationships and associated feedbacks, is the defining quality of geomicrobiology. Indeed, the more we learn about how life interacts with the planet, the more we realize that it is the feedbacks and drivers between the two that are the key agents of change. For example, one can ask: How did microbial genetic innovation in the past lead to biological reactions that modified Earth's surface? Environmental change, through processes such as tectonics, in turn influences biology, which can create new opportunities for evolutionary innovation – sometimes in unexpected ways. In this regard, geomicrobiologists are at the forefront of studies that strive for a more complete understanding of the Earth as a dynamic, interrelated system, to know our origins, to predict our future, and to explore for life beyond our planet and solar system.

One of the greatest challenges in any emerging field is the exchange of relevant information – specifically methodologies – among different subdisciplines. In part, this problem reflects historical gaps among the disciplines that contribute to geomicrobiology, and specifically the highly specialized training that each subgroup receives. Although expertise in any given subdiscipline in geomicrobiology requires specialized training, it is the cross-fertilization among these seemingly disparate fields that defines geomicrobiology – and highlights the need for a book that integrates the tools being utilized by our community. *Analytical Geomicrobiology: A Handbook of Instrumental Techniques* does just this. It presents a summary of the advanced instrumentation used today to determine how organics, metals, and/or minerals interact with bacterial surfaces and their environment. It is designed to give researchers a broad knowledge of the available analytical techniques and will provide for the first time a way for researchers to standardize their methodologies.

The chapters begin with an introduction of the standard methods used to study bacteria and their secondary mineralization products, including commonly used techniques in general microbiology and geochemistry. Following are two chapters on isothermal titration calorimetry and potentiometric titrations, tools that are key to characterizing the surface chemistry and thermodynamics of metal binding to bacterial cell envelopes, microbial exudates, and biogenic minerals. Chapter 4 covers the use of multi-collector inductively coupled plasma-mass spectrometry (ICP-MS) to measure stable isotopes, a technique of increasing importance in geomicrobiology used to understand the biotic and abiotic cycling of metals in marine and terrestrial environments. Chapters 5–8 focus on imaging techniques, including atomic force microscopy, scanning tunneling microscopy, transmission and scanning electron microscopy, and various fluorescence *in situ* hybridization (FISH) techniques. All of these techniques are commonly used to visualize materials at the micro to nano scale, and the authors of these chapters have put particular effort into providing information specific to sample preparation for geomicrobiological samples. Spectroscopic techniques comprise the next five chapters, including benchtop techniques such as infrared, Mössbauer, and X-ray photoelectron spectroscopies, the use of X-ray diffraction techniques to study biogenic minerals, and a chapter on the use of synchrotron-based X-ray absorption spectroscopy. Finally, Chapters 14 and 15 discuss advanced microbiological tools to study microbial communities and their functions, including signature lipid biomarkers, and provide a survey of phylogenetic techniques available to the geomicrobiology community.

The layout of the chapters makes this book an easy reference tool for both experienced and novice researchers. In each chapter, the authors first provide a succinct and insightful introduction to the instrumental technique at hand, along with key references for further reading. Following this, an applied example of the technique is discussed in detail. This allows the reader to immediately appreciate the sampling methods, tools, materials, and expertise required to use the technique, and to decide whether the instrumental method is appropriate to answer a particular scientific question. While studies in peer-reviewed publications often use the advanced tools discussed in this book to determine how metals, organics, and minerals interact with microbes, those journal publications necessarily provide little information about how to use the technique, or for that matter, why it was even chosen. This book, written by worldwide experts in geomicrobiology, provides an entry point for novice and experienced scientists to select the most appropriate instrumental techniques to study their system. Perhaps most importantly, the text brings together in one place analytical tools used in the many subgroups of geomicrobiology. The aim of the book is simple: to promote more collaboration and integration among scientists in our field by informing them of the analytical possibilities.

Kurt O. Konhauser, FRSC
University of Alberta

PART I

STANDARD TECHNIQUES IN GEOMICROBIOLOGY

1 General Geochemistry and Microbiology Techniques

SARRAH M. DUNHAM-CHEATHAM AND YAQI YOU

Abstract
Geomicrobiological investigations benefit from knowledge of geochemical and biological systems at different scales, including information about both the abiotic and the biotic components. Gathering this information requires analysis and characterization of both abiotic and biotic components of the target system. The techniques presented in this chapter were selected to cover a variety of needs in geomicrobiological studies, including general sample collection and storage, organic and inorganic compound quantification, and best practices for cultivation, observation, and analysis of microorganisms and microbial communities. In this chapter, introductions and discussions for common techniques provide the reader with a basic understanding of the technique itself, which samples can be analyzed using the technique, and how to prepare samples for analysis. Detailed methods are provided for select techniques, and citations to standard methods are provided for techniques whenever available. For techniques that are rapidly evolving, recent developments and applications are discussed.

1.1 Field Sampling and Sample Collection, Preservation, and Storage

Field samples are important resources to answer many geomicrobiological research questions. They can provide information on target systems under real conditions that laboratory-synthesized samples cannot. Before samples can be collected from the field, a sampling design must be established. For design planning, care must be taken to prevent introduction of bias, misrepresentation, and insufficient data in the final dataset. If, for example, a research question is aimed at understanding the relationship between rhizosphere microbial communities and tree roots in a forest ecosystem, collecting samples exclusively from the rhizosphere of oak trees might bias the results. In another example, when trying to understand the soil carbon (C) content of a 100 acre agricultural field, it would be a poor sampling design to collect all replicate samples from the area within a square half acre, because data from one specific area does not accurately and/or sufficiently represent the entire field; likewise, collecting one replicate sample from each edge of the field would also be a poor sampling design, because edge effects (e.g., transition from agricultural field to nonagricultural field) may be captured in the samples, and the results may not accurately represent the conditions in the agricultural field. A thorough discussion of sampling strategies for field sample collection is presented by Thompson (2012).

The following sections discuss common strategies and methods for sample collection, preservation, and storage for geochemical and microbiological analyses. Special considerations for each analysis type are examined.

1.1.1 Samples for Geochemical Analyses

For geochemical analyses, once a sampling design is established, the next step is to decide *how* the samples will be collected. To decide how to collect samples, many variables must be considered: (1) Is the sample solid or liquid? (2) What analyte(s) will be measured? (3) Is the analyte sensitive to light? (4) Is the analyte sensitive to atmospheric gases (e.g., O_2, CO_2)? and (5) When is the best time to sample? Liquid samples are commonly collected in high-density polyethylene (HDPE) or glass bottles that have been cleaned (e.g., acid-washed to remove contaminant metals, combusted to remove residual carbon) prior to sample collection. If the target analyte is known to react with or sorb to either of these materials, alternative materials (e.g., Teflon) should be used for sample collection. Solid samples (e.g., soils, rocks, and minerals) can be collected in plastic storage bags, paper bags, or geological sample bags, including plastic-lined bags; if C is a target analyte, paper bags and unlined geological sample bags should not be used. If an analyte reacts with light, the sample can be collected in a UV-blocking or opaque container (e.g., amber glass, brown HDPE, plastic bag placed inside larger opaque container, or aluminum foil-wrapped container). For analytes that react with atmospheric gases, care must be taken to prevent interaction with and exchange of atmospheric gases into the sample during and after collection. Glass sample containers fitted with airtight lids provide the best protection from potential interactions and exchanges. Further, whenever possible, samples should be sealed with no ambient headspace in the container to avoid interactions with atmospheric gases (e.g., CO_2, H_2, or S-bearing gases) that might contaminate the sample; small bottles of inert gas (e.g., Ar or N_2) can be brought to the field to purge headspace in containers for sensitive samples. Another consideration for sample collection is when to sample; some systems have natural cycles (e.g., diurnal or seasonal), which can contribute to sampling errors if not considered carefully.

Ideally, samples are collected and analyzed immediately to most accurately capture the information present at the time of sample collection, though this is rarely possible. The next best option is to analyze the samples as soon as possible. When this is not possible either, preservation methods may be used to stabilize analytes by preventing physical, chemical, and/or biological reactions that would otherwise alter those analytes between the time of sample collection and analysis. Common preservation methods for aquatic samples include sterilization, reduction in temperature, filtration, and acidification. Sterilization inactivates biological processes and prevents alterations to analytes, especially analytes that are nutrients for organisms (e.g., C-, N-, and S-bearing compounds). It can be achieved with radiation, though this process is expensive and has the potential to alter some analytes. Alternatively, a reduction in biological activity can be achieved with a reduction in temperature (≤ 4 °C). Caution should be used when reducing the temperature of samples where the analyte may form mineral precipitates under lower-temperature conditions. Filtration removes suspended particles from the sample, which can either sorb or desorb analytes and affect their concentrations, and allows the determination of the "dissolved" fraction of analytes; typically, 0.45 μm filters are used, but 0.2 μm filters may be used to remove smaller particles. Filters and filtration methods should be carefully chosen to avoid significant analyte loss during filtration (Batley, 1989; Horowitz et al., 1992). Acidification reduces interferences from atmospheric CO_2 and stabilizes dissolved metals, preventing their sorption to the container material and/or precipitation from solution. Concentrated acid (e.g., HCl) is added to bring the pH of the sample down to 1.5 ± 0.5. Table 1 in Uzoukwu (2000) provides a detailed list of parameters and suitable container materials and preservation methods for common geochemical analytes in aquatic samples. For solid samples, the most common preservation method is drying. Moist solid samples are susceptible to continued physical, chemical, and biological reactions, which are minimized upon drying. Most analytes are not significantly affected by drying, with the notable exception of those that are easily oxidized under ambient atmospheric conditions (e.g., Fe(II)) (Bates, 1993; Tan, 1996). Drying should be achieved at room temperature using forced air flow (e.g., a fan or exhaust hood) to hasten the process; oven drying should be avoided if possible, and an oven above 35 °C should never be used. Depending on the analyte and the analytical technique (and potential interferences), preservation of samples may not be required. However, in all cases, methods that alter, destroy, or contaminate analytes should be avoided entirely.

Storage time between sample collection and analysis should be minimized as much as possible. For some analytes (e.g., pH, Fe(II), Mn, nitrate, organics, and nutrients), time is of the essence, and despite best efforts to preserve the sample, analyte alterations will occur over time. Other analytes (e.g., total metal content), however, may be stable for long periods of time. As a rule of thumb, solid samples can be stored longer than water samples, though exceptions exist in both cases. The method of storage (e.g., room temperature, chilled, or frozen) will depend on the analyte(s) of interest; if unsure, default to chilled (~4 °C) but reduce fluctuations in temperature over time (i.e., either keep chilled or keep unchilled). Methods for long-term storage of samples should prevent contamination to and alteration of the sample, and eliminate temperature and moisture fluctuations as much as possible.

1.1.2 Samples for Microbiological Analyses

In addition to the considerations discussed for geochemical analyses (see Section 1.1.1), special considerations must be made when collecting samples for microbiological analyses. The first consideration is how best to prevent contamination of the sample with microbial cells and biomolecules from the sample collector(s), collection gear and tools, and surrounding environment (e.g., surfaces and air). Conditions for collecting and handling samples in the field are not as ideal as working in a controlled laboratory environment, and achieving and maintaining an aseptic working environment in the field is a constant challenge. Working carefully and frequently (re)sterilizing surfaces are common approaches to minimizing potential contamination risks. To minimize contamination risks from sample collectors, collectors should wear sterile gloves during sample collection and frequently sterilize their gloves with either 70% ethanol or isopropyl alcohol in water. If multiple samples are being collected, gloves should be thoroughly rinsed and sterilized, or preferably changed, between samples to prevent cross-contamination of samples. Collectors should also be careful to avoid breathing directly on the sample and not to touch the sample with any part of their body, including hair and clothing that has not been sterilized. It should be noted that ribonuclease (RNase) activity is abundant in bodily fluids, such as perspiration, skin oils, and saliva. Therefore, ungloved hands can introduce RNase contamination that compromises sample accuracy. To minimize contamination risks from collection gear and tools, all surfaces that will contact the sample should be thoroughly and frequently sterilized. Field gear can be sterilized with 70% ethanol or isopropyl alcohol, or by heat sterilization using a small, portable torch. Tools and gear that cannot withstand these treatments, or those with scratches or porous surfaces, should not be used to collect samples for microbiological analyses, as they are difficult to sterilize and increase risks of contamination. Alternatively, manufacturer-sterilized disposable tools can be used. When working with sensitive, valuable paleoecological samples (e.g., ancient glacial ice and permafrost sediments), every precaution should be employed to prevent contamination of the sample (e.g., wearing sterile caps and facemasks in addition to sterile gloves, and spiking drilling equipment with recognizable microorganisms to identify contamination). Microbes and/or their biomolecules are ubiquitous in all environments, resulting in an extremely high risk of contamination of paleoecological samples by contemporary microbial signatures. Even trace amounts of contaminant contemporary microbes or their biomolecules can cause inaccurate microbiological analysis results, especially when nonspecific cultivation or molecular amplification is used. A detailed discussion of precautions, controls, and criteria for reducing the risk of contamination of ice and permafrost samples, both in the field and in the laboratory, has been given by Willerslev et al. (2004).

The second consideration that must be made when collecting samples for microbiological analyses is selecting the appropriate container to store and transport the sample once it is collected. Commonly used containers include manufacturer-sterilized disposable sampling bags (e.g., Whirlpak brand bags), as well as sterilized vials, tubes, and bottles, which are rigid and thus provide structural protection for samples that are sensitive to compaction during transport. These container options are inexpensive and available in multiple sizes or materials to accommodate different sample volumes and analysis requirements. Special attention should be paid when the sample will be analyzed for biomolecules such as nucleic acids and proteins. If nucleic acids are to be analyzed, field samples should be collected in sterile, nuclease-free, nonpyrogenic containers for best performance. If peptides and proteins are to be analyzed, adsorption of the target biomolecules to the container surface should be considered, as these amphiphilic biomolecules can easily sorb to most surfaces. Choosing the optimal container helps avoid inaccurate measurements and false conclusions due to unpredictable peptide/protein loss (Goebel-Stengel et al., 2011). Other containers can also be used, but above all else, the selected container must be sterile prior to use to prevent contamination of the sample by the container. In addition to sterilization, labware used to collect samples for nucleic acid analyses should be pretreated to eliminate nucleic acids and nucleases for best performance. Labware can either be heated (e.g., 250+ °C for 2+ hours) or treated with commercially available surface decontaminants that can eliminate nucleic acids and nucleases.

The third consideration when collecting field samples for microbiological analyses is to maintain the sample under appropriate conditions both in the field and during transport. Due to the sensitivity of microbes to environmental changes and the subsequent shifts in cellular activity and community composition and structure, field samples should ideally be processed and analyzed immediately upon collection. If immediate processing is not feasible, samples should be preserved immediately upon collection. If field samples are to be analyzed using culture-dependent methods (e.g., cultivation and isolation), it is imperative to maintain the microbial community in the original sample as close to native as possible. This is difficult to achieve, as living organisms respond to small changes in the surrounding environment (e.g., temperature, humidity, and O_2 level) at the molecular, cellular, and population levels. Chilling samples can slow microbial metabolism and changes to the community, but will not prevent changes to the community. Usually, samples are kept on ice (~4 °C) in a cooler or at room temperature during collection and transport, and are processed for culture-dependent analyses as soon as feasibly possible to minimize potential changes and biasing of analysis results. If field samples are to be analyzed using culture-independent analyses (e.g., nucleic acid and protein analyses), samples are commonly frozen in dry ice or liquid nitrogen as quickly as possible and kept in the dark in the field to avoid degradation of biomolecules by enzymes such as nucleases and proteases, as well as by hydrolysis, oxidation, UV exposure, and other physicochemical processes. This is especially critical for samples that will be analyzed for ribonucleic acid (RNA); RNA is more susceptible than deoxyribonucleic acid (DNA) to degradation, because RNases are ubiquitous in the environment and the ribose sugar in RNA contains a 2′-hydroxyl group, which acts as a nucleophile in reactions with nucleases. Additionally, samples can be transferred to specific RNA storage and stabilization solutions (e.g., commercially available RNAlater solutions) before freezing to minimize degradation. When dry ice or liquid nitrogen is unavailable, samples should be kept on ice in the dark and transported to the laboratory as quickly as possible, although some degradation is expected to occur under these conditions. While there are commercially available reagents designed to simultaneously stabilize nucleic acids and proteins in tissue samples at room temperature, those reagents may or may not be suitable for environmental samples, given the complicated physical and biochemical characteristics of many environmental samples. The field-frozen samples should be transferred to a −20 °C freezer, or preferably a −80 °C freezer, upon arrival at the laboratory and should remain frozen until sample processing begins. Freeze-thaw cycles must be avoided as much as possible, as biomolecules will rapidly degrade during repeated freeze-thaw cycles.

1.1.3 Samples for Both Geochemical and Microbiological Analyses

Collecting field samples that will be used for both geochemical and microbiological analyses requires some preplanning. For example, considerations need to be made regarding whether strategies and methods for sample collection, preservation, and storage for one type of analyses will also accommodate the other type of analyses. Ideally, the same field sample would be used for both types of analyses, but extracting multiple aliquots from the same sample for multiple analyses has the potential to introduce contamination into the sample, especially if the tools used to extract the aliquots are not sterile or do not meet the needs of specific analyses (e.g., using tools containing nucleases to prepare aliquots for nucleic acid analyses). If all requirements for both geochemical and microbiological analyses have been met, once in the laboratory, a small aliquot of the sample can be archived appropriately and used strictly for microbiological analyses to minimize contamination, and the remaining sample can be subsampled for geochemical analyses. However, in cases where the preservation and/or storage methods required for geochemical and microbiological analyses differ (e.g., room temperature vs. frozen, acidified vs. non-acidified, wet vs. dried), this approach is not feasible. Instead, replicate field samples should be collected and each replicate sample preserved and stored according to the requirements for the analyses to be performed on that replicate sample.

1.2 Geochemical Techniques

Investigating the large variety of materials and analytes in geochemical studies requires an equally large variety of techniques, from simple benchtop wet chemistry to extremely sophisticated approaches, and most often a combination of multiple approaches. In this section, basic approaches to geochemical investigations are presented. Considerations for collecting, preserving, and storing samples are discussed, followed by techniques applied for quantification of inorganic and organic compounds.

1.2.1 Inorganic Compounds

Geochemical studies rely heavily on the analysis of inorganic compounds to elucidate information and trends from samples under investigation. In this section, common analytical techniques used to quantify and characterize inorganic components are presented, including spectrophotometric and atomic spectroscopic techniques. Technique overviews, analysis considerations and limitations, sample preparation requirements, and analysis-specific data processing procedures are covered. For common analytes, step-by-step methods are also provided.

1.2.1.1 Spectrophotometric Techniques

Spectrophotometric techniques allow users to quantify the concentration of an analyte by measuring the absorbance of energy (i.e., light) by the analyte or the analyte after reaction with a photosensitive compound. The principles of spectrophotometric techniques are based on adaptations to the Beer–Lambert Law, which states that the absorbance is linearly proportional to the concentration of the absorbing species and the pathlength of the sample (Kaur, 2007; Tan, 1996). Because it is difficult to measure absorbance directly, most techniques directly measure transmittance and calculate the corresponding absorbance. Today, capabilities exist to perform spectrophotometric analyses in the ultraviolet, visible, and infrared regions of the electromagnetic spectrum. Infrared spectroscopy is discussed in more detail in Chapter 12.

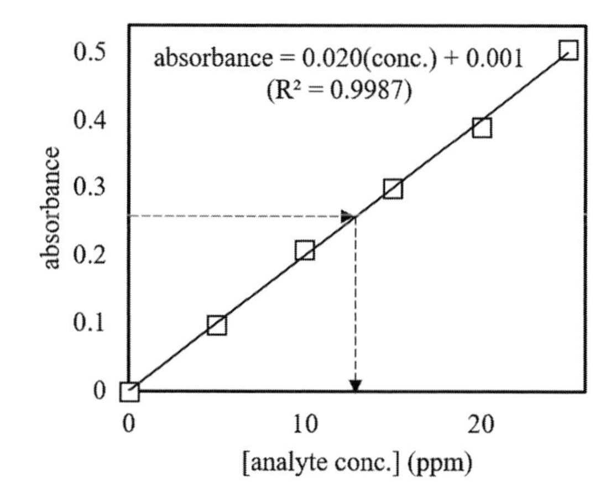

Figure 1.1 Example calibration curve for an analyte ranging in concentration from 0 to 25 ppm. The resulting calibration equation is included for illustration. The arrows indicate the calculation performed to determine the analyte concentration using the absorbance value from the sample and the calibration curve. For some techniques, absorbance will be replaced by the concentration-dependent metric (e.g., intensity, peak height, peak area) specific to the technique.

This section introduces colorimetric assay and ion chromatography (IC) methods, which are common spectrophotometric techniques used in analyzing geochemical samples. Colorimetric assays quantify concentrations of a single analyte using a reagent to produce a reagent–analyte complex, which absorbs light at a characteristic wavelength. Generally speaking, these assays are selective for the target analyte and are not sensitive to complex sample matrices. However, potential interfering compounds for each assay are discussed later. IC quantifies multiple analytes from a single sample by passing the mixture through sorbent-packed columns that separate the analytes based on chemical properties of both the sorbent and the analytes. For both colorimetric assay and IC analyses, analyte concentrations are calibrated by treating a set of standards using the same method used to treat samples. The standards are analyzed, and the resulting calibration curve is used to calculate the concentration of the analyte in a sample (Figure 1.1). The r^2 values for all analyses included in this section should be ≥ 0.99; if the resulting r^2 value is lower, create a fresh set of standards and start over.

1.2.1.1.1 Colorimetric Assay – Iron

Due to its prevalence in geomedia and biological tissues and its importance in redox cycles, iron (Fe) is an element of common interest in geochemical investigations. In some cases, researchers are interested in understanding the redox cycling of Fe and want to measure both the oxidized ferric (Fe(III)) and reduced ferrous (Fe(II)) valences in their sample. In other cases, researchers are interested in quantifying Fe extracted from Fe minerals. Common Fe mineral extraction methods include the dithionite-citrate-bicarbonate (DCB) method for extracting long- and short-range order Fe minerals (Jackson, 2005; Mehra and

Jackson, 1960; Pansu and Gautheyrou, 2006), the HCl method for extracting short-range order Fe minerals (Lovley and Phillips, 1986), and the Tamm's reagent method for extracting colloidal Fe (Blakemore, 1968). Whatever the case, the ability to quantify Fe is a useful geochemical technique.

Aqueous and extracted Fe concentrations can be quantified using simple colorimetric assays (Verschoor and Molot, 2013), including the ferrozine method (Stookey, 1970; Viollier et al., 2000). The ferrozine method is widely used and a preferred Fe assay for several reasons. First, the method is rapid, requires few chemicals, and is relatively inexpensive to perform. Second, the ferrozine method uses a non-reducing reagent to generate the color compound and results in more accurate quantification of ferrous ions, unlike other methods (e.g., TPTZ) that use strong reducing agents and overestimate ferrous content. Third, the method is easily adapted. A common method adaptation is selecting a reagent buffer that is appropriate for the experimental samples. The most common reagent buffer is ammonium acetate, but sodium acetate, 2-[4-(2-hydroxyethyl)piperazin-1-yl] ethanesulfonic acid (HEPES) and 1,4-piperazinediethanesulfonic acid (PIPES) buffers are also commonly used. The method can also be easily adapted to accurately quantify Fe within the ppb-to-ppm range.

The ferrozine reagent forms a color compound with ferrous iron only. Thus, ferrous Fe can be directly quantified using the ferrozine method. Ferric Fe, however, must first be reduced to ferrous iron before quantification using this method. To reduce ferric iron, ascorbic acid or hydroxylamine hydrochloride reductants are routinely used. Ascorbic acid is often preferred because it is an efficient reducing agent at room temperature. However, it loses its reducing ability shortly after being mixed with the sample (~2 h), and reduced Fe can reoxidize; if samples are prepared and immediately analyzed, this is not a concern. Hydroxylamine does not lose its reducing ability over time and has a long shelf-life once the reagent is prepared, but it must be purified to remove contaminant Fe before it can be used in the assay. If more than ~2 h will elapse between sample preparation and analysis, hydroxylamine is the preferred reductant.

There are several considerations to be made when performing the ferrozine method. First, the ferrozine method is best performed in an anoxic environment, such as in a glovebox or glovebag with an N_2 atmosphere, to eliminate the risk of oxidation of ferrous Fe. When working with low concentrations of Fe, it is safest to also sparge the ferrozine reagent with N_2 to remove dissolved oxygen. Second, adaptations made to the method may alter the ferrozine and/or reductant reaction. For each adaptation, new method controls should be performed to verify and optimize that method. For example, changing a buffer system may yield a longer color development period, necessitating a kinetics control to determine the reaction time required to reach color optimization. Third, color development for the ferrous–ferrozine compound is achieved between pH 3 and 8, and is most stable between pH 4 and 6 (Schilt and Hoyle, 1967). If an alternative buffer system is selected, it should be adjusted to and stable within this pH range and should not absorb energy at 562 nm. Lastly, Si in the sample can interfere with the ferrozine method. The orthophenanthroline (phen) method (Fortune and Mellon, 1938; O'Connor et al., 1965) is the appropriate Fe assay to use if high concentrations of Si are expected in a sample (e.g., in samples extracted from Si-rich materials). Other interfering ions include Mn^{2+}, MoO_4^{2-}, F^-, $C_2O_4^{2-}$, UO_2^+, NO_2^{2-}, Cr^{3+}, Hg^{2+}, Cu^+, Cu^{2+}, Ni^{2+}, Co^{2+}, CN^-, and oxalate, though high concentrations are required to result in significant interferences. A detailed, step-by-step ferrozine method is provided later.

X-ray absorption spectroscopy (Chapter 10) and Mössbauer (Chapter 13) are additional techniques commonly used to characterize Fe. Both techniques provide information about the Fe valence state(s) and can be used to determine the Fe speciation and mineralogy in samples. For more information about these techniques, please refer to Chapters 10 and 13.

Ferrozine Method (for determination of Fe(II) in the range of 0.1–10 ppm)

Reagents

- Ferrozine Reagent – 1 g/L ferrozine (3 - (2 - pyridyl) - 5,6 - bis(4 - phenylsulfonic acid) - 1,2,4 - triazine) with 50 mM buffer (e.g., ammonium acetate, PIPES, or HEPES). Adjust solution pH to 7.0 using dilute NaOH or HCl.
- Reducing Reagent – 10% w/w reductant (e.g., ascorbic acid or hydroxylamine hydrochloride).
- Fe Standards – ferrous ammonium sulfate prepared with N_2-sparged water.

Procedure

Notes: This procedure has been adapted from the original method and optimized for the determination of ferrous and ferric iron in environmental water samples. The volume of the final sample fits in a 1×1 cm^2 cell but can be increased to accommodate larger cells. All reagents and samples should be prepared with Type II (deionized) water or Type I (ultrapure reagent grade) water if available.

Acid-wash all labware in a hydrochloric acid bath and rinse thoroughly with Type II water before drying. If quantifying low levels of Fe, sparge all reagents with N_2 to remove dissolved oxygen. Transfer all materials required for the assay into an anaerobic chamber and allow the oxygen level to drop to 0 ppm before proceeding.

Determine which type(s) of Fe will be quantified. Ferrous Fe and total Fe quantification require one calibration curve each, whereas ferric Fe quantification requires two calibration curves (both ferrous and total Fe) and two sets of ferrozine solutions (with and without reducing reagent).

For ferrous iron quantification, pipet 2 mL of ferrozine reagent into a clean test tube. For Fe(II) concentrations between 0.1 and 10 ppm Fe(II), pipet 20 μL of filtered sample into the test tube. For Fe(II) concentrations below 0.1 ppm, pipet up to 200 μL of sample into the test tube. If Fe(II) concentrations are above 10 ppm, dilute the sample with N_2-sparged Type II water and then pipet 20 μL of diluted sample into the test tube. Swirl the test tube to mix the reagent and the sample. If Fe(II) is present, a magenta color should form rapidly. To the mixture, add Type II water to a sample+water volume of 200 μL; if 20 μL of sample was added, add 180 μL of water. Swirl the test tube again to create a homogeneous mixture. Allow the mixture to react for 10 minutes inside the chamber to reach optimal color development.

For total iron quantification, repeat this procedure for ferrous iron in a separate test tube. Additionally, add 500 μL of reducing reagent to the ferrozine before adding the sample.

Ferrous iron standards should be prepared to bracket the expected concentration range. Prepare a stock solution of ferrous ammonium sulfate at the highest concentration in the range to be tested and transfer to an anaerobic chamber. Dilute the stock solution with N_2-sparged Type II water to create standards to cover the experimental concentration range. Prepare ferrozine standard solutions for each Fe concentration by adding 2 mL ferrozine reagent to a test tube, followed by 20 µL of ferrous Fe standard and 180 µL water. Swirl to homogenize the mixture and allow to react for 10 minutes. For total iron quantifications, a set of ferrous Fe standards should be prepared including 500 µL of reducing reagent.

Reagent blanks should be used to quantify background absorbance resulting from the reagents and cell material. To prepare a reagent blank, mix 2 mL ferrozine reagent with 200 µL water, swirl, and allow to react for 10 minutes. For total Fe quantification, the reagent blank should be prepared by mixing 2 mL ferrozine reagent, 200 µL water, and 500 µL reducing reagent.

Standards, reagent blanks, and samples should be analyzed on a UV-Vis spectrophotometer at 562 nm, and absorbance values recorded. Solutions with absorbance values above 1.0 should be prepared again with a dilution of the initial sample. If the absorbance value for a reagent blank is non-zero, the value should be subtracted from all standards and samples prior to calibration calculations. To calibrate Fe(II), apply the Beer–Lambert Law as discussed earlier. If samples were concentrated or diluted, the appropriate concentration or dilution factor, respectively, should be applied to the final calculation. Note that the ferrous Fe reagent blank and calibration curve (without reducing reagent) should only be applied to samples without reducing agent. Similarly, the total Fe reagent blank and calibration curve should only be applied to samples with reducing agent. To calculate ferric iron concentrations in samples, simply subtract the ferrous iron concentration from the total iron concentration of a sample.

1.2.1.1.2 Colorimetric Assay – Chromium

Chromium (Cr) is the focus of many geochemical and microbiological studies due to its toxicity and carcinogenicity. Though Cr occurs naturally, high concentrations in natural samples are usually a result of human activity and indicate contamination from industrial wastes. The contamination can spread to soil, surface waters, groundwater, ocean water, and air, potentially leading to exposure and risk to wildlife and humans.

The toxicity of Cr depends upon its form. The two most abundant Cr oxidation states under environmental conditions are trivalent Cr and hexavalent Cr. The other oxidation states of Cr (e.g., Cr(0), Cr(I), Cr(II), Cr(IV), and Cr(V)) are rare. Trivalent Cr (Cr(III)) compounds are relatively stable and have low solubility in water, but can exist in natural waters as Cr–colloid complexes or $Cr(OH)_2^+ \cdot 4H_2O$. Cr(III) is a micronutrient and is required for life. Hexavalent Cr, Cr(VI), and its complexes (e.g., chromate and dichromate) are highly soluble and mobile in water, strong oxidants, and toxic. Due to their mobility and risk potential, Cr(VI) complexes are a major focus of geochemists and microbiologists.

Colorimetric assays provide a rapid technique for quantifying Cr(VI) in aqueous samples. In the assay, aqueous Cr(VI) is reacted with excess diphenylcarbazide (DPC) under acidic (pH ~2) conditions to yield a colored Cr–DPC complex. The intensity of the complex color is proportional to the concentration of Cr(VI) and can be quantified using a calibration curve with standards of known Cr(VI) concentrations. Aqueous Cr(III) concentrations can be determined by first oxidizing all Cr(III) to Cr(VI) and repeating the assay on the oxidized sample (Rice et al., 2017); the calculated difference in Cr(VI) concentrations between paired non-oxidized and oxidized samples is the Cr(III) concentration in the original sample. Interferences for this method are not common; however, mercury salts with concentrations exceeding 200 ppm, hexavalent molybdenum, extremely high concentrations of vanadium (up to 10× the Cr concentration), and ferrous iron concentrations above 1 ppm may all interfere with the analysis. These interferences can be removed with an extraction procedure (Rice et al., 2017). Detailed, step-by-step methods for the diphenylcarbazide chromium assay can be found in Rice et al. (2017) and United States Environmental Protection Agency Method 7196A (US EPA, 1995).

1.2.1.1.3 Colorimetric Assay – Phosphate

Phosphorus is important in a range of geochemical and microbiological investigations. Phosphorus mined from phosphate minerals is widely used in industry, including as fertilizers in agriculture, additives to foods, and ingredients in pharmaceuticals and household products (e.g., detergents). Ecologists are particularly interested in phosphate because of its potential to cause eutrophic conditions in surface and subsurface waters if present in excess. Biologists may be more concerned with mediation of energy storage and release within organisms through the phosphorylation and dephosphorylation of proteins. Regardless of the motivation, the quantification of phosphorus in samples is a useful technique to many investigators.

Orthophosphate (PO_4^{3-}), sometimes referred to as reactive phosphate, is routinely quantified using a molybdate blue colorimetric assay. Under acidic conditions, orthophosphate ions in an aqueous sample are reacted with a molybdate reagent to form a complex. The complex is then reduced using ascorbic acid – or stannous chloride in older methods – resulting in a blue-colored solution. The intensity of the color is proportional to the orthophosphate concentration, and the Beer–Lambert Law can be applied to quantify the concentration.

Sample pretreatment allows the quantification of phosphorus fractions other than aqueous orthophosphate. Total phosphorus is quantified by digesting the sample with persulfate prior to orthophosphate determination. Acid-hydrolyzable phosphorus, including metaphosphates, pyrophosphate, polyphosphates, and some organic phosphates, is determined by first hydrolyzing the phosphate with sulfuric acid to break the complex structures into the component orthophosphate ions; the resulting orthophosphate concentration can then be measured. Organic phosphates are calculated as the difference between total phosphorus and the sum of acid-hydrolyzable phosphorus plus orthophosphate (with no pretreatment). If the sample has not been filtered, the term "total" can be applied to each phosphorus fraction (e.g., total acid-hydrolyzable phosphorus). However, if it has been filtered, the term "dissolved" should be added to indicate that particulate phosphorus fractions have been removed.

There are a couple of potential interferences to be aware of for the colorimetric phosphate assay. First, silica forms a blue complex with molybdate that absorbs energy at the same wavelength as the orthophosphate–molybdate complex. Fortunately, relatively high silica concentrations (~30 mg SiO_2/L) are necessary before significant interferences are observed (Lachat, 2008). Second, certain metals (e.g., ferrous iron and calcium) will react with orthophosphate to form precipitates, decreasing the analyte in the sample. Bisulfite can be added to the sample as a pretreatment to prevent this interference. In many cases, no interference is noted for low metal concentrations. A detailed, step-by-step method for the orthophosphate assay is provided in the following. For a complete discussion of phosphate chemical and physical properties, sample preparation and storage, pretreatments, and analytical methods, refer to McKelvie (2000).

Method (for determination of orthophosphate in the range of 0.01–2.0 mg P/L)

Reagents

- Ammonium Molybdate–Antimony Potassium Tartrate Reagent – 8 g/L ammonium molybdate and 0.2 g/L antimony potassium tartrate.
- Ascorbic Acid Reagent – 60 g/L ascorbic acid and 0.2% acetone. Prepare fresh reagent every 10–14 days.
- Sulfuric Acid Reagent – 11 N sulfuric acid.
- Stock Orthophosphate Solution – 0.4393 g/L monopotassium phosphate. Final concentration is 100 mg P/L.
- Ammonium persulfate – for total phosphorus determination only.

Procedure

Notes: This method follows the US Environmental Protection Agency's Standard Method 365.3 (US EPA, 1978) and the methods for orthophosphate and total phosphorus determination established by Lachat Instruments, with slight adaptations. Samples with low (0.01–0.3 mg/L) P concentrations should be analyzed using a large cell to increase the pathlength, and the appropriate correction made in the final calculations; samples with higher concentrations can be analyzed using a standard 1×1 cm^2 cell. All reagents and samples should be prepared with Type II (deionized) water or Type I (ultrapure reagent grade) water if available.

Acid-wash all glassware in a hydrochloric acid bath and rinse thoroughly with Type II water. If samples are likely to have low phosphorus concentrations, it is best to treat all glassware with all reagents, using Type II water in place of the sample, and rinse the glassware thoroughly before allowing to dry.

For orthophosphate: Add 50 mL of sample to a ~100 mL flask, followed by 1 mL sulfuric acid and 4 mL molybdate reagent, and swirl to mix. Pipette 2 mL of ascorbic acid reagent into the flask, and mix. A blue color should develop if orthophosphate is present. Allow the mixture to react for 5 minutes to reach optimum color development and then determine the absorbance at 880 nm with a UV-Vis spectrophotometer. Analyze samples within 1 hour after ascorbic acid addition.

For total phosphorus: Add 50 mL of sample to an ~100 mL flask, followed by 1 mL sulfuric acid and 0.4 g ammonium persulfate. Mix the solution, and boil until the final volume has reduced to ~10 mL. Bring the volume of the solution back to 50 mL and filter through a phosphorus-free filter. Add 4 mL of molybdate reagent and 2 mL of ascorbic acid reagent, and mix. Follow the orthophosphate procedure to quantify the orthophosphate in solution once the blue color has developed.

For acid-hydrolyzable phosphorus: Repeat the total phosphorus method without the addition of ammonium persulfate.

Create an orthophosphate calibration curve by diluting the stock solution to make several standards of varying concentrations ranging from 0.01 to 2 mg P/L. Treat each standard using the orthophosphate method. Plot the concentration in each standard against the absorbance value to achieve a linear calibration curve ($R^2 > 0.99$). Use the curve equation to calculate the concentration of P in each sample from the measured absorbance. Concentrations are usually reported as concentration of P, not as PO_4^{3-}.

1.2.1.1.4 Ion Chromatography – Anions and Cations

IC is a separation technique that uses chemical properties of mixture components to separate ions from the mixture. First, an eluent solution is passed through the separation column, which contains a solid, stationary phase. The eluent ions saturate the charged sites on the stationary phase material. The sample is then injected onto the column, and ions in the sample interact with the stationary phase. Each ion interacts differently; ions with a high affinity for the stationary phase material will adsorb strongly, whereas ions with a low affinity will adsorb weakly. Eluent is continuously pumped through the column to elute the ions. Analytes separate based on their affinity for the stationary phase as they flow through the column, allowing the detection of isolated analytes as they are sequentially eluted (Figure 1.2). Elution of ions with very high adsorption affinities for the stationary phase can be hastened by increasing the salt concentration and/or adjusting the pH of the eluent.

Both anions and cations can be analyzed using IC. Historically, anions were the main analytes of interest in IC analysis, but cations are increasingly being analyzed by IC. One of the major advantages of IC is its capability to simultaneously analyze a large assortment of anions (or cations). Common anion determinations include bromide, chlorate, chloride, fluoride, nitrate, phosphate, selenite, sulfate, tungstate, molybdate, and carbonate. Common cation determinations include lithium, sodium, potassium, ammonium, magnesium, calcium, and zinc. However, IC can be used to quantify almost any ion and is widely utilized for the determination of organic ions (e.g., proteins and amino acids).

There are two types of IC analysis: anion and cation. The selection of stationary phase material and eluent composition depends on the analysis type. The stationary phase material should be oppositely charged from the analyte(s) of interest. Thus, for anion determinations, the stationary phase material, typically a resin, should be positively charged. The stationary phase material particle size will affect the analysis resolution and flow rate (analysis time), with decreasing particle size resulting in higher resolution and lower flow rates (longer analysis). Eluent compositions should be selected such that the analyte(s) is soluble in the eluent, the analyte(s) does not interact with the eluent, and the eluent has a low electrical conductivity (EC). Further, eluent pH should be within 0.5–1.5 pH units above (anions) or below (cations) the

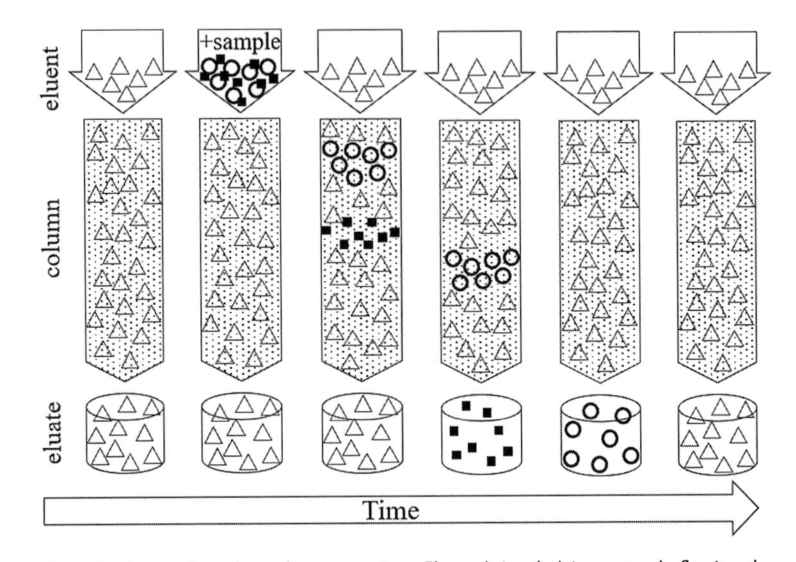

Figure 1.2 Schematic of sample elution through a column over time. Eluent (triangles) is constantly flowing through the column. Sample, a mixture of two components (circles and squares) in this example, is injected and separated in the column based on chemical and physical interactions with the column stationary solid phase. Each sample component is eluted from the column at different time points and is detected in the eluate.

isoelectric point of the analyte to optimize adsorption efficiency on the stationary phase material. Common eluent buffers include pyridine (pH 4.9–5.6), histidine (5.5–6.0), 2-[bis(2-hydroxyethyl)amino]-2-(hydroxymethyl)propane-1,3-diol (bis-tris) (5.8–7.2), 2-amino-2-(hydroxymethyl)propane-1,3-diol (tris) (7.5–8.0), and 2-(bis(2-hydroxyethyl) amino)acetic acid (bicine) (7.6–9.0) for anions, and lactic acid (3.6–4.3), acetic acid (4.8–5.2), PIPES (6.1–7.5), and HEPES (6.8–8.2) for cations, though many others are frequently used. It is important to keep in mind that concentrations of many analytes are affected by changes in pH (e.g., bicarbonate, carbonate, and phosphate); thus, selecting an eluent buffer that is at the same pH as the samples is critical to minimize changes in the analyte concentration.

EC detectors are used to quantify analytes in IC analysis. As analytes are eluted from the separation column, the EC of the eluate increases. The relative increase in EC is proportional to the concentration of the analyte, and measured changes in EC can be related to analyte concentration by using a calibration curve. UV-Vis detectors can be used to increase the sensitivity for analytes that absorb light energy, and amperometric detectors are best used for analytes that form deposits on the electrode surfaces (e.g., sugars). For analyses using an eluent with a high EC, a dual-column system can be used to suppress the EC signal from the eluent buffer, reducing the background signal.

Sample requirements for IC analysis are straightforward: the sample must be liquid (filtered is best) and have a low ionic strength. A low ionic strength is required in order not to oversaturate the surface sites of the stationary phase material; oversaturation results in analyte ions eluting through the column without separation and underestimation of the analyte concentration. Dilution of samples with high ionic strengths may bring the ionic strength within the operating range, as long as the analyte does not become too dilute to be accurately detected. Detection limits for many analytes are in the ppb range, with some in the low ppm range, though detection limits can change based on the complexity of the matrix. Lastly, and importantly, the pH of the sample and the eluent should match so that pH-sensitive analytes are accurately quantified.

For a complete discussion of IC, refer to Crompton (2001), Fritz and Gjerde (2009), Nollet (2000), and Rice et al. (2017), who provide standard IC methods for a variety of water and soil analyses.

1.2.1.2 Atomic Spectroscopic Techniques

Atomic spectroscopy techniques characterize samples by taking advantage of the fact that each element and molecule has a characteristic response to electromagnetic radiation. The excitation of outer-shell valence electrons results in the absorption of discrete quantities of energy by the electron and its transition from a stable ground state to the excited state. The measurement of the energy absorbed by the atom is the basis for atomic absorption spectrometry techniques. The photon energy emitted from the excited electron when it relaxes back to its stable ground state is the basis for atomic emission spectrometry techniques. By exciting an analyte and measuring the absorption or emission, the concentration of the analyte can be quantified.

A variety of atomic spectroscopic techniques exist, including inductively coupled plasma mass spectrometry (Chapter 4), X-ray absorption spectroscopy (Chapter 10), and X-ray fluorescence spectroscopy, among others. Though a variety of atomic spectroscopy instruments exist, many are expensive to purchase and operate, require extensive training to operate independently and interpret the data appropriately, or can be challenging to gain access to. Some techniques, however, are inexpensive, accessible, and require minimal training to operate. Two such techniques will be presented in this chapter: atomic absorption spectrometry and inductively coupled plasma-atomic emission spectrometry. Both techniques can analyze many elements with high sensitivity.

1.2.1.2.1 Atomic Absorption Spectrometry

Atomic absorption spectrometry (AAS) has been the "gold standard" for quantifying trace levels of metals since the late 1950s, with methods fully developed for 24 metals and metalloids in environmental samples. The appeal of the technique stems from its simplicity, ease of use, sensitivity for metals, and rapid sample processing time. But this is not to say that the instrumentation itself is basic.

AAS instruments are divided into three segments: sample introduction, sample excitation, and detection. The sample introduction system introduces the sample into the atom cell and creates a fine spray of sample. The spray is then passed into the excitation system – a flame. The flame is typically an oxygen acetylene, or nitrous oxide acetylene, plasma, but other flames can be used for specific applications. Droplets from the sample spray are heated, generating an atomic vapor. Electromagnetic radiation at a wavelength(s) characteristic to the analyte of interest is emitted through the vapor, and the analyte absorbs some of the radiative energy. The detection system detects and quantifies the energy absorbance from the analyte by measuring the difference between energy emitted by the radiation source and energy received at the detector. Analyte concentrations can be calculated by comparing the sample absorbance value with the calibration curve. Modern AAS instruments utilize monochromators and photomultipliers to reduce background noise from the flame, reduce interferences from the matrix, and improve sensitivity for analytes, effectively improving limits of detection.

Samples are typically introduced into AAS instruments in the aqueous phase. Filtered environmental water samples (e.g., surface waters, groundwater, soil extracts, and wastewater) can be directly introduced into the instrument without additional sample preparation. If potential interferences are expected for the analyte of interest, then pretreatment steps may be necessary, though this is uncommon. Solid samples can be prepared using a variety of methods. For example, if the water-extractable fraction of the analyte in the sample is desired, simply add water to the sample and allow the phases to equilibrate before analyzing the filtrate on AAS. The weak-acid-extractable fraction can be determined by adding a dilute acid to the sample, allowing the mixture to equilibrate, and analyzing the filtrate. For determination of the total analyte concentration, digestion of the sample may be required (consult Hu and Qi, 2014 for standard methods for all extractions and digestions). If samples cannot be extracted or digested, graphite furnace AAS may be used; this technique introduces solid samples to a furnace at around 1 000 °C, atomizing the sample, and the absorbance signals associated with the analyte are detected and quantified. Graphite furnace AAS has many complicating factors and should not be used for all samples.

Though AAS is the standard for the analysis of many metals, it is not without its limitations. The first limitation is its sensitivity for metals; though the limits of detection for many metals are in the ppb–ppm range, accurate quantification below this range is not possible without a pretreatment method to concentrate the analyte. Though concentration methods exist, they can be lengthy and introduce errors to the analysis. The second major limitation of AAS is that many instruments have the capacity to analyze only one analyte at a time due to the specificity of the excitation source for the analyte. Modern AAS instruments are often adapted with excitation sources that allow simultaneous analysis of multiple analytes, typically up to nine, but they are predetermined. Quantification of additional analytes requires additional excitation sources and sample reanalysis.

For more detailed discussion of AAS theory and instrument options and adaptations, refer to Jackson and Jackson (1999), Welz and Sperling (1999), and Welz et al. (2005). Crompton (2001), Nollet (2000), and Rice et al. (2017) provide standard methods for quantifying metals in water, soil, and sludge samples using AAS.

1.2.1.2.2 Inductively Coupled Plasma-Atomic Emission Spectrometry

Like AAS, inductively coupled plasma-atomic emission spectrometry (ICP-AES) is used for the rapid quantification of metals in aqueous samples. Methods have been developed to use this technique to quantify 50 total elements, including 30 metals and metalloids. An ICP-AES instrument is similar to an AAS instrument in that it has sample introduction, excitation, and detection systems. The key difference is that the detector measures the energy (photons) emitted during the relaxation of excited electrons, not the energy absorbed during the excitation of electrons. A calibration curve is used to calculate the concentration of the analyte based on atomic emission.

Samples for ICP-AES analysis must be aqueous and must be filtered to prevent clogs in the lines and to maximize exposure of the analyte to the excitation source. Especially for metal analytes, samples should be acidified to 1–2% acid content to stabilize the analyte in solution phase and prevent its sorption to instrument materials (e.g., tubing). In cases where acidification will destroy or alter the analyte, sample acidification should not be used. Additionally, standards should be prepared in the same matrix and to the same acid content as the sample to minimize "matrix effects" resulting from interfering chemicals. For more complex samples, achieving true matrix-matched standards may be difficult; in such instances, alternative techniques (e.g., AAS) should be used.

There are several advantages to using ICP-AES over AAS for metal analysis. First, ICP-AES can analyze multiple analytes in the same run, reducing the total analysis duration and sample volume needed. Second, ICP-AES has a larger linear dynamic range for analytes, up to 1 000 times larger than AAS for most analytes. It also has lower detection limits for several analytes (e.g., Al, B, Ba, Ti, and Zr), though the detection limits of many analytes are similar between the techniques; detection limits for ICP-AES depend on the sample matrix and can be higher in complex matrices. Third, several non-metals (e.g., P, S, I, Br, and Cl) can be quantified using ICP-AES but not AAS. Fourth, the stability and precision over long analysis durations is better for ICP-AES than for AAS. Lastly, argon gas is commonly used to generate the plasma in an ICP-AES instrument, as opposed to the flammable gases used in AAS instruments; thus, there are fewer hazards associated with operating the instrument.

For a thorough discussion of ICP-AES, refer to Huang et al. (2000) and Thompson and Walsh (1989). Van Loon and Barefoot (1989) provide a user-friendly discussion on how to handle spectral and non-spectral interferences and calibrate samples using internal references.

1.2.2 Organic Compounds

Geochemical techniques would be incomplete without the ability to quantify and characterize organic compounds. In this section, common analytical techniques used to quantify total C in both aqueous and solid samples, as well as individual organic molecules in various types of samples, are presented. Detailed methods are presented for total C quantification in aqueous and solid phases, and technique overviews, analysis considerations and limitations, sample preparation requirements, and analysis-specific data processing procedures are covered for techniques used to separate organic molecules from fluid matrices.

1.2.2.1 Solution-Phase Carbon – TOC

Carbon (C) is ubiquitous in the environment and present in many geochemical samples. In aqueous samples, C exists as either inorganic carbon (IC) or organic carbon (OC). Inorganic carbon consists of dissolved inorganic ions (carbonic acid, bicarbonate, and carbonate) as well as particulate inorganic carbon. Every sample exposed to a CO_2-containing atmosphere will contain some amount of inorganic carbon. Organic carbon consists of purgeable organic carbon (POC) – organic carbon removed from the sample by purging with an inert gas – and non-purgeable organic carbon (NPOC). NPOC can further be divided into dissolved and particulate fractions, where the dissolved fraction is the NPOC that passes through a 0.45 μm filter and the particulate fraction is the NPOC retained by the filter.

Total organic carbon (TOC) analyzers quantify C concentrations in aqueous samples via combustion. A small aliquot of sample is passed through a combustion chamber at around 700 °C, converting the C to CO_2 gas; often, the combustion is catalyzed to increase oxidation efficiency. The CO_2 generated from the sample is then carried to a nondispersive infrared (NDIR) detector by a carrier gas, and the volumetric CO_2 concentration is detected and recorded as a peak. The peak area for each sample is then used to quantify the OC in the original sample. Inorganic C is quantified in much the same way, except that the aliquot is acidified, not combusted, to convert all inorganic carbon to CO_2.

Care must be taken when collecting and preparing samples for IC and/or POC analysis. The IC concentration in an aqueous sample changes readily when changes to gaseous CO_2 concentrations occur, with increasing CO_2 resulting in increased IC and vice versa. If the sample is well buffered, these changes may be minor. Also, addition of preservatives and acidification of samples may drastically change the IC concentration. POC, also referred to as volatile organic carbon, can be lost from aqueous samples if there is any headspace above the sample. To avoid changes in IC and POC during sample collection and preparation, it is best to collect the sample in C-free, airtight glass containers to prevent gas exchange, not to add any preservative to the sample, and to seal the container such that it contains absolutely no headspace.

A detailed, step-by-step method for TOC determination is provided in the following. For a complete discussion of TOC analyses and alternative methods (e.g., persulfate oxidation), refer to standard methods (Rice et al., 2017).

Method (for determination of Total C and IC in the range of 0.1–100 ppm)

Materials

- Total C Stock Solution – 2.125 g/L desiccated (110 °C, 1 h) reagent grade potassium hydrogen phthalate. The resulting C concentration is 1 000 ppm C.
- IC Stock Solution – 3.500 g/L desiccated (2+ h in desiccator) reagent grade sodium bicarbonate and 4.410 g/L desiccated (280 °C, 1 h) reagent grade sodium carbonate. The resulting C concentration is 1 000 ppm C, all IC.

Procedure

Notes: The method presented here is adapted from standard operating procedures for a Shimadzu TOC (Shimadzu, 2013). TOC instruments need to be calibrated before each analysis, as they can drift significantly over time. If many samples are to be analyzed at one time, it is wise to add standard checks throughout and at the end of the analysis to monitor drift. Additional reagents (e.g., hydrochloric acid or phosphoric acid) are required to operate the TOC instrument. Please consult the instrument manual for specific details regarding operation prior to analysis. If the TOC instrument is equipped with a total nitrogen (TN) detector, the TN content of samples can be simultaneously quantified. All reagents and samples should be prepared with Type I (ultrapure reagent grade) water.

To reduce risk of C contamination from residues on vials, wrap glass sample vials in aluminum or tin foil and combust in a 550 °C oven for 2–4 h; allow vials to return to room temperature slowly, then store in a sealed container to prevent dust and other debris from contaminating the vials. Ensure that vial septa are also stored in a dust-free location.

A separate set of standards is required for total C and IC. To prepare total C standards, dilute the 1 000 ppm total C stock solution with Type I water that has been sparged with N_2 to reduce contaminant IC. After preparation, cap and seal each standard with no headspace; to verify that no gas is trapped, check that bubbles are not present in the vial when flipped upside down. Repeat the process using the 1 000 ppm IC stock solution to create a separate set of IC standards. The final range of both sets of standards should bracket the expected range of C in the samples to be analyzed; typical ranges for operation are 1–100 ppm for total C and 1–25 ppm for IC, though some samples may require either narrower or broader calibration ranges. Fill prepared vials with standard solutions and immediately cap, ensuring there is no headspace captured in the vial. Caps should be open-topped to allow the sampling needle to pierce into the sample, and a clean, unpierced septum should be used to seal the vessel.

Prepare samples by transferring into prepared sample vials and capping, ensuring no headspace is captured in the vials. If only the DOC fraction is to be measured, filter the sample (0.45 μm) into the vial. For all other C fractions, filtering is not required; however, visible particles must be removed from the sample (by, e.g., filtration or centrifugation) prior to analysis to prevent clogs in the lines. If C concentrations are expected to be higher than 100 ppm, dilute the sample with N_2-sparged Type I water. Analyze the samples as soon as possible to prevent sample alteration (e.g., precipitation) and biasing of results. Keep samples and standards cold (e.g., in a refrigerator or on ice) until analysis.

There are two methods for quantifying organic carbon fractions: TOC determination and NPOC determination. Depending on the capabilities of the instrument, both methods may be performed automatically by the instrument, or they may need to be performed manually. For TOC determination, the total carbon (IC + OC) in the sample is quantified first. An aliquot of the sample is then acidified to convert all IC into CO_2. The IC concentration is then subtracted from the total C concentration to calculate the total OC concentration in the sample. Most instruments perform this method and calculation automatically. Caution should be used when using this method for samples with relatively high IC concentrations compared with OC concentrations; in this case, pretreatment for IC removal should be performed to reduce error associated with IC determination. If a manual TOC determination is required, acidify an aliquot of the sample with HCl to achieve pH 1–2. Measure C in the acidified sample; this value represents the TOC in the initial sample. To quantify NPOC, both the IC and the POC must be removed from the sample prior to C measurement. Many instruments perform this method and calculation automatically by purging the sample with an inert gas prior to analysis to remove the POC fraction. This step can be manually performed prior to capping the sample vials for analysis if the instrument does not have the capability.

Most TOC software will automatically calculate the IC, TC, and TOC (and NPOC, if this method was used) concentrations and will display the concentrations of each fraction. Concentrations in samples are based on the standard calibration curves generated from the standard solutions prepared earlier. Each C concentration corresponds to a CO_2 peak, with increasing C concentrations resulting in larger CO_2 peaks. The C concentration in each standard is plotted against the area under the CO_2 peak, and a linear calibration curve ($R^2 > 0.99$) is generated and used to calculate the C in samples. If samples were diluted, the appropriate dilution correction factor should be applied to determine the C concentration in the original sample.

1.2.2.2 Solid Phase Carbon – Elemental Analysis

Elemental analysis (EA) is an accurate, rapid, inexpensive technique used to detect elements within a solid sample and quantify their mass fractions. In environmental samples, C, H, N, and S are commonly investigated using this technique, but other elements (e.g., halogens) can also be analyzed. Applications of this technique include qualitatively characterizing the composition of an unknown sample, quantitatively characterizing the mass ratios of elements present in a sample, and determining the purity of a known sample. Elemental analyzers equipped with a mass spectrometer can additionally determine the stable isotopic composition of the sample.

Quantification of elements using EA is achieved using concepts of the Pregl–Dumas method by combusting solid samples in an O_2-rich environment and quantifying the resulting gaseous compounds. Through combustion, each element is oxidized to its corresponding gaseous form (e.g., C to CO_2, N to N_2 and N oxides, H to H_2O, and S to SO_2). The resulting gas mixture is passed through a sequence of traps, to remove elements and compounds that are not being quantified, and columns, to separate the compounds in the mixture based on mass and chemical properties. The gases are additionally passed through a column packed with copper to remove excess, unreacted O_2 gas and to convert any N oxides to N_2. The separated compounds are carried to a thermal conductivity (TCD) and/or NDIR detector with a carrier gas (e.g., He) and detected in sequence (Thompson, 2008).

Sample preparation for EA analysis is straightforward: samples should be dry and homogeneous. Air-dried samples are sufficient unless humidity is high, in which case oven-dried (60 °C) samples are preferred to reduce mass contribution from water. If H or O is to be analyzed, oven drying samples is required to remove contamination from hygroscopic water in the sample. Homogenization is achieved by physical mixing. If the sample contains large or uneven-sized particles, pulverization should be performed to increase the surface area and homogenize the sample. Samples are transferred to the instrument in capsules (e.g., tin) that are compactly folded over the sample to prevent loss of the material. Analyte concentrations in each sample are calibrated using an analyte-specific standard calibration curve. Standard calibration curves should be prepared using a commercial micro-analytical standard compound (e.g., acetanilide or benzoic acid) to ensure maximum accuracy of the analysis.

Sample combustion in most EA instruments is achieved in a 1 000 °C oven. If samples contain thermally resistant compounds, they may not be fully oxidized, causing the analyte mass fraction to be underestimated. Accelerants (e.g., vanadium pentoxide) can be added to the sample to elevate the local temperature around the sample, increasing the combustion and oxidation efficiencies. Caution should be used when selecting an accelerant to ensure it is compatible with the sample and the instrument. Additionally, many accelerants are hazardous and should be handled with care.

A detailed, step-by-step method for determination of C and N using EA is provided in the following.

Method (for determination of 0.01+ mg of total C or N)

Materials

- Tin capsules (or other sample holders).
- Micro-analytical Standard Compounds – high-purity, certified standard containing each analyte.
- Accelerant (if using).

Procedure

Notes: The method presented here is adapted from several standard methods (ECS, 2001; ISO, 1995; Zimmerman et al., 1997). Elemental analyzer instruments require carrier gases and O_2 for operation, and can be arranged in various configurations, depending on which analyte is being analyzed. Please consult the instrument manual and/or instrument supervisor for instrument-specific training prior to analysis. All weighing should be performed using an analytical balance capable of accurately weighing microgram masses.

First, prepare the standards. Consult the instrument manual to determine the detection range for the analyte of interest, and be sure to work within this range. Determine the highest concentration to be used in the calibration curve, and calculate the mass of the standard compound that must be used to achieve the concentration. For example, if the highest desired C concentration is 1 mg C and the standard compound contains 71% C by mass, then 1.408 mg (1/0.71) of the standard compound will need to be used. Gravimetrically add the standard compound to a tin capsule to achieve the calculated mass, and record the actual mass of the compound. Carefully fold the tin capsule into a tight ball, making sure that the open end of the capsule is completely sealed and that the standard compound cannot escape. Make several more standards, adding progressively less standard compound to each tin capsule, recording the mass of standard compound in each and tightly folding each capsule into a sealed ball. A standard blank should also be made by folding an empty tin capsule; the blank will be used to determine whether the tin capsule contains the analyte and is contributing background to standards and samples.

Next, prepare samples. If the analyte concentration (mass analyte per mass sample) in the sample is approximately known, calculate the mass of sample that must be analyzed to achieve a concentration within the calibration range. For example, if the sample contains approximately 50 mg C/g sample (5%), and the target mass is 0.50 mg C, then 10 mg (0.50/

50) of sample should be weighed. If the concentration is truly unknown, make an educated guess or weigh out a few different masses of the same sample. Weigh out the calculated mass of the sample into a tin capsule, record the mass, and fold the capsule into a sealed ball. If an accelerant will be used, add the accelerant to the weighed sample prior to folding. Though this technique is highly accurate, replicates of each sample should be analyzed to minimize errors associated with sample preparation, weighing, and analysis.

After analysis, use the reported peak area values plotted against the mass of analyte in the standard to create a calibration curve for the analyte. Use the calibration curve to calculate the mass of analyte in each sample. To determine the analyte mass in the solid sample (mass analyte per mass sample), divide the analyte mass in the combusted sample by the mass of the solid sample added to the tin capsule. Note: if the peak area for the standard blank is non-zero, subtract the peak area value from all standards and samples prior to the data processing.

1.2.2.3 Chromatography – HPLC and GC

High performance liquid chromatography (HPLC) and gas chromatography (GC) are two separation techniques used to quantify a variety of organic compounds. Both techniques separate compounds in a mixture based on physical (e.g., mass) properties of mixture constituents. The mixture is introduced to the instrument and the constituent compounds are carried by a mobile phase through a separation column containing a stationary phase that retards the movement of larger compounds and compounds that have higher affinity for the stationary phase. Each isolated constituent exits the column and is detected. The time between sample introduction and detection of a compound is characteristic for individual compounds, assuming all other conditions (e.g., stationary phase, operating temperature, pressure, etc.) are consistent. Standard solutions must be prepared in a matrix similar to the samples and analyzed to identify the time-to-peak for the analyte. A linear calibration curve of the peak area (or height) plotted against the known analyte concentration for each standard solution is used to calculate the analyte concentration in the unknown samples. In modern instruments, the accompanying software automatically performs this calculation and reports the concentration. These techniques are commonly used for quantification of pesticides, phospholipids, hormones, amino acids, peptides, sugars, and phosphates (Tabatabai and Frankenberger, 1996), though application to other compounds is possible. Neither technique can positively identify an unknown compound.

HPLC requires samples to be in the liquid or dissolved phase and is best suited for analytes that cannot be volatilized at low temperatures (e.g., high-molecular-mass organics, inorganics). In HPLC, samples are carried by a liquid mobile phase under high pressure (ca. 5 000 psi) at room temperature and are passed through a small, densely packed column containing a solid stationary phase. Compounds in the mobile phase are separated based on physical (e.g., size) and/or chemical (e.g., hydrophilicity) characteristics. For example, small compounds pass through the column more easily than large compounds and exit the column first, or nonpolar compounds are retained in a C18 (a common silica-based stationary phase) column and polar compounds exit the column first. For mixtures containing analytes with similar physical and/or chemical characteristics, specialized (e.g., longer) columns can be used to improve separation. UV detectors are commonly used to detect analytes, but additional detectors (e.g., photodiode array, fluorescence, refractive index, or conductivity) may be installed for more sensitive detection.

GC differs from HPLC in that the analytes are detected in the gas phase and carried via a gaseous mobile phase. Thus, GC is best suited to analytes that are easily volatilized (e.g., low-molecular-mass organics, inorganic gases) and do not degrade at the temperatures reached in the oven (if used). Samples can be introduced to the GC as a liquid and passed through an oven (<450 °C) to volatilize the analyte(s), or they can be directly introduced as a gas. An inert mobile phase (e.g., He or N_2) carries the analyte(s), and separation is achieved by passing the mobile phase through a polymer or liquid-covered solid phase or through a long (ca. 3 m) capillary column. Organic analytes are typically detected by a flame ionization detector (FID) and inorganic gases by a TCD, though specialized detectors can be used to increase sensitivity for some analytes (e.g., nitrogen phosphorus detector, flame photometric detector for S and organophosphorus gases, photoionization detector for aromatics, and electron capture detector for halogenated gases).

Sample preparation for HPLC and GC depends on the analyte and its concentration. Often, samples can be directly analyzed without additional sample preparation. If the concentration of the analyte is low, additional techniques (e.g., solid phase extraction) may be required to concentrate the analyte so that it can be accurately detected and quantified. If the analyte concentration is too high, then dilution may be required. HPLC samples are easily concentrated or diluted to adjust analyte concentrations. However, for GC analysis, the concentration of a gaseous analyte is determined by the vapor pressure of the gas, and thus, concentration or dilution may not be possible.

In many laboratories where both techniques are available, samples containing analytes with characteristics suited to each technique are often analyzed on both HPLC and GC. The technique that results in the best selectivity for the analyte and fastest analysis is used for further analyses. Additional discussions about HPLC and GC techniques can be found in Tabatabai and Frankenberger (1996), with detailed methods for a variety of compounds provided by Crompton (2001), Nollet (2000), and Rice et al. (2017).

1.3 Microbiological Techniques

Microorganisms are present in all ecosystems, driving a myriad of biogeochemical reactions. Geomicrobiological investigations often need to deal with a variety of microorganisms at the molecular, cellular, population, and community levels. In this section, common culture-dependent and culture-independent microbiological techniques are presented, and their advantages and disadvantages are discussed. It should be stressed that avoiding contamination during experiments is highly critical for microbiological analyses. Culture-dependent methods should avoid contamination by microbes from the laboratory environment, labware, and researchers. Culture-independent methods should avoid contamination by both microbial cells and biomolecules from the same sources. Additionally, culture-independent methods should avoid conditions that can degrade the quantity and quality of target biomolecules. Detailed discussions on how to prevent contamination are presented in the following sections wherever appropriate.

1.3.1 Culture-dependent Methods

Culture-dependent methods are conventional and have been used in microbiology since its origin. Current knowledge of many perspectives of microbiology is obtained from studying pure cultures of model microorganisms (e.g., cellular structure and physiology, the biochemical basis of cell proliferation and function, and nutrient cycling). Reference genomes of model microorganisms also pave the way for understanding the genetic basis of metabolic pathways involved in biogeochemical cycles. Even though the majority of environmental microorganisms are uncultured or unculturable in the laboratory, cultivating a microorganism remains the only way to fully characterize its properties and predict its impact on an environment (Madigan et al., 2014). The availability of pure cultures of a single organism or a group of organisms (i.e., a consortium) can serve the purpose of studying these organisms under controlled laboratory conditions, particularly for analysis of microbial morphology, physiology, and metabolism.

1.3.1.1 Cultivation

Microorganisms can be cultivated from their habitats using the enrichment approach. Using a selective growth medium and a set of incubation conditions, enrichment selects for the desired organisms and counterselects for undesired organisms. There are two approaches to cultivating microorganisms in pure cultures under controlled laboratory conditions: batch culture and continuous culture. A batch culture uses a defined medium to which a fixed amount of substrate is added, while a continuous culture requires a steady inflow of growth medium containing substrate and an outflow of spent medium containing microbial cells. Batch cultures are usually grown in a flask or a similar container, while continuous cultures are grown in a more sophisticated vessel called a bioreactor or chemostat (Figure 5.12 in Madigan et al., 2014). Both approaches rely on the choice of an appropriate medium for the target organism(s) and on the maintenance of suitable environmental parameters. A single species can be purified from an enrichment culture by isolation, usually on a streak plate. A pure culture of that species can be established by restreaking a well-isolated colony multiple times. By monitoring growth of pure cultures in conjunction with mathematical modeling, researchers can measure physiological parameters such as generation time, growth rate, and death rate under different conditions.

1.3.1.1.1 Enrichment

To bring organisms of interest into laboratory cultures, appropriate media and incubation conditions must be used that can duplicate as closely as possible the resources and conditions of those organisms' ecological niches (Hurst et al., 2007). Because of the extreme microbial diversity, hundreds of different enrichment strategies have been developed (Madigan et al., 2014). Factors that should be considered include resources (C source, N source, other macronutrients, and micronutrients such as trace metals and growth factors) and conditions (e.g., temperature, pH, and osmotic conditions). A summary of common enrichment methods for environmental microorganisms can be found in Table 18.1 and Table 18.2 in Madigan et al. (2014). For many organisms whose pure cultures have been established in the laboratory, there are defined and/or complex media available, and a more complete list of all known media with detailed recipes can be found in Atlas (2010). Information about cultivation media for organisms that have been archived by American Type Culture Collection (e.g., 3-(N-morpholino)propanesulfonic acid (MOPS) buffered medium for *Dehalococcoides* spp.) can be found on their website (www.atcc.org). Cold Spring Harbor protocol (http://cshprotocols.cshlp.org) is another source for enrichment media recipes. Commonly used media, such as Luria-Bertani medium and nutrient medium, are commercially available as powders that are easily dissolved in water. Occasionally, specific media are required for cultivating mutant strains. For example, osmotically protected media have been used to grow *Bacillus subtilis* without a cell wall (Maier et al., 2009). All culture media must be sterilized before use, which is usually achieved by exposing the medium to high-pressure saturated steam at 121 °C for a sufficient period of time in an autoclave. For temperature-sensitive media and medium ingredients, the medium is filter-sterilized through a 0.22 μm membrane.

Enrichment starts with inoculation; that is, placing an inoculum containing the organism of interest into the prepared sterile medium. Such an inoculum can be any environmental sample, such as fresh water, seawater, a hot spring microbial mat, sediments, soils, and decomposing leaf litter, among others. Inoculation procedures should be conducted aseptically to avoid potential cross-contamination, including airborne contaminants. In the laboratory, bench space is usually cleaned with 70% ethanol before use; utensils are sterilized by autoclaving or radiation; open vessels are operated close to a Bunsen burner flame; and container openings and metal inoculation loops are resterilized by passing through a flame ("flaming") in between inoculations or, alternatively, new sterile plastic loops are used for each inoculation. After inoculation, the culture is incubated under conditions that support the organism's growth. Methods to cultivate microbial consortia and communities using a bioreactor or chemostat have been described by Hurst et al. (2007).

1.3.1.1.2 Isolation

For many purposes, researchers isolate a single species of microorganism from an enrichment culture. This can be achieved by a variety of methods, including streaking, agar dilution, and others.

A streak plate can be used for purifying aerobes and also for anaerobes with proper incubation facilities such as anoxic jars or chambers. Streak plates consist of a solid agar medium, made by adding 1–2% agar to the liquid medium. The resulting mixture is autoclaved, cooled to 50–60 °C (essential if temperature-sensitive components, such as antibiotics, need to be added), and poured into a sterile Petri dish. After the molten agar medium has solidified, the following streaking procedures can be performed. A sterile loop,

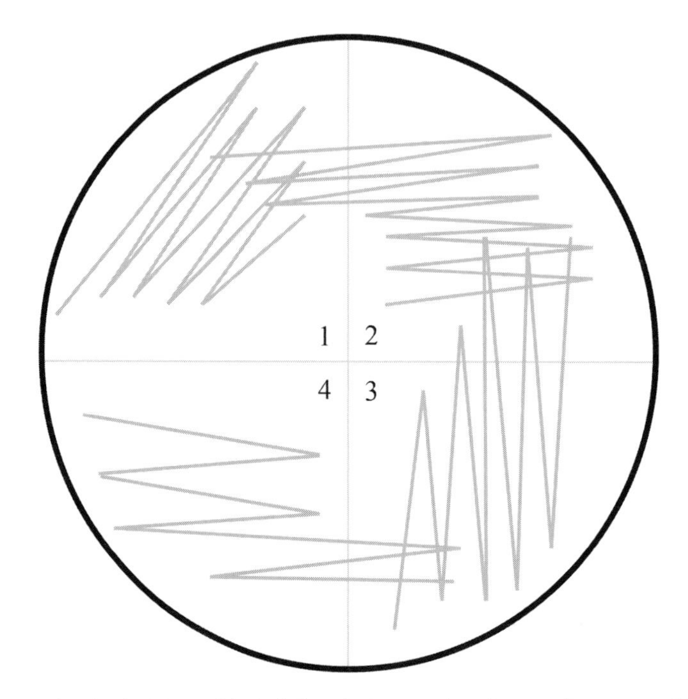

Figure 1.3 Schematic diagram of a streak plate used for isolation of microorganisms from a culture on a solid agar medium. Numbers indicate sequential streaking in four quadrants of the plate.

either a presterilized plastic loop or a metal loop flamed on a Bunsen burner, is used to remove a loopful of inoculum from an enrichment culture. The loop carrying the inoculum is placed at one corner of the agar plate and spread over one quadrant of the plate using close parallel streaks. Subsequently, the plate is turned through a 90° angle; a second streaking is made by lightly sweeping the loop through the previous quadrant of the plate and streaking it into the next quadrant of the plate without overlapping the previous streaks. This step is repeated until the final streaking covers the fourth quadrant of the plate (Figure 1.3). It should be noted that for each streaking step, a new presterilized plastic loop should be used, or the metal loop should be flamed again for resterilization. The plate is then inverted and incubated at the required temperature. Streaking allows microbial cells to grow on a solid medium and form visible colonies. After multiple rounds of streaking, single and separate colonies are formed on the plate at the end of streaks, whose shapes, sizes, and colors can vary depending on the organisms and the cultivation conditions. To obtain a pure culture, a single colony is selected and restreaked repeatedly on solid medium using the aforementioned procedure, until only one kind of colony morphology is observed on the plate.

The agar dilution method is common for purifying anaerobes. By inoculating a mixed culture into a tube containing molten agar medium, one can obtain colonies embedded in the agar. Cell suspensions in the first tube are successively diluted in new tubes of molten agar medium until pure cultures are obtained.

Similarly to inoculation, using aseptic techniques is critical to avoid contamination for isolation. It is always recommended to open tube caps and plate covers just enough to permit transfer and streaking operations, and to work close to a Bunsen burner flame so that flaming can be conducted if necessary.

During the past decade, there have been many efforts to renew isolation methods to capture more microbial diversity, particularly for as-yet-uncultivated species. These novel methods are directed to better reflect natural environments where microorganisms dwell, taking into account nutrients (composition and concentration), oxygen levels, pH, and natural associations between microorganisms (e.g., coculturing with helper organisms). They also take advantage of new techniques, including flow cytometry and cell sorting, optical tweezers, microfluidics, multiwall microbial culture chips, and high-throughput microbioreactors (Cardenas and Tiedje, 2008; Pham and Kim, 2012; Vartoukian et al., 2010).

1.3.1.1.3 Growth Measurements and Kinetics

Microbial growth at the cellular level is a complex process, involving many anabolic and catabolic reactions and molecular processes that eventually result in cell division. At the population level, microbial growth results in an increase in the number of cells or the amount of biomass. Methods used to measure cell numbers and monitor population increase include the measure of optical density using spectrophotometry, the measure of viable cells using plate counts, the measure of total cell numbers using a microscope, the measure of gene copy numbers using real-time quantitative polymerase chain reaction (qPCR; see Chapter 15), and many others.

Measurement of Optical Density Using Spectrophotometry

This method is based on the fact that cells scatter light in proportion to cell mass, resulting in the turbidity of a culture suspension. A measure of the light scattered from a culture suspension thus provides a rapid and easy estimate of cell numbers in the suspension. This is usually conducted using a spectrophotometer at a specific wavelength, commonly at 480 nm, 540 nm, 600 nm, or 660 nm, depending on the color of the microorganism(s). The estimate of cell numbers is presented as optical density (OD) at the wavelength (e.g., OD_{600}). This estimate can be related to total or viable cell numbers by creating a standard curve for OD readings versus the desired cell counting. Culture growth and OD measurement can also be conducted in microliter volumes using a 96-well microtiter plate in line with a microplate reader. While OD measurement can be highly accurate for pure cultures of a single organism, it may be inappropriate for consortia containing multiple organisms. This is because different organisms usually differ in size and shape, and therefore, equal numbers of different organisms will not necessarily yield the same OD readings. Moreover, OD measurement may not be accurate for microorganisms that tend to form clumps or even biofilms, although continuous shaking could alleviate cell aggregation to some extent. Furthermore, OD measurement does not distinguish between live and dead cells.

Measurement of Viable Cells Using Plate Count

Viable cells divide and form colonies on an agar plate. Assuming that the growth and division of each viable cell generates a single colony on the plate, colony numbers thus reflect the number of viable cells in a sample. Procedures of plate counting are similar to streaking used in isolation. A small volume (usually 200 μL or less) of an appropriately diluted culture is spread over the surface of an agar plate using a sterile spreader (the spread-plate method). Alternatively, a volume of the culture is pipetted into a sterile Petri dish, and the molten agar medium, cooled to just above gelling temperature, is added and mixed with the culture by gently swirling the plate (the pour-plate method). A total number of 30 to 300 colonies developing on or in the plate is usually considered to be a reliable count in practice. To obtain these numbers, a sample is usually serially diluted, often by 10-fold and in sterile phosphate buffered saline solution or 10% sodium chloride solution. The calculation of total viable cells in the sample considers the dilution factor that generates the appropriate colony number and is presented as colony-forming units (CFU), because a clump of multiple cells may form a single colony. As in enrichment and isolation, the choice of appropriate media and incubation conditions largely affects plate counts. This is particularly the case for environmental samples, as microorganisms in these vary a lot in terms of growth requirements. Moreover, microorganisms in a mixed culture, such as a soil suspension, have different growth strategies and may compete with each other on agar plates. All these factors make plate counting unreliable for assessing environmental samples, although it is possible to achieve reliable results for a known group of organisms by using a highly selective medium.

Measurement of Total Cell Numbers Using a Microscope

For complex samples, direct microscopic counting typically yields higher numbers of microorganisms than plate counting. This is because both live and dead cells, and sometimes also cell debris, are observed and counted under a microscope. With special staining techniques, dead and/or live cells can be distinguished and determined accurately using this method. A detailed description of microscopic counting is given in Section 1.3.1.2.2.

Measurement of Microbial Growth Using Other Methods

Microbial growth can also be measured by monitoring a variety of physiological activities. For example, the growth of heterotrophs can be measured by monitoring CO_2 evolution, which is detailed in Section 1.3.1.3.1. Microbial growth can further be measured using molecular biology techniques. For example, nucleic acid synthesis can be tracked by quantifying particular genes using qPCR, which is detailed in Section 1.3.2.2.2.

Growth Kinetics for a Batch Culture

Using the aforementioned measurements, microbial growth under both batch and continuous conditions has been well characterized. For a batch culture, the increase in cell number or biomass is measured as a function of time, which is called a growth curve (Figure 1.4). A growth curve consists of several distinct growth phases, including the lag phase, the exponential or log phase, the stationary phase, and the death phase. The lag phase represents the time duration between when inoculation is completed and when culture growth begins. The exponential phase represents a period when the microbial population in the culture, as reflected by the cell concentration (cell number or biomass), doubles at regular intervals. At any time within the exponential phase, the rate of the increase in cell concentration is proportional to the cell concentration:

$$\frac{dN}{dt} = \mu N \qquad \text{(Eq.1.1)}$$

where N is the cell concentration (cell number or biomass per volume), t is time, and μ is the specific growth rate (time^{-1}). Solving Eq. (1.1) gives an expression of the cell concentration after a period of time:

$$N = N_0 e^{\mu t} \qquad \text{(Eq.1.2)}$$

where N_0 is the initial cell concentration after inoculation. The time it takes for a cell division to occur and thereby, the cell concentration to double ($N/N_0 = 2$) is called the doubling time or the generation time (g):

$$g = \frac{\ln 2}{\mu} \qquad \text{(Eq.1.3)}$$

Because within the exponential phase the microbial population is doubling, the cell concentration after n divisions or generations can also be expressed as

$$N = N_0 2^n \qquad \text{(Eq.1.4)}$$

where $n = t/g$. Both the specific growth rate (μ) and the generation time (g) of the culture can be determined from the linear portion of the semilogarithmic growth curve that corresponds to the exponential phase (Figure 1.4). The slope of the linear portion is proportional to the reciprocal of the generation time (g):

$$\text{slope} = \frac{\log N - \log N_0}{t} = \frac{\log(2^n)}{t} = 0.301\frac{n}{t} = \frac{0.301}{g} \qquad \text{(Eq.1.5)}$$

The specific growth rate (μ) is then determined based on Eq. (1.3). The depletion of essential nutrients and/or the accumulation of waste growth products eventually cause the

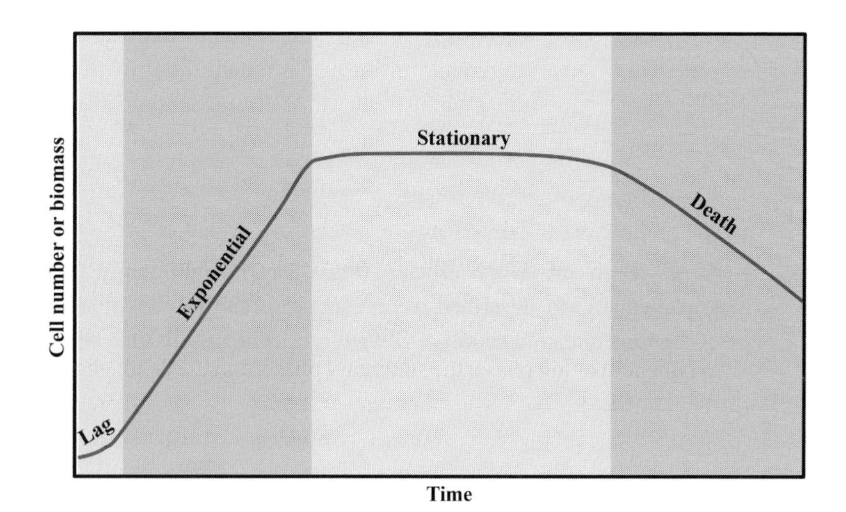

Figure 1.4 Growth curve for a batch culture.

batch culture to end the exponential phase and enter the stationary phase. In the stationary phase, the processes of cell growth and cell death are balanced out, such that the microbial population size remains constant with no net increase or decrease in cell number or cell mass:

$$\frac{dN}{dt} = 0 \qquad \text{(Eq.1.6)}$$

Some stationary phase cells may still have metabolic activities, utilizing resources provided by dead and lysed cells (endogenous metabolism). But such metabolic activities are usually at reduced rates as compared with the exponential phase, resulting in relatively smaller cells in the stationary phase. The final phase of the growth curve is the death phase, when the microbial population experiences a net loss of viable cells. This phase often occurs exponentially, although the rate of population decrease due to cell death is typically slower than the rate of population increase in the exponential phase:

$$\frac{dN}{dt} = -k_d N \qquad \text{(Eq.1.7)}$$

where k_d is the specific death rate (time^{-1}). Despite the net population decrease, individual viable cells may remain in the culture throughout the death phase.

For a batch culture, one mathematical description of the entire growth curve comprising all the four growth phases is the Monod equation:

$$\mu = \frac{\mu_{max}\ S}{K_S\ +\ S} = \frac{dN}{dt} \qquad \text{(Eq.1.8)}$$

where S is the substrate concentration (mass per volume), μ_{max} is the maximum specific growth rate constant (time^{-1}), and K_S is the half-saturation constant or the affinity constant of the substrate (mass per volume). The Monod equation reflects a relationship between the specific growth rate of the culture and the substrate concentration available to the culture. The two constants μ_{max} and K_S both depend on intrinsic physiological properties of the culture and on the substrate and the growth temperature. Considering cell yield in the culture along with substrate utilization, the Monod equation can be expressed as

$$\frac{dS}{dt} = -\frac{1}{Y}\frac{dN}{dt} = -\frac{1}{Y}\frac{\mu_{max}\ S}{K_S + S} \qquad \text{(Eq.1.9)}$$

where Y is the cell yield coefficient (mass/mass). Additionally, the entire growth curve can be mathematically described using equations that only involve the cell concentration but not the substrate concentration (Zwietering et al. 1990). One such example is the Logistic equation:

$$\frac{dN}{dt} = kN(1 - \frac{N}{N_{max}}) \qquad \text{(Eq.1.10)}$$

where N_{max} is the maximum cell concentration at the end of batch growth (cell number or biomass per volume), and k is the logistic growth rate constant (time^{-1}). The Logistic equation and other similar equations are special cases of the Monod equation with certain approximations, and the parameters in the Logistic and other equations are mathematically related to the parameters in the Monod equations.

Growth Kinetics for a Continuous Culture

A continuous culture is an open system where fresh medium is continuously added and spent medium containing cells continuously removed at a rate called the dilution rate. The growth rate and the growth yield of the continuous culture are independently controlled through the dilution rate and the concentration of a growth-limiting nutrient in the incoming medium (e.g., a C or N source) (Figure 3.10 in Maier et al., 2009). By varying the dilution rate, different growth rates can be achieved, which is useful for maintaining the microbial population in the exponential growth phase for long periods, optimizing the generation of primary or secondary growth products, and removing metabolites from the system.

1.3.1.2 Light Microscopy

Microscopes are powerful tools for visualizing cells. Coupled with other techniques, microscopy has been applied for studying microbial structure, motility, and *in situ* activity. Among many types of microscopes, light microscopes are usually available in most laboratories. Light microscopes use visible light to illuminate cell structures. Based on their contrast methods, there are different types of light microscopes, including bright-field microscopes, phase-contrast microscopes, dark-field microscopes, differential interference contrast (DIC) microscopes, and fluorescence microscopes (Table 9.1 in Maier et al., 2009). Each of these microscopes operates at a maximum practical magnification of 2000×, commonly with a resolution of 0.2 μm.

Working on these microscopes requires similar operation, and general operation steps are presented here. Sample preparation for light microscopy is simple when using wet mounts. Viable microorganisms are suspended in water, phosphate buffered saline solution, or any other appropriate liquid medium, which maintains cell viability and allows cell locomotion. A drop of the specimen suspension is placed on a precleaned glass slide with a coverslip. The glass slide is then mounted on the microscope stage and viewed. When using a 100× objective oil-immersion lens, a drop of optical grade oil should be placed between the microscope slide and the objective lens, which allows the light rays emerging from the specimen at angles to be collected and viewed, thus increasing the light-gathering ability of the lens.

While wet mounts are easy to prepare, they often yield poor contrast between the specimen and the surrounding medium under light microscopes, such that microbial morphology cannot be well observed. This is often the case for bright-field microscopy. One way to overcome this obstacle is through the fixation and staining of specimens on glass slides. However, these processes kill organisms and therefore, only allow the observation of nonviable cells. For fixation, a thin film of a wet mount on a slide is dried in air. This generates a smear, which is then fixed on the slide by gently heating it over a flame for a few seconds. A variety of dyes are used for staining, such as organic compounds that have different affinities for specific cellular materials. In general, dyes can be classified as basic dyes (positively charged) or acidic dyes (negatively charged). Commonly used basic dyes include crystal violet, methylene blue, malachite green, and safranin. They bind strongly to negatively charged cell components, such as nucleic acids and acidic polysaccharides. Since cell surfaces tend to be negatively charged, basic dyes also bind to high-affinity sites on cell surfaces. This property makes basic dyes useful for general-purpose stains. Commonly used acidic dyes include acid fuchsin, eosin, rose Bengal, and others. These bind to cellular components such as positively charged proteins. Stains using two different dyes, designated as the primary dye and the counterstain, can lend different colors to different types of cells and are therefore called differential stains. One important differential stain is the Gram stain. Regardless of which dye is used, the stain procedure is similar. The next section will use the Gram stain as an example to describe the general stain procedure in more detail.

In addition to light microscopes, there are many more advanced microscopes that can provide higher magnifications and resolutions, as well as different information about microorganisms (e.g., surface imaging). They will be presented later in this book, including atomic-force microscopy/scanning tunneling microscopy (Chapter 5), scanning electron microscopy (Chapter 6), and transmission electron microscopy (Chapter 7).

1.3.1.2.1 Gram Staining for Identification

The Gram stain is the most commonly used stain in microbiology, often used as the first step to characterize a newly isolated bacterium. Two dyes, crystal violet–iodine complex and safranin, are used in the Gram stain. Based on a bacterium's reaction with these two dyes, the bacterium is characterized as Gram-positive or Gram-negative. Gram-positive bacteria retain the primary dye complex (crystal violet–iodine) and appear purple-violet, whereas Gram-negative bacteria lose the primary dye during the rinse and are stained by the counterstain (safranin) to appear pink. This differential staining arises from differences in the cell wall structure between these two groups of bacteria. The Gram-positive cell wall mostly consists of a thick layer of peptidoglycan, whereas the Gram-negative cell wall consists of a lipopolysaccharide outer membrane and a thin layer of peptidoglycan (Madigan et al., 2014). During the Gram stain, an insoluble crystal violet–iodine complex forms inside the cells of both groups of bacteria. For Gram-positive bacteria, the alcohol rinse dehydrates the cell well and prevents the insoluble crystal violet–iodine complex from leaching out. For Gram-negative bacteria, the alcohol penetrates the lipid-rich outer membrane and rinses out the insoluble crystal violet–iodine complex. The nearly colorless Gram-negative cells are then counterstained by safranin. It should be noted that for many environmental isolates, the Gram stain could be inconclusive, such that both purple-violet cells and red cells may appear afterwards. The organisms that exhibit this trait are called Gram-variable (e.g., *Arthrobacter* spp. in soils) (Maier et al., 2009). A detailed, step-by-step method for Gram staining is given in the following (US Food and Drug Administration), which serves as an example for all stains. Reagents used for the Gram stain can also be purchased as commercial kits.

Method

Materials

- Crystal violet solution:
  - Solution A – 2 g crystal violet dissolved in 20 mL 95% ethanol.
  - Solution B – 0.8 g ammonium oxalate dissolved in 80 mL Type II (deionized) water.
  - Mix Solutions A and B. Store for 24 hours, then filter through coarse filter paper.
- Gram's iodine solution – 1 g iodine and 2 g potassium iodine dissolved in 300 mL Type II water. Store in an amber bottle.
- Decolorizer solution – 95% ethanol.
- Safranin stock solution – 2.5 g safranin dissolved in 100 mL 95% ethanol.
- Safranin solution – 10 mL safranin stock solution diluted to 100 mL with Type II water.

Procedure

First, prepare a specimen smear on a microscope slide. For a liquid culture, aseptically transfer a loopful of the liquid culture onto a precleaned glass slide and spread over a small area. For colonies grown on solid media, place a small drop of distilled water or phosphate buffered saline solution on a precleaned glass slide; aseptically transfer a loopful of colonies to the drop, and emulsify. This step should result in a thin suspension film for better staining outcomes. Allow the suspension thin film to air dry, and fix the dried film by gently heating the glass slide over a Bunsen burner flame two or three times (1–2 seconds) without exposing the dried film directly to the flame. Heat-fixation is essential for the Gram stain. Loosely fixed specimens are easily washed off during staining steps, whereas over-heating could distort cellular morphology.

Next, differentially stain the specimen smear on the microscope slide. Flood the slide with crystal violet solution and allow it to stand for 60 seconds. Gently rinse with distilled water for 5 seconds and drain the slide. Flood the slide with Gram's iodine solution and allow it to stand for 60 seconds, then rinse and drain as before. Remove excess water from the slide by blotting dry with lint-free, absorbent paper or bibulous paper. Flood the slide with the decolorizer solution and allow it to stand for 5–10 seconds, then rinse and drain. Note that excessively thick specimen smears may require longer decolorization. Sufficient and appropriate decolorization is essential for the Gram stain. Flood the slide with safranin solution and allow it to stand for 30 seconds, then rinse and drain. Gently remove excess water from the slide by blotting dry with absorbent or bibulous paper. Place a coverslip over the stained smear and press firmly on the coverslip to remove trapped air bubbles and ensure good contact between the coverslip and the smear. The slide can then be placed on a microscope, and the stained smear can be observed.

1.3.1.2.2 Cell Counting with Hemocytometer

Microbial cells in a culture or natural sample can be directly counted under a microscope (i.e., microscopic cell count), which is quick and easy and requires minimal sample preparation. This measurement can be applied to samples dried and stained on slides, as well as to liquid samples prepared as wet mounts.

For liquid samples, a device called a counting chamber (either a Neubauer hemocytometer or a Petroff-Hausser counter) can be used. A counting chamber is a specialized glass slide, consisting of squares of known area etched in a grid on the surface (Figure 1.5). The entire grid can be seen under a light microscope at 40× (10× ocular lens and 4× objective lens). To avoid the necessity for staining, a phase-contrast or DIC microscope is typically used for good visualization. When a coverslip is placed on the counting chamber, each of the squares on the grid has a precise volume. After loading a small volume of the liquid sample onto the counting chamber, one can count the number of cells per unit area in the squares and calculate the number of cells per small chamber volume accordingly. For larger organisms, cells are usually counted in squares with a surface area of 1 mm^2; for smaller organisms, cells are counted in squares with a surface area of 1/25 mm^2 (Figure 1.5). The height of the chamber formed with the coverslip is typically 0.1 mm for a Neubauer hemocytometer and 0.02 mm for a Petroff-Hausser counter. For a Petroff-Hausser counting chamber, each large square represents a volume of 2×10^{-5} mL; each small square represents a volume of 8×10^{-7} mL. Cells are counted in several large or small squares and the numbers are averaged to yield the final count, presented as number of cells per unit volume.

Certain limitations of this measurement should be noted. First, for soils or other environmental samples that contain considerable amounts of colloids and particles, interference from matrix materials is substantial. Second, microorganisms vary in their sizes and shapes, and it may be difficult to observe some small organisms or directly count organisms with atypical morphologies under light microscopes. Third, debris from dead and partially lysed cells could be mistaken for microbial cells. Additionally, for samples containing a small number of microorganisms, it is often necessary to concentrate the sample before microscopic cell counting. Without special staining techniques, dead cells cannot be distinguished from live cells; thus, microscopic counts give the total number of cells, alive or dead. A variety of fluorescent dyes can be used to distinguish live and dead cells. For example, the nucleic acid stain propidium iodide only penetrates cells with damaged membranes and is used to stain dead cells, whereas the nucleic acid stain SYTO 9 stains all cells. With such dyes, microscopic counts can be used to measure live and/or dead cells.

1.3.1.3 Physiological Measurements

In geomicrobiology, one important question is "what" activities a microbial community carries out. This is of particular interest when assessing possible impacts of certain disturbances, either natural (e.g., fire or flooding) or anthropogenic (e.g., contamination or deforestation), on functions of the microbial community. Physiological measurements also allow one to estimate the contribution of microbial activities to nutrient cycling within and between reservoirs in an ecosystem. Numerous measurements have been developed for studying microbial activities. A more comprehensive summary of commonly used methods, as well as advantages and disadvantages of each method, has been given by Maier et al. (2009). Some measurements can be performed in the field (e.g., respiration measurement using a nondispersive infrared gas sensor), but many need special reagents and equipment

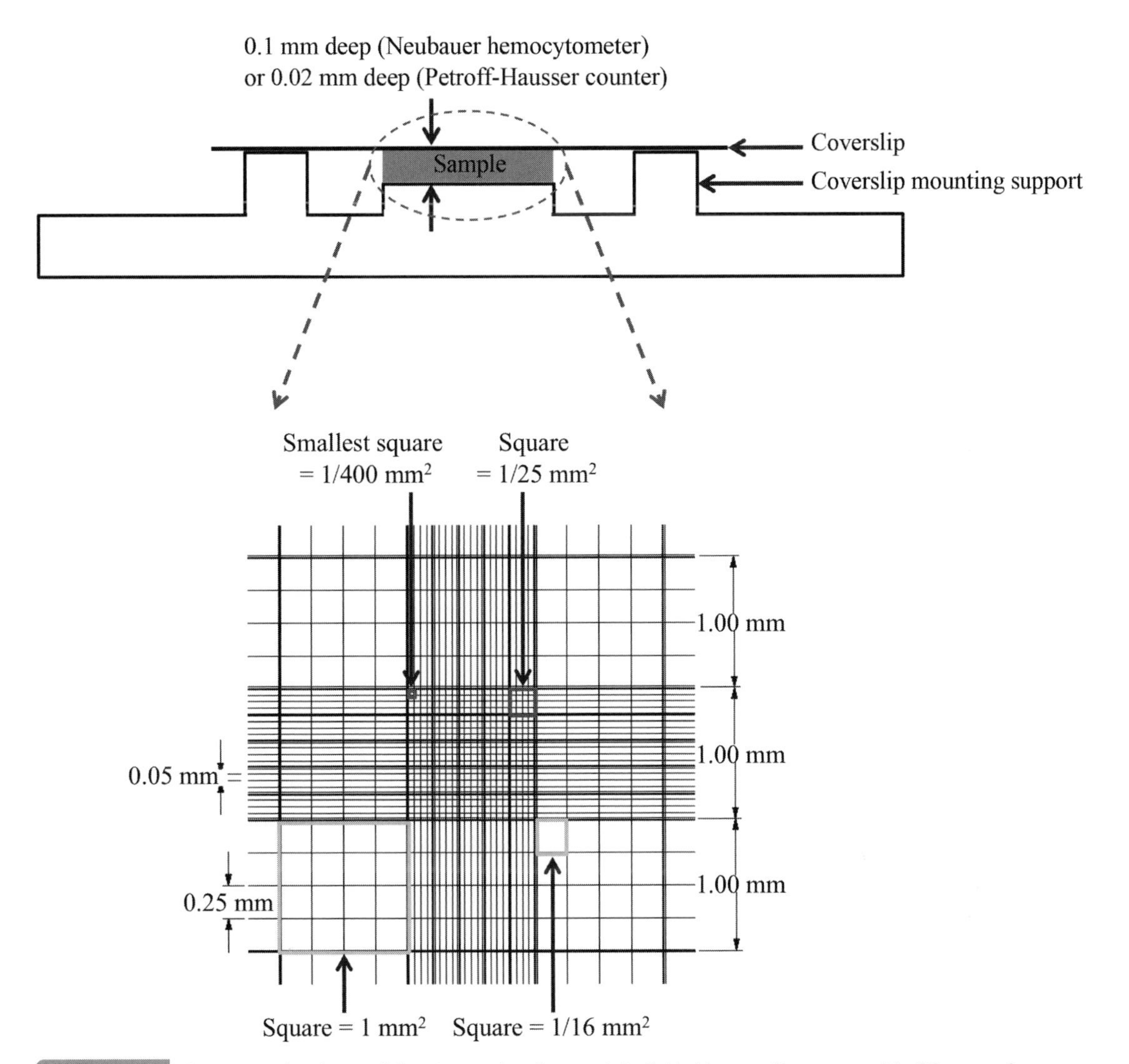

Figure 1.5 A counting chamber used for microscopic cell count. It is divided into smaller squares with different surface areas. Larger cells are counted in squares with a surface area of 1 mm^2; smaller cells are counted in squares with a surface area of 1/25 mm^2.

and therefore, are performed in the laboratory. The next sections present some physiological measurements that are commonly used and relatively feasible for many laboratories.

1.3.1.3.1 Respiration Measurement (CO_2)

One primary activity of microbial communities is the utilization of organic materials as a carbon and energy source followed by the release of CO_2, known as heterotrophic respiration. There are various approaches to measure CO_2 evolved from microbial respiration, such as CO_2 trapping, using infrared gas analyzers, and introducing ^{14}C-labeled

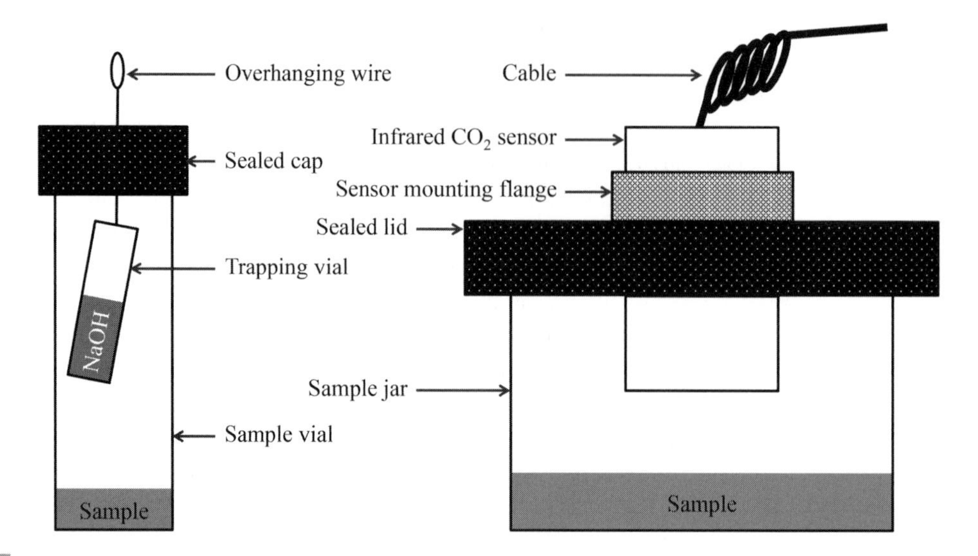

Figure 1.6 Measurement of CO_2 evolved from microbial respiration. Left schematic shows a setup using the CO_2 trapping approach; right schematic shows a setup using a nondispersive infrared CO_2 sensor. Both solid and liquid samples can be measured in these setups. Anaerobic conditions can also be achieved by appropriate modifications of these setups.

C substrates followed by isotopic analysis of $^{14}CO_2$. With appropriate experimental design, these approaches can be adapted to measure microbial respiration under both aerobic and anaerobic conditions. Methods involving chromatography are not examined here; some relevant information is presented earlier in this chapter (Section 1.3.2.2) and in Chapter 4.

The CO_2 trapping approach can be applied to both liquid and solid samples in various setups, such as a biometer flask, a glass vial with a test tube attached underneath the cap, or a plastic jar with a glass vial placed inside (Figure 1.6). This approach uses NaOH or another strong base to transform CO_2 into bicarbonate (HCO_3^-):

$$CO_2 + OH^- \rightarrow HCO_3^- \qquad \text{(Eq.1.11)}$$

The carbonate can be quantified using titrimetric, gravimetric, or conductimetric methods (Maier et al., 2009). In the titrimetric method, CO_3^{2-} is first precipitated with $BaCl_2$, and the solution is titrated with HCl in the presence of a pH indicator (e.g., phenolphthalein). The amount of CO_2 trapped is calculated from the reduction of the NaOH concentration in the trapping solution (Liu et al., 2009; Zhang et al., 2005). In the gravimetric method, CO_3^{2-} is precipitated as $BaCO_3$ by the addition of excessive $BaCl_2$. Afterwards, the $BaCO_3$ precipitate is collected, washed, dried, and weighed (Fontaine et al., 2007). In the conductimetry method, the conductance of ions in the solution is directly measured, considering the sum of total ions present (Critter et al., 2004). Often, the CO_2 trapping method is used in conjunction with substrate-induced respiration involving ^{14}C-labeled C substrates. In that case, the radioactivity of the trapped $^{14}CO_2$ is quantified using a liquid scintillation counter (Hanson et al., 2000; Hu and van Bruggen, 1997).

The evolved CO_2 can also be measured by infrared gas analyzers or sensors (Anderson and Domsch, 1993; Hartley et al., 2008; Mikan et al., 2002; Valentini et al., 2000). NDIR gas sensors (Neethirajan et al., 2009), incorporated into soil gas flux chambers or used alone as portable devices, are particularly useful for measuring soil respiration in the field (Luo and Zhou, 2010). Portable NDIR sensors are commercially available; these come with special adapter kits for horizontal or vertical installation in the soil. They perform well under a relatively wide range of operating conditions, with a wide measurement range and high accuracy. It is also convenient to use portable NDIR sensors for small samples such as a pure liquid culture or a soil microcosm in the laboratory (Figure 1.6).

During long-term incubations in the laboratory, oxygen in the headspace of sealed, airtight vessels would gradually deplete. Therefore, experimental setups should allow oxygen in the headspace to be replenished. This can be done by fluxing CO_2-free air through the system or by periodically exchanging the headspace with CO_2-free air. When measuring soil samples, nonmicrobial contribution of CO_2 (Hanson, 2000), as well as abiotic contribution of CO_2 (e.g., acid soils containing high amounts of $CaCO_3$), should be carefully considered, and the appropriate adjustments should be made.

1.3.1.3.2 Bioassays for Metabolic Activity

In the environment, intercellular and extracellular enzymatic activities of microorganisms drive a myriad of biogeochemical reactions, significantly contributing to nutrient cycling in an ecosystem (Burns, 1982; Tabatabai, 1994). Bioassays are used to determine the enzymatic activity of a pure culture or a complex sample. An enzymatic assay typically uses a surrogate substrate that is transformed by a specific class of enzymes. By monitoring the resulting product, the activity of a specific class of enzymes is quantified. It should be noted that the added surrogate substrate might not well represent the substrate spectra of all microorganisms present in the sample. Hence, the result from such an assay should be considered as an index of the general level of enzymatic activity of the microbial community. It should also be noted that such an assay usually measures the gross enzymatic activity in a sample, which could come from live microorganisms or extracellular enzymes. The measured enzymatic activity may also include contributions from microscopic eukaryotic organisms. Presented next are several common enzymatic assays, though many more have been developed (Maier et al., 2009; Nannipieri et al., 2003; Taylor et al., 2002).

Dehydrogenase Assay

Dehydrogenases are intracellular enzymes required for the respiration of organic compounds and are inactive when outside the cell. They catalyze the transfer of hydrogen and electrons between organic compounds, and their activity reflects the intracellular flux of electrons to O_2 (Nannipieri et al., 2003). Dehydrogenase assays are based on the utilization of synthetic electron acceptors, typically a water-soluble, almost colorless tetrazolium salt. When reduced, the tetrazolium salt forms a reddish formazan dye, which is detected in a variety of ways. One popularly used tetrazolium salt is INT (2-(*p*-iodophenyl)-3-(*p*-nitrophenyl)-5-phenyltetrazolium chloride), which is reduced to water-insoluble, intensively reddish INT–formazan in the assay. Environmental samples are suspended in a buffered solution (e.g., 1 M tris buffer, pH 7.0) containing INT and incubated at an appropriate temperature for a certain period of time, depending on the nature of the microbial community. A uniform rate of INT–formazan production should occur during this time, and cellular proliferation, particularly an enrichment of certain microbial groups, should not occur (Casida, 1977). Afterwards, a microscope can be used to examine the production of red INT–formazan deposits within microbial cells and to count the total cell number. The proportion of metabolically active cells can be calculated accordingly. Alternatively, INT–formazan can be extracted from microbial cells using a solvent (e.g., methanol or a 1:1 mixture of dimethylformamide and ethanol). Total INT–formazan production can then be quantified using a spectrophotometric method (464 nm) based on a standard curve (Taylor et al., 2002; von Mersi and Schinner, 1991).

Phosphatase Assay

Phosphatases are a broad group of enzymes that catalyze the hydrolysis of both esters and anhydrides of phosphoric acid (Tabatabai, 1994). These enzymes are produced and secreted by organisms, including microbes. Phosphatase activity is usually measured based on the use of *p*-nitrophenol phosphate, which is transformed to *p*-nitrophenol in the assay. The produced *p*-nitrophenol is extracted in aqueous solution and quantified using a spectrophotometric method (400–420 nm, maximum absorbance at 400 nm) with a standard curve (Tabatabai and Bremner, 1969). The pH condition of sample incubation should be carefully chosen to optimize the activity of acid or alkaline phosphatases.

Fluorescein Diacetate Hydrolysis

Fluorescein diacetate (FDA) (3′,6′-diacetylfluorescein) can be hydrolyzed, both inside and outside cells, by a variety of different enzymes including esterases, lipases, and proteases. Thus, FDA hydrolysis activity is an indicator of total microbial activity in an environmental sample, providing information on organic matter turnover in the habitat. In the assay, environmental samples are usually suspended in a buffered solution containing FDA and incubated. After incubation, the sample is centrifuged and filtered, and the supernatant of the suspension is analyzed using a spectrophotometric method (490 nm). For samples containing substantial colloids, 50% acetone (vol/vol with water) can be used to terminate FDA hydrolysis, followed by centrifugation, filtration, and spectrometric analysis (Schnürer and Rosswall, 1982).

Other Bioassays for Metabolic Activity

Assays targeting other enzymes, such as aminopeptidases, endocellulases, β-glucosidases, peroxidases, phenol oxidases, proteases, ureases, and others, have been widely used in research on functions of microbial communities in the environment (Böhme et al., 2005; Kourtev et al., 2002; Taylor et al., 2002). They are similar to the assays described earlier in terms of principles and experimental procedures.

The adenine-containing nucleotides adenosine triphosphate (ATP), adenosine diphosphate (ADP), and adenosine monophosphate (AMP) are responsible for coupling intracellular energy-producing and energy-requiring metabolic reactions (Karl and Holm-Hansen, 1978). In particular, ATP is synthesized by actively growing cells, and the ATP content of a microbial cell or a microbial population reflects the level of metabolic activity in that cell or population at that moment. ATP can be quantified using a luciferin–luciferase substrate–enzyme system. Luciferases utilize the energy supplied by free ATP and convert luciferin to oxyluciferin. The oxidation reaction generates radiant energy, which is detected and quantified using a luminometer, with light intensity proportional to the ATP concentration. Because the luciferase assay measures free ATP, it requires a cell/matrix extraction step, which could be more challenging for environmental samples than for pure cultures (Ciardi and Nannipieri, 1990; Jenkinson and Oades, 1979; Martens, 1995). Commercial kits are available for the luciferase assay, which provide enhanced quantification of kinetics, enzymatic turnover, and light intensity. In addition to ATP, ADP and AMP can be quantified similarly after enzymatic conversion to ATP (Karl and Holm-Hansen, 1978). A parameter called the adenylate energy charge (AEC) can then be calculated:

$$\mathrm{AEC} = \frac{\mathrm{ATP} + 1/2\ \mathrm{ADP}}{\mathrm{ATP} + \mathrm{ADP} + \mathrm{AMP}} \qquad \text{(Eq.1.12)}$$

AEC is a linear measure of the amount of metabolically available energy that is momentarily stored in the adenylate system. High AEC values (>0.8) reflect cells in an active state, intermediate AEC values (0.4–0.8) reflect a resting state, and low AEC values (<0.4) reflect a dead or moribund state (Maier et al., 2009).

Finally, metabolic profiles of heterotrophic microbial populations or communities can be determined using commercially available redox-based sole carbon source utilization tests, such as the BIOLOG system (Choi and Dobbs, 1999; Garland and Mills, 1991; Girvan et al., 2003). Such assays are based on tetrazolium dye reactions and allow 95 carbon sources to be simultaneously tested in a 96-well microtiter plate. The color responses are continuously monitored at 590 nm using a microplate reader.

1.3.2 Culture-independent Methods

Culture-dependent methods have a prerequisite that the species under study must be culturable in the laboratory, which already limits their usefulness in studying the majority of environmental microorganisms. This is particularly the case for microorganisms living in extreme environments, whose physiological niches have not been perceived or duplicated experimentally (Hurst et al., 2007; Parmar and Singh, 2013). This is also the case for slow-growing microorganisms in subsurface terrestrial environments (Maier et al., 2009).

Culture-independent methods are much less biased in general because they do not rely on the growth of microorganisms in the laboratory. By implementing molecular biology, they target biological macromolecules such as nucleic acids, proteins, and lipids that can provide information on microbial community structure and function. Their successful application to samples from diverse environments has unveiled the immense diversity of environmental microorganisms (Ward et al., 1990). It is now recognized that less than 1% of the total bacterial population in the environment has been culturable (Amann et al., 1995). The application of culture-independent techniques has also uncovered unprecedented genetic diversity of microbes in the environment, shedding light on microbial functions in different ecosystems (Venter et al., 2004).

Culture-independent methods use biochemical or biophysical techniques to detect, monitor, and quantify biological macromolecules. The successful recovery of target biomacromolecules is the first step in culture-independent methods, which requires efficient extraction of target molecules from samples and effective purification to remove coexisting matrix inhibitors that can interfere in downstream analyses. Depending on the target molecules, a variety of culture-independent methods can be used, including methods for DNA, RNA, and protein, which are presented in this chapter, and methods for lipid biomarkers (Chapter 14) and phylogenetic techniques (Chapter 15), among others.

While minimizing contamination risks is essential for all these methods, it is especially important for methods involving PCR, as trace amounts of contaminant DNA will be exponentially amplified during PCR, leading to biased results. Furthermore, while degradation should be avoided for all target molecules, certain molecules are more susceptible than others. As discussed earlier in this chapter (Section 1.2.2), RNA degrades more easily than DNA because of its molecular structure and the ubiquity of RNases in the environment. For culture-independent analyses, common contamination sources include the laboratory environment (e.g., untreated surfaces and dust particles), researchers (e.g., saliva, hair, perspiration, and skin oils), labware (e.g., contaminated tips and tubes), and experimental reagents (e.g., low-quality water). Ideally, laboratories designed for molecular biology analyses should consist of separate areas for sample extraction, reaction setup, and post-reaction analyses. Whenever possible, RNA work should be conducted in an area separate from other work. PCR assays should ideally be set up in a workstation that combines an International Standards Organization (ISO) 5 clean air environment with UV light sterilization for protection from background and cross-contamination. Decontamination of laboratory surfaces should be achieved using 70% ethanol and 10% bleach before and after experiments, and an RNase decontamination solution should additionally be used to decontaminate surfaces when working with RNA. Disposable labware and filter tips that are certified free of detectable RNase, DNase, DNA, and pyrogens are highly recommended. Culture-independent analyses should also use commercially available molecular biology grade water and buffers free of nucleic acids and nucleases, since autoclaving does not completely remove nucleic acids or nucleases.

1.3.2.1 Extraction of Biological Macromolecules

Nucleic acids (DNA and RNA) and proteins are two types of biomacromolecules commonly analyzed in geomicrobiological investigations. Analysis of DNA, the genetic basis of organisms other than RNA viruses, can provide researchers with taxonomic information on a microbial community, such as species diversity (richness and evenness) and community composition, and the population size of specific microbial groups. Analysis of DNA can also inform researchers of the functional potential of the microbial community and the abundance of specific functional genes in the community. Analysis of RNA, especially messenger RNA (mRNA) produced during transcription, and proteins generated during translation can provide researchers with qualitative and quantitative information on gene expression in the microbial community, including the activity of specific enzymes, under particular environmental conditions.

Efficient and effective extraction of these biomacromolecules is important for culture-independent methods (Tan and Yiap, 2009). It can be easily achieved for pure cultures but can be challenging for environmental samples (Daniel, 2005). This is particularly the case where concentrations of biomacromolecules are low (e.g., deep subsurface with a small microbial population) or where environmental matrices contain extensive inhibiting chemicals (e.g., sediments or soils with substantial humic substances). Many conventional methods can be used for extraction of biomacromolecules from both pure cultures and environmental samples. Additionally, a variety of commercial kits have been developed for pure cultures and to meet specific needs of challenging samples such as sediments and soils. These commercial kits come with manufacturer protocols, which can be further optimized to accommodate particular samples. Automated systems are increasingly used to streamline extraction tasks, making it easier to handle a large volume of samples within a relatively short time. Regardless of the extraction method used, care should be taken to achieve unbiased representation of the microbial community, which is critical to all downstream culture-independent analyses.

1.3.2.1.1 Nucleic Acids

Two nucleic acids, DNA and RNA, are commonly analyzed in geomicrobiology. While it may be possible to separate microbial cells from matrix materials first and then extract nucleic acids from separated cells, direct extraction of total nucleic acids from the environment is more widely used because it is less biased and more efficient. Direct extraction of nucleic acids from various environmental samples usually comprises similar steps, including cell lysis, by chemical/biochemical approaches (e.g., using lysis buffer containing surfactants such as sodium dodecyl sulfate (SDS) or Triton X-100 and lysozyme) or physical approaches (e.g., bead beating or ultrasonication), followed by purification through ethanol or isopropanol precipitation, phenol/chloroform/isoamyl alcohol extraction, or spin column-based solid-phase extraction (Hurt et al., 2001; Tan and Yiap, 2009).

In the first step, physical approaches can better disrupt cellular structure and desorb nucleic acids from particles, but mechanical shearing may cause fragmentation of nucleic acids, which could be a problem for certain downstream analyses. The crude extract contains both DNA and RNA, and an endonuclease called deoxyribonuclease I (DNase I) is used to nonspecifically degrade DNA. This allows the removal of DNA from the crude extract and only RNA to be purified afterwards. Additionally, recovery of RNA from environmental samples requires special attention, as RNA is highly susceptible to degradation by RNases during extraction procedures. RNases are ubiquitously present in the environment. Therefore, RNA extraction methods or commercial kits include a special treatment to inactivate RNases, often using β-mercaptoethanol.

In the second step, nucleic acids are separated from coextracted biological molecules such as proteins and coexisting matrix chemicals such as humic acids and salts. In ethanol or isopropanol precipitation, an acetate salt is added with ethanol or isopropanol (sodium acetate with ethanol and ammonium acetate with isopropanol) to the crude extract; the mixture forms precipitates of nucleic acids at low to normal temperatures (Green and

Sambrook, 2016). In phenol/chloroform/isoamyl alcohol extraction, coextracted proteins are denatured and accumulate in the organic phase or at the interphase, while nucleic acids remain in the aqueous phase, allowing further collection. In spin column-based methods, nucleic acids are selectively bound to a silica-based membrane under optimal binding conditions, and coextracted molecules are washed off through the mini column. Nucleic acids are then released from the membrane and eluted in an appropriate solution. Alternatively, nucleic acids are selectively bound to magnetic particulate materials, which are separated from the crude extract by the application of a magnetic field.

Conventional methods for nucleic acid extraction usually comprise various combinations of cell lysis by chemical/biochemical approaches and nucleic acid purification by ethanol precipitation or phenol/chloroform extraction. An example protocol of this kind is in Wilson (1997). Commercial extraction kits mostly utilize bead beating to facilitate cell lysis and solid-phase extraction to purify nucleic acids, thereby avoiding the incomplete phase separation commonly associated with conventional methods. Some commercial kits are designed particularly for samples containing a high content of humic substances, using reagents to precipitate contaminant organic and inorganic matter. It is also often worth combining conventional methods and spin column-based methods for better performance (Miller et al., 1999; Zhou et al., 1996). However, it should be noted that contaminant DNA is likely present in commonly used extraction kits and other laboratory reagents, and caution should be advised when applying sequence-based analyses to study microbial communities in low-biomass environments (Salter et al., 2014).

Extraction methods vary in performance, and the quality and quantity of extracted nucleic acids could influence microbial community profiling (Carrigg et al., 2007; Feinstein et al., 2009; Leff et al., 1995; Willner et al., 2012). Different methods are used to examine the quality and quantity of extracted nucleic acids. One method is to measure absorbance using a microvolume spectrophotometer (e.g., the NanoDrop) or a regular spectrophotometer if sufficient sample volume is available. In particular, the ratio of absorbance at 260 nm and 280 nm (A_{260}/A_{280}) serves as an indicator of purity. A_{260}/A_{280} of ~1.8 and ~2.0 is considered acceptable for DNA and RNA, respectively. Lower A_{260}/A_{280} could indicate contamination from protein, phenol, or other molecules that absorb strongly at ~280 nm. The ratio of absorbance at 260 nm and 230 nm (A_{260}/A_{230}) serves as another indicator, with an acceptable value of 2.0–2.2. Lower A_{260}/A_{230} may indicate the presence of contaminants that strongly adsorb at ~230 nm, such as carbohydrates and phenolic solutions. The quality and quantity of nucleic acids can also be examined by gel electrophoresis (Kasibhatla et al., 2006; Rio et al., 2010). After gel electrophoresis, gel-trapped nucleic acids of a particular size range can be purified from the gel using conventional methods or by commercial kits (Nilsen, 2013). Additionally, more sensitive and accurate quantification of double-strand DNA (dsDNA) can be accomplished using dsDNA-specific fluorescent stains (e.g., PicoGreen assay).

Finally, most direct extraction methods do not distinguish between intracellular and extracellular DNA, although some methods have been developed to capture particular portions (Taberlet et al., 2012). Because extracellular DNA can persist in the environment (Barnes et al., 2014; Levy-Booth et al., 2007), information gained from analysis of the total microbial DNA in a sample reflects intact cells (live or dead), cell debris, and extracellular

DNA fragments. RNA analysis gives more accurate information on active microbial populations in the environment.

1.3.2.1.2 Proteins

Similarly to nucleic acid extraction, recovering proteins from environmental samples mainly comprises cell lysis followed by protein purification, either after separating cells from the environmental matrix or by direct extraction from the bulk sample, with the latter having advantages in unbiased representation of complete proteomes (Scopes, 1993).

A critical step of direct extraction is cell lysis, which is usually accomplished through chemical approaches, such as surfactant treatment (e.g., SDS or Triton X-100) or strong chemical treatment (e.g., NaOH or urea), or through physical approaches such as ultrasonication or boiling. To better break down the cell wall of Gram-positive bacteria, lysozyme can be added to the lysis buffer. To minimize protease-catalyzed protein degradation, a commercially available protease inhibitor cocktail can be added to the lysis buffer. After extraction, proteins can be purified using conventional approaches such as phenol extraction coupled with ammonium acetate precipitation, or trichloroacetic acid precipitation coupled with an acetone wash. A detailed step-by-step protocol of such conventional approaches is outlined in Isaacson et al. (2006). Individual extraction and purification approaches can be applied in various combinations to suit the need of a particular sample (Benndorf et al., 2007; Chourey et al., 2010; Keiblinger et al., 2012). Currently, there are commercially available lysis buffers designed for optimal performance on special sample types. There are also commercially available filters for removing cell debris, proteases, and other contaminants from the cell lysis process. They can be used individually or in combination to achieve the purpose of a particular study.

Extraction and purification methods also depend on the types of target proteins and the analyses to be performed on the proteins. For research on membrane proteins, extraction of the functional protein from its natural lipid membrane is often required. Although detergents in conventional extraction methods could solubilize the target membrane protein or protein complex from the membrane phospholipid bilayer (Caldwell et al., 1981; Filip et al., 1973), such methods have certain disadvantages in maintaining protein structure and activity. New methods are continuously being developed to overcome this challenge. For example, a recently developed method utilizes styrene maleic acid copolymer to isolate membrane proteins, allowing the proteins to remain with the native lipids (Lee et al., 2016).

After extraction, the concentration of proteins can be measured using a variety of methods, most commonly by the Lowry assay based on the Folin reaction (Lowry et al., 1951), the Bradford assay based on the Coomassie Blue reaction (Bradford, 1976), or the BCA assay based on the bicinchoninic acid–copper reaction (Smith et al., 1985). These assays are all colorimetric methods, with the Lowry assay monitoring a blue color at 750 nm, the Bradford assay monitoring a blue color at 595 nm, and the BCA assay monitoring a purple color at 562 nm. It should be noted that the BCA assay is not suitable for samples containing reducing agents, as they interfere with the reaction.

Extracted proteins can be separated using sodium dodecyl sulfate polyacrylamide gel electrophoresis (SDS-PAGE) and analyzed by two-dimensional PAGE (Görg et al., 2004;

O'Farrell, 1975). Additionally, spots and bands can be excised from the gel with individual proteins identified by liquid chromatography in line with mass spectrometry (MS) via electrospray ionization source (LC-ESI-MS) (Benndorf et al., 2007). Currently, new MS techniques have driven the development of MS-based shotgun proteomics, in which proteins are first digested to peptides (Schneider and Riedel, 2010). Particular sample preparation methods have been developed to meet the requirements of MS-based proteomics (Wiśniewski et al., 2009).

1.3.2.2 Nucleic Acid-based Analysis

A variety of methods can be employed using nucleic acids to study microbial populations and communities, such as gene probing techniques (e.g., microarrays and fluorescent *in situ* hybridization; Chapter 8), PCR fingerprinting (e.g., denaturing gradient gel electrophoresis [DGGE], terminal restriction enzyme fragment length polymorphism [T-RFLP], or automated ribosomal intergenic spacer analysis [ARISA]), reverse transcription PCR (RT-PCR) and real-time qPCR, and next-generation sequencing (NGS)-based metagenomics and metatranscriptomics. A typical flowchart of nucleic acid-based analysis is shown in Figure 1.7. This section does not aim to provide an exhaustive elucidation of these techniques, but focuses on several techniques, including microarray, qPCR, metagenomics, and metatranscriptomics, that are of increasing interest to geomicrobiological investigations due to their ability to provide quantitative and/or high-throughput data.

1.3.2.2.1 Microarray

A microarray is a collection of thousands to hundreds of thousands of gene probes (oligonucleotides) arrayed on a chip. It exploits complementary base pair hybridization between gene probes and nucleic acid samples to detect sequences of interest. Nucleic acid samples are fluorescently labeled and hybridized to gene probes on the microarray. Successful hybridization is detected using a microarray reader that is equipped with a laser fluorescent imaging device. Gene probe construction has been described in detail in manuals such as that by Sambrook and Russell (2001).

Depending on research purposes, the oligonucleotides are designed to probe genetic markers of specific microbial groups (phyloarrays or phylochips) or functional genes encoding proteins with biogeochemical significance (functional gene microarrays). Phylochips are used for biodiversity studies, allowing researchers to track population dynamics and community profile changes across a variety of species on the same chip (Brodie et al., 2006; DeSantis et al., 2007). Functional gene microarrays are used for studying the functional diversity of microbial communities, allowing researchers to examine a broad range of metabolic capacities on the same chip. For example, the *GeoChip* contains ~50000 gene sequences from more than 290 gene categories, encompassing metabolic functions such as the production and consumption of methane, dissimilative metal reduction, heavy metal resistance, degradation of recalcitrant chlorinated pollutants, and common redox reactions in the C, N, and S cycles (He et al., 2007).

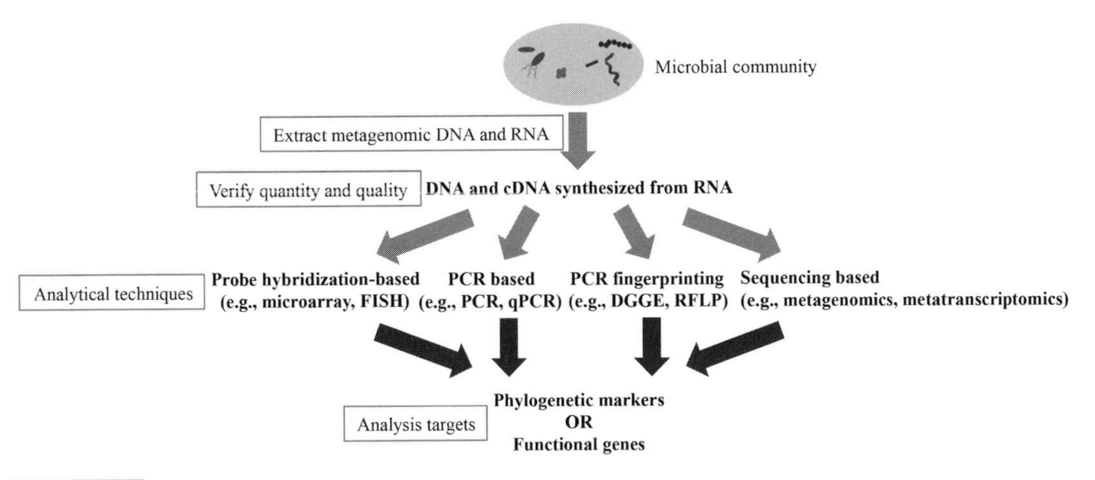

Figure 1.7 Typical flow chart of nucleic acid-based analyses for studying a microbial community in the environment.

Microarrays are also used for transcriptome analysis, in which mRNA extracted from a sample is fluorescently labeled and hybridized to DNA probes on the microarray. Alternatively, mRNA is converted through reverse transcription to complementary DNA (cDNA), which is then fluorescently labeled and hybridized to DNA probes on the microarray (Schena et al., 1995).

Compared with PCR and PCR-based fingerprinting techniques, microarrays can generate much more intensive data within a short time. However, gene variants that are highly similar in sequence may not be resolved on a microarray due to nonspecific hybridization, leading to false positive results.

1.3.2.2.2 Real-time Quantitative PCR

The development of qPCR by Higuchi et al. (1992) has been a great advance for PCR-based techniques (Kubista et al., 2006). qPCR assays can be designed for different purposes. If targeting phylogenetic markers, qPCR can quantify specific groups in a microbial community, such as all eubacteria, ammonium-oxidizing bacteria, ammonium-oxidizing archaea, and others. If targeting functional genes, qPCR can quantify genes involved in specific metabolic and/or biogeochemical processes, such as ammonium oxidases, arsenate reductases, mercury methyltransferases, and others. Similarly to conventional PCR, qPCR exploits DNA polymerases to amplify sequence fragments between primer binding sites. Specifically, a qPCR reaction mixture contains fluorescent molecules that bind to amplification products as they are generated during thermal cycles, resulting in an increase in fluorescence. At the end of each PCR cycle, the fluorescence level is quantified by a fluorescence reader integrated with the thermocycler. This allows an almost immediate result, avoiding the need for post-amplification DNA visualization.

Two approaches are used to introduce fluorescence molecules in qPCR. The first uses a general fluorescent dye, SYBR Green I, which nonspecifically intercalates into dsDNA (Morrison et al., 1998). Because SYBR Green I present in the qPCR reaction mixture becomes fluorescent only when bound, an increase in fluorescence indicates the generation of amplicons. The second approach is a combination of PCR and gene probe techniques. It uses sequence-specific fluorescent probes made by attaching a fluorescent dye to a short DNA probe that matches the target sequence. The dye fluoresces only when dsDNA of the correct sequence accumulates. TaqMan or 5'-exonuclease probes are the most commonly used sequence-specific fluorescent probes (Heid et al., 1996; Holland et al., 1991). With sequence-specific fluorescent probes and multiple dyes, multiplex qPCR reactions can be performed using similar setups as for conventional PCR. However, consideration should be given to fluorescence excitation and emission wavelengths in order to minimize the interference between samples.

qPCR generates an amplification curve for each reaction (Figure 1.8A). When the fluorescence signal of the sample increases to higher than the background or threshold value, a positive result is indicated. The threshold fluorescence value can be set manually or using a default algorithm that considers the average fluorescence of a number of early cycles (i.e., 6–10 cycles). For each sample, the cycle number at which the amplification curve crosses the threshold value is defined as the sample's threshold cycle value (*Ct* or *Cq* depending on the instrument). *Ct* values are used to calculate the concentration of initial template present in the qPCR reaction mixture, which is then converted to the concentration of template in the sample.

For qPCR reactions that use the nonspecific dsDNA dye SYBR Green I, a melt curve analysis should be performed afterwards to confirm the specificity of the reaction (Ririe et al., 1997). A melt curve measures fluorescence change as a function of temperature change (Figure 1.8B). As temperature gradually increases from below the annealing temperature to above the denaturing temperature of the amplicon, SYBR Green I is released and the system's fluorescence signal gradually decreases. When the amplicon's denaturing temperature is reached, a rapid decrease in fluorescence results in a sharp peak in the melt curve. The shape and position of this melt curve are functions of the GC (guanine–cytosine) content, length, and sequence of the amplicon and are used to differentiate desired and undesired products, eliminating the need for verification by gel electrophoresis. Using high-resolution instruments, melting analysis can permit simultaneous mutation scanning and genotyping (Montgomery et al., 2007).

Quantification based on qPCR data is usually conducted using two methods: absolute quantification, which utilizes a standard curve, and relative quantification, which makes use of an internal reference gene. In absolute quantification, standard curves are generated using serial dilutions of plasmid DNA or genomic DNA (including cDNA) (Heid et al., 1996). While genomic standards are straightforward, the construction of plasmid standards requires amplification and cloning of the target sequence. Standard curves are plotted as *Ct* values against logarithms of standard concentrations in copy number or mass (Figure 1.9). The slope of the standard curve is used to determine the amplification efficiency (*E*):

$$E = 10^{\frac{-1}{\text{slope}}} - 1 \qquad \text{(Eq.1.13)}$$

Under ideal conditions, PCR products will double during each cycle in the exponential phase, giving an efficiency of 100%. It is therefore useful to check the PCR efficiency and assess the assay's performance. In relative quantification, the *Ct* value of a target gene is

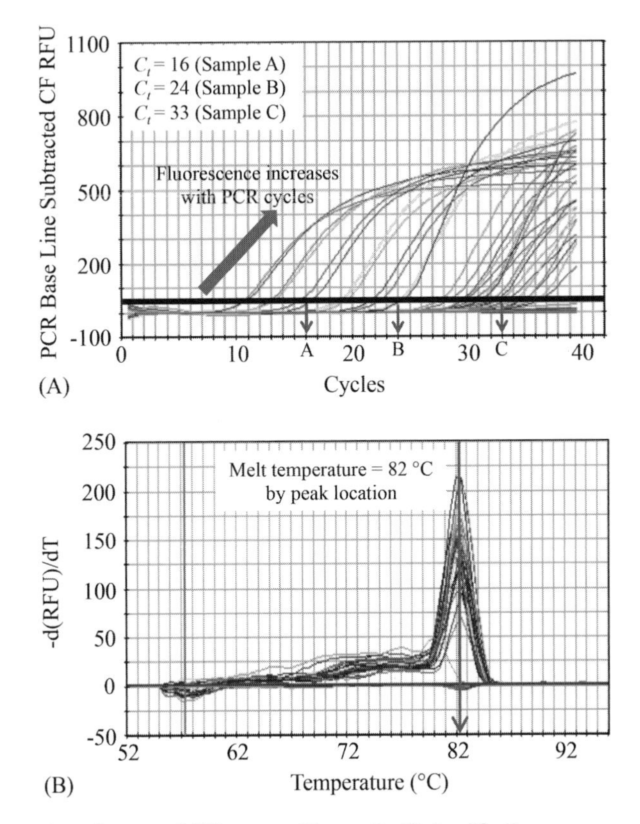

Figure 1.8 qPCR of a target gene in environmental DNA extracted from soils. (A) Amplification curves generated using 10-fold serial dilutions of soil DNA samples, along with positive and negative controls. The horizontal line represents the algorithm-defined threshold value used to determine *Ct* values. (B) Melt curves of the amplicons generated from soil DNA samples, with positive and negative controls.

compared with an internal reference gene (Livak and Schmittgen, 2001; Pfaffl, 2001). One common method for relative quantification is the $2^{-\Delta\Delta Ct}$ method, which makes the assumptions that the PCR efficiency is close to 100% and that the efficiency of the target gene is similar to that of the internal reference gene (Livak and Schmittgen, 2001). A detailed equation derivation and instructions for the $2^{-\Delta\Delta Ct}$ method have been given by Schmittgen and Livak (2008). An alternative method for relative quantification considers differences in the PCR efficiency of target and reference genes but is otherwise similar to the $2^{-\Delta\Delta Ct}$ method (Pfaffl, 2001). In general, absolute quantification is more useful for determining absolute concentrations of genes, whereas relative quantification is often used for assessing gene expression changes after treatment.

Compared with conventional PCR, qPCR has advantages of high sensitivity (it is possible to detect the template down to a single-molecule level), time savings by avoiding gel electrophoresis, and accurate quantification. qPCR is also used to analyze transcripts, in which mRNA is first converted to cDNA through reverse transcription (Bustin, 2000). Coupled with appropriate primers, qPCR can be applied to quantify both phylogenetic

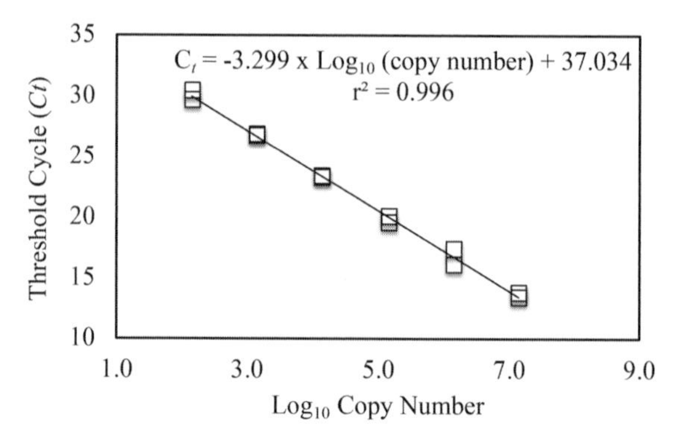

Figure 1.9 Example of a standard curve used for quantification in real-time quantitative PCR (qPCR). Two independent serial dilutions of plasmid clones are used as standards. In this example, PCR efficiency is calculated as 101%, based on the slope of −3.299 and Eq. (1.13). The correlation coefficient is determined as 0.996.

markers and functional genes. However, many factors should be carefully considered for robust quantitative results (Bustin et al., 2009). For example, primers used in qPCR are usually designed for short amplicons (75–300 bp) in order to achieve optimal assay efficiencies. The presence of primer dimers or any short nonspecific products could influence the fluorescence signal and thereby, the quantitative results. This is particularly the case for SYBR Green I-based assays. Potential contamination by bioaerosols during processing should be avoided, as this not only leads to false positive results but also lowers the sensitivity of an assay. For environmental samples, several issues need additional attention, including the presence of PCR inhibitors (e.g., humic substances and metals coextracted with nucleic acids) and nonspecific primer binding due to the vast sequence diversity of environmental metagenomes, among others. These issues all affect the sensitivity, accuracy, and efficiency of qPCR reactions. In particular, the removal of inhibiting substances could be challenging, depending on the nature of the sample, and so improved nucleic acid extraction and purification should be carried out for those samples. Serial dilution of environmental DNA is another convenient approach to test and mitigate the inhibiting effects of matrix substances (Bustin et al., 2009). Additionally, modified DNA polymerases with increased tolerance to inhibitors are now commercially available, which can enhance the sensitivity and overall performance of qPCR under various experimental conditions.

1.3.2.2.3 Metagenomics and Metatranscriptomics

Metagenomics is the analysis of all genomes in a microbial community inhabiting a particular environmental niche. Through the sequencing and analysis of total DNA directly extracted from the environmental sample, metagenomics holds the promise to provide a comprehensive picture of the structure and function of the microbial community. Original metagenomics used cloning methods to capture fragments of environmental DNA and create a clone library for sequencing (Handelsman, 2004; Tringe et al., 2005; Tyson et al.,

2004). The introduction and widespread use of NGS techniques greatly accelerated the advance of metagenomics, given its significantly lower per-base cost and higher throughput within a shorter time period as compared with conventional sequencing (Schuster, 2008).

A variety of commercial NGS systems are now widely used for metagenomics, including the Roche 454 system, Applied Biosystems SOLiD (Sequencing by Oligo Ligation Detection) system, Illumina GA (Genome Analyzer) system/HiSeq system/MiSeq system, Life Technologies Ion Torrent PGM (Personal Genome Machine) system/S5 system/S5 XL system, Helicos Biosciences HeliScope system, Pacific Bioscience (PacBio) system, and Nanopore system (Liu et al., 2012; Shendure and Ji, 2008). These second- and third-generation sequencing technologies comprise a number of methods that can be grouped broadly based on template preparation, and sequencing and imaging (Metzker, 2010). Two methods are used for template preparation in NGS: clonal amplification of templates from random fragments or mate-pair targets by emulsion PCR or solid-phase amplification (e.g., Roche 454 and Illumina systems), and single-molecule templates prepared by an immobilizing primer, template, or polymerase on solid supports (e.g., PacBio system). Sequencing and imaging of NGS is conducted through single nucleotide addition (pyrosequencing) (Roche 454), cyclic reversible termination (e.g., Illumina systems), sequencing by ligation (e.g., SOLiD system), hydrogen ion detection (e.g., Ion Torrent systems), or real-time sequencing (e.g., PacBio system). The sequencing techniques of NGS can be categorized into sequencing-by-synthesis (most systems) and sequencing by oligo ligation (SOLiD system). NGS imaging is based on bioluminescence (Roche 454), fluorescence (most systems), or chemical (Ion Torrent systems) signals. Each technique has its own advantages and disadvantages. More detailed descriptions, including principle, technique, sample preparation, and output (read length, run time, etc.), and a comparison of these systems in terms of performance, usage, and cost have been outlined by Liu et al. (2012), Mardis (2008), Metzker (2010), Quail et al. (2012), and Shokralla et al. (2012). Figure 1.10 shows the historical development of NGS techniques and their major features (Shokralla et al., 2012).

NGS-based metagenomics is currently applied in two types of research, either focusing on phylogenetic markers such as 16S ribosomal RNA (rRNA) genes for a comprehensive picture of a microbial community's structure, or targeting all genes and genomes present in the community for a better understanding of the community's metabolic function or functional potential. A full presentation of the first type of metagenomics is given in Chapter 16. A brief introduction of the second type of metagenomics, sometimes also called whole genome shotgun (WGS) metagenomics, is given here. WGS metagenomics allows researchers to probe the metabolism of all microorganisms, both cultivated and uncultivated, through directly studying their genomes without a reliance on cultivation or reference genomes. Comparison of the collective information on microbial communities, assisted by niche space metadata, has been a powerful approach to explore microbial functions across different ecosystems (Dinsdale et al., 2008). Occasionally, complete or near-complete genomes of uncultivated organisms can be *de novo* assembled from the environmental metagenomic DNA, which sheds light on the metabolic capacities of uncultivated and little-known organisms that drive transformations central to biogeochemical cycles (Wrighton et al., 2012). New genome-resolved information on numerous uncultivated organisms has allowed researchers to infer a dramatically expanded version

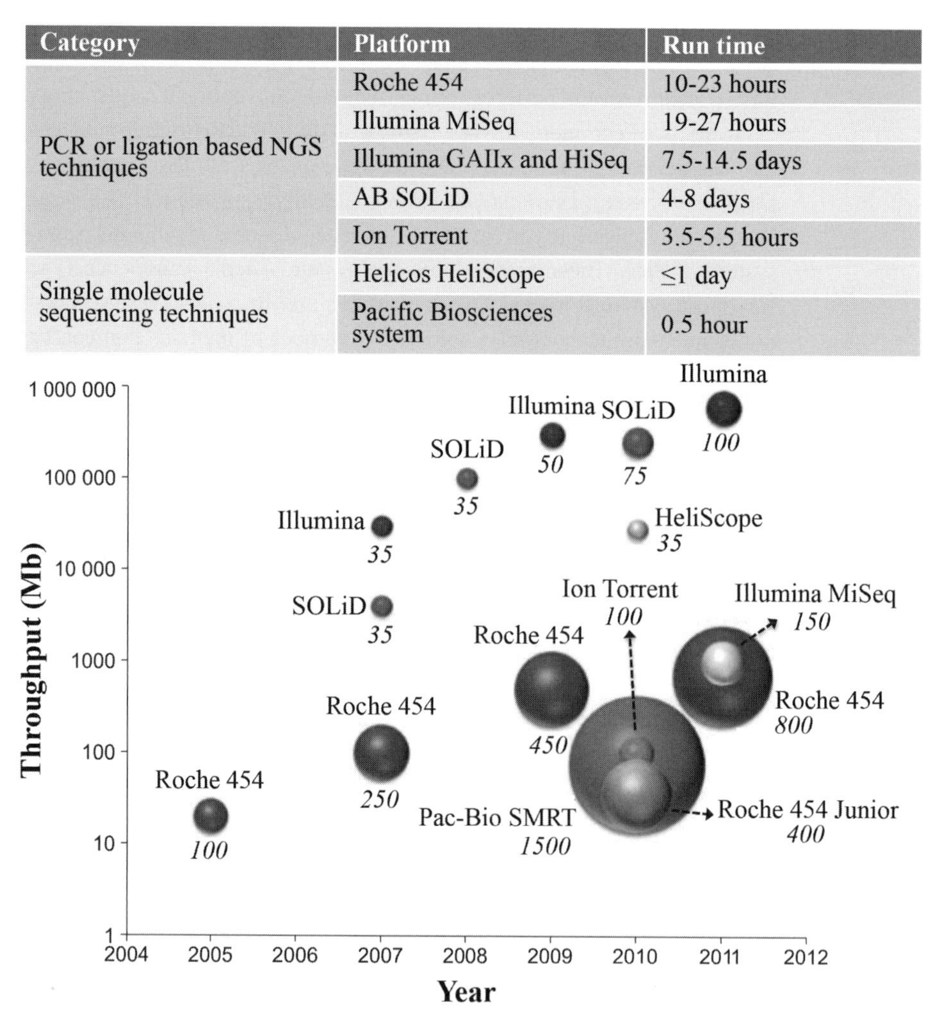

Category	Platform	Run time
PCR or ligation based NGS techniques	Roche 454	10-23 hours
	Illumina MiSeq	19-27 hours
	Illumina GAIIx and HiSeq	7.5-14.5 days
	AB SOLiD	4-8 days
	Ion Torrent	3.5-5.5 hours
Single molecule sequencing techniques	Helicos HeliScope	≤1 day
	Pacific Biosciences system	0.5 hour

Figure 1.10 Historical development of next-generation sequencing (NGS) techniques. The diameter of each bubble represents the sequencing read length of the platform (in base pairs [bp]). Adapted from Shokralla et al. (2012).

of the tree of life, highlighting major lineages currently underrepresented in biogeochemical models (Brown et al., 2015; Hug et al., 2016).

Metagenomic data analysis is usually more challenging than data generation itself, as modern NGS techniques easily generate large volumes of data within days. Analysis of short reads generated by NGS requires special bioinformatics efforts (Thomas et al., 2012). A typical workflow of metagenomic data analysis consists of quantity control of sequence data (Patel and Jain, 2012), alignment of short reads to a reference (Li and Homer, 2010) or *de novo* assembly from paired or unpaired reads (Magoc et al., 2013; Miller et al., 2010), and metagenome annotation by classifying sequences to known functions or taxonomic units based on homology searches against annotated databases (Markowitz et al., 2011; Meyer et al., 2008). In addition, composition-based or similarity-based binning facilitates recovery of more complete and higher-fidelity microbial genomes from metagenomes by sorting sequences into groups that might represent an individual genome or genomes from closely related organisms (Thomas et al., 2012).

While metagenomics unveils a microbial community's metabolic capacity, metatranscriptomics deciphers which genes in the community are actually expressed and the relative level of that expression. Thereby, metatranscriptomics provides a snapshot of the global transcriptional activities of a microbial community in its environmental niche at the time of sampling. Technically, metatranscriptomics is analogous to metagenomics in many regards, but it analyzes the collective RNA sequences (mRNA and noncoding RNA species) from a microbial community rather than DNA (Wang et al., 2009). A schematic for data generation and analysis steps in NGS-based transcriptomics is shown in Martin and Wang (2011) and in Ozsolak and Milos (2011). For microbial communities, noncoding RNA species can constitute a substantial proportion of total RNA. Depending on research purposes, noncoding RNA can be removed before cDNA synthesis (He et al., 2010). However, these depletion methods may bias the quantification of highly abundant transcripts (Martin and Wang, 2011). Alternatively, analysis of both mRNA and rRNA allowed simultaneous assessment of community structure and function in a single experiment (Urich et al., 2008). Focusing on mRNA, metatranscriptomes can reveal gene-specific and taxon-specific expression patterns as well as gene categories previously undetected in metagenome annotation (Frias-Lopez et al., 2008).

1.4 Conclusions

Traditional and modern geochemical and microbiological techniques applicable to geomicrobiology investigations were presented in this chapter. Background information, sample preparation techniques, and limitations are provided for each technique, with additional information and complete methods provided where appropriate. While this chapter does not go into depth on many of the techniques, relevant references are provided. The reader should refer to the references within this chapter for standard or detailed methods and for more complete discussions of technique development, advantages and disadvantages, and specific requirements. Furthermore, many of these techniques are rapidly evolving, so the reader should learn as much as possible about the current status of the technique prior to use.

1.5 References

Amann, RI, Ludwig, W and Schleifer, KH 1995, 'Phylogenetic identification and *in situ* detection of individual microbial cells without cultivation', *Microbiological Reviews*, vol. 59, no. 1, pp. 143–169.

Anderson, TH and Domsch, KH 1993, 'The metabolic quotient for CO_2 (qCO_2) as a specific activity parameter to assess the effects of environmental conditions, such as pH, on the microbial biomass of forest soils', *Soil Biology and Biochemistry*, vol. 25, no. 3, pp. 393–395.

Atlas, RM 2010, *Handbook of microbiological media*, CRC Press, Florida.

Barnes, MA, Turner, CR, Jerde, CL, et al. 2014, 'Environmental conditions influence eDNA persistence in aquatic systems', *Environmental Science & Technology*, vol. 48, no. 3, pp. 1819–1827.

Bates, TE 1993, 'Soil handling and preparation', in MR Carter, (ed), *Soil sampling and methods of analysis*, CRC Press, Florida.

Batley, GE (ed) 1989, *Trace element speciation analytical methods and problems*, CRC Press, Florida.

Benndorf, D, Balcke, GU, Harms, H and von Bergen, M 2007, 'Functional metaproteome analysis of protein extracts from contaminated soil and groundwater', *The ISME Journal*, vol. 1, no. 3, pp. 224–234.

Blakemore, LC 1968, 'Determination of iron and aluminum in Tamm's soil extracts', *New Zealand Journal of Agricultural Research*, vol. 11, no. 2, pp. 515–520.

Böhme, L, Langer, W and Böhme, F 2005, 'Microbial biomass, enzyme activities and microbial community structure in two European long-term field experiments', *Agriculture, Ecosystems & Environment*, vol. 109, no. 1, pp. 141–152.

Bradford, MM 1976, 'A rapid and sensitive method for the quantitation of microgram quantities of protein utilizing the principle of protein-dye binding', *Analytical Geochemistry*, vol. 72, no. 1–2, pp. 248–254.

Brodie, EL, DeSantis, TZ, Joyner, DC, et al. 2006, 'Application of a high-density oligonucleotide microarray approach to study bacterial population dynamics during uranium reduction and reoxidation', *Applied and Environmental Microbiology*, vol. 72, no. 9, pp. 6288–6298.

Brown, CT, Hug, LA, Thomas, BC, et al. 2015, 'Unusual biology across a group comprising more than 15% of domain Bacteria', *Nature*, vol. 523, no. 7559, pp. 208–211.

Burns, RG 1982, 'Enzyme activity in soil: location and a possible role in microbial ecology', *Soil Biology and Biochemistry*, vol. 14, no. 5, pp. 423–427.

Bustin, SA 2000, 'Absolute quantification of mRNA using real-time reverse transcription polymerase chain reaction assays', *Journal of Molecular Endocrinology*, vol. 25, no. 2, pp. 169–193.

Bustin, SA, Benes, V, Garson, JA, et al. 2009, 'The MIQE guidelines: minimum information for publication of quantitative real-time PCR experiments', *Clinical Chemistry*, vol. 55, no. 4, pp. 611–622.

Caldwell, HD, Kromhout, J and Schachter, J 1981, 'Purification and partial characterization of the major outer membrane protein of *Chlamydia trachomatis*', *Infection and Immunity*, vol. 31, no. 3, pp. 1161–1176.

Cardenas, E and Tiedje, JM 2008, 'New tools for discovering and characterizing microbial diversity', *Current Opinion in Biotechnology*, vol 19, no. 6, pp. 544–549.

Carrigg, C, Rice, O, Kavanagh, S, Collins, G and O'Flaherty V 2007, 'DNA extraction method affects microbial community profiles from soils and sediment', *Applied Microbiology and Biotechnology*, vol. 77, no. 4, pp. 955–964.

Casida, LE 1977, 'Microbial metabolic activity in soil as measured by dehydrogenase determinations', *Applied and Environmental Microbiology*, vol. 34, no. 6, pp. 630–636.

Choi, KH and Dobbs, FC 1999, 'Comparison of two kinds of Biolog microplates (GN and ECO) in their ability to distinguish among aquatic microbial communities', *Journal of Microbiological Methods*, vol. 36, no. 3, pp. 203–213.

Chourey, K, Jansson, J, VerBerkmoes, N, et al. 2010, 'Direct cellular lysis/protein extraction protocol for soil metaproteomics', *Journal of Proteome Research*, vol. 9, no. 12, pp. 6615–6622.

Ciardi, C and Nannipieri, P 1990, 'A comparison of methods for measuring ATP in soil', *Soil Biology and Biochemistry*, vol. 22, no. 5, pp. 725–727.

Critter, SAM, Freitas, SS and Airoldi, C 2004, 'Comparison of microbial activity in some Brazilian soils by microcalorimetric and respirometric methods', *Thermochimica Acta*, vol. 410, no. 1, pp. 35–46.

Crompton, TR 2001, *Determination of metals and anions in soils, sediments and sludges*, Spon Press, London.

Daniel, R 2005, 'The metagenomics of soil', *Nature Reviews Microbiology*, vol. 3, no. 6, pp. 470–478.

DeSantis, TZ, Brodie, EL, Moberg, JP, et al. 2007, 'High-density universal 16S rRNA microarray analysis reveals broader diversity than typical clone library when sampling the environment', *Microbial Ecology*, vol. 53, no. 3, pp. 371–383.

Dinsdale, EA, Edwards, RA, Hall, D, et al. 2008, 'Functional metagenomic profiling of nine biomes', *Nature*, vol. 452, no. 7187, pp. 629–632.

ECS 2001, 'EN 13137:2001 Characterization of waste: determination of total organic carbon (TOC) in waste, sludges and sediments', European Committee for Standardization.

Feinstein, LM, Sul, WJ and Blackwood, CB 2009, 'Assessment of bias associated with incomplete extraction of microbial DNA from soil', *Applied and Environmental Microbiology*, vol. 75, no. 16, pp. 5428–5433.

Filip, C, Fletcher, G, Wulff, JL and Earhart, CF 1973, 'Solubilization of the cytoplasmic membrane of *Escherichia coli* by the ionic detergent sodium-lauryl sarcosinate', *Journal of Bacteriology*, vol. 115, no. 3, pp. 717–722.

Fontaine, S, Barot, S, Barré, P, et al. 2007, 'Stability of organic carbon in deep soil layers controlled by fresh carbon supply', *Nature*, vol. 450, no. 7167, pp. 277–280.

Fortune, WB and Mellon, MG 1938, 'Determination of iron with o-phenanthroline: a spectrophotometric study', *Industrial and Engineering Chemistry, Analytical Edition*, vol. 10, no. 2, pp. 60–64.

Frias-Lopez, J, Shi, Y, Tyson, GW, et al. 2008, 'Microbial community gene expression in ocean surface waters', *Proceedings of the National Academy of Sciences*, vol. 105, no. 10, pp. 3805–3810.

Fritz, JS and Gjerde, DT 2009, *Ion chromatography*, 4th completely revised and enlarged ed, Wiley, New Jersey.

Garland, JL and Mills, AL 1991, 'Classification and characterization of heterotrophic microbial communities on the basis of patterns of community-level sole-carbon-source utilization', *Applied and Environmental Microbiology*, vol. 57, no. 8, pp. 2351–2359.

Girvan, MS, Bullimore, J, Pretty, JN, Osborn, AM and Ball, AS 2003, 'Soil type is the primary determinant of the composition of the total and active bacterial communities in arable soils', *Applied and Environmental Microbiology*, vol. 69, no. 3, pp. 1800–1809.

Goebel-Stengel, M, Stengel, A, Taché, Y and Reeve, JR 2011, 'The importance of using the optimal plasticware and glassware in studies involving peptides', *Analytical Biochemistry*, vol. 414, no. 1, pp. 38–46.

Görg, A, Weiss, W and Dunn, MJ 2004, 'Current two-dimensional electrophoresis technology for proteomics', *Proteomics* vol. 4, no. 12, pp. 3665–3685.

Green, MR and Sambrook, J 2016, 'Precipitation of DNA with ethanol', *Cold Spring Harbor Protocols*, Available from: http://cshprotocols.cshlp.org/content/2016/12/pdb.prot093377.full.pdf+html [August 18, 2017].

Handelsman, J 2004, 'Metagenomics: application of genomics to uncultured microorganisms', *Microbiology and Molecular Biology Reviews*, vol. 68, no. 4, pp. 669–685.

Hanson, PJ, Edwards, NT, Garten, CT and Andrews, JA 2000, 'Separating root and soil microbial contributions to soil respiration: a review of methods and observations', *Biogeochemistry*, vol. 48, no. 1, pp. 115–146.

Hartley, IP, Hopkins, DW, Garnett, MH, Sommerkorn, M and Wookey, PA 2008, 'Soil microbial respiration in arctic soil does not acclimate to temperature', *Ecology Letters*, vol. 11, no. 10, pp. 1092–1100.

He, S, Wurtzel, O, Singh, K, et al. 2010, 'Validation of two ribosomal RNA removal methods for microbial metatranscriptomics', *Nature Methods*, vol. 7, no. 10, pp. 807–812.

He, Z, Gentry, TJ, Schadt, CW, et al. 2007, 'GeoChip: a comprehensive microarray for investigating biogeochemical, ecological and environmental processes', *The ISME Journal*, vol. 1, no. 1, pp. 67–77.

Heid, CA, Stevens, J, Livak, KJ and Williams, PM 1996, 'Real time quantitative PCR', *Genome Research*, vol. 6, no. 10, pp. 986–994.

Higuchi, R, Dollinger, G, Walsh, PS and Griffith, R 1992, 'Simultaneous amplification and detection of specific DNA sequences', *Nature Biotechnology*, vol. 10, no. 4, pp. 413–417.

Holland, PM, Abramson, RD, Watson, R and Gelfand, DH 1991, 'Detection of specific polymerase chain reaction product by utilizing the 5′-3′ exonuclease activity of *Thermus aquaticus* DNA polymerase', *Proceedings of the National Academy of Sciences*, vol. 88, no. 16, pp. 7276–7280.

Horowitz, AJ, Elrick, KA and Colberg, MR 1992, 'The effect of membrane filtration artifacts on dissolved trace element concentrations', *Water Research*, vol. 26, no. 6, pp. 753–763.

Hu, S and van Bruggen, AHC 1997. 'Microbial dynamics associated with multiphasic decomposition of ^{14}C-labeled cellulose in soil', *Microbial Ecology*, vol. 33, no. 2, pp. 134–143.

Hu, Z and Qi, L 2014, 'Sample digestion methods' in H Holland, and K Turekian, (eds), *Treatise on geochemistry*, 2nd ed, Elsevier Science, Amsterdam.

Huang B, Ying H, Yang P, Wang, X and Gu, S 2000, *An atlas of high resolution spectra of rare earth elements for ICP-AES*, Royal Society of Chemistry, London.

Hug, LA, Baker, BJ, Anantharaman, K, et al. 2016, 'A new view of the tree of life', *Nature Microbiology*, vol. 1, no. 16048.

Hurst, CJ, Crawford, RL, Garland, JL and Lipson, DA (eds) 2007, *Manual of environmental microbiology*, 3rd ed, American Society for Microbiology Press, Washington, DC.

Hurt, RA, Qiu, X, Wu, L, et al. 2001. 'Simultaneous recovery of RNA and DNA from soils and sediments', *Applied and Environmental Microbiology*, vol. 67, no. 10, pp. 4495–4503.

Isaacson, T, Damasceno, CM, Saravanan, RS, et al. 2006, 'Sample extraction techniques for enhanced proteomic analysis of plant tissues', *Nature Protocols*, vol. 1, no. 2, pp. 769–774.

ISO 1995, 'ISO 10694:1995 Soil quality: Determination of organic and total carbon after dry combustion (elementary analysis)', International Organization for Standardization.

Jackson, KW and Jackson, E 1999, *Electrothermal atomization for analytical atomic spectrometry*, Wiley, New Jersey.

Jackson, ML 2005, *Soil chemical analysis: advanced course*, Parallel Press, Wisconsin.

Jenkinson, DS and Oades, JM 1979, 'A method for measuring adenosine triphosphate in soil', *Soil Biology and Biochemistry*, vol. 11, no. 2, pp. 193–199.

Karl, DM and Holm-Hansen, O 1978, 'Methodology and measurement of adenylate energy charge ratios in environmental samples', *Marine Biology*, vol. 48, no. 2, pp. 185–197.

Kasibhatla, S, Amarante-Mendes, GP, Finucane, D, et al. 2006, 'Analysis of DNA fragmentation using agarose gel electrophoresis', *Cold Spring Harbor Protocols*, Available from: http://cshprotocols.cshlp.org/content/2006/1/pdb.prot4429.short [August 18, 2017].

Kaur, K 2007, *Handbook of water and wastewater analysis*, Atlantic Publishers & Distributors (P) Ltd., New Delhi.

Keiblinger, KM, Wilhartitz, IC, Schneider, T, et al. 2012, 'Soil metaproteomics–comparative evaluation of protein extraction protocols', *Soil Biology and Biochemistry*, vol. 54, pp. 14–24.

Kourtev, PS, Ehrenfeld, JG and Häggblom, M 2002, 'Exotic plant species alter the microbial community structure and function in the soil', *Ecology*, vol. 83, no. 11, pp. 3152–3166.

Kubista, M, Andrade, JM, Bengtsson, M, et al. 2006, 'The real-time polymerase chain reaction', *Molecular Aspects of Medicine*, vol. 27, no. 2, pp. 95–125.

Lachat 2008, Data Pack Lachat Applications in Standard Methods 21st Edition, Lachat Instruments, Milwaukee, Wisconsin.

Lee, SC, Knowles, TJ, Postis, VL, et al. 2016, 'A method for detergent-free isolation of membrane proteins in their local lipid environment', *Nature Protocols*, vol. 11, no. 7, pp. 1149–1162.

Leff, LG, Dana, JR, McArthur, JV and Shimkets, LJ 1995, 'Comparison of methods of DNA extraction from stream sediments', *Applied and Environmental Microbiology*, vol. 61, no. 3, pp. 1141–1143.

Levy-Booth, DJ, Campbell, RG, Gulden, RH, et al. 2007, 'Cycling of extracellular DNA in the soil environment', *Soil Biology and Biochemistry*, vol. 39, no. 12, pp. 2977–2991.

Li, H and Homer, N 2010, 'A survey of sequence alignment algorithms for next-generation sequencing', *Briefings in Bioinformatics*, vol. 11, no. 5, pp. 473–483.

Liu, F, Ying, GG, Tao, R, et al. 2009, 'Effects of six selected antibiotics on plant growth and soil microbial and enzymatic activities', *Environmental Pollution*, vol. 157, no. 5, pp. 1636–1642.

Liu, L, Li, Y, Li, S, et al. 2012, 'Comparison of next-generation sequencing systems', *Journal of Biomedicine and Biotechnology*, vol. 2012, no. 251364, pp. 1–11.

Livak, KJ and Schmittgen, TD 2001, 'Analysis of relative gene expression data using real-time quantitative PCR and the 2−ΔΔCT method', *Methods*, vol. 25, no. 4, pp. 402–408.

Lovley, DR and Phillips, EJ 1986, 'Availability of ferric iron for microbial reduction in bottom sediments of the freshwater tidal Potomac River', *Applied and Environmental Microbiology*, vol. 52, no. 4, pp. 751–757.

Lowry, OH, Rosebrough, NJ, Farr, AL and Randall, RJ 1951, 'Protein measurement with the Folin phenol reagent', *Journal of Biological Chemistry*, vol. 193, no. 1, pp. 265–275.

Luo, Y and Zhou, X 2010, *Soil respiration and the environment*, Academic Press, Burlington, Massachusetts.

Madigan, MT, Martinko, JM, Bender, KS, Buckley, DH and Stahl, DA (eds) 2014, *Brock biology of microorganisms*, 14th ed, Pearson Education, New York.

Magoc, T, Pabinger, S, Canzar, S, et al. 2013, 'GAGE-B: an evaluation of genome assemblers for bacterial organisms', *Bioinformatics*, vol. 29, no. 14, pp. 1718–1725.

Maier, RM, Pepper, IL and Gerba, CP (eds) 2009, *Environmental microbiology*, 2nd ed, Academic Press, Burlington, MA.

Mardis, ER 2008, 'Next-generation DNA sequencing methods', *Annual Review of Genomics and Human Genetics*, vol. 9, pp. 387–402.

Markowitz, VM, Chen, IMA, Palaniappan, K, et al. 2011, 'IMG: the integrated microbial genomes database and comparative analysis system', *Nucleic Acids Research*, vol. 40, no. D1, pp. D115–D122.

Martens, R 1995, 'Current methods for measuring microbial biomass C in soil: potentials and limitations', *Biology and Fertility of Soils*, vol. 19, no. 2, pp. 87–99.

Martin, JA and Wang, Z 2011, 'Next-generation transcriptome assembly', *Nature Reviews Genetics*, vol. 12, no. 10, pp. 671–682.

McKelvie, ID 2000, 'Phosphates' in LML Nollet, (ed), *Handbook of water analysis*, Marcel Dekker, Inc., New York.

Mehra, OP and Jackson, ML 1960, 'Iron oxide removal from soils and clays by a dithionite-citrate system buffered with sodium bicarbonate', *Proceedings 7th National Conference on Clays and Clay Minerals*, pp. 317–327.

Metzker, ML 2010, 'Sequencing technologies—the next generation', *Nature Reviews Genetics*, vol. 11, no. 1, pp. 31–46.

Meyer, F, Paarmann, D, D'Souza, M, et al. 2008, 'The metagenomics RAST server—a public resource for the automatic phylogenetic and functional analysis of metagenomes', *BMC Bioinformatics*, vol. 9, no. 386, pp. 1–8.

Mikan, CJ, Schimel, JP and Doyle, AP 2002, 'Temperature controls of microbial respiration in arctic tundra soils above and below freezing', *Soil Biology and Biochemistry*, vol. 34, no. 11, pp. 1785–1795.

Miller, DN, Bryant, JE, Madsen, EL and Ghiorse, WC 1999, 'Evaluation and optimization of DNA extraction and purification procedures for soil and sediment samples', *Applied and Environmental Microbiology*, vol. 65, no. 11, pp. 4715–4724.

Miller, JR, Koren, S and Sutton, G 2010, 'Assembly algorithms for next-generation sequencing data', *Genomics*, vol. 95, no. 6, pp. 315–327.

Montgomery, J, Wittwer, CT, Palais, R and Zhou, L 2007, 'Simultaneous mutation scanning and genotyping by high-resolution DNA melting analysis', *Nature Protocols*, vol. 2, no. 1, pp. 59–66.

Morrison, TB, Weis, JJ and Wittwer, CT 1998, 'Quantification of low-copy transcripts by continuous SYBR Green I monitoring during amplification', *Biotechniques*, vol. 24, no. 6, pp. 954–958.

Nannipieri, P, Ascher, J, Ceccherini, M, et al. 2003, 'Microbial diversity and soil functions', *European Journal of Soil Science*, vol. 54, no. 4, pp. 655–670.

Neethirajan, S, Jayas, DS and Sadistap, S 2009, 'Carbon dioxide (CO_2) sensors for the agri-food industry – a review', *Food and Bioprocess Technology*, vol. 2, no. 2, pp. 115–121.

Nilsen, TW 2013, 'Gel purification of RNA', *Cold Spring Harbor Protocols*, Available from: http://cshprotocols.cshlp.org/content/2013/2/pdb.prot072942.full.pdf [August 18, 2017].

Nollet, LML (ed) 2000, *Handbook of water analysis*, Marcel Dekker, Inc., New York.

O'Connor, JT, Komolrit, K and Engelbrecht, RS 1965, 'Evaluation of the orthophenanthroline method for ferrous-iron determination', *American Water Works Association*, vol. 57, no. 7, pp. 926–934.

O'Farrell, PH 1975, 'High resolution two-dimensional electrophoresis of proteins', *Journal of Biological Chemistry*, vol. 250, no. 10, pp. 4007–4021.

Ozsolak, F and Milos, PM 2011, 'RNA sequencing: advances, challenges and opportunities', *Nature Reviews Genetics*, vol. 12, no. 2, pp. 87–98.

Pansu, M and Gautheyrou, J 2006, *Handbook of soil analysis: mineralogical, organic and inorganic methods*, Springer-Verlag, Berlin, Heidelberg, New York.

Patel, RK and Jain, M 2012, 'NGS QC Toolkit: a toolkit for quality control of next generation sequencing data', *PloS One*, vol. 7, no. 2, pp. e30619.

Pfaffl, MW 2001, 'A new mathematical model for relative quantification in real-time RT-PCR', *Nucleic Acids Research*, vol. 29, no. 9, pp. 2002–2007.

Pham, VH and Kim, J 2012, 'Cultivation of unculturable soil bacteria', *Trends in Biotechnology*, vol. 30, no. 9, pp. 475–484.

Quail, MA, Smith, M, Coupland, P, et al. 2012, 'A tale of three next generation sequencing platforms: comparison of Ion Torrent, Pacific Biosciences and Illumina MiSeq sequencers', *BMC Genomics*, vol. 13, no. 1, p. 341.

Rice, EW, Baird, RB and Eaton, AD 2017, *Standard methods for the examination of water and wastewater*, 23rd ed, American Public Health Association, American Water Works Association, Water Environment Federation, Maryland.

Rio, DC, Ares, M, Hannon, GJ and Nilsen, TW 2010, 'Polyacrylamide gel electrophoresis of RNA', *Cold Spring Harbor Protocols*, Available from: http://cshprotocols.cshlp.org/content/2010/6/pdb.prot5444.full.pdf+html [August 18, 2017].

Ririe, KM, Rasmussen, RP and Wittwer, CT 1997, 'Product differentiation by analysis of DNA melting curves during the polymerase chain reaction', *Analytical Biochemistry*, vol. 245, no. 2, pp. 154–160.

Salter, SJ, Cox, MJ, Turek, EM, et al. 2014, 'Reagent and laboratory contamination can critically impact sequence-based microbiome analyses', *BMC Biology*, vol. 12, no. 1, p. 87.

Sambrook, J and Russell, DW (eds) 2001, *Molecular cloning: a laboratory manual*, 3rd ed, Cold Spring Harbor Laboratory Press, Cold Spring Harbor, NY.

Schena, M, Shalon, D, Davis, RW and Brown, PO 1995, 'Quantitative monitoring of gene expression patterns with a complementary DNA microarray', *Science*, vol. 270, no. 5235, pp. 467–470.

Schilt, AA and Hoyle, WC 1967, 'Improved sensitivity and selectivity in the spectrophotometric determination of iron by use of a new Ferroin-type reagent', *Analytical Chemistry*, vol. 39, no. 1, pp. 114–117.

Schmittgen, TD and Livak, KJ 2008, 'Analyzing real-time PCR data by the comparative CT method', *Nature Protocols*, vol. 3, no. 6, pp. 1101–1108.

Schneider, T and Riedel, K 2010, 'Environmental proteomics: analysis of structure and function of microbial communities', *Proteomics*, vol. 10, no. 4, pp. 785–798.

Schnürer, J and Rosswall, T 1982, 'Fluorescein diacetate hydrolysis as a measure of total microbial activity in soil and litter', *Applied and Environmental Microbiology*, vol. 43, no. 6, pp. 1256–1261.

Schuster, SC 2008, 'Next-generation sequencing transforms today's biology', *Nature Methods*, vol. 5, no. 1, pp. 16–18.

Scopes, RK 1993, *Protein purification: principles and practice*, 3rd ed, Springer, New York.

Shendure, J and Ji, H 2008, 'Next-generation DNA sequencing', *Nature Biotechnology*, vol. 26, no. 10, pp. 1135–1145.

Shimadzu 2013, TOC Application Handbook: Document SCA-130–101-604, Shimadzu Europa GmbH, Duisburg, Germany. <www.shimadzu.eu>

Shokralla, S, Spall, JL, Gibson, JF and Hajibabaei, M 2012, 'Next-generation sequencing technologies for environmental DNA research', *Molecular Ecology*, vol. 21, no. 8, pp. 1794–1805.

Smith, PK, Krohn, RI, Hermanson, GT, et al. 1985, 'Measurement of protein using bicinchoninic acid', *Analytical Biochemistry*, vol. 150, no. 1, pp. 76–85.

Stookey, LL 1970, 'Ferrozine – a new spectrophotometric reagent for iron', *Analytical Chemistry*, vol. 42, no. 7, pp. 779–781.

Tabatabai, MA 1994, 'Soil enzymes' in RW Weaver, JS Angel and PS Bottomley, (eds), *Methods of soil analysis: Part 2 – microbiological and biochemical properties*, pp. 775–834. Soil Science Society of America, Wisconsin.

Tabatabai, MA and Bremner, JM 1969, 'Use of *p*-nitrophenyl phosphate for assay of soil phosphatase activity', *Soil Biology and Biochemistry*, vol. 1, no. 4, pp. 301–307.

Tabatabai, MA and Frankenberger Jr., WT 1996, 'Liquid chromatography' in DL Sparks, AL Page, PA Helmke, and RH Loeppert, (eds), *Methods of soil analysis: Part 3 – chemical methods*, Soil Science Society of America, Wisconsin.

Taberlet, P, Prud'homme, SM, Campione, E, et al. 2012, 'Soil sampling and isolation of extracellular DNA from large amount of starting material suitable for metabarcoding studies', *Molecular Ecology*, vol. 21, no. 8, pp. 1816–1820.

Tan, KH 1996, *Soil sampling, preparation, and analysis*, Marcel Dekker, Inc., New York.

Tan, SC and Yiap, BC 2009, 'DNA, RNA, and protein extraction: the past and the present', *Journal of Biomedicine and Biotechnology*, vol. 2009, no. 574398, pp. 1–10.

Taylor, JP, Wilson, B, Mills, MS and Burns, RG 2002, 'Comparison of microbial numbers and enzymatic activities in surface soils and subsoils using various techniques', *Soil Biology and Biochemistry*, vol. 34, no. 3, pp. 387–401.

Thomas, T, Gilbert, J and Meyer, F 2012, 'Metagenomics – a guide from sampling to data analysis', *Microbial Informatics and Experimentation*, vol. 2, no. 3, pp. 1–12.

Thompson, M 2008, 'CHNS elemental analysers', Royal Society of Chemistry Analytical Methods Committee, no. 29.

Thompson M and Walsh JN 1989, *Handbook of inductively coupled plasma spectrometry*, 2nd ed, Springer US, New York.

Thompson, SK 2012, *Sampling*, 3rd ed, Wiley, New Jersey.

Tringe, SG, von Mering, C, Kobayashi, A, et al. 2005, 'Comparative metagenomics of microbial communities', *Science*, vol. 308, no. 5721, pp. 554–557.

Tyson, GW, Chapman, J, Hugenholtz, P and Allen, EE 2004, 'Community structure and metabolism through reconstruction of microbial genomes from the environment', *Nature*, vol. 428, no. 6978, pp. 37–43.

Urich, T, Lanzén, A, Qi, J, et al. 2008, 'Simultaneous assessment of soil microbial community structure and function through analysis of the meta-transcriptome', *PloS One*, vol. 3, no. 6, pp. e2527.

US EPA 1978, 'Method 365.3: Phosphorous, All Forms (Colorimetric, Ascorbic Acid, Two Reagent)', United States Environmental Protection Agency, Washington, DC.

US EPA 1995, 'Method 7196A: Chromium, hexavalent (colorimetric)', Test methods for evaluating solid waste, physical/chemical methods. SW-846, 3rd ed, US Environmental Protection Agency, Washington, DC.

US Food and Drug Administration, Laboratory Methods. <www.fda.gov/Food/FoodScienceResearch/LaboratoryMethods/ucm062229.htm>.

Uzoukwu, BA 2000, 'Methods of water preservation' in LML Nollet, (ed), *Handbook of water analysis*, Marcel Dekker, Inc., New York.

Valentini, R, Matteucci, G, Dolman, AJ, et al. 2000, 'Respiration as the main determinant of carbon balance in European forests', *Nature*, vol. 404, no. 6780, pp. 861–865.

Van Loon, JC and Barefoot, RR 1989, *Analytical methods for geochemical exploration*, Academic Press, Burlington, Massachusetts.

Vartoukian, SR, Palmer, RM and Wade, WG 2010, 'Strategies for culture of "unculturable" bacteria', *FEMS Microbiology Letters*, vol. 309, no. 1, pp. 1–7.

Venter, JC, Remington, K, Heidelberg, JF, et al. 2004, 'Environmental genome shotgun sequencing of the Sargasso Sea', *Science*, vol. 304, no. 5667, pp. 66–74.

Verschoor, MJ and Molot, LA 2013, 'A comparison of three colorimetric methods of ferrous and total reactive iron measurement in freshwaters', *Limnology and Oceanography: Methods*, vol. 11, pp. 113–125.

Viollier, E, Inglett, PW, Hunter, K, Roychoudhury, AN and Van Cappellen, P 2000, 'The ferrozine method revisited: Fe(II)/Fe(III) determination in natural waters', *Applied Geochemistry*, vol. 15, pp. 785–790.

von Mersi, W and Schinner, F 1991, 'An improved and accurate method for determining the dehydrogenase activity of soils with iodonitrotetrazolium chloride', *Biology and Fertility of Soils*, vol. 11, no. 3, pp. 216–220.

Wang, Z, Gerstein, M and Snyder, M 2009, 'RNA-Seq: a revolutionary tool for transcriptomics', *Nature Reviews Genetics*, vol. 10, no. 1, pp. 57–63.

Ward, DM, Weller, R and Bateson, MM 1990, '16S rRNA sequences reveal numerous uncultured microorganisms in a natural community', *Nature*, vol. 345, no. 6270, pp. 63–65.

Welz, B and Sperling, M 1999, *Atomic absorption spectrometry*, 3rd ed, Wiley-VCH, Weinheim.

Welz, B, Becker-Ross, H, Florek, S and Heitmann, U 2005, *High-resolution continuum source AAS: the better way to do atomic absorption spectrometry*, Wiley-VCH, Weinheim.

Willerslev, E, Hansen, AJ and Poinar, HN 2004, 'Isolation of nucleic acids and cultures from fossil ice and permafrost', *Trends in Ecology & Evolution*, vol. 19, no. 3, pp. 141–147.

Willner, D, Daly, J, Whiley, D, et al. 2012, 'Comparison of DNA extraction methods for microbial community profiling with an application to pediatric bronchoalveolar lavage samples', *PloS One*, vol. 7, no. 4, pp. e34605.

Wilson, K 1997, 'Preparation of genomic DNA from bacteria' in FM Ausubel, R Brent, RE Kimston, DD Moore, JG Seidman, JA Smith and K Struhl, (eds), *Current protocols in molecular biology*, pp. 2.4.1–2.4.5. John Wiley & Sons, Inc., New York, NY.

Wiśniewski, JR, Zougman, A, Nagaraj, N and Mann, M 2009, 'Universal sample preparation method for proteome analysis', *Nature Methods*, vol. 6, no. 5, pp. 359–362.

Wrighton, KC, Thomas, BC, Sharon, I, et al. 2012, 'Fermentation, hydrogen, and sulfur metabolism in multiple uncultivated bacterial phyla', *Science*, vol. 337, no. 6102, pp. 1661–1665.

Zhang, W, Parker, KM, Luo, Y, et al. 2005, 'Soil microbial responses to experimental warming and clipping in a tallgrass prairie', *Global Change Biology*, vol. 11, no. 2, pp. 266–277.

Zhou, J, Bruns, MA and Tiedje, JM 1996, 'DNA recovery from soils of diverse composition', *Applied and Environmental Microbiology*, vol. 62, no. 2, pp. 316–322.

Zimmerman, CF, Keefe, CW and Bashe, J 1997, 'Method 440.0 Determination of carbon and nitrogen in sediments and particulates of estuarine/coastal waters using elemental analysis', National Exposure Research Laboratory, Office of Research and Development, U.S. Environmental Protection Agency, Cincinnati, Ohio.

Zwietering, MH, Jongenburger, I, Rombouts, FM and van't Riet, K 1990, 'Modeling of the bacterial growth curve', *Applied and Environmental Microbiology*, vol. 56, no. 6, pp. 1875–1881.

PART II

ADVANCED ANALYTICAL INSTRUMENTATION

2 The Application of Isothermal Titration Calorimetry for Investigating Proton and Metal Interactions on Microbial Surfaces

DREW GORMAN-LEWIS

Abstract

Isothermal titration calorimetry combined with surface complexation modeling is an ideal technique to provide further characterization of microbial surface reactivity towards protons and metal ions. This technique can produce enthalpies of protonation and metal ion coordination of acidic functional groups on microbial surfaces. This information is critical for understanding the thermodynamic driving force of surface complexation and provides key information for the indirect identification of surface ligands. Topics covered in this chapter include how this technique complements traditional methods of microbial surface reactivity, necessary system characterization prior to performing calorimetric experiments, how to prepare biomass and solutions for calorimetric titrations, difficult aspects of this technique, and data analysis and interpretation.

2.1 Introduction

Microbial life is implicated in both the formation and the dissolution of minerals (Banfield and Nealson, 1997; Ehrlich, 1996). For example, microbes can influence mineral dissolution through surface reactions by lowering the activity of free ions in solution with the formation of surface complexes (Wightman and Fein, 2004). In contrast to that behavior, the concentration of ions on microbial surfaces can promote the occurrence of precipitation (Ehrlich, 2002). Surface reactions can be important for microbial attachment to minerals, while exudates attack the solid phase to release nutrients locked up in the solid phase (Sheng et al., 2008). Consequently, the interaction of metal ions and microbial surfaces is a fundamental process that affects both the environment and cellular function. For example, microbial surfaces exert influences on mineral dissolution and weathering (Barker et al., 1998; Wightman and Fein, 2004) and surface–metal ion interactions are critical to the bioavailability of metal ions to cells (Flynn et al., 2014; Nell et al., 2016). To understand why these spontaneous surface reactions occur from a thermodynamic perspective necessitates delving into the thermodynamic driving force of surface complexation reactions.

There are myriads of techniques used to directly study or support studies of metal–microbial surface reactions. These can include, but are not limited to, potentiometric titrations, metal ion adsorption experiments, surfaces complexation modeling, and various spectroscopic techniques (Alessi et al., 2010; Boyanov et al., 2003; Ginn and Fein, 2009;

Mishra et al., 2010; Ngwenya et al., 2003). These techniques support investigations of why metal ion–microbial surface reactions occur from a thermodynamic perspective, but are limited in the information they can provide. For example, a critical parameter needed to quantify why reactions occur is the Gibbs energy of a reaction. Deriving the Gibbs energy of metal ion adsorption onto microbial surfaces can be broken down into two parts: quantifying protonation of functional groups and quantifying metal ion adsorption. Quantitatively determining acidity constants and acidic functional group concentrations on microbial surfaces is performed by measuring protonation of surface functional groups with potentiometric titrations followed by applying surface complexation modeling to the potentiometric data (see Chapter 3). Gibbs energies of reaction for surface protonation can easily be determined from the surface site acidity constants with Eq. (2.1), where ΔG_r is Gibbs energy of reaction, R is the universal gas constant, T is absolute temperature, and K is an equilibrium constant.

$$\Delta G_r = 2.303 \times R \times T \times \log K \quad \text{(Eq.2.1)}$$

A quantitative understanding of protonation is necessary to quantify metal ion adsorption, since the proton active sites on microbial surfaces are also sites of metal adsorption. Once microbial surfaces are sufficiently characterized with respect to protons, quantitative investigations of metal ion adsorption can be performed by applying surface complexation modeling to the results of bulk metal ion adsorption experiments. Surface complexation models will produce stability constants for metal ion surface complexes, and similarly to microbial surface protonation, Eq. (2.1) can be used to calculate the Gibbs energy of metal ion complexation from the stability constant. The results of surface complexation modeling can give indirect indications of the coordination environment based on the values of acidity constants and stability constants of surface complexes. Spectroscopic investigations can provide direct information on average metal ion coordination environments, which provides contextual information for surface complexation modeling results. All these lines of investigation support thermodynamic investigations of metal ion complexation on microbial surfaces, yet they do not provide enough information to determine the thermodynamic driving force of reactions. This requires a closer look at Gibbs energy of reaction.

The decomposition of Gibbs energies of reaction is necessary in order to understand the thermodynamic driving force behind reactions. Gibbs energies are composed of two parameters: enthalpy and entropy. Equation (2.2) describes the relationship between Gibbs energy (ΔG_r), enthalpy (ΔH_r), and entropy (ΔS_r) of reaction.

$$\Delta G_r = \Delta H_r - T\Delta S_r \quad \text{(Eq.2.2)}$$

It becomes clear that if one can determine Gibbs energies of reaction, as described earlier, and enthalpies of reaction for surface complexation reactions, then entropies of reaction can be calculated. With the full suite of thermodynamic parameters in Eq. (2.2), determining whether reactions are enthalpy or entropy driven becomes a matter of comparing ΔH_r with $T\,\Delta S_r$.

There are two common methods for determining enthalpies of reaction: 1) calorimetry and 2) determining stability constants at two temperatures, assuming the relative heat capacities of the products and reactants are zero for the reaction, and calculating the enthalpies with the van't Hoff equation. The second option can prove experimentally challenging for obtaining

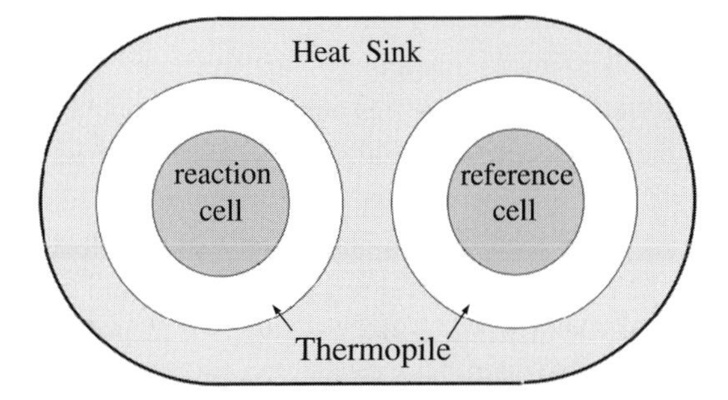

Figure 2.1 Schematic of a twin channel calorimeter. Heat flow between the reaction and reference cell is effectively monitored over time according to Eq. (2.3).

well-defined stability constants with low errors at elevated temperatures (Ginn and Fein, 2009; Wightman et al., 2001), and the assumptions related to the heat capacities may not be valid for microbial reactions (Gorman-Lewis, 2011). Calorimetry provides the means to directly measure heats of reaction without making assumptions about heat capacities and is a more reliable means to determine enthalpies of reaction.

Isothermal calorimetry measurements monitor the heat being evolved (or absorbed) from a reaction over time. That heat is conducted to (or from) a heat sink surrounding the reaction vessel. The heat flow is recorded by allowing the heat to pass through a thermopile positioned between the reaction vessel and the heat sink. A twin channel calorimeter design is depicted in Figure 2.1. Heat flow measurements in a twin channel design are effectively monitoring heat flow between a reaction and a reference cell as described by Eq. (2.3), where T denotes of the temperature of the reaction cell (s), reference cell (r), and heat sink (o). Advantages of this design include increased sensitivity due to cancellation of thermal effects from background disturbances, increased accuracy and precision, and long-term stability. Isotherm titration calorimetry involves the titrant being delivered to the reaction cell via cannula, while the heat flow is monitored over time.

$$\Delta T = (T_s - T_o) - (T_r - T_o) = T_s - T_r \qquad \text{(Eq.2.3)}$$

This chapter describes the techniques for applying isothermal titration calorimetry (ITC) to microbial surfaces. The information presented here will outline growing and preparing the biomass, preliminary data acquisition, performing calorimetric experiments, data reduction, and interpretation.

2.2 Methods

2.2.1 Biomass Preparation

ITC experiments are facilitated by using highly concentrated biomass suspensions. Typically, 2 to 4 mL of suspension is sufficient for calorimetric experiments. If possible, initial experiments should aim to contain 10 to 25 g/L biomass dry weight, which usually corresponds to approximately 10^9–10^{10} cells/mL. These cell densities may not be achievable, depending on the species being investigated; consequently, obtaining a suspension of the highest concentration realistically possible is advised for initial experiments. Highly concentrated suspensions are necessary to ensure that the heat flow signal from the reactions being measured is above the baseline heat flow noise of the instrument for the sample being measured. Therefore, the minimum biomass concentration will depend on your particular instrument and sample.

Biomass is typically grown in a medium designed to facilitate maximum growth and ease of harvesting. Separation of biomass from the medium by centrifugation is ideal; however, filtration can also be used to harvest biomass from the medium. Once the biomass is harvested from the medium, it needs be rinsed to eliminate any residual medium. The biomass should be rinsed with the same electrolyte that will be used in the experiments to avoid the introduction of unnecessary ions into the suspension. Repeated centrifugation, decanting, and resuspension in the electrolyte can be used as a rinsing procedure. Alternatively, the biomass can be rinsed with an electrolyte during filtration if the biomass will not pellet sufficiently with centrifugation. This alternative approach is often necessary with-low concentration suspensions. Cell viability should be checked after the rinse procedure with a differential stain, such as LIVE/DEAD Baclight kit (Life Technologies), though previous work shows that 80 to 90% of cells are viable after these treatments (Gorman-Lewis, 2011; Gorman-Lewis et al., 2014).

Once biomass is harvested, suspensions can be prepared for calorimetric and auxiliary experiments. The biomass should be suspended in an inert electrolyte with an appropriate concentration to prevent changes in ionic strength during the experiment. Common electrolytes for creating bacterial suspensions include sodium perchlorate, sodium nitrate, and sodium chloride. However, the electrolyte should be chosen based on specific experimental needs.

2.2.2 Titrant Preparation

The background composition of the titrant should match the background composition of the electrolyte in the biomass suspension. For example, in an experiment measuring functional group protonation of biomass suspended in sodium perchlorate, the titrant used should be perchloric acid. Matching titrant and electrolyte compositions helps to lessen heats of dilution during experiments.

Titrant concentration also needs to be carefully considered for many reasons. Depending on the instrument in use, the titrant volume with each injection may be fixed or flexible. One needs to ensure that the volume of titrant delivered is sufficient to cause a measurable reaction and that the number of injections during the experiment will produce enough data points. A 250 μL syringe is commonly used for ITC; consequently, there are volume limitations on the amount of titrant that can be delivered during the course of one experiment. The higher the concentration of titrant, the smaller the titrant dose can be, and one can increase the number of data points gathered in each experiment. However, increasing titrant concentration will also affect heats of dilution. Care should also be taken to avoid solution conditions that would encourage precipitation of components in the titrant.

2.2.3 Background Data

There are a number of background data needed to establish initial calorimetric conditions. These data are heats of dilution, heats of neutralization, extent of adsorption, and kinetics of adsorption. Heats of dilution should be determined by adding the titrant to the supporting biomass-free electrolyte under the same conditions (i.e., same volumes, same number of injections, same stirring rate, etc.). If calorimetric experiments are occurring above approximately pH 8 and there is the potential for hydroxide neutralization, one should predict the extent of hydroxide neutralization to calculate its heat contribution. Hydroxide neutralization is very exothermic and could complicate high-pH measurements; therefore, having an initial sense of the extent of neutralization heat will be helpful for planning experimental conditions. Initial experimental conditions should be chosen to encourage the maximum adsorption possible under the conditions. This means that the experimental system being investigated should be thoroughly characterized with traditional surface techniques (i.e., proton/metal adsorption measured and modeled with a surface complexation model) so that one can choose conditions that would encourage measurable heat flows. Kinetics experiments determining the length of time necessary for adsorption reactions to come to equilibrium should be undertaken prior to calorimetric experiments. This will provide an initial injection interval for the calorimetric experiments. The interval should provide enough time for the heat flow signal to return to baseline prior to the next injection. This background data is crucial to planning experimental conditions that will produce a signal above background after correction for extraneous heats.

2.2.4 Calorimetric Experiments – Heats of Protonation

Calorimetric experiments for measuring heats of protonation are set up similarly to standard potentiometric titrations of microbial suspensions (Chapter 3). Parameters that need to be considered are the concentration of acid, injection volume, number of injections to achieve measurable heat flow, and a sufficient number of data points. Predicting the extent of protonation with a surface complexation model can be very helpful in planning the experimental conditions.

Suspension contamination with CO_2 needs to be avoided if experiments are carried out above approximately pH 6. Experimental suspensions should be prepared with CO_2-free electrolyte and the pH adjusted with CO_2-free base. Electrolytes can be sparged with N_2 for at least 30 min and stored in sealed serum vials under an N_2 atmosphere to maintain a CO_2-free solution prior to preparing the suspension. CO_2-free base can be prepared from 50% w/w NaOH and water sparged with N_2 for at least 30 min and stored in sealed serum vials under an N_2 atmosphere to maintain a CO_2-free solution. Ideally, the suspension and pH adjustments are carried out in a glove box with a CO_2-free atmosphere to prevent reabsorption of CO_2 into the solution. If such facilities are not available, then the suspension should be sparged once it is prepared and prior to loading in the calorimeter. The reaction vessel should be flushed with inert gas (e.g., N_2 or Ar) and the experiment performed under conditions that prevent absorption of CO_2 into the solution. Protonation of carbonate species is exothermic; therefore, CO_2 contamination would cause the measured heat to be more exothermic than in a CO_2-free suspension.

The configuration of a titration calorimeter does not typically accommodate simultaneous pH measurements within the reaction vessel. Consequently, options for obtaining the necessary pH data include 1) modifying the calorimeter, 2) collecting pH measurements on a parallel titration under the same conditions running outside the calorimeter, and 3) predicting the pH after each addition of acid using a surface complexation model of protonation. Modifying a titration calorimeter to allow for a pH probe in the reaction vessel may be possible; however, it will likely be accompanied by additional cost and perhaps an increase in baseline noise. Options 2 and 3 are likely more realistic for most investigators, and both have been successfully applied to multiple systems (Gorman-Lewis, 2009, 2014; Gorman-Lewis et al., 2006; Harrold and Gorman-Lewis, 2013). For both options 2 and 3, the initial and final pH of the calorimetric titration suspension should be measured for comparison with either the external measurements or the predicted pH to ensure that results are consistent.

2.2.5 Calorimetric Experiments – Heats of Metal Complexation

Measuring heats of metal complexation on microbial surfaces is similar to performing a concentration-dependent adsorption experiment at a fixed pH. Aliquots of a concentrated metal solution are delivered to the reaction vessel with the heat flow being monitored over time. Parameters that need to be considered when setting up this type of experiment are the titrant concentration, titrant volume delivered, extent of adsorption, and cell exudates. Preliminary adsorption experiments should be performed prior to calorimetric experiments to ensure that experiments are performed under appropriate conditions, where adsorption is maximized and reversible.

Heats of metal complexation tend to produce a smaller heat flow than protonation experiments. Consequently, titrant injections tend to be larger for metal absorption experiments than for protonation experiments. The ideal experimental conditions would produce 100% adsorption in order to maximize the heat flow signal and minimize the influence of cellular exudates. Depending on the system being investigated, 100% adsorption cannot always be achieved. Therefore, if cellular exudates are present in the experimental system, the heat flow being measured may be influenced by metal–exudate solution complexes forming. To test whether the heat flow being measured in the system is influenced by metal–exudate complexation, a titration of the metal solution into the supernatant of the microbial suspension can be compared with the heats of dilution. A significant difference between the heats of dilution and the heats from titration into the microbial suspension could indicate that metal–exudate complexes are formed. If consistent amounts of exudates are released from bacterial suspensions (e.g., verified by dissolved organic carbon measurements of supernatants), a correction can be applied to the experimental data by performing heat of dilution experiments with bacterial supernatant produced from a bacterial suspension at the same pH as the experiments. These data can be used in lieu of standard heat of dilution measurements with electrolyte alone.

Once background data are collected, experimental conditions determined, and heats of dilution measured, calorimetric data collection can begin. During the calorimetric titration, the pH in the reaction vessels needs to remain constant, or a parallel external experiment needs to be performed to determine pH with each addition. Often, adjusting the suspension and the titrant solution to the same value will hold the pH sufficiently constant. The pH of the reaction vessel should be measured, either to verify that the pH has not substantially drifted from the initial value or to determine its consistency with the external parallel experiment. Total metal adsorption should be measured at the end of the titration to ensure that the results are consistent with expected uptake/modeling predictions.

2.2.6 Potential Problems

Determining experimental conditions that produce heat flows large enough to reliably measure is the primary challenge associated with measuring heats of microbial surface complexation reactions. Maximizing adsorption of the ion of interest and minimizing heats of noninterest (e.g., hydroxide neutralization, carbonate protonation) are the best techniques to ensure a measurable signal. This is best done by working with highly concentrated biomass suspensions, making injections that deliver enough titrant to produce a clear signal, working in appropriate pH ranges, and ensuring that carbon dioxide contamination is minimal. To handle microbial cultures that only achieve cell densities of approximately 10^6 or 10^7 cells/mL, multiple cultures will need to be concentrated and combined. For example, Harrold and Gorman-Lewis (2013) grew and combined the endospore harvest from approximately 10 L of *Bacillus subtilis* cultures to produce sufficient endospore biomass for their work. Titrant concentration and injection volume can be adjusted to achieve optimal adsorption with each injection. Fewer data points per experiment may be needed if larger injection volumes are necessary to avoid precipitation in titrant solutions or large heats of dilution upon injection. This approach will allow the delivery of the number of moles of titrant necessary to produce a reliably measurable heat flow signal. Selecting a pH range that promotes adsorption and avoids unwanted reactions is essential for metal ion adsorption measurements. Investigators may want to avoid metal ion hydrolysis if the complexed ions are not in the reaction being investigated. Achieving these conditions may require performing experiments at lower than optimal pH for maximum adsorption. By maximizing biomass concentration and optimizing titrant additions as described here, less-than-optimal pH conditions can be used to gather data.

2.2.7 Experimental Data

Figures 2.2 and 2.3 illustrate typical raw calorimetric data for measuring heats of protonation and heats of metal complexation. These raw data should consist of several distinct exothermic or endothermic peaks that correspond to each titrant injection. Often, the first injection is anomalous due to diffusion of titrant into the reaction vessel while the instrument returns to baseline after vessel insertion. Integrating the area under each peak in the

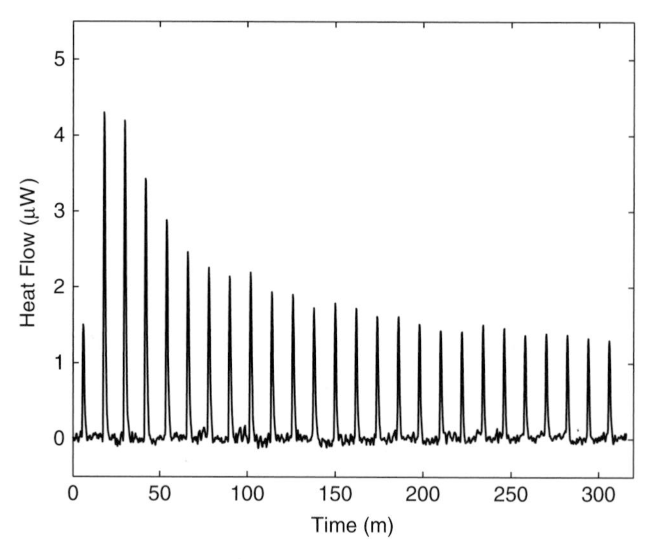

Figure 2.2 *Nitrosopumilis maritimus* strain SCM1 titrated from pH 9.4 to 4.6. Each titrant addition (0.01 M $HClO_4$) caused a strong exothermic signal. Figure used with permission from Gorman-Lewis et al., 2014.

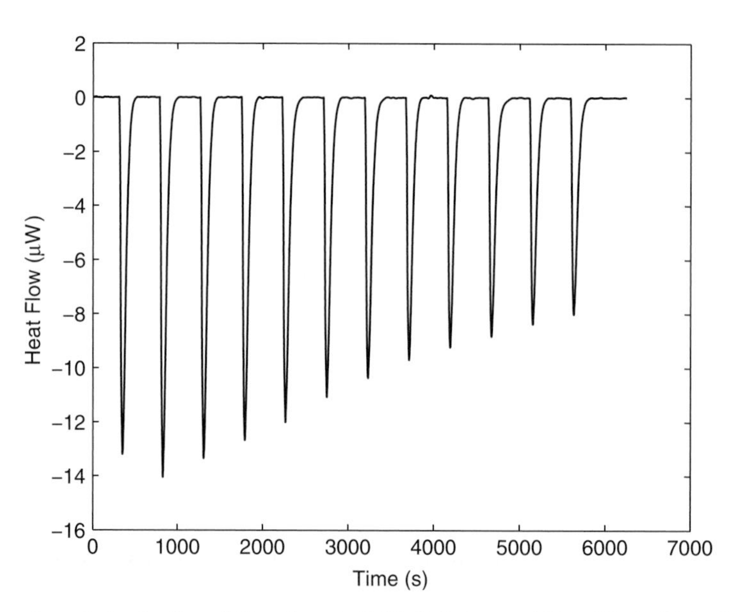

Figure 2.3 Zn^{2+} complexation onto the *Bacillus subtilis* surface at pH 5.8. Each titrant addition (10 μL of 3.6 mM $Zn(ClO_4)_2$) caused a strong endothermic signal.

heat flow signal produces the heats of protonation or metal complexation that correspond to each titrant injection. These heats are deemed "experimental heats" (Q^{exp}). Experimental heats need to be corrected for heats of dilution and other heats intrinsic to the titration process by subtracting out background heats (Q^{bkg}) measured during the heat of dilution experiments. Equation 2.4 describes the background correction that produces corrected heats (Q_i^{corr}), where *i* corresponds to the titrant injection.

$$Q_i^{corr} = Q_i^{exp} - Q_i^{bkg} \tag{Eq.2.4}$$

Experimental heats in the absence of modeling can provide some useful information about the system under study. The experimental heats are a summation of heats from various reactions occurring in the system (e.g., heats of dilution, heats of protonation/deprotonation, heats of metal adsorption if metal ions are present, etc.). Considering whether the experimental heats are exothermic or endothermic provides information about the enthalpy of the reactions occurring that dominate the system. For example, endothermic reactions dominate Zn^{2+} complexation onto *B. subtilis*, as shown in Figure 2.3, while protonation of *Nitrosopumilis maritimus* strain SCM1 surface is dominated by exothermic reactions, as shown in Figure 2.2. These experimental heats include heats (e.g., heats of dilution) not associated with the reaction of interest; therefore, the dominant enthalpy evident from experimental heats may not be the same sign as the heats of reactions that are being targeted for study. With microbial surfaces, protonation and metal complexation can occur on a variety of sites simultaneously; consequently, the corrected heats are a summation of all the various protonation/complexation reactions occurring. The magnitudes of corrected heats are also helpful for determining whether heat flow during reactions is strongly exothermic or endothermic. More specific information about the enthalpies of reactions requires combining corrected heats with surface complexation modeling to determine site-specific thermodynamics.

2.2.8 Derivation of Site-specific Parameters

Combining surface complexation modeling with calorimetric measurements provides much more information about the reactivity of the microbial surface than the raw calorimetric data alone. It should be noted that any analysis involving surface complexation modeling will produce model-dependent parameters. Models are needed to describe proton adsorption and metal ion surface complexation. Quantifying heats of protonation is the first step in site-specific analysis of calorimetric data.

As the microbial surface is protonated, multiple types of sites can be protonated simultaneously. Consequently, the corrected heats are a sum of heats produced by protonation of multiple types of sites as described for individual corrected heats in Eq. (2.5) or total heat produced in Eq. (2.6). In Eq. (2.5) and (2.6), corrected heats of protonation are related to the enthalpies of protonation of site Lk (ΔH_{HL_k}), where $\delta n^i_{HL_k}$ is the change in the number of moles of the protonated sites caused by the *i*th addition of titrant.

$$-Q_i^{corr} = \Delta H_{HL_k} \times \delta n^i_{HL_k} \tag{Eq.2.5}$$

$$-\Sigma Q_i^{corr} = \sum (\Delta H_{HL_k} \times \delta n^i_{HL_k}) \tag{Eq.2.6}$$

The $\delta n^i_{HL_k}$ values are a function of pH, protonation constant, site concentrations, and volume of acid added. Using a suitable chemical speciation program (e.g., FITEQL

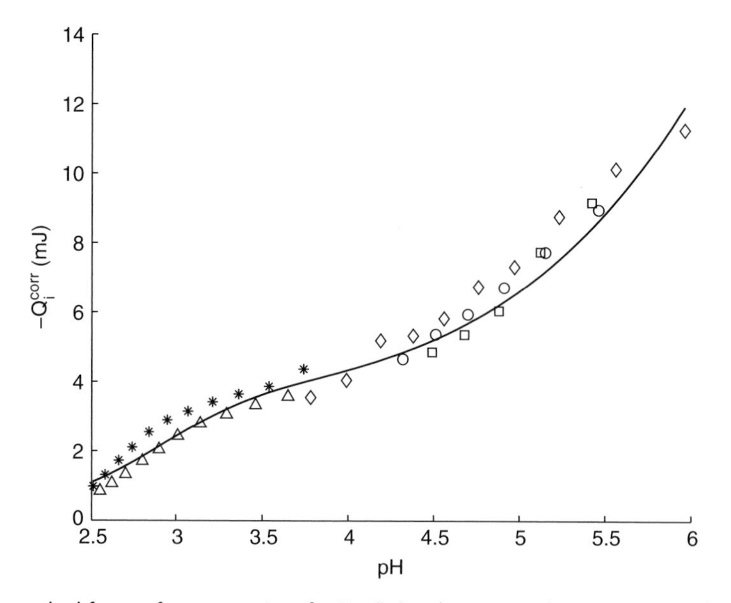

Figure 2.4 Corrected heat evolved from surface protonation of a 40 g/L *Pseudomonas putida* suspension as a function of pH (open circles). The solid curve represents the fit of the four-site nonelectrostatic model developed by Borrok and Fein (2005). Figure used with permission and modified from Gorman-Lewis (2009).

(Westall, 1982) or HYSS (Alderighi et al., 1999)), the surface complexation model, experimental parameters, $\delta n^i_{HL_k}$ values can be calculated for each addition of titrant. Derivation of the site-specific enthalpies is achieved through optimization by minimizing the sum-of-squares difference between the measured corrected heats (Q_i^{corr}) and the model-derived heats Qi^{calc}. Figure 2.4 illustrates the fit of Eq. (2.5) to corrected heats of protonation of *Pseudomonas putida*.

The derivation of enthalpies of metal complexation is more complicated, because the corrected heats also include heats from the redistribution of protons on the microbial surface. Consequently, the heat of proton redistribution must be taken into consideration when determining site-specific heats of metal complexation. Equation 2.7 describes the relationship between corrected heats of metal complexation, proton redistribution, and enthalpies of metal complexation, where ΔH_{ML_k} is the enthalpy of metal M complexing with microbial surface ligand L_k and $\delta n^i_{ML_k}$ is the change in complexed metal with each titrant addition. Proton and metal surface complexation models are necessary for calculating the changes in protonation and complexed metal at each step in the titration.

$$-\Sigma Q_i^{corr} = \Sigma(\Delta H_{HL_k} \times \delta n^i_{HL_k} + \Delta H_{ML_k} \times \delta n^i_{ML_k}) \qquad \text{(Eq.2.7)}$$

A suite of thermodynamic parameters can be determined for ion interactions on the microbial surface using equilibrium constants and derived enthalpies. Gibbs energies of reactions are calculated with Eq. 2.1 and entropies of protonation or complexation are calculated using Eq. 2.2.

2.2.9 Interpretation of Enthalpies of Protonation

Using the characteristic thermodynamic parameters of various ligand groups to identify and quantify acidic sites in large, complex molecules is a method used on macromolecules that is also applicable to the microbial surface (Fein et al., 2005; Jespersen and Jordan, 1970; Ngwenya et al., 2003; Perdue, 1978; Plette et al., 1995). Acidic functional groups on microbial surfaces react similarly to free organic acids in solution (Fein et al., 1997). Consequently, comparing the thermodynamic parameters of functional groups on microbial surfaces to organic acids and interpreting metal complexation in terms of metal binding by aqueous ligands provides a framework for understanding reactions occurring on microbial surfaces.

Protonation of the microbial surface is due to the presence of acidic functional groups within surface structures, and these are also the sites of metal complexation (Beveridge, 1994, 1988; Beveridge and Murray, 1980; Ferris and Beveridge, 1985; Hoyle and Beveridge, 1983). Functional groups known within the microbial cell surface include carboxylic acids, phosphates, amines, thiols, and hydroxyl groups. These functional groups have been confirmed spectroscopically by X-ray absorption spectroscopy and Fourier-transform infrared spectroscopy (Boyanov et al., 2003; Culha et al., 2008; Guine et al., 2006; Kelly et al., 2002; Mishra et al., 2009; Wei et al., 2004; Yee et al., 2004). Thermodynamic parameters determined from the combination of ITC and surface complexation modeling have been consistent with spectroscopic investigations of surface functional groups and provide characterization of these sites in terms of enthalpies and entropies of protonation, which cannot effectively be determined through other methods (Gorman-Lewis, 2011, 2009; Gorman-Lewis et al., 2006; Harrold and Gorman-Lewis, 2013).

Protonation of microbial surfaces tends to produce exothermic heats (Gorman-Lewis, 2011, 2009; Gorman-Lewis et al., 2014; Harrold and Gorman-Lewis, 2013). This suggests that enthalpies of protonation are dominated by exothermic reactions but recall that these heats are a cumulative quantity as described in Eq. 2.5 and 2.6; therefore, endothermic heats may contribute to the cumulative heats. An example of a microbial surface that has both endothermic and exothermic enthalpies of protonation is *Nitrosomonas europaea*, which has a functional group with a pKa of 4.8 and an endothermic enthalpy of protonation, while functional groups with higher pKa values have thermoneutral to exothermic enthalpies of protonation. The sign and magnitude of enthalpies and the magnitude of entropies protonation of are diagnostic to their identity.

The protonation of carboxylic acids can be both endothermic and exothermic depending on the structure of the acid (Martell, 2001; Pettit and Powell, 2005). Monofunctional carboxylic acids tend to be endothermic to thermoneutral, while the protonation of multifunctional acids begins endothermic to thermoneutral and becomes progressively more exothermic with subsequent protonations (Martell, 2001; Pettit and Powell, 2005). Characteristic entropies of protonation for monofunctional carboxylic acids are large and positive on the order of +80 to +120 J/mol K while the initial protonation of multifunctional acids is similar to monofunctional acids with subsequent protonations becoming less positive (Christensen et al., 1967, 1968, 1976; Martell, 2001; Pettit and Powell, 2005).

Previous investigations of thermodynamic parameters of functional groups with pKa values consistent with carboxylic acids have been consistent with both monofunctional and multi-functional carboxylic acids (Gorman-Lewis, 2009, 2011; Gorman-Lewis et al., 2014). This suggests that subtleties in functional group conformation may be detectable with calorimetric investigations.

Phosphate-bearing functional groups are an important component of microbial surface reactivity (Beveridge and Murray, 1980; Boyanov et al., 2003; Guine et al., 2006). The first deprotonation of phosphate-bearing functional groups is typically less than pH 4 and the second deprotonation in the circumneutral range. Enthalpies of protonation are typically exothermic and become less exothermic with subsequent protonation. For example, methylenediphosphonic acid has enthalpies of protonation of –10.36 and –3.17 kJ/mol for the first and second protonation, respectively. Entropies of protonation are typically large for the first protonation (e.g., +120 J/mol K) and decrease with subsequent protonations (Jensen et al., 2000; Martell, 2001; Nash et al., 1995; Pettit and Powell, 2005). Calorimetric investigations of the microbial surface have produced thermodynamic parameters consistent with the presence of phosphate-bearing functional groups, which have also been consistent with spectroscopic investigations (Boyanov et al., 2003; Gorman-Lewis, 2009; Gorman-Lewis et al., 2006).

More recent work has identified the presence of thiol groups on the microbial surface (Mishra et al., 2009, 2010). Thiol groups have pKa values of approximately 8, which is similar to that of the second protonation of phosphate-bearing functional groups. However, thiols have enthalpies of protonation that are distinct from phosphate-bearing functional groups, with thiols being more exothermic (e.g., –25 to –45 kJ/mol). Entropies of protonation are typically between +25 and +45 J/mol K, which is smaller than the first protonation of phosphate-bearing functional groups and may be in the range of the second protonation (Christensen et al., 1976; Martell, 2001; Pettit and Powell, 2005).

Proton ionization on the microbial surface at high pH (approximately 9 to 11) has been attributed to the presence of hydroxyl and primary amine functional groups (Beveridge and Murray, 1980; Fein et al., 1997). These functional groups have similar pKa values and similar enthalpies of protonation, in the range of approximately –30 to –50 kJ/mol (Christensen et al., 1976; Martell, 2001; Pettit and Powell, 2005). The similarity in pKa values and enthalpies of primary amines and hydroxyl groups makes it difficult to distinguish between them based on potentiometric titrations and site-specific enthalpies. In addition, these functional groups are not thought to heavily contribute to cation binding, making spectroscopic investigations of these sites difficult. However, a distinctive thermodynamic parameter that can aid in the identification of these basic functional groups is their entropies of protonation. Primary amines and hydroxyl groups have distinguishable entropies of protonation, with hydroxyl groups having much larger entropies (e.g., +100 to +130 J/mol K) than primary amines (e.g., +50 to +80 J/mol K) (Martell, 2001; Pettit and Powell, 2005).

2.2.10 Interpretation of Enthalpies of Metal Complexation

The interpretation of metal complexation on bacterial surfaces is considered in the context of metal complexation by aqueous ligands; therefore, the interpretation of metal complexation on the microbial surface is facilitated by comparison to metal binding by aqueous ligands. This discussion will begin by first considering metal complexation by aqueous ligands and then metal complexation by microbial surfaces.

The sign and magnitude of enthalpies of complexation are related to the two dominant processes that occur during complexation: the metal–ligand bond formation and changes in hydration of the central ion and ligand. Breaking metal ion–water bonds is an endothermic process, while making metal ion–ligand bonds is an exothermic process. Enthalpically driven complexation is characteristic of "soft" donors and "soft" metal ions (Ahrland et al., 1958; Beck and Nagypal, 1990; Martell and Hancock, 1996; Pearson, 1968a, 1968b). When complexation occurs between "soft" donors and "soft" metal ions, the bond is largely covalent, which often includes some degree of back-bonding between d orbitals of the metal ion and vacant d or $\pi*$ orbitals of the ligand. "Soft" metal ions typically have low oxidation states and d10 electron configurations. This electronic configuration facilitates the formation of π bonds to donors that have vacant d orbitals by sharing the t2g6 electrons. Exothermic enthalpies of complexation result when complexes with substantial covalent character are formed. Ligands like the S in thiol groups can participate in this type of enthalpy-driven complexation with soft metals. In contrast to "soft" ligand donors, anionic oxygen ligands are less polarizable and drive complexation through the net loss of water from the primary and secondary hydration spheres of the metal ion and ligand. This dehydration results in a large positive entropy, driving the complexation reaction and endothermic to thermoneutral enthalpy of complexation (Ahrland et al., 1958; Beck and Nagypal, 1990; Martell and Hancock, 1996; Pearson, 1968a, 1968b). Anionic oxygen ligands are found in carboxylic and phosphate-bearing functional groups on microbial surfaces.

Considering the sign and magnitude of enthalpies and entropies of complexation along with the pKa values of the functional groups provides essential information to understand the nature of complexation reactions occurring at microbial surfaces. For example, Figure 2.5 illustrates results of Cd complexation onto *B. subtilis*, which produced enthalpies of complexation of –0.2 and +14.4 kJ/mol onto carboxyl and phosphoryl surface sites, respectively (Gorman-Lewis et al., 2006). The $-T\Delta S$ values for these reactions are –17.0 and –38.2 kJ/mol, respectively, which illustrates that these reactions are entropically driven. Spectroscopic

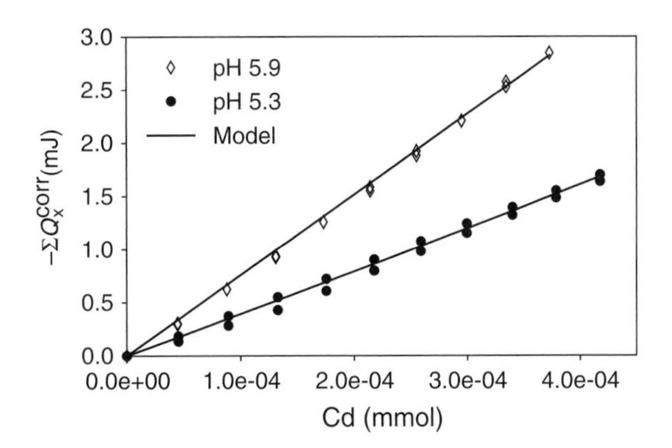

Figure 2.5 Cumulative background corrected heat generated by titration of 5.3 mM $Cd(ClO_4)_2$ into 80 g/L *B. subtilis* starting at pH 6 or 5.3, with the curve representing the fit to Eq. (2.7) using parameters from Gorman-Lewis (2006). Negative Q represents endothermic heats. Figure used with permission and modified from Gorman-Lewis (2006).

investigations of Cd adsorption onto *B. subtilis* are consistent with the interpretation of complexation to carboxyl and phosphoryl surface sites. Enthalpically driven metal complexation on microbial surfaces would likely result from thiol ligands participating in binding. Gorman-Lewis (2014) investigated Cd and Zn adsorption onto *B. licheniformis* and derived exothermic enthalpies (–15.6 and –15.5 kJ/mol, respectively) of complexation onto a postulated thiol site. While spectroscopic corroboration of thiol groups on *B. licheniformis* is lacking under similar experimental conditions, there is mounting evidence that thiol groups are present on microbial surfaces and are capable of complexing metal ions (Culha et al., 2008; Guine et al., 2006; Mishra et al., 2009, 2010; Song et al., 2012).

2.3 Conclusions

ITC of microbial surfaces combined with surface complexation modeling provides the means to decompose the Gibbs energy of complexation reactions and understand the driving force behind complex formation. This is the sole technique that can measure enthalpies of surface complexation reactions with sufficient accuracy and precision to afford a meaningful comparison of the enthalpic contribution to the Gibbs energy of reaction. It is that comparison that allows one to answer the question of "what drives complexation on microbial surfaces?" from a thermochemical perspective. The strength of the technique lies in integrating the information gained from calorimetric measurements with more traditional techniques of investigating surface reactivity. The very limited amount of data from this technique makes it difficult to identify trends among different microbes or specific microbe–metal complexes. However, as the body of research grows and is considered in the context of multiple lines of investigation, trends may emerge between the subtleties of the thermodynamic driving force of reactions and how those reactions influence cells and/or the environment.

2.4 References

Alderighi, L., Gans, P., Ienco, A., et al., 1999. Hyperquad simulation and speciation (HySS): a utility program for the investigation of equilibria involving soluble and partially soluble species. *Coord Chem Rev* 184, 311–318.

Ahrland, S., Chatt, J., Davies, N.R., 1958. The relative affinities of ligand atoms for acceptor molecules and ions. *Q Rev Chem Soc* 12, 265–276.

Alessi, D.S., Henderson, J.M., Fein, J.B., 2010. Experimental measurement of monovalent cation adsorption onto *Bacillus subtilis* cells. *Geomicrobiol J* 27, 464–472.

Banfield, J.F., Nealson, K.H., 1997. *Geomicrobiology: Interactions between Microbes and Minerals, Reviews in Mineralogy*. Mineralogical Society of America, Washington, DC.

Barker, W.W., Welch, S.A., Chu, S., Banfield, J.F., 1998. Experimental observations of the effects of bacteria on aluminosilicate weathering. *Am Mineral* 83, 1551–1563.

Beck, M.T., Nagypal, I., 1990. *Chemistry of Complex Equilibria*. Ellis Horwood, Chichester.

Beveridge, T.J., 1988. The bacterial surface: general considerations towards design and function. *Can J Microbiol* 34, 363–372.

Beveridge, T.J., 1994. Bacterial S-layers. *Curr Opin Struct Biol* 4, 204–212.

Beveridge, T.J., Murray, R.G.E., 1980. Sites of metal deposition in the cell wall of *Bacillus subtilis*. *J Bacteriol* 141, 876–887.
Borrok, D.M., Fein, J.B., 2005. The impact of ionic strength on the adsorption of protons, Pb, Cd, and Sr onto the surfaces of Gram negative bacteria: testing non-electrostatic, diffuse, and triple-layer models. *J Colloid Interface Sci* 286, 110–126.
Boyanov, M.I., Kelly, S.D., Kemner, K.M., et al., 2003. Adsorption of cadmium to *Bacillus subtilis* bacterial cell walls: a pH-dependent X-ray absorption fine structure spectroscopy study. *Geochim Cosmochim Acta* 67, 3299–3311.
Christensen, J.J., Hansen, L.D., Izatt, R.M., 1976. *Handbook of Proton Ionization Heats and Related Thermodynamic Quantities*. Wiley-Interscience, New York.
Christensen, J.J., Izatt, R.M., Hansen, L.D., 1967. Thermodynamics of proton ionization in dilute aqueous solution. VII. DH and DS values for proton ionization from carboxylic acids at 25C. *J Am Chem Soc* 89, 213–222.
Christensen, J.J., Wrathall, D.P., Izatt, R.M., 1968. Calorimetric determination of log K, ΔH, and ΔS from thermometric titration data. *Anal Chem* 40, 175–181.
Culha, M., Adiguzel, A., Yazici, M.M., et al., 2008. Characterization of thermophilic bacteria using surface-enhanced Raman scattering. *Appl Spectrosc* 62, 1226–1232.
Ehrlich, H.L., 1996. How microbes influence mineral growth and dissolution. *Chem Geol* 132, 5–9.
Ehrlich, H.L., 2002. *Geomicrobiology*, 4th ed. Marcel Dekker, Inc., New York, NY.
Fein, J.B., Boily, J.-F., Yee, N., Gorman-Lewis, D., Turner, B.F., 2005. Potentiometric titrations of *Bacillus subtilis* cells to low pH and a comparison of modeling approaches. *Geochim Cosmochim Acta* 69, 1123–1132.
Fein, J.B., Daughney, C.J., Yee, N., Davis, T.A., 1997. A chemical equilibrium model for metal adsorption onto bacterial surfaces. *Geochim Cosmochim Acta* 61, 3319–3328.
Ferris, F.G., Beveridge, T.J., 1985. Functions of bacterial cell surface structures. *Bioscience* 35, 172–177.
Flynn, S., Szymanowski, J., Fein, J., 2014. Modeling bacterial metal toxicity using a surface complexation approach. *Chem Geol* 374–375, 110–116.
Ginn, B., Fein, J.B., 2009. Temperature dependence of Cd and Pb binding onto bacterial cells. *Chem Geol* 259, 99–106.
Gorman-Lewis, D., 2009. Calorimetric measurements of proton adsorption onto *Pseudomonas putida*. *J Colloid Interface Sci* 337, 390–395.
Gorman-Lewis, D., 2011. Enthalpies of proton adsorption onto *Bacillus licheniformis* at 25, 37, 50, and 75°C. *Geochim Cosmochim Acta* 75, 1297–1307.
Gorman-Lewis, D., 2014. Enthalpies and entropies of Cd and Zn adsorption onto *Bacillus licheniformis* and enthalpies and entropies of Zn adsorption onto *Bacillus subtilis* from isothermal titration calorimetry and surface complexation modeling. *Geomicrobiol J* 31, 383–395.
Gorman-Lewis, D., Fein, J.B., Jensen, M.P., 2006. Enthalpies and entropies of proton and cadmium adsorption onto *Bacillus subtilis* bacterial cells from calorimetric measurements. *Geochim Cosmochim Acta* 70, 4862–4873.
Gorman-Lewis, D., Martens-Habbena, W., Stahl, D.A., 2014. Thermodynamic characterization of proton-ionizable functional groups on the cell surfaces of ammonia-oxidizing bacteria and archaea. *Geobiology* 12, 157–171.
Guine, V., Spadini, L., Sarret, G., et al., 2006. Zinc sorption to three Gram-negative bacteria: combined titration, modeling, and EXAFS study. *Environ Sci Technol* 40, 1806–1813.
Harrold, Z.R., Gorman-Lewis, D., 2013. Thermodynamic analysis of *Bacillus subtilis* endospore protonation using isothermal titration calorimetry. *Geochim Cosmochim Acta* 109, 296–305.
Hoyle, B., Beveridge, T.J., 1983. Binding of metallic ions to the outer membrane of *Escherichia coli*. *Appl Environ Microbiol* 46, 749–752.
Jensen, M.P., Beitz, J.V., Rogers, R.D., Nash, K.L., 2000. Thermodynamics and hydration of the europium complexes of a nitrogen heterocycle methane-1,1-diphosphonic acid. *J Chem Soc Dalton Trans*, 3058–3064.

Jespersen, N.D., Jordan, J., 1970. Thermometric enthalpy titration of proteins. *Anal Lett* 3, 323–334.
Kelly, S.D., Kemner, K.M., Fein, J.B., et al., 2002. X-ray absorption fine structure determination of pH-dependent U-bacterial cell wall interactions. *Geochim Cosmochim Acta* 66, 3855–3871.
Martell, A.E., 2001. NIST Critically Selected Stability Constants of Metal Complexes. U.S. Department of Commerce, Technology Administration, National Institute of Standards and Technology.
Martell, A.E., Hancock, R.D., 1996. *Metal Complexes in Aqueous Solution*. Modern Inorganic Chemistry. Plenum Press, New York.
Mishra, B., Boyanov, M., Bunker, B.A., et al., 2010. High- and low-affinity binding sites for Cd on the bacterial cell walls of *Bacillus subtilis* and *Shewanella oneidensis*. *Geochim Cosmochim Acta* 74, 4219–4233.
Mishra, B., Boyanov, M.I., Bunker, B.A., et al., 2009. An X-ray absorption spectroscopy study of Cd binding onto bacterial consortia. *Geochim Cosmochim Acta* 73, 4311–4325.
Nash, K.L., Rao, L.F., Choppin, G.R., 1995. Calorimetric and laser induced fluorescence investigation of the complexation geometry of selected europium-gem-diphosphonate complexes in acidic solutions. *Inorg Chem* 34, 2753–2758.
Nell, R. M., Szymanowski, J. E. S., Fein, J. B., 2016. Divalent metal cation adsorption onto *Leptothrix cholodnii* SP-6SL bacterial cells. *Chem Geol* 439, 132–138.
Ngwenya, B.T., Sutherland, I.W., Kennedy, L., 2003. Comparison of the acid-base behavior and metal adsorption characteristics of a Gram-negative bacterium with other strains. *Appl Geochem* 18, 527–538.
Pearson, R.G., 1968a. Hard and soft acids and bases, HSAB, part I: fundamental principles. *J Chem Educ* 45, 581–587.
Pearson, R.G., 1968b. Hard and soft acids and bases, HSAB, part II: underlying theories. *J Chem Educ* 45, 643–648.
Perdue, E.M., 1978. Solution thermochemistry of humic substances – I.Acid-base equilibria of humic acid.*Geochim Cosmochim Acta* 42, 1351–1358.
Pettit, L.D., Powell, K.J., 2005. IUPAC Stability Constants Database (SC-Database). Acad. Softw. Data Version 4.82.
Plette, A.C.C., van Riemsdijk, W.H., Benedetti, M.F., Van der Wal, A., 1995. pH dependent charging behavior of isolated cell walls of a Gram-positive soil bacterium. *J Colloid Interface Sci* 173, 354–363.
Sheng, X.F., Zhao, F., He, L.Y., Qiu, G., Chen, L., 2008. Isolation and characterization of silicate mineral-solubilizing *Bacillus globisporus* Q12 from the surfaces of weathered feldspar. *Can J Microbiol* 54, 1064–1068.
Song, Z., Kenney, J.P.L., Fein, J.B., Bunker, B.A., 2012. An X-ray absorption fine structure study of Au adsorbed onto the non-metabolizing cells of two soil bacterial species. *Geochim Cosmochim Acta* 86, 103–117.
Wei, J., Saxena, A., Song, B., et al., 2004. Elucidation of functional groups on Gram-positive and Gram-negative bacterial surfaces using infrared spectroscopy. *Langmuir* 20, 11433–11442.
Westall, J.C., 1982. FITEQL, a Computer Program for Determination of Chemical Equilibrium Constants from Experimental Data. Version 2.0. Report 82–02. Department of Chemistry, Oregon State University.
Wightman, P.G., Fein, J.B., 2004. The effect of bacterial cell wall adsorption on mineral solubilities. *Bact Geochem Speciat Met* 212, 247–254.
Wightman, P.G., Fein, J.B., Wesolowski, D.J., et al., 2001. Measurement of bacterial surface protonation constants for two species at elevated temperatures. *Geochim Cosmochim Acta* 65, 3657–3669.
Yee, N., Benning, L.G., Phoenix, V.R., Ferris, F.G., 2004. Characterization of metal-cyanobacteria sorption reactions: A combined macroscopic and infrared spectroscopic investigation. *Environ Sci Technol* 38, 775–782.

3 Potentiometric Titrations to Characterize the Reactivity of Geomicrobial Surfaces

DANIEL S. ALESSI, SHANNON L. FLYNN, MD. SAMRAT ALAM, LESLIE J. ROBBINS, AND KURT O. KONHAUSER

Abstract

Potentiometric titrations are a widely used technique to quantify proton-active functional groups on microorganisms, their exudates, biominerals, and biofilms. In this chapter, we provide a step-by-step introduction to the preparation of microbial cells for the determination of proton buffering capacity using modern autotitration systems. Following a discussion of how to process titration data and plot titration curves, we review commonly used thermodynamic approaches to model titration curves in order to calculate cell wall functional group acidity constants and corresponding site concentrations. In geomicrobiology, protonation models are primarily used as a basis for the development of surface complexation models that can predict the adsorption of charged species such as metals to biomass. The case example that follows outlines the development of a surface complexation model for the adsorption of cadmium to a species of the marine cyanobacterium *Synechococcus*, using a protonation model developed from titration data as its basis. In the last section, we introduce the reader to other analytical tools that are complementary to titration results, and discuss a few common complications to the titration approach when it is applied to natural materials.

3.1 Introduction

The reactivity of chemical functional groups on microbial cell walls, cell exudates, and biofilms often controls the mobility of metals and nutrients in natural systems, and may modulate the formation or dissolution of (bio)minerals (Fein, 2006). Potentiometric titrations are among the most widely used experimental techniques to quantify how the charge densities of these surfaces develop as a function of pH, a key parameter in understanding their reactivity towards charged species in aqueous solution (Lützenkirchen et al., 2012). Titration experiments are conceptually simple. A biomass of interest is suspended in an electrolyte at a known concentration (g of biomass per L of electrolyte), and precise volumes of acid or base are added in a series of small aliquots. After each addition, the solution proton (H^+) condition is perturbed, so that the solution is in disequilibrium with proton-active functional groups on the biomass surface. Thus, after each addition, the solution is allowed to reach equilibrium with the biomass before a subsequent aliquot of acid or base is added. The resulting data form a titration curve, which represents the buffering capacity of the biomass suspension across the tested pH range. Cell damage may occur at highly acidic or moderately alkaline pH (Fein et al., 2005), and so a typical titration pH range for bacteria would be approximately 3 to 10.

The advent of modern automated potentiometric titration systems has considerably increased the application of titrations in the field of geomicrobiology. Autotitrators use high-precision piston-driven systems to add sub-microliter volumes of acid or base to cell suspensions during the titration process. As a result, far less sample volume is needed to obtain a high-quality titration curve. Furthermore, the software that runs modern autotitration systems has numerous tools to optimize the titration process. For example, a pH stability threshold can be set to ensure that equilibrium has been achieved prior to the addition of the next aliquot of acid or base. The software can also calculate the instantaneous buffering capacity at any point in the titration process by comparing the pH change with the amount of acid or base added during the previous addition. Using this information, the software can make an intelligent guess at an appropriate acid or base amount for the subsequent addition. These advances have resulted in titrations that are generally accurate, rapid, and reproducible.

Biomass titration curves provide the basis for the development of surface complexation models, which are described in more detail in Section 3.3. The surface complexation model (SCM) approach has considerable advantages over empirical modeling approaches such as sorption isotherms, because SCMs can make predictions about metal speciation in a system outside the experimental conditions used to develop the model. Detailed commentaries on the advantages and disadvantages of these modeling approaches can be found in Koretsky (2000) and Bethke and Brady (2000). In this chapter, we introduce commonly used methods to prepare geomicrobial samples for potentiometric titration experiments, including instrument setup considerations for automated titration systems. The remainder of the chapter discusses data processing techniques and the fundamental theory behind protonation models, and provides an introduction to methods used to model titration data. Finally, we discuss the coupling of protonation models to microbe–metal adsorption data to construct an SCM, complementary analytical techniques to titration analyses, and issues that should be considered when titrating complex, multi-component natural materials.

3.2 Conducting Potentiometric Titrations

3.2.1 Sample Preparation

Potentiometric titrations can be performed on a variety of environmentally relevant solids, including minerals (e.g., Duc et al., 2005a, 2005b; Dzombak and Morel, 1990), microbes (e.g., Alessi et al., 2010; Cox et al., 1999; Fein et al., 1997, 2005), microbial exudates (e.g., Baker et al., 2010; Kenney and Fein, 2011; Petrash et al., 2011a), carbon-bearing materials (e.g., Alam et al., 2018; Gorgulho et al., 2008), plant and organic matter (e.g., Driver and Perdue, 2015; Ginn et al., 2008), worm mucus (Lalonde et al., 2010; Petrash et al., 2011b), and admixtures of these materials (e.g., Alessi and Fein, 2010; Davis et al., 1998;

Pagnanelli et al., 2006). Here, we focus on the preparation of microbial cells for acid–base titrations to collect data essential in developing a protonation model of the titrated biomass. In preparation for titrations and corresponding metal adsorption experiments, planktonic cells (e.g., *Synechococcus*) are typically grown from pure culture or as a consortium in a growth medium such as lysogeny broth (LB), tryptic soy broth (TSB), or various minimal media, although growth media vary widely depending on the microorganism of interest. Growth of cells is normally monitored by measuring the optical density of the cell suspension at 600 nm (OD_{600}); by comparing the OD_{600} with a predetermined growth curve (see Chapter 1), the density of cells in solution can be determined. Researchers often target the exponential or stationary phases of cell growth, depending on the experiment to be conducted; in either case, cells used for potentiometric titrations should be at the same growth phase as those used for other parts of the experimental plan (e.g., metal adsorption experiments, metal redox experiments) to ensure that the functional group identities and densities are matched.

Cells are normally collected by centrifugation at 5 000–10 000 *g* for 5–10 min. Following the initial centrifugation, the resulting cell pellet is repeatedly washed (3–5×) with an electrolyte solution, often sodium chloride, sodium nitrate, or sodium perchlorate, at a concentration that is generally between 0.01 and 0.1 M and at circumneutral pH. A common practice is to match the ionic strength and composition of the wash solution to that which will be used in the metal adsorption experiments, which in effect accounts for the effect of ionic strength on the electric field at the microbial surface. After each wash, the cells are pelleted as described in preparation for the subsequent wash. The use of the electrolyte wash solution both removes residues from the growth medium and ensures that cells do not lyse, which may occur if deionized water is used. Cells may be washed in electrolytes of composition or concentration that differ from those mentioned if experiments are to be conducted at those conditions; for example, cell washing may be done in a higher-ionic-strength medium if the research study focuses on constraining the impact of ionic strength on the magnitude of cell wall electrostatic effects, or on the adsorption of metals in marine systems (e.g., Borrok and Fein, 2005; Liu et al., 2015). Once the washing process is complete, cells should be repeatedly centrifuged and the supernatant decanted each time to remove as much excess water from the cell pellet as possible. The pellet is then precisely weighed to determine the bacterial wet weight, which has been determined to be approximately 10 times the dry weight (Fein and Delea, 1999). Following the wet weight determination, cells should be resuspended in the electrolyte of interest. The cell concentration used (often given in g of cells per L of electrolyte) varies widely depending on the cell size and surface chemistry; a reasonable trial range for initial potentiometric titration experiments would be between 1 and 10 g/L as wet weight. A typical solution volume for a titration would be between 10 and 50 mL, but this varies depending on the titrator model and solution type.

3.2.2 Conducting Automated Potentiometric Titrations

Modern potentiometric titrations are conducted using automated systems (Figure 3.1), where precisely known volumes of acid and/or base are added to a solution of known volume containing the material being titrated. Prior to starting the titrations, both the stock

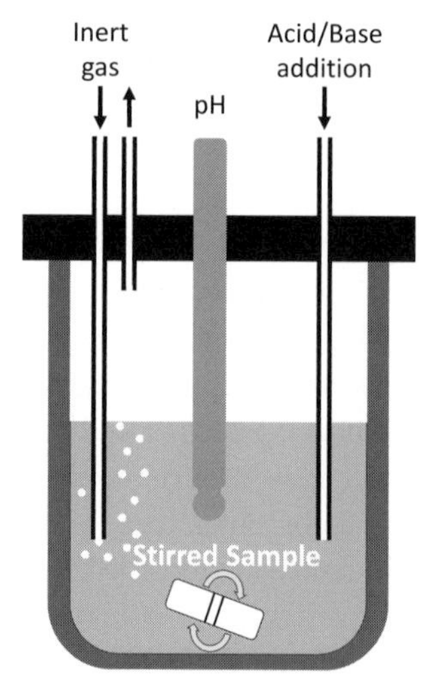

Figure 3.1 Basic setup of a potentiometric titration. A sealed cup containing the sample is purged with an inert gas (commonly N_2 or Ar) to eliminate aqueous carbonate species. During titration, the suspension is continuously stirred while precisely measured, small aliquots of acid or base are added, and the resulting change in pH is measured after equilibrium is achieved.

titrants (acid and base) and the cell suspension should be purged with an inert gas such as nitrogen (N_2) for at least 30 minutes to remove inorganic carbonate species that may add proton buffering capacity to the solutions. The beaker or cup containing the cell suspension to be titrated should be gently bubbled with the same gas throughout the titration process; for this reason, it is standard practice to cover the container with a piece of perforated Parafilm® or a custom cover to better isolate it from the atmosphere. Prior to initiating a titration, the following steps should be taken:

1. Purge the burette(s) with the acid and/or base solution(s) to be used in the titrations, often 1 M solutions of HCl and NaOH, both to ensure that fresh titrant is available and to remove any bubbles in the burettes.
2. Calibrate the pH meter connected to the autotitrator using certified pH reference solutions; ensure that the slope of the calibration curve is within recommended limits.
3. The suspension being titrated should be stirred during the titration process; add either a magnetic stir bar of appropriate size or a stirring impeller to the solution.
4. Just prior to initiating the titration, immerse the pH meter, the gas line for N_2 or another inert gas, and the tip(s) of the acid/base burette(s) into the cell suspension, and then ensure the cup is covered with Parafilm® or an appropriate custom lid.
5. Adjust the flow of inert gas to ensure that the solution is bubbled throughout the titration process, and the rate of stirring is such that the solution remains visibly homogeneous and the tip(s) of the burette(s) and the pH meter remain submerged.
6. The temperature of the solution should also be monitored if there is concern about it changing during the period of the titration. The pK_a values are grounded in equilibrium thermodynamics and are therefore temperature dependent (see Section 3.2.3).

Titrations can be performed either alkalimetrically (from low pH to high pH; often called a 'forward' titration) or acidimetrically (from high pH to low pH; often called a 'reverse' titration). Some autotitrator models can control two burettes simultaneously, allowing the automation of forward and reverse titrations without a burette change between the two. Either the software that controls the autotitrator can add fixed volumes of acid or base to the solution being titrated until the designated terminal pH is achieved, or the addition of the titrant can be based on the instantaneous buffering capacity of the solution during the titration process. The latter choice is often better, because it ensures a more evenly distributed set of titration data points as a function of pH, regardless of changes in buffering capacity across the pH range tested. In this case, the software uses the volume of acid or base, and the corresponding change in pH obtained from the previous addition, to calculate the volume of the subsequent addition. In the autotitrator software, the user can usually set the minimum and maximum titrant addition volume per step (often ranging between a single and hundreds of microliters, respectively), the electrode stability required for the next addition of titrant (normally 0.1–0.2 mV/s), and the desired pH change per addition (often 0.1–0.2 pH units) (Alam et al., 2018; Liu et al., 2016). Times to complete titrations depend on these parameters, but a typical forward or reverse titration on a modern auto-titration system may take 30 to 60 minutes.

In principle, a single forward or reverse titration provides sufficient data to produce a protonation model (i.e., to calculate pK_a values and corresponding site concentrations, which define how surface charge density develops as a function of pH). In practice, both forward and reverse titrations of the cells should be conducted, as hysteresis between forward and reverse curves is diagnostic of either damage to the substrate being titrated (e.g., at low or high pH) or some issue with the titration process itself. Figure 3.2 shows an example of a forward and reverse titration of a suspension of *Synechococcus* sp. PCC 7002 cyanobacterial cells conducted between pH 4.0 and 10.0 using an autotitration system (Liu et al., 2016). One of the more common ways of plotting titration data is as moles of excess charge (due to deprotonated surface functional groups) per gram of bacteria, according to

$$C_A - C_B = [-Q] + [H^+] - [OH^-] \tag{3.1}$$

where $[-Q]$ is the negative excess charge per gram of bacteria due to deprotonated surface functional groups, $[H^+]$ and $[OH^-]$ are the concentrations of protons and hydroxyl ions in solution, respectively, and $C_A - C_B$ is the concentration difference between acid and base added during the titration process. Quantifying the initial proton condition of the cell wall, prior to the beginning of the titration, is a nontrivial matter. Bacterial surfaces are often net negatively charged to pH <2 (Fein et al., 2005), a fact that can be confirmed by zeta potential measurements, implying that there are proton-active sites at highly acidic pH. In Figure 3.2 (which is consistent with Eq. 3.1), the net excess charge is assigned a value of zero at approximately pH 4, the pH at which the forward titration was initiated; however, because there are certainly deprotonated cell wall functional groups below pH 4 that contribute excess charge to the system, it is more correct to say that Figure 3.2 shows excess charge relative to the pH at which the forward titration started. Fein et al. (2005) point out that defining a zero-proton condition for cell wall functional groups is essential to

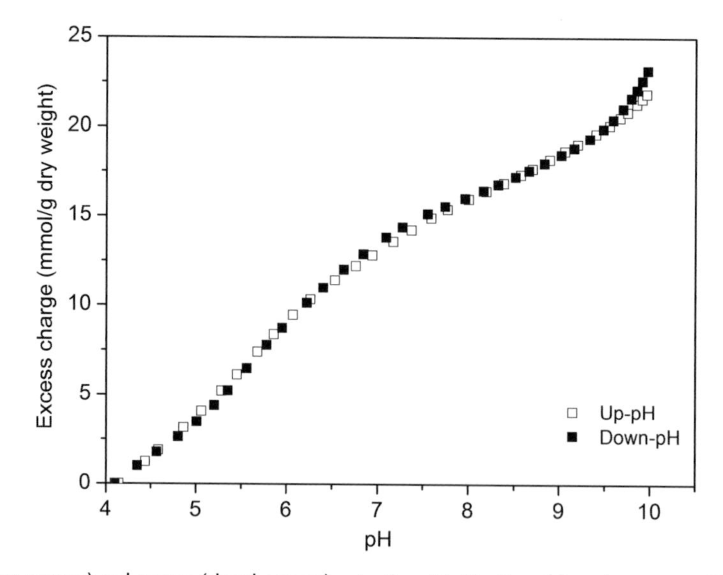

Figure 3.2 Forward (open squares) and reverse (closed squares) potentiometric titration of *Synechococcus* sp. PCC 7002 cells. The curves show little hysteresis, indicating that cell wall functional group reactions with protons are reversible and that the microbial cells were not significantly damaged during the titration process. Source: Liu et al. (2016) Geochim. Cosmochim. Acta 187: 179–194.

calculate the hydrogen mass balance of the system. In the case of minerals, this problem is solved by defining the initial hydrogen concentration of the system as zero at the point of zero charge (PZC). The PZC itself can be determined by titrating the mineral at several different ionic strengths to determine the common intersection point (CIP) of the titration curves (see, e.g., Hao et al., 2018; Zhao et al., 2011). Because the point of zero charge is often <2 for bacteria, it is not possible to titrate to the PZC without damaging the cells, and so the CIP cannot be directly measured by titration. Fein et al. (2005) note that two practical solutions are possible: to assign either the fully protonated or the fully deprotonated case as the condition at which the total initial hydrogen concentration is zero, and to perform all subsequent hydrogen mass balance calculations to that reference state. Westall et al. (1995) and Fein et al. (2005) provide more detailed explanations of the proper consideration of hydrogen mass balances in potentiometric titrations.

In the example given in Figure 3.2, negligible hysteresis between the forward and reverse curves is observed, indicating that protonation reactions at cell wall functional groups are reversible and that the cells themselves were not damaged as a result of the solution pH changes induced by the titration process. If cell damage is a concern, further confirmation of cell viability can be conducted using live/dead staining kits such as BacLight in consort with microscopy (e.g., Cox et al., 1999). Additionally, it is common practice to conduct titrations in duplicate or triplicate, using a new cell suspension each time. These practices ensure the accuracy of the data collected and provide information about variability in the protonation constants and corresponding site concentrations calculated from the data (discussed later).

In addition to titrating the cell suspension or other proton-active substrate of interest, titrations must also be conducted to quantify the proton buffering capacity of the background electrolyte solution used to suspend the cells. This titration serves as a baseline by quantifying proton and hydroxyl consumption due to reactions between the two to form water, and aqueous complexation reactions of H^+ and OH^- with the monovalent ions in the electrolyte solution itself or ions that are added as counterions to the acids and bases used for the titration, e.g., Na^+ (from NaOH), Cl^- (from HCl), or NO_3^- (from HNO_3). The blank titration is particularly important at relatively high or low pH, where a considerable fraction of the measured buffering capacity is due to solution H^+ or OH^-. Once collected, the blank titration is subtracted from the titration of the cell suspension to determine the true buffering capacity attributable to the cells across the tested pH range. As described for the cell titration, repeated forward and reverse titrations of the background electrolyte are the normal practice to collect a rigorous data set.

3.2.3 Data Processing and Modeling

Potentiometric titration data are typically used to determine the acidity constant ($-\log K_a$, or pK_a) values and corresponding site densities of proton-active functional groups on the cell surface. To do so, the excess charge determined in the blank electrolyte titration should be subtracted from that of the titrations of the cells to determine the potential due to the cells themselves. Doing so can be nontrivial, because the positions of titration data points along the pH axis in the blank titration often do not align precisely with those of the titration of the cell suspension. To overcome this, curve fitting and subtraction can be performed using a linear programming approach in software such as MATLAB to generate a set of corrected titration data that can subsequently be modeled (see, e.g., Lalonde et al., 2008a, b). In some cases, the buffering capacity of the cell suspension is so great in comparison to that of the electrolyte that the subtraction can be ignored within the pH range of interest. However, the titration data should be carefully compared with those of the blank electrolyte solution to ensure that this is the case, particularly at relatively low and high pH values, where elevated H^+ or OH^- concentrations, respectively, may represent a considerable fraction of the buffering capacity of the overall system.

A variety of approaches are used to model potentiometric titration data, including linear programming optimization methods (LPM) (e.g., Brassard et al., 1990; Cox et al., 1999; Martinez and Ferris, 2001; Martinez et al., 2002), which often result in relatively more binding sites, and least-squares error (LSE) approaches (e.g., Fein et al., 1997; Hetzer et al., 2006; Ngwenya et al., 2003), such as those that are used in titration data modeling software such as FITEQL (Herbelin and Westall, 1999) or ProtoFit (Turner and Fein, 2006). In the latter approach, the operator fixes the number of discrete sites to be solved for and guesses at the initial pK_a values for each of those sites, and the model with the lowest variance is accepted as the one having the appropriate number of sites (Martinez and Ferris, 2001). Because smaller differences are often discarded in the LSE approach, these models tend to have fewer sites (i.e., discrete pK_a values) than do LPM models. Both modeling approaches are well accepted in the geobiology community, and the LSE approach, as executed in FITEQL, will be discussed in the example that follows.

Many protonation models of the bacterial surface invoke a set of monoprotic functional groups to describe the proton buffering behavior observed in the potentiometric titrations (Figure 3.2). In the most general sense, the deprotonation reaction can be described as

$$\mathrm{R-A(H)^0 \leftrightarrow H^+ + R-A^-} \tag{3.2}$$

where R represents the cell wall macromolecule to which functional group A is bound. The acidity constant, K_a, for Eq. 3.2 is then expressed as

$$\mathrm{K_a = \frac{[R-A^-]a_{H^+}}{[R-A(H)^0]}} \tag{3.3}$$

where $[R-A^-]$ and $[R-AH^0]$ represent the concentrations of deprotonated and protonated functional groups, respectively, and a_{H^+} represents the activity of H^+ in solution. The K_a values, and all mass action constants, are based in equilibrium thermodynamics, according to

$$\Delta_r G = \Delta_r G^0 + RT\ln\prod_{i=1}^{n}(a_i^{\gamma_i}) \tag{3.4}$$

where $\Delta_r G^O$ and $\Delta_r G$ represent the Gibbs energy of reaction at the reference condition and the actual chemical free energy at any point in the reaction, respectively, R is the gas constant, T is the temperature in degrees Kelvin, and $\prod_{i=1}^{n}(a_i^{\gamma_i})$ is the reaction quotient (e.g., the right side of Eq. 3.3). By definition, at equilibrium $\Delta_r G = 0$, and, at this unique condition, the reaction quotient is defined as the mass action constant (K), so that Eq. 3.4 simplifies to

$$\Delta_r G^0 = -RT\ln K \tag{3.5}$$

As discussed earlier, when using LSE approaches, the user must suggest a discrete number of sites (often between one and four) and make initial guesses at the K_a values at these sites. In some instances, these initial guesses can result in the assignment of K_a values at local minima in the modeled data (i.e., not at the true K_a values; Martinez and Ferris, 2001), and so testing a range of initial guesses is good practice.

A final major consideration during the modeling of protonation data is the choice of a nonelectrostatic or an electrostatic model. In electrostatic models, an additional term is added to the mass action expression (e.g., Eq. 3.3) to account for changes in the extent of the surface electric field of the bacteria, for example according to

$$K_{apparent} = K_{intrinsic} \times e^{\frac{-\Delta z \psi F}{RT}} \tag{3.6}$$

where $K_{intrinsic}$ is the true mass action constant (i.e., it represents the chemical free energy of the surface protonation reaction of interest), $K_{apparent}$ is the effective mass action constant, and $e^{\frac{-\Delta z \psi F}{RT}}$ is the electrostatic (or Coulombic) term, in which Δz is the change in surface charge due to the sorption of a proton (or other charged species), Ψ is the potential of the surface, and F is Faraday's constant. In cases where the ionic strength is relatively high, the electrostatic correction is critical, because the observed mass action constant is far from the intrinsic value. In cases where the ionic strength of the system is modest, a nonelectrostatic model is often acceptable, because the values of $K_{intrinsic}$ and $K_{apparent}$ are similar. A wide variety of electrostatic SCM approaches exist, including the constant capacitance model (CCM), the double layer model (DLM), and the triple layer model (TLM). Davis and Kent (1990) provide a thorough description of the theory underpinning commonly used electrostatic SCMs. Most modeling software packages, including FITEQL, have a number of electrostatic models that can be employed when modeling titration data.

3.3 An Example of the Use of Protonation Models in Surface Complexation Modeling

Protonation models developed using the approaches described in Section 3.2 can be used as the basis of an SCM to predict the adsorption behavior of charged species (often metals) to the surface of bacteria across a wide range of system conditions. In order to develop a SCM, the metal binding constant (K_M) values must be determined for each metal–surface functional group pair in the system of interest. As an example of this workflow, we describe here the results of our earlier study on the adsorption of Cd(II) to the marine cyanobacterium *Synechococcus* sp. PCC 7002 (Liu et al., 2015). In this study, the bacteria were titrated in electrolytes of different ionic strengths and a nonelectrostatic SCM was used to calculate pK_a values and corresponding site concentrations. In the case of the 0.56 M NaCl electrolyte, it was determined that a three-site protonation model, with pK_a values of 5.07 ± 0.03 (Site 1), 6.71 ± 0.07 (Site 2), and 8.54 ± 0.15 (Site 3), provided the best fit to the titration data, as determined by the variance, or V(Y), value given by FITEQL. Following the titration modeling, experiments of Cd(II) adsorption to *Synechococcus* cells were conducted as a function of pH in the same electrolyte. Cadmium is typically used in metal adsorption studies to bacterial surfaces because it is not complicated by the precipitation of carbonate or hydroxide solids in most experimental conditions. Moreover, because of its common use, comparisons can readily be made with other metal adsorption studies involving humic acids and other bacterial species, where Cd is the usual metal chosen (e.g., Borrok et al., 2004; Yee and Fein, 2001).

In general, adsorption of cationic metals to the cells increases markedly with increased pH (Figure 3.3), because the cell wall surface functional groups are increasingly deprotonated and therefore negatively charged as pH increases. For example, Cd^{2+} binding to any of the three proton-active sites can occur according to this reaction:

$$R - A^- + Cd^{2+} \leftrightarrow R - A - Cd^+ \tag{3.7}$$

The corresponding mass action constant (K_M) is

$$K_M = \frac{[R - A - Cd^+]}{[R - A^-]a_{Cd^{2+}}} \tag{3.8}$$

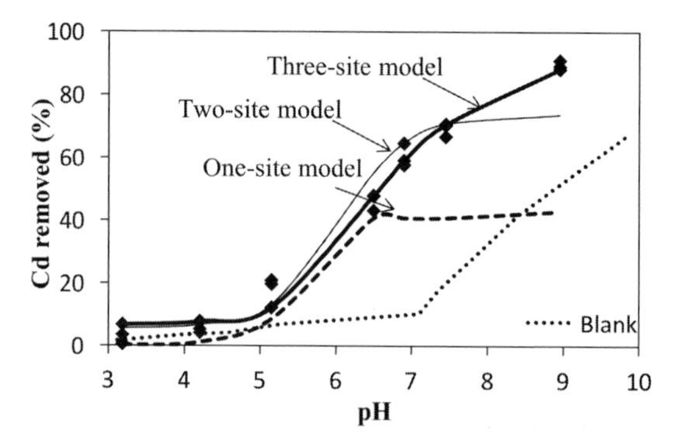

Figure 3.3 pH adsorption edge of Cd(II) to *Synechococcus* sp. PCC 7002 cells. Diamonds (◆) indicate experimental data, and the dashed, thin, and thick lines represent models that invoke Cd(II) adsorption to one, two, or three of the proton-active sites on the bacterial surface, respectively. The Cd(II) concentration is 8.9×10^{-6} M, electrolyte concentration 0.56 M NaCl, and the biomass concentration 10 g/L (wet weight). Source: Liu et al. (2015) Geochim. Cosmochim. Acta 169: 30–44.

Combining Eq. 3.3 and 3.8 in a system of nonlinear equations, which is the basis of an SCM, places H^+ and Cd^{2+} in direct competition for adsorption to functional groups of interest on the bacterial cell wall. The proton binding constants (pK_a values) and corresponding site concentrations are known from prior modeling of the potentiometric titration data. This second modeling step uses those parameters and the metal adsorption data (e.g., Figure 3.3) to calculate the metal binding constants to each proton-active site.

Opinion varies on how best to model metal adsorption data using the SCM approach. An attractive option is to invoke the smallest number of sites needed to explain the metal adsorption behavior observed. In the example here, three proton-active sites are available to adsorb Cd species from solution. Since at any given pH the site with the lowest pK_a (Site 1, $pK_a = 5.07$) will have a higher degree of deprotonation than the sites with higher pK_a values (Sites 2 and 3), it is logical to start by invoking Cd^{2+} adsorption to Site 1. The dashed line in Figure 3.3 shows the result of this fitting exercise. The data is well fitted between pH 3 and approximately 6.5, after which Cd uptake plateaus at approximately 40%. In order to account for observed Cd adsorption at higher pH, Cd adsorption to the second site ($pK_a = 6.71$) is allowed, resulting in a reasonable fit to approximately pH 7.5 (thin solid line). In this case, invoking Cd adsorption to all three proton-active sites provides the best overall fit (thick solid line). The corresponding metal binding constants (K_M values) for Cd^{2+} adsorption to sites 1, 2, and 3 were calculated to be 1.15 ± 0.67, 1.78 ± 0.07, and 2.74 ± 0.03, respectively. In principle, these metal binding constants, in consort with the pK_a values and site concentrations, can be used to predict Cd adsorption to *Synechococcus* across a wide range of metal-to-bacteria ratios. Properly constructed SCMs also consider the aqueous speciation of the metal of interest, Cd(II) in this example, the possibility of metal precipitation (often as carbonate or hydroxide species), and the possibility that multiple aqueous species of the metal of interest may compete for adsorption to bacterial surface functional groups.

3.4 Final Considerations

It is well accepted that proton-active functional groups on microbial cell surfaces and their exudates are primarily comprised of carboxyl, hydroxyl, phosphoryl, amino, and sulfhydryl groups (Beveridge and Murray, 1980; Fein et al., 1997; Liu et al., 2016; Yu and Fein, 2016). To confidently assign functional group identities to sites calculated during the modeling of potentiometric titration data, spectroscopic approaches are also required. Other chapters in this book provide detailed discussions of common techniques to determine the local environment of metal binding on bacterial cell walls, including synchrotron-based X-ray absorption spectroscopy (Chapter 10) and Fourier-transform infrared spectroscopy (Chapter 12). Additionally, the use of calorimetry (Chapter 2) coupled to surface complexation modeling provides an opportunity to determine the thermodynamic driving force behind adsorption reactions (e.g., Gorman-Lewis et al., 2006).

The buffering capacity of microbial cells across a range of pH values is relatively easy to determine by potentiometric titration. Conversely, the development of protonation models and their extension to SCMs that consider metal binding is nontrivial. In order to accurately capture metal binding behavior in a system, the key aqueous complexation and adsorption reactions must be included in the SCM. In complex systems, such as soils, sediments, and biofilms, accurately assessing these reactions may require considerable experimental work and an expert-level knowledge of geochemistry. Furthermore, for some natural systems, potentiometric titrations are not possible, because the sample contains material such as carbonate minerals, which irreversibly dissolve (e.g., Flynn et al., 2017; Warchola et al., 2017), or because individual sorbents in complex mixtures may interact with each other, resulting in site blockage (e.g., Alessi and Fein, 2010; Davis et al., 1998). For these reasons, in some cases, empirical approaches to adsorption data modeling, including isotherms, may be more practical. However, the utility of the SCM approach is its ability to make predictions about equilibrium metal distribution across wide ranges of pH, ionic strength, water chemistry, and sorbent-to-sorbate ratios. Potentiometric titrations are the first step in developing such models.

3.5 Acknowledgments

DSA and KOK received Natural Sciences and Engineering Research Council of Canada (NSERC) Discovery grants that supported several of the experimental studies that are discussed in this chapter.

3.6 References

Alam, M. S., Gorman-Lewis, D., Chen, N., et al. (2018) Thermodynamic analysis of nickel(II) and zinc(II) adsorption to biochar. *Environ. Sci. Technol.* 52(11): 6246–6255.

Alessi, D. S., Fein, J. B. (2010) Cadmium adsorption to mixtures of soil components: Testing the component additivity approach. *Chem. Geol.* 270(1–4): 186–195.

Alessi, D. S., Henderson, J. M., Fein, J. B. (2010) Experimental measurement of monovalent cation adsorption onto *Bacillus subtilis* cells. *Geomicrobiol. J* 27(5): 464–472.

Baker, M. G., Lalonde, S. V., Konhauser, K. O., Foght, J. M. (2010) Role of extracellular polymeric substances in the surface chemical reactivity of *Hymenobacter aerophilus,* a psychrotolerant bacterium. *Appl. Environ. Microbiol.* 76(1): 102–109.

Bethke, C. M., Brady, P. V. (2000) How the Kd approach undermines ground water cleanup. *Ground Water* 38(3): 435–443.

Beveridge, T. J., Murray, R. G. E. (1980) Sites of metal deposition in the cell wall of *Bacillus subtilis. J. Bacteriol.* 141(2): 876–887.

Borrok, D. M., Fein, J. B. (2005) The impact of ionic strength on the adsorption of protons, Pb, Cd, and Sr onto the surfaces of Gram negative bacteria: testing non-electrostatic, diffuse, and triple-layer models. *J. Colloid Interface Sci.* 286: 110–126.

Borrok, D. M., Fein, J. B., Kulpa, C. F. (2004) Proton and Cd adsorption onto natural bacterial consortia: testing universal adsorption behavior. *Geochim. Cosmochim. Acta* 68: 3231–3238.

Brassard, P., Kramer, J. R., Collins, P. V. (1990) Binding site analysis using linear programming. *Environ. Sci. Technol.* 24(2): 195–201.

Cox, J. S., Smith, D. S., Warren, L. A., Ferris, F. G. (1999) Characterizing heterogeneous bacterial surface functional groups using discrete affinity spectra for proton binding. *Environ. Sci. Technol.* 33(24): 4514–4521.

Davis, J.A., Kent, D. (1990) Surface complexation modeling in aqueous geochemistry. *Rev. Mineral. Geochem.* 23(1): 177–260.

Davis, J. A., Coston, J. A., Kent, D. B., Fuller, C. C. (1998) Application of the surface complexation modeling concept to complex mineral assemblages. *Environ. Sci. Technol.* 32(19): 2820–2828.

Driver, S. J., Perdue, E. M. (2015) Acid-base chemistry of natural organic matter, hydrophobic acids, and transphilic acids from the Suwannee River, Georgia, as determined by direct potentiometric titration. *Environ. Engineer. Sci.* 32(1): 66–70.

Duc, M., Gaboriaud, F., Thomas, F. (2005a) Sensitivity of the acid-base properties of clays to the method of preparation and measurement: 1. Literature review. *J. Colloid Interface Sci.* 289: 139–147.

Duc, M., Gaboriaud, F., Thomas, F. (2005b) Sensitivity of the acid-base properties of clays to the method of preparation and measurement: 2. Evidence from continuous potentiometric titrations. *J. Colloid Interface Sci.* 289: 139–147.

Dzombak, D. A., Morel, F. M. M. (1990) *Surface Complexation Modeling: Hydrous Ferric Oxide.* Wiley-Interscience, New York, NY, 393 pp.

Fein, J. B. (2006) Thermodynamic modeling of metal adsorption onto bacterial cell walls: Current challenges. *Adv. Agron.* 90: 179–202.

Fein, J. B., Delea, D. E. (1999) Experimental study of the effect of EDTA on Cd adsorption by *Bacillus subtilis:* A test of the chemical equilibrium approach. *Chem. Geol.* 161: 375–383.

Fein, J. B., Daughney, C. J., Yee, N., Davis, T. A. (1997) A chemical equilibrium model for metal adsorption onto bacterial surfaces. *Geochim. Cosmochim. Acta* 61: 3319–3328.

Fein, J. B., Boily, J.-F., Yee,N., Gorman-Lewis, D., Turner, B. F. (2005) Potentiometric titrations of Bacillus subtilis cells to low pH and a comparison of modeling approaches. *Geochim. Cosmochim. Acta* 69: 1123–1132.

Flynn, S. L., Gao, Q., Robbins, L. J., et al. (2017) Measurements of bacterial mat metal binding capacity in alkaline and carbonate-rich systems. *Chem. Geol.* 451: 17–24.

Ginn, B. R., Szymanowski, J. S., Fein, J. B. (2008) Metal and proton binding onto the roots of *Fescue rubra. Chem. Geol.* 253: 130–135.

Gorgulho, H. F., Mesquita, J. P., Gonçalves, F., Pereira, M. F. R., Figueiredo, J. L. (2008) Characterization of the surface chemistry of carbon materials by potentiometric titrations and temperature-programmed desorption. *Carbon* 46(12): 1544–1555.

Gorman-Lewis, D., Fein, J. B., Jensen, M. P. (2006) Enthalpies and entropies of proton and cadmium adsorption onto *Bacillus subtilis* bacterial cells from calorimetric measurements. *Geochim. Cosmochim. Acta* 70: 4862–4873.

Hao, W., Flynn, S. L., Alessi, D. S., Konhauser, K. O. (2018) Change of the point of zero net proton charge (pH_{PZNPC}) of clay minerals with ionic strength. *Chem. Geol.* 493: 458–467.

Herbelin, A. L., Westall, J. C. (1999) FITEQL: A Computer Program for Determination of Equilibrium Constants from Experimental Data. Department of Chemistry, Oregon State University, Corvallis, OR, Report 99–01.

Hetzer, A., Daughney, C. J., Morgan, H. W. (2006) Cadmium ion biosorption by the thermophilic bacteria *Geobacillus stearothermophilus* and *G. thermocatenulatus*. *Appl. Environ. Microbiol,* 72 (6): 4020–4027.

Kenney, J. P. L., Fein, J. B. (2011) Importance of extracellular polysaccharides in proton and Cd binding to bacteria: A comparative study. *Chem. Geol.* 286(3–4): 109–117.

Koretsky, C. (2000) The significance of surface complexation reactions in hydrologic systems: A geochemist's perspective. *J. Hydrol.* 230(3): 127–171.

Lalonde, S. V., Smith, D. S., Owttrim, G. W., Konhauser, K. O. (2008a) Acid-base properties of cyanobacterial cell surfaces. I: Influences of growth phase and nitrogen metabolism on cell surface reactivity. *Geochim. Cosmochim. Acta* 72: 1257–1268.

Lalonde, S. V., Smith, D. S., Owttrim, G. W., Konhauser, K. O. (2008b) Acid-base properties of cyanobacterial cell surfaces. II: Silica as a chemical stressor influencing cell surface reactivity. *Geochim. Cosmochim. Acta* 72: 1269–1280.

Lalonde, S. V., Dafoe, L., Pemberton, S. G., Gingras, M. K., Konhauser, K. O. (2010) Investigating the geochemical impact of burrowing animals: Proton and cadmium adsorption onto the mucus-lining of Terebellid polychaete worms. *Chem. Geol.* 271: 44–51.

Liu, Y., Alessi, D. S., Owttrim, G. W., et al. (2015) Cell surface reactivity of *Synechococcus* sp. PCC 7002: Implications for metal sorption from seawater. *Geochim. Cosmochim. Acta* 169: 30–44.

Liu, Y., Alessi, D. S., Owttrim, G. W., et al. (2016) Cell surface acid-base properties of cyanobacterium *Synechococcus:* Influences of nitrogen source, growth phase, and N:P ratios. *Geochim. Cosmochim. Acta* 187: 179–194.

Lützenkirchen, J., Preočanin, T., Kovačević, D., et al. (2012) Potentiometric titrations as a tool for surface charge determination. *Croat. Chem. Acta* 85(4): 391–417.

Martinez, R. E., Ferris, F. G. (2001) Chemical equilibrium modeling techniques for the analysis of high-resolution bacterial metal sorption data. *J. Colloid Interface Sci.* 243: 73–80.

Martinez, R. E., Smith, D. S., Kulczycki, E., Ferris, F. G. (2002) Determination of intrinsic bacterial acidity constants using a Donnan shell model and a continuous pK_a distribution method. *J. Colloid Interface Sci.* 253(1): 130–139.

Ngwenya, B. T., Sutherland, I. W., Kennedy, L. (2003) Comparison of the acid-base behaviour and metal adsorption characteristics of a gram-negative bacterium with other strains. *Appl. Geochem.* 18(4): 527–538.

Pagnanelli, F., Bornoroni, L., Moscardini, E., Toro, L. (2006) Non-electrostatic surface complexation models for protons and lead(II) sorption onto single minerals and their admixtures. *Chemosphere* 63(7): 1063–1073.

Petrash, D. P., Raudsepp, M., Lalonde, S. V., and Konhauser, K. O. (2011a) Assessing the importance of matrix materials in biofilm chemical reactivity: Insights from proton and cadmium adsorption onto the commercially-available biopolymer alginate. *Geomicrobiol. J.* 28: 266–273.

Petrash, D. A., Lalonde, S. V., Gingras, M. K., and Konhauser, K. O. (2011b) A surrogate approach to studying the chemical reactivity of burrow mucus linings. *Palaios* 26: 595–602.

Turner, B. F., Fein, J. B. (2006) Protofit: A program for determining surface protonation constants from titration data. *Comput. Geosci.* 32: 1344–1356.

Warchola, T., Flynn, S. L., Robbins, L. J., et al. (2017) Field- and lab-based potentiometric titrations of microbial mats from the Fairmont Hot Spring, Canada. *Geomicrobiol. J.* 34(10): 851–863.

Westall, J. C., Jones, J. D., Turner, G. D., Zachara, J. M. (1995) Models for association of metal ions with heterogeneous environmental sorbents. 1. Complexation of Co(II) by leonardite humic acid as a function of pH and $NaClO_4$ concentration. *Environ. Sci. Technol.* 29: 951–959.

Yee, N., Fein, J. (2001) Cd adsorption onto bacterial surfaces: A universal adsorption edge? *Geochim. Cosmochim. Acta* 65: 2037–2042.

Yu, Q., Fein, J. B. (2016) Sulfhydryl binding sites within bacterial extracellular polymeric substances. *Environ. Sci. Technol.* 50(11): 5498–5505.

Zhao, Z., Jia, Y., Xu, L., Zhao, S. (2011) Adsorption and heterogeneous oxidation of As(III) on ferrihydrite. *Water Res.* 45(19): 6496–6504.

4 Use of Multi-collector ICP-MS for Studying Biogeochemical Metal Cycling

KAI LIU, LINGLING WU, AND SHERRY L. SCHIFF

Abstract

Multi-collector inductively coupled plasma mass spectrometry (MC-ICP-MS) is a powerful technique for the study of biogeochemical cycling of a variety of metals. The advantages of this technique include high ionization efficiency, low detection limits, and rapid analysis. It can produce highly precise and accurate elemental isotope compositions of natural and experimental samples, which can provide insights into the mechanisms of both biological and abiological processes in in natural environments. In this chapter, the operating principles of the instrument, purification of samples, interferences encountered, correction methods to eliminate the instrumental mass discrimination, and data analysis with respect to reliability and reproducibility are discussed. A case study is included that highlights the capability of MC-ICP-MS to infer mechanisms of Fe redox processes in an acidic oligotrophic lake using natural abundance of stable Fe isotopes.

4.1 Introduction

Isotopic analysis of metals (e.g., Li, Mg, Cu, Zn, Mo, Fe) has been performed by geochemists for several decades in order to explore metal cycling in both natural and perturbed environments. With new innovations and developments in analytical methods, the isotope compositions of these metals have been validated in a variety of settings, including both low-temperature (e.g. marine, soil, riverine, lake) and high-temperature environments. The isotopic systematics of some metals is described here to provide a general understanding of the behavior of metals influenced by biogeochemical processes.

Lithium (Li) and magnesium (Mg) isotopes are promising tracers to study key biogeochemical processes. Both Mg and Li are fluid-mobile elements that are not involved in redox reactions (Teng et al., 2017). Li isotopes are a useful tracer to study both low-temperature (e.g., silicate weathering; Li et al., 2015) and high-temperature processes (fluid–rock interactions; Wunder et al., 2007). Natural variation of Li isotope compositions (Table 4.1) spans a large range (from –20 to 40‰) in terrestrial environments due to the substantial mass difference (~16%) between ^{6}Li and ^{7}Li. Low-temperature weathering could induce Li isotope fractionation between fluids (river, groundwater) and solids (sediment, rock) due to the enrichment of ^{7}Li in fluids (Burton and Vigier, 2011).

Table 4.1 Isotopes, standard materials, and natural isotopic compositions for selected elements

Element	Stable isotopes and natural relative abundances (%)	Standard material		Data reporting	Natural isotopic variation
		Name	Composition		
Li	^{6}Li (7.6) ^{7}Li (92.4)	L-SVEC[a] IRMM016[b]	Li carbonate (solid) Li carbonate (solid)	δ^{7}Li	−20 to 40‰
Mg	^{24}Mg (79.0) ^{25}Mg (10.0) ^{26}Mg (11.0)	NIST-SRM980[c] DSM3	Mg chips (solid) Mg in nitric acid (solution)	δ^{26}Mg	−6 to 2‰
Cu	^{63}Cu (69.2) ^{65}Cu (30.8)	NIST-SRM976[a] ERM-AE633	Cu disk (solid) Cu in nitric acid (solution)	δ^{65}Cu	Biotic sample: −1.3 to 1.5‰ Abiotic sample: 0.0 to 0.5‰
Zn	^{64}Zn (49.2) ^{66}Zn (27.8) ^{67}Zn (4.0) ^{68}Zn (18.4) ^{70}Zn (0.6)	JMC3-Lyona IRMM3702	Mg in nitric acid (solution) Mg in nitric acid (solution)	δ^{66}Zn	Biotic sample: −1.0 to 1.5‰ Abiotic sample: 0.5 to 1.0‰
Mo	^{92}Mo (14.8) ^{94}Mo (9.3) ^{95}Mo (15.9) ^{96}Mo (16.7) ^{97}Mo (9.6) ^{98}Mo (24.1) ^{100}Mo (9.6)	NIST-SRM3134	Mo in hydrochloric acid (solution)	δ^{98}Mo	−1.5 to 2.4‰
Fe	^{54}Fe (5.8) ^{56}Fe (91.8) ^{57}Fe (2.1) ^{58}Fe (0.3)	IRMM014a IRMM634	Fe wire (solid) Fe in hydrochloric acid (solution)	δ^{56}Fe	−3.6 to 1.2‰

[a] Discontinued: Standard material is no longer available.

[b] IRMM016 has the identical isotope composition to L-SVEC.

[c] NIST-SRM980 was widely used before 2003 and was then replaced by DSM3, a standard developed at the University of Cambridge.

Magnesium is the fifth most abundant element in continental crust and thus, a major component of minerals in different rock types (Teng, 2017). A variation of >7‰ in Mg isotope compositions (Table 4.1) is observed in terrestrial samples (Wombacher et al., 2011). Low-temperature weathering (e.g., silicate weathering) can produce significant Mg isotope fractionation, which is due to the partitioning of light Mg (^{24}Mg) into the aqueous phase, and thus, the Mg isotope has been used as a tracer to study weathering processes and water–rock interaction (Li et al., 2010; Teng et al., 2010).

Copper (Cu) and zinc (Zn) are redox-sensitive elements affected by biogeochemical processes, making them good tracers to understand complex biological pathways and the fate of contaminants. Cu and Zn either form complexes with organic matter or adsorb onto the surface of oxyhydroxides in near-surface environments such as rivers, oceans, and soils (Moynier et al., 2017). The variation of both Cu and Zn isotopic compositions (Table 4.1) in natural samples that are involved in biological processes span a range of 2.5‰, while the $\delta^{66}Zn$ and $\delta^{65}Cu$ of inorganic samples are less variable, with a range less than 0.5‰ (Cloquet et al., 2008; Moynier et al., 2017). Zn and Cu isotopes can also be used as tracers to study processes in high-temperature environments (Moynier et al., 2017). As the most incompatible chalcophile element, Cu is highly related to the fusion of sulfides during partial melting of the mantle, making it a good tracer to study the role of sulfides during igneous differentiation processes. The moderate volatility and high abundance of Zn in bulk silicate earth makes it a good tracer to understand the volatility processes in rocks. The Cu and Zn isotopic compositions in bulk silicate earth are reported as 0.07 ± 0.10‰ (Savage et al., 2015) and 0.16 ± 0.06‰ (Sossi et al., 2018), respectively.

Molybdenum (Mo) can be used to study redox conditions in near-surface environments due to its redox sensitivity, role in microbiological processes, and relatively large isotopic range in nature (from −1.5‰ to 2.4‰) (Kendall et al., 2017). The largest isotope fractionation (~3‰) is observed from the adsorption of Mo to Mn oxides in oxic seawater by preferentially accumulating lighter isotopes of Mo onto the surface of Mn oxides (Barling and Anbar, 2004; Wasylenki et al., 2008). Mo is an essential cofactor for enzymes that play an important role in biological nitrate reduction and nitrogen fixation (Glass et al., 2009; Kendall et al., 2017). Bacterial uptake prefers the lighter isotopes of Mo, inducing isotope fractionation with the range of ~0.2 to 1.0‰ (Barling and Anbar, 2004). Some microorganisms can also utilize oxidized Mo species as electron acceptors in catabolic metabolism when generating energy from chemiosmosis (Lovley, 1993). Thus, Mo isotopes are a good tracer to study redox conditions as well as the evolution of life on Earth (Kendall et al., 2017) and euxinic environments because of the large differences in isotopic fractionation between processes that sequester Mo in sedimentary minerals.

Mass spectrometry is a powerful instrumental technique used to determine the isotopic composition of elements. Types of mass spectrometry used for isotopic analyses are listed in Table 4.2. Thermal ionization mass spectrometry (TIMS) was first developed for isotopic analysis of nontraditional isotopes and was widely used before the introduction of inductively coupled plasma mass spectrometry (ICP-MS) (Vanhaecke and Degryse, 2012). The approaches of introducing and ionizing samples are the major differences between the TIMS and ICP-MS methods (Albarède and Beard, 2004). The ICP-MS was developed after 1983 and contains three essential components: an ion source, an ion separator, and

Table 4.2 Different types of mass spectrometry for measurement of stable isotope ratios

Mass spectrometry	Ionization	Sample form	Elements of interest
Dual inlet or continuous flow IRMS	Bombard sample with e-beam	Gas	Light elements (e.g., H, C, N, O, S, Cl, Br)
TIMS	Heat by electric current on a filament	Solid or solution	Heavy elements (e.g., Ca, Ti, Cr, Ni, Sr, Pb, Nd)
MC-ICP-MS	Introduce into an inductively coupled argon plasma	Solid or solution	Light and heavy elements (e.g., Li, B, Mg, Si, Ca, Fe, Cu, Zn, Mo, Hf, Hg)
Noble gas MS	Bombard sample with e-beam in the ion gun	Gas	He, Ne, Ar, Kr, Xe

a detection system. The ion separator used in commercial ICP-MS instrumentation can be classified into three types: quadrupole mass analyzer, double-focusing sector, and time-of-flight mass analyzer. The double-focusing sector can produce higher unit mass resolution, which can result in high precision of isotope ratios, compared with the other two types. An enhanced version of the ICP-MS, the multi-collector (MC)-ICP-MS, was developed on the double-focusing sector field ICP-MS over the following decade (Vanhaecke and Degryse, 2012) and allows the study of isotope abundances of all elements (Wieser et al., 2012). TIMS is not suitable for elements with high first ionization potentials, whereas MC-ICP-MS is capable of analyzing a wider range of elements (e.g., Mo, Fe, Hf, Hg) (Becker, 2007; Vanhaecke and Degryse, 2012). More recently, MC-ICP-MS is increasingly used for isotope analysis in many fields (e.g., biogeochemistry, cosmochemistry) due to the advantages of MC-ICP-MS over TIMS (e.g., higher sensitivity, higher ionization efficiency, faster sample throughput, superior precision and accuracy). Common MC-ICP-MS instruments that are commercially available include the Isoprobe from GV Instruments, the Neptune (Plus) from Thermo Fisher Scientific, and the Nu Plasma II and the Nu Plasma 1700 from Nu Instruments (Wieser et al., 2012).

The availability of TIMS or MC-ICP-MS (gradually becoming a widely used instrumental technique) has led to new discoveries in metal cycling. For example, Fe biogeochemistry has been investigated to understand the Fe source, binding environment, and electron transfer and atom exchange pathways. The development of the double spike technique allows high-precision Fe isotope measurements to be achieved (Beard and Johnson, 1999; Beard et al., 1999; Johnson and Beard, 1999). The double spike technique is employed to correct and account for instrumental mass fractionation during mass spectrometric analysis. This well-established method (Dodson, 1963) is applicable to any element that has four or more isotopes. The double spike technique is being increasingly applied in nontraditional stable isotope research (Albarède and Beard, 2004; Fantle and Bullen, 2009), where it is ideally suited to distinguish between natural and instrumental mass fractionation. MC-ICP-MS is more suitable for nontraditional stable isotope measurements, since its inherent instrumental mass bias is constant and several elements, including

Fe, have high ionization efficiency (Beard et al., 2003a). However, MC-ICP-MS also introduces some new disadvantages, such as isobaric interferences produced by the Ar plasma and interference from other elements (e.g., Ca, Cr, Ni) (Beard et al., 2003a). Great efforts have been made to minimize these interferences (Anbar et al., 2000; Beard and Johnson, 1999; Belshaw et al., 2000). Over the past decade, MC-ICP-MS has been predominantly deployed to study isotope fractionation of a wide range of elements, and to produce high-precision isotope measurements that capture the true naturally occurring mass-dependent variations. Since isotopic analysis for each element and type of sample necessitates detailed method development, this chapter will focus on the application of MC-ICP-MS for studying Fe isotope fractionation in general with specific application in a case study. The overall methods and sample preparation for measuring Fe isotopes are broadly similar for studying other metal isotopes.

4.2 Methodology

4.2.1 Instrumentation

Mass resolution(R) is an important parameter of MC-ICP-MS. It estimates the ability of MC-ICP-MS to identify the signals of separated ion beams with slightly different masses:

$$R = m/\Delta m \qquad \text{(Eq.4.1)}$$

where Δm is the mass difference between two separated ion beams with masses m and $m + \Delta m$ (Becker, 2007), and is indicative of the capability of MC-ICP-MS to separate the neighbor peaks in a mass spectrum. The mass resolution can be calculated in two different ways. In the 10% valley definition, two neighbor peaks (at mass m_1 and mass m_2 in a mass spectrum) with the same height (ion intensity) are separated by a valley whose lowest point is 10% of the peak height. In the peak width definition, the Δm_x is the width of a single peak with mass m at a desired peak height (x% (e.g., 5%) of maximum peak height) (Becker, 2007; Vanhaecke, 2012):

$$\mathrm{R}_{10\%\ \text{valley definition}} = \left(\frac{m_1+m_2}{2}\right)/(m_1+m_2) \qquad \text{(Eq.4.2)}$$

$$\mathrm{R}_{\text{peak width definition}} = m/\Delta m_x \qquad \text{(Eq.4.3)}$$

The two definitions are equivalent when the mass resolution is taken at 5% of peak height in peak width definition.

The MC-ICP-MS instrument used for discussion in this book chapter is a Nu Plasma II, which is a double-focusing sector field mass spectrometer (Figure 4.1). The sample is introduced into the system as an aerosol by a nebulizer (CETAC Aridus II Desolvating Nebulizer System) to the inductively coupled argon plasma for ionization. The ions are accelerated first and then separated from other ions in space based on their different mass to charge ratio and introduced into the magnetic sector field of the mass spectrometer.

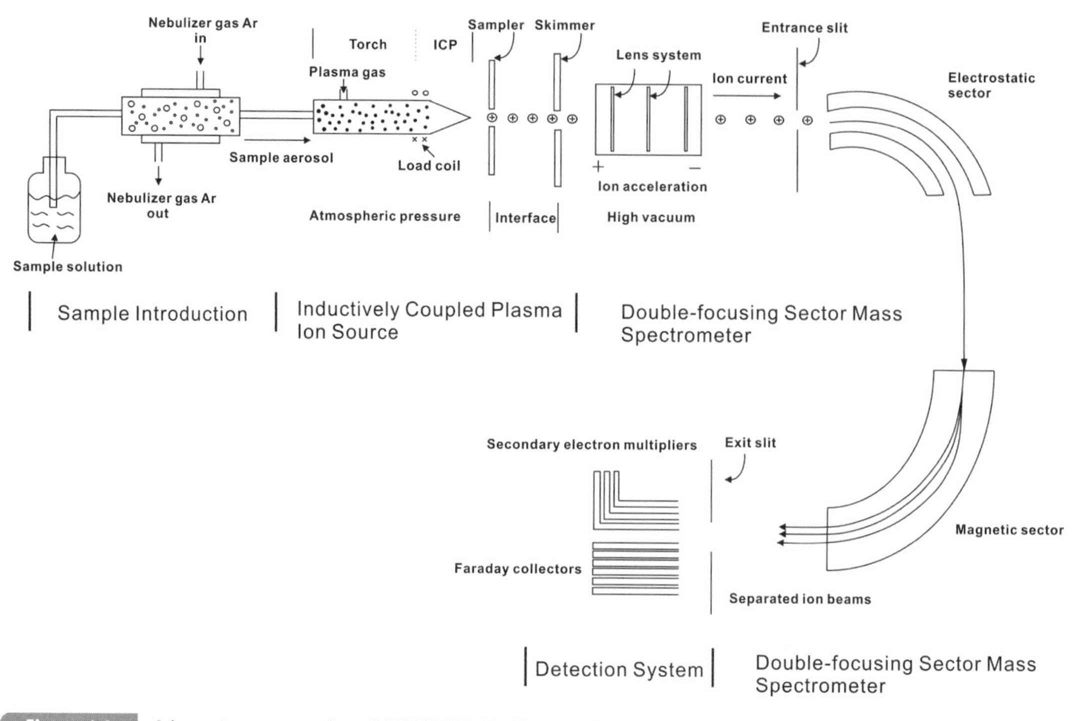

Figure 4.1 Schematic representation of MC-ICP-MS: Nu Plasma II (coupled with sample introduction system). Nu Plasma II contains three distinct parts: an inductively coupled plasma ion source, a double-focusing sector mass spectrometer, and a detection system consisting of 16 Faraday collectors (also called Faraday cups), five ion counters (discrete dynode secondary electron multiplier), and a high-abundance sensitivity filter. Modified from Vanhaecke, 2012, with permission of John Wiley and Sons.

The separated ion beam is subsequently converted into a measurable electrical signal by the detection system (Vanhaecke, 2012). There is a strong negative correlation between the ion transmission efficiency and mass resolution for MC-ICP-MS: the higher the mass resolution, the lower the ion transmission efficiency (Vanhaecke, 2012). For example, reducing the width of the exit slit can increase the mass resolution but decreases the intensity of the ion beam, resulting in low ion transmission efficiency (Becker, 2007). With high mass resolution, the MC-ICP-MS can separate ions of interest (e.g., Fe) from interfering ions to get accurate and precise isotope ratios (Wieser et al., 2012). The double-focusing sector, including magnetic and electrostatic sectors, has been developed for the purpose of producing high mass resolution without significant damage to the ion transmission efficiency. This includes a magnetic sector, which separates ions by the strength of the magnetic field based on their mass to charge ratio (m/z), and an electrostatic sector, which separates ions by acceleration voltage based on their kinetic energy (Vanhaecke, 2012). The double-focusing (energy and directional focusing) sector in spectrometers similar to the Nu Plasma II follows the Nier–Johnson geometry, in which the electrostatic sector is followed by the magnetic sector (Becker, 2007). The detection system is a combination of Faraday collectors and ion counters. The Faraday collectors are deep buckets that are installed behind the exit slits for collecting the separated ion beams that pass through the exit slits. The ion counters in the Nu Plasma II are discrete dynode secondary electron multipliers, which can improve the sensitivity of mass spectrometers by several orders of magnitude compared with Faraday collectors (Becker, 2007). The combination of Faraday collectors and ion counters can extend the dynamic range of the detection system and produce accurate and precise isotopic ratios (Wieser et al., 2012).

4.2.2 Reagents

Ultrapure deionized water (18.2 MΩ, 4 ppb total organic carbon [TOC] at 25 °C) is used for the preparation of all samples, standards, and test solutions. Dilute nitric acid solutions (0.1% and 2% by volume) are prepared with ultrapure nitric acid (70% w/w, Omni*Trace* Ultra®). Hydrochloric acid is distilled twice, often using the Savillex DST-1000 Acid Purification System, from analytical reagent-grade HCl (Baker Analyzed® A. C. S. Reagent HCl). For the analysis of Fe isotopes, two stock Fe standard solutions are used: IRMM634 (the certified standard replacing IRMM014, from the Institute for Reference Materials and Measurements [IRMM]) and HPS Fe (High-Purity Standards from Delta Scientific Laboratory Products Ltd., 1000 μg/mL Fe solution [source: Fe metal, purity: 99.999%] in 2% HNO_3). The mission of the IRMM is to promote a common and reliable European measurement system in support of EU policies. The standard stock solutions are specific depending on the elemental isotope of interest. For example, the IRMM3702 is the certified reference material for $\delta^{64}Zn$ in 1 M nitric acid. The dilute standard solutions are prepared by dilution of the concentrated standard solutions in 2% nitric acid. Synthetic solutions are prepared that contain the same concentrations of major ions as are present in the samples to be analyzed.

4.2.3 Interferences

MC-ICP-MS is usually coupled with ion exchange chromatography in order to eliminate spectral (or isobaric) interferences and nonspectral interferences (Albarède and Beard, 2004). Spectral interferences can be further classified into elemental interferences and molecular interferences. Matrix elements such as Cr and Ni can introduce elemental interferences during the Fe isotope measurement (Beard et al., 2003a). Therefore, care must be taken to limit their possible interference during analysis. For example, ^{54}Fe suffers from the interference of ^{54}Cr and ^{58}Ni because the latter two overlap the ^{58}Fe signal. Some matrix elements such as Ca can form oxide (or hydroxide) ions such as $^{40}Ca^{16}O$ at ^{56}Fe, $^{40}CaOH^+$ at ^{57}Fe, and $^{42}CaO^+$ at ^{58}Fe during analysis and result in molecular interferences (Beard et al., 2003a). These interfering ions will produce significant bias if their concentrations in solutions are an appreciable fraction of the concentration of the target isotope. Albarède and Beard (2004) suggest that an interfering ion can bias the true abundance of the target species to result in inaccurate isotopic compositions. In order to erase the effect of these interferences, a strict Fe–matrix separation protocol is applied in the preliminary stages to purify the samples. A correction is also applied to erase the signal of interfering ions based on a standard method (Albarède and Beard, 2004).

The MC-ICP-MS instrumental technique also introduces spectral molecular interferences by ionized plasma gas (Ar). The Ar-containing polyatomic ions, $^{40}Ar^{14}N$, $^{40}Ar^{16}O$, and $^{40}Ar^{16}OH$, can overlap signals corresponding to ^{54}Fe, ^{56}Fe, and ^{56}Fe, respectively. However, several approaches (e.g., addition of H_2 gas in collision cell, cool plasma conditions) have been used to remove the interfering effect of these Ar-containing polyatomic ions (Beard et al., 2003a). Furthermore, the double-focusing sector with which machines similar to Nu Plasma II are equipped can produce both energy-centralized and direction-controlled ion beams, thus reducing the width of source and exit slits in the detection system and further increasing the mass resolution (Vanhaecke, 2012). Similarly, with high mass resolution, the Fe ion beams can be distinguished from other Ar-containing polyatomic ions.

Nonspectral interferences result from the space-charge effect or plasma condition effect (Albarède and Beard, 2004). The space-charge effect can cause heavier ions to be more favorably transmitted than lighter ions, resulting in a compositional change of the ion beams (a defocusing effect) (Vanhaecke, 2012). The nonspectral interferences are generally coupled with a sensitivity shift of the target ion caused by other elements (Albarède and Beard, 2004). Like the spectral interferences, the nonspectral interferences can also produce significant bias in isotopic compositions.

4.2.4 Sample Material

MC-ICP-MS has been used to investigate isotopic signatures useful to geomicrobiologists, such as Fe, in various environments such as rivers, oceans, groundwater, and lakes. The sample material in these environmental settings includes pore water and dissolved and particulate phases in the water column, groundwater and sediment. Sediment samples can be dissolved with HF for bulk isotope analysis or treated with sequential acid extraction to extract different metal phases. Several studies have used sedimentary rock samples to study biogeochemical Fe cycling (e.g., Li et al., 2015; Planavsky et al., 2012; Tsikos et al., 2010). Sedimentary rock such as black shales, carbonate rocks, or mudrocks can be effectively dissolved in a mixture of HF–HNO_3–HCl (2:1:1) (HF 29 M: HCl 8 M: HNO_3 14 M) (Li et al., 2015) (the optimal ratio may differ for different types of sedimentary rock).

4.2.5 Sample Preparation

Before the samples are introduced into the MC-ICP-MS, they must be purified to separate the ions for isotopic analysis from the background matrix. The matrix contains interfering ions that can induce spectral interferences. In the case of Fe, a strict Fe–matrix chemical separation is required by the standard–sample–standard bracketing approach (Albarède and Beard, 2004) for mass bias correction.

For Fe isotope analysis, ion exchange chromatography is utilized for the purpose of Fe–matrix separation. Ion exchange resin is an insoluble polymer with exchangeable ions that can exchange with other ions from an aqueous solution, either positively charged ions (cation exchange resin) or negatively charged ions (anion exchange resin). An anion exchange resin (Bio-Rad AG® 1-X4 Resin, Analytical grade, 4% cross-linking, 200–400 mesh) has been used in ion exchange chromatography to separate Fe(III) from other elements in a hydrochloric acid medium (Beard et al., 2003a; Strelow, 1980). The anion exchange resin is strongly basic, having quaternary ammonium groups ($-N^+(CH_3)_3$) that can absorb hydrochloric acid ($-N^+(CH_3)_3OH^- + Cl^- \leftrightarrows -N^+(CH_3)_3Cl^- + OH^-$) (Cullum, 1994). Fe(III) has high affinity for resin conditioned with concentrated hydrochloric acid and thus forms chloride complexes (e.g., $FeCl_4^-$) with the anion exchange resin (Strelow, 1980). The stabilities of Fe(III)–chlorine complexes increase with increasing hydrochloric acid concentration (Borrok et al., 2007; Strelow, 1980). Therefore, Fe(III) can be adsorbed to resin in 7 M HCl and eluted from resin in 0.5 M HCl. Prior to use, the anion exchange resin is washed twice using 8 M HCl (distilled twice from analytical reagent-grade HCl) and 10 times using H_2O to rid the column of the finest resin particles (Marechal et al., 1999) and finally, is filled with H_2O before the sample is loaded onto the column. In order to minimize contamination problems, all procedures are conducted in a laminar flow cabinet located in a metals-free clean room. The detailed procedure is described in Table 4.3 (Beard et al., 2003a). It has been proven that no isotopic fractionation occurs during the ion exchange chromatography following these procedures (Beard and Johnson, 1999; Beard et al., 2003a). However, it is crucial to obtain an Fe yield quantitatively close to 100% through chemical separation to avoid mass-dependent fractionation (Albarède and Beard, 2004).

Test solutions are carried through ion exchange chromatography to test the accuracy of chemical separation. A certain amount (μg) of high purity standard® (HPS) Fe (in-house standard) is added to synthetic samples, which contain the same major ions as the original samples. The amount of HPS Fe should match the range of iron concentrations of the samples to be measured.

4.2.6 MC-ICP-MS Analysis

The required minimal concentrations of measurement solutions introduced into MC-ICP-MS could vary depending on the Fe intensities of different MC-ICP-MS instruments. Samples and Fe standards are prepared at (e.g.) 600 ppb Fe concentration in weak nitric acid (e.g., 0.01%) at the time of the measurement. It is important to keep all samples and standards matched in Fe concentration and acid matrix, since the apparent mass discrimination could change due to any changes of the Fe concentration or acid strength (Malinovsky et al., 2003). Different concentrations of HPS Fe (400, 500, 700, and 800 ppb) are also prepared in 0.01% nitric acid for the correction of mass discrimination caused by variation in Fe concentration. These diluted Fe standards are included in each run-set and analyzed as unknown samples. In order to minimize contamination problems, all the sample dilution processes are operated in a laminar flow cabinet that is situated as close to the MC-ICP-MS instrument as possible. The MC-ICP-MS is computer controlled, and each isotopic composition is the average of 40 (varies as desired) 10 s on-peak measurements following a 1 min on-peak measurement of blank acid (0.01% nitric acid) (Skulan et al., 2002). The on-peak measurements of blank acid are conducted in order to use the on-peak zero method (OPZ). This method can eliminate the variation in the background from both the instrumental memory effect and small operational variations, which can cause the accuracy

Table 4.3 Procedure for ion exchange chromatography

1	Measure Fe concentrations of sample
2	Calculate the volume of sample to contain 50 μg (vary from 15 μg to 100 μg) and pipette into preleached (using 1× 8 M HCl) labeled Teflon beakers
3	Dry out samples on hot plates
4	Add four drops ultrapure HNO_3
5	Dry out samples on hot plates
6	Add four drops ultrapure HNO_3
7	Dry out samples on hot plates
8	Add 500 μL 2× 7 M HCl. Dry out, preserve in closed beakers overnight
9	Add 100 μL 2× 7 M HCl immediately before column chemistry
10	Run through column
a	Fill column with water and make sure there are no bubbles in it. Then load resin onto column
b	Use a pipette to pass 800 μL 2× 0.5 M HCl twice
c	Use a pipette to pass 800 μL 2× 7 M HCl twice
d	Load sample in 100 μL 2× 7 M HCl
e	Pass 200 μL of 2× 7 M HCl
f	Pass 600 μL of 2× 7 M HCl twice
g	Collect 2.1 mL of 2× 0.5 M HCl (700 μL three times) in new preleached (using 1× 8 M HCl) labeled Teflon beakers
11	Dry samples on hot plates
12	Add 100 μL 2× 7 M HCl immediately before column chemistry
13	Run through column
a	Fill column with water and make sure there are no bubbles in it. Then load resin onto column
b	Use a pipette to pass 800 μL 2× 0.5 M HCl twice
c	Use a pipette to pass 800 μL 2× 7 M HCl twice
d	Load sample in 100 μL 2× 7 M HCl
e	Pass 200 μL of 2× 7 M HCl
f	Pass 783 μL of 2× 7 M HCl six times
g	Collect 2.1 mL of 2× 0.5 M HCl (700 μL three times) in new preleached (using 1× 8 M HCl) labeled Teflon beakers
14	Use 38 μL collected sample for Ferrozine[a] and compare with value from beginning (yield should be close to 100%)
15	Dry out samples on hot plates
16	Add four drops ultrapure HNO_3; once Fe is dissolved, add four drops H_2O_2
17	Dry out samples on hot plates
18	Add four drops ultrapure HNO_3; once Fe is dissolved, add four drops H_2O_2
19	Dry out samples on hot plates
20	Dissolve samples to 25 ppm Fe in 2% nitric acid matrix (using concentrated ultrapure HNO_3)

[a] Iron concentration is measured spectrophotometrically by the Ferrozine method. Ferrozine solution is prepared by dissolving 0.2 g ferrozine reagent, 12 g HEPES, and 10 g hydroxylamine hydrochloride in 1 L deionized water (pH ~7). Thirty-eight microliters of sample is added to 1 mL ferrozine solution for the measurement of iron concentration.

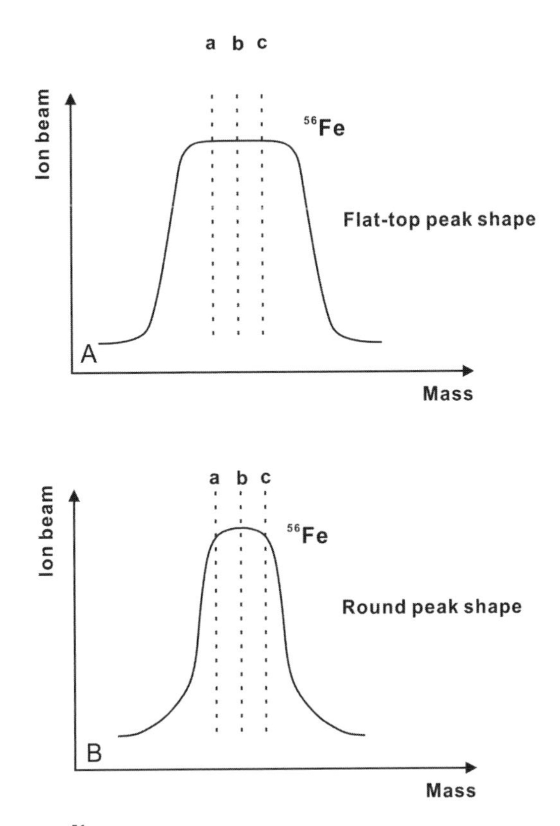

Figure 4.2 Magnet scans for iron isotope ^{56}Fe in two peak shapes: (A) flat-top peak shape; (B) round peak shape. The dashed lines (a, b, c) show different analysis mass positions caused by small shifts. In (B), the ion beams (intensity) of ^{56}Fe change with the small variations of mass position (a, b, c), but they remain constant in (A).

and precision of isotopic ratios to deteriorate (~0.1‰ range) (Albarède and Beard, 2004). In order to get precise Fe isotope ratios, the peak shapes of Fe isotopes (^{54}Fe, ^{56}Fe, ^{57}Fe, and ^{58}Fe) during a mass scan need to be "flat top" (Figure 4.2). If not, the collector setting will bias the isotopic ratios (Albarède and Beard, 2004). Some instruments, such as Neptune and Nu Plasma II, can produce high-quality "flat top" peak shapes by adjusting independently the width of both source and exit slits. Reducing the slit width results in high mass resolution by separating analyte peaks from other interferences, therefore producing a large flat-top peak area.

An instrumental mass bias (also called mass discrimination effect) between the measured isotope composition and the corresponding true value is created due to the shift of ion beams during transmission from the ion source to the mass spectrometer (Albarède and Beard, 2004). Although this bias in isotope measurements is not fully understood, it is likely caused by some processes in the inductively coupled plasma ion source, such as supersonic expansion of ions and space-charge effect (Meija et al., 2012). These processes can cause heavier ions to be more favorably transmitted than lighter ions, which can produce large biases to the measured isotope ratios (Meija et al., 2012).

The variations in the transmission of samples containing different matrices can also introduce matrix-dependent nonspectral discrimination effects (Meija et al., 2012). Therefore, a correction must be applied to correct for the drift, which impacts the instrumental mass discrimination.

Because the mass bias changes slightly during an analysis by MC-MS-ICP, a standard–sample–standard bracketing approach is used for mass bias correction (Beard et al., 2003a; Skulan et al., 2002). This approach is designed to erase the drift in measured isotope ratio data due to the instrumental mass bias in order to reveal the natural Fe isotope fractionation inherent in the samples (Rehkamper et al., 2012). Before and after each measurement of a sample, the isotopic ratio of Fe standard solution is measured and used to correct the raw data of a sample (Rehkamper et al., 2012). This approach can only be used when the matrix difference between samples and Fe reference standards is very small. The matrix difference can introduce drifts in the instrumental mass bias, which can cause inaccurate results during correction (Rehkamper et al., 2012). Therefore, a strict chemical separation during the preliminary stage and proper preparation of matrix-specific standards are required to eliminate the matrix effect.

4.2.7 Data Reporting

Natural mass-dependent variation in Fe isotopes covers a range of ~6‰ (Table 4.4). Early studies have used the epsilon notation (ε) to report metal isotopic values (ε^{56}Fe, ε^{57}Fe) (e.g., Matthews et al., 2001). However, now isotope compositions in most studies are reported as δ values (e.g.: δ^{56}Fe, δ^{57}Fe) given in ‰:

$$\delta^{56}\text{Fe} = [(^{56}\text{Fe}/^{54}\text{Fe}_{\text{sample}})/(^{56}\text{Fe}/^{54}\text{Fe}_{\text{standard}}) - 1] \times 10^3 \qquad \text{(Eq.4.4)}$$

$$\delta^{57}\text{Fe} = [(^{57}\text{Fe}/^{54}\text{Fe}_{\text{sample}})/(^{57}\text{Fe}/^{54}\text{Fe}_{\text{standard}}) - 1] \times 10^3 \qquad \text{(Eq.4.5)}$$

δ^{56}Fe and δ^{57}Fe values are typically normalized to the average of igneous rocks (Beard et al., 2003a). However, some studies report δ^{56}Fe and δ^{57}Fe values relative to the international standard IRMM014 (δ^{56}Fe = −0.09 ± 0.05‰ relative to the igneous Fe baseline; e.g., Anbar, 2004; Azrieli-Tal et al., 2014; Beard et al., 2003a).

4.2.8 Reference Standards

IRMM certified reference materials, provided by EC-JRC-IRMM (Institute for Reference Materials and Measurements, Belgium), are widely used for interlaboratory calibration in stable isotope studies. Examples of available reference standards include IRMM014 for Fe and IRMM3702 for Zn (Table 4.1). When standards like IRMM014 are no longer available (e.g., due to the exhaustion of IRMM014), alternative standards such as IRMM634 are used for Fe isotopic measurements. Other in-house standards (such as HPS Fe; see details in Section 4.2.2), alongside IRMM634, can be additionally employed to test the reliability and reproducibility of data by MC-ICP-MS analysis.

Table 4.4 Summary of ranges of Fe isotopic compositions in natural environments

Environments		$\delta^{56}Fe^{a}$ (‰)		Reference
		From	To	
Aquifer				
	Groundwater	−3.4	0.58	(Dekov et al., 2014; Guo et al., 2013; Teutsch et al., 2005; Xie et al., 2014)
	Sediment	−1.1	0.75	
Lake				
	Pore water	−1.81	0.64	(Busigny et al., 2014; Malinovsky et al., 2005; Percak-Dennett et al., 2013; Schiff et al., 2017; Teutsch et al., 2009)
	Sediment	−0.72	0.34	
	Water column	−2.14	0.57	
Continental shelf				
	Pore water	−4	1.22	(Chever et al., 2015; Homoky et al., 2009, 2013; John et al., 2012; Scholz et al., 2014; Severmann et al., 2008, 2010; Staubwasser et al., 2006)
	Sediment	−0.89	0.15	
	Water column	−3.45	0.04	
Marine				
	Seawater	−0.9	0.71	(Gelting et al., 2010; Labatut et al., 2014; Nishizawa et al., 2010; Radic et al., 2011; Rouxel and Auro, 2010)
	Sediment	−1.8	1	
Soil		−0.64	0.41	(Brantley et al., 2004; Buss et al., 2010; Collins and Waite, 2009; Guelke et al., 2010; Ilina et al., 2013; Mansfeldt et al., 2011; Pinheiro et al., 2014; Poitrasson et al., 2008; Song et al., 2011; Wiederhold et al., 2006)
River				
	Dissolved Fe	−0.60	0.51	(Bergquist and Boyle, 2006; Chen et al., 2014; Escoube et al., 2009; Ingri et al., 2006; Pinheiro et al., 2014)
	Suspended Fe	−0.90	0.31	
Human body		−3.6	−0.9	(Hotz and Walczyk, 2013; Van Heghe et al., 2013; Walczyk and Blanckenburg, 2002)
Total range		−3.6	1.22	

4.2.9 Precision and Accuracy

The precision of this technique using the described procedures is ±0.05‰ for $\delta^{56}Fe$ and ±0.07‰ for $\delta^{57}Fe$ (see Appendix in Beard et al., 2003b). To estimate the external reproducibility, at least 10% of the samples are randomly selected to be carried through the entire analytical procedure of sample preparation (different aliquots of the same sample carried through both the ion exchange chromatography and MC-ICP-MS analysis) for replicate measurement. The Fe isotope composition of test solutions should be identical to that of pure standard solutions to ensure that no bias is produced during chemical separations and confirm the accuracy of isotope measurements. The precision and accuracy of isotopic data are evaluated by the standard deviation (2σ) calculated from the repeated analyses of pure standards, test solutions, and samples. The measured external precision is calculated by the average standard deviation of all samples, including the replicates (e.g., Liu et al., 2015).

4.3 Case Study: Iron Redox Transformations in an Oligotrophic Softwater Lake

Iron is a major crustal element on Earth and plays an important or even dominant role in the cycling of many other elements, such as nitrogen, sulfur, carbon, and phosphorus, due to its sensitivity to redox processes and the difference in solubility of oxidized and reduced forms (Melton et al., 2014). Iron is also an essential nutrient for organisms and may control, for example, the microbial community or the domination of phytoplankton communities by cyanobacteria (Molot et al., 2014). Some microorganisms utilize oxidized or reduced Fe species as electron acceptors or electron donors in catabolic metabolism when generating energy from chemiosmosis (Burdige, 2006; Byrne et al., 2015; Davison, 1993; Melton et al., 2014). These microorganisms are referred to as Fe-oxidizing and Fe-reducing bacteria. The phototrophic Fe(II)-oxidizing bacteria can utilize light energy while using Fe(II) as electron donor (Caiazza et al., 2007; Jiao et al., 2005). This anoxygenic phototrophic Fe(II) oxidation has been postulated to be one of the most important lithoautotrophic metabolisms in the ancient Fe-based ecosystems (Caiazza et al., 2007). It may also play an important role in the Archean and early Proterozoic banded iron formations (Caiazza et al., 2007; Ehrenreich and Widdel, 1994; Kappler et al., 2005).

Natural abundance Fe isotopes are useful tracers for biogeochemical processes involving Fe. At low temperature, the Fe isotopes experience mass-dependent isotopic fractionation resulting from the very small different efficiencies involved in biophysicochemical reactions for respective isotopes differing in mass (Vanhaecke and Kyser, 2012). The Fe isotopic values of an Fe species fundamentally reflect the binding environments (i.e., the Fe(III)–O bond is stronger than the Fe(II)–O bond; thus, the heavier Fe isotopes preferentially remain in the Fe(III)–O bond in order to lower the overall energy of the system where both bonds coexist). The Fe isotopic fractionations are regulated by different physical and chemical reactions occurring in different binding environments (Vanhaecke and Kyser, 2012). There are two Fe isotopic fractionations that occur in natural environments: equilibrium isotopic fractionation, which occurs during reversible reactions in chemical equilibrium between reactant and product, and kinetic isotopic fractionation, which occurs during a unidirectional reaction or by reversible reactions that do not reach equilibrium. The largest fractionation (about −3‰) between aqueous Fe(II) and Fe(III) oxyhydroxides is observed in dissimilatory Fe reduction and produces an isotopically light Fe(II) (Beard et al., 1999; Crosby et al., 2005, 2007; Percak-Dennett et al., 2011; Tangalos et al., 2010). Abiotic processes such as reduction of Fe(III) by H_2S could also induce Fe isotopic fractionation between ferric and ferrous Fe. Furthermore, the sorption of Fe(II) to Fe(III) oxyhydroxides can produce Fe isotopic fractionation with a magnitude in the range of 0.2 to 0.9‰ between sorbed Fe(II) and aqueous Fe(II) (Brantley et al., 2004; Bullen et al., 2001; Crosby et al., 2005; Wu et al., 2009, 2010).

A range in Fe isotopic compositions of about 5‰ (Table 4.1) have been reported in a variety of subsurface environments, including marine, soil, riverine, groundwater, and lake settings (Table 4.4). A number of papers also report Fe isotopic compositions in the human body (Table 4.4). The variation of δ^{56}Fe in the sediments from Archean oceans (Severmann and Anbar, 2009) has been hypothesized to reflect microbial Fe(III) reduction during diagenesis and/or Fe(II) oxidation potentially dominated by phototrophic Fe(II) oxidizers in the water column (Gauger et al., 2016; Johnson et al., 2008).

Iron isotopic compositions measured by MC-ICP-MS can be used to study the Fe biogeochemical cycling in natural lacustrine environments. The major pathways of Fe redox transformations in lake systems can lead to Fe isotope fractionation between various Fe phases in lake sediments; thus, stable isotope measurements can be used to shed light on Fe biogeochemical cycling. In this case study, Fe isotopes of pore water and sediment samples from Lake Tantaré (47° 04′ N, 71° 33′ W), an acidic (pH 5.4–6.0) oligotrophic lake located 40 km northwest of Québec city, Québec, Canada (Couture et al., 2010b), are measured by MC-ICP-MS (Liu et al., 2015).

4.3.1 Methods

Particular attention should be paid during the sampling process if the target metal is redox sensitive. In the case of Fe, significant effort is required to avoid artificial oxidation during sampling. Natural Fe sample material in near-surface environments includes pore water and dissolved and particulate phases in the water column and sediment. The methods for collection of pore water can be divided into an *in situ* method and an *ex situ* method. The *in situ* method requires installing a device (membrane dialysis [e.g., peepers; Couture et al., 2010a]) or suction samplers (Bufflap and Allen, 1995) directly into the sediment to the desired depth. Pore water samples are acidified immediately following sampling. This treatment can prevent Fe(II) oxidation by atmospheric O_2, which is rapid at normal lake pH but very slow at pH <3.0. The *ex situ* method requires sediment cores to be removed from the natural environment. The sediment cores are stored under anoxic conditions until the pore water is collected by either centrifugation (e.g., Tangalos et al., 2010) or Rhizon samplers (e.g., Busigny et al., 2014). Water column samples are filtered and acidified for dissolved Fe. Particulate Fe can be obtained from filters. All samples are stored at 4 °C prior to analysis. In this case study (Liu et al., 2015), pore water samples are acidified to a 0.5 M HCl matrix using 6 M ultrapure HCl (Seastar grade) immediately following sampling. Sediment samples are treated by sequential acid extractions following the procedures given in Tangalos et al. (2010) to separate different Fe phases. Solid-phase speciation was characterized in Liu et al. (2015). Iron concentrations (Fe(II) and Fe(tot) [total Fe]) of pore water and sediment extractions are measured by the Ferrozine method (Stookey, 1970).

Wet chemistry is performed on samples to separate Fe from other cations by passing them through an anion exchange resin following the standard procedures (e.g., Beard et al., 2003a). Additionally, test solutions are prepared to evaluate the accuracy of Fe isotope analyses as

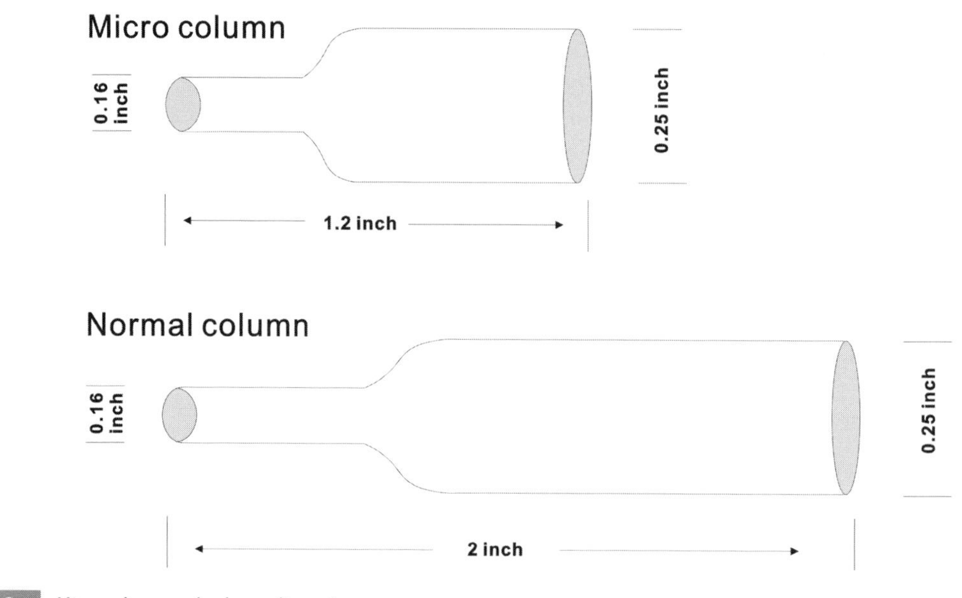

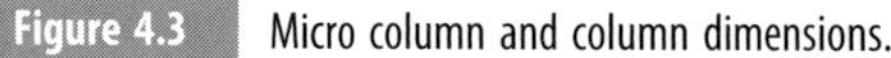
Figure 4.3 Micro column and column dimensions.

described in the sample preparation section. Although the standard procedures are well established, some problems may be encountered during the chemical separation processes. The problems listed here are some general problems when dealing with natural samples.

Low iron concentrations: the pore water samples from oxic environments normally have low Fe concentrations and are operationally limited in volume. In this case study (Liu et al., 2015), the relatively low iron concentrations (2.2 to 30.7 μM Fe(tot)) and the limited volume (~6 mL) of pore water samples did not allow the collection of 10 μg Fe, the lower limit for loading onto the ion exchange resin. Therefore, a micro column was used to replace the column to analyze the wet chemistry of the pore water samples (Figure 4.3).

Dissolved organic matter: the sediment extractions can contain a large amount of organic matter. After the sediment extracts are dried, this organic matter often cannot be redissolved in 7 M HCl (step 9 in Table 4.3). This organic matter may bind Fe ions, resulting in Fe loss in the extracts that are analyzed, and thereby may produce an artificial Fe isotopic fractionation during the wet chemistry preparatory steps. To solve this problem, H_2O_2 can be added to the sediment extractions to oxidize the organic matter. After H_2O_2 treatment, the sediment extract can be easily dissolved in 7 M HCl after being dried.

Insolubility of extraction matrix in strong acid: similarly to the problem with organic matter, other substances in extractions may become insoluble in 7 M HCl (step 9 in Table 4.3). A recent study (Shi et al., 2016) used 1 M NaH_2PO_4 solution to extract the edge-bound Fe(II) from clay minerals. During drying, this NaH_2PO_4 solution will precipitate as a white crystalline substance and cannot be redissolved in 7 M HCl. This white crystalline substance may bind Fe ions and cause Fe loss during wet chemistry preparatory steps. To solve this problem,

4 mL 7 M HCl (500 μL × eight times) was added to dissolve the samples and then loaded onto the column. In order to test whether any Fe ions are eluted during this process, a collection beaker was placed under the column to collect the 7 M HCl during step 10(d–f). Then, the 7 M HCl solutions were reduced to a small volume (approx. 0.5 mL) and measured by the Ferrozine method to check whether there were any Fe ions in the solutions.

4.3.2 Interpretation of Results

Iron isotope compositions of samples are measured by MC-ICP-MS and usually reported as δ^{56}Fe. The ratio for the variation of δ^{57}Fe–δ^{56}Fe in the mass-dependent fractionation line is around 1.5:1 (Beard and Johnson, 2004), and therefore, the δ^{57}Fe can be derived from δ^{56}Fe ($\delta^{57}Fe \approx 1.5 \times \delta^{56}Fe$). Some studies have reported both δ^{56}Fe and δ^{57}Fe. The purpose of presenting the δ^{57}Fe value is to show that there are no unresolved analytical artifacts (Dauphas et al., 2017). The δ^{58}Fe is not reported for iron isotope compositions because of its low abundance. The range of differences in δ^{56}Fe between different types of natural-abundance samples can be up to 5‰. In this case study (Liu et al., 2015), isotopic differences in δ^{56}Fe occur within and between different sediment sequential extractions and sediment pore waters. The different Fe pools were –1.33 ± 0.42‰ for aqueous Fe (pore water Fe); –0.92 ± 0.18‰ for Fe pool extracted by 0.1 M HCl; 0.42 ± 0.08‰ for Fe pool extracted by 0.5 M HCl; and 0.56 ± 0.17‰ for the Fe pool extracted by 7 M HCl (Figure 4.4) (Liu et al., 2015). The δ^{56}Fe values increased from the most negative in pore waters to the Fe pool extracted by the highest concentration of acid (Liu et al., 2015).

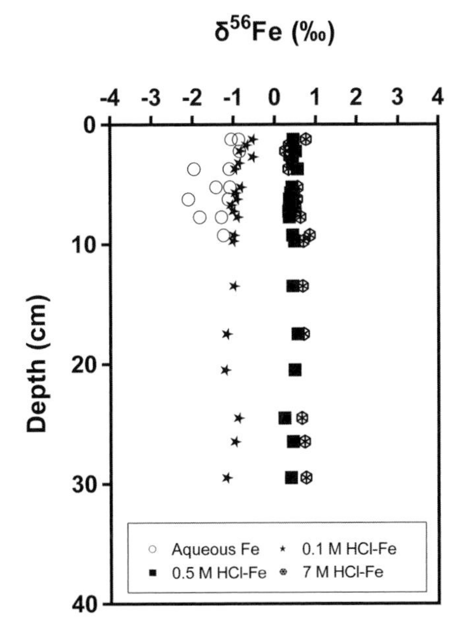

Figure 4.4 Measured δ^{56}Fe for pore water and sediment extractions. Precision is 0.08‰ at 2 SD. Modified from Liu et al., 2015, with permission of Elsevier.

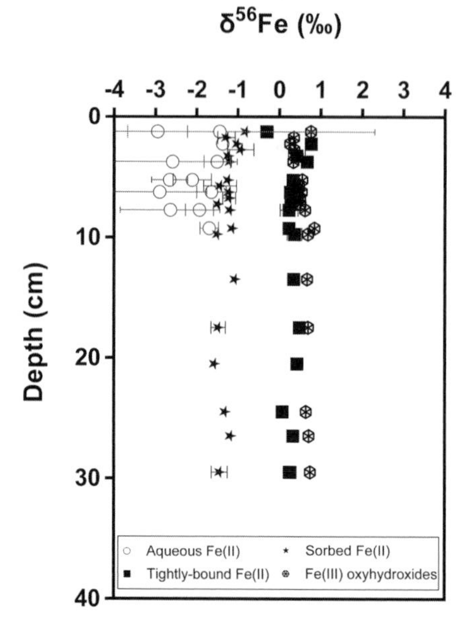

Figure 4.5 Calculated δ^{56}Fe for aqueous Fe(II), sorbed Fe(II), tightly bound Fe(II), and Fe(III) oxyhydroxides from measurements of δ^{56}Fe (Figure 4.4) and the percentage of Fe(II) and Fe(III) concentrations in the pore water and extracts 1 and 2. Modified from Liu et al., 2015, with permission of Elsevier.

It should be noted that measured Fe isotope composition is the δ^{56}Fe for all Fe species in sample solutions. Therefore, measured Fe isotope compositions can be further separated into δ^{56}Fe of Fe(II) and δ^{56}Fe of Fe(III) if sample solutions are mixtures of Fe(II) and Fe(III) species. In this case study (Liu et al., 2015), Fe pools, including pore water and extracts 1 and 2, are mixtures of Fe(II) and Fe(III) species. Fe species in the sediment are distributed among sorbed Fe(II) (extract 1), tightly bound Fe(II) (extract 2), amorphous Fe oxyhydroxides (extracts 1 and 2), and crystalline Fe(III) phases (extract 3). One of the common methods for isotopic mass balance is to use the percentage of Fe(II) and Fe(III) along with the measured isotope composition to determine the δ^{56}Fe for Fe(II) and Fe(III) separately. Some Fe components in different Fe pools may be assumed to have the same Fe isotope compositions if the related isotopic fractionations are small or negligible. In this case study (Liu et al., 2015), Fe isotope compositions for different Fe(II) phases were constrained from the Fe(III) component (Figure 4.5) using two assumptions. We first assumed that all the Fe(III) components in acid extractions had the same Fe isotope composition. This is supported by previous studies showing that partial dissolution by HCl does not introduce Fe isotopic fractionation (Skulan et al., 2002; Wiederhold et al., 2006). Since Fe(III) in pore water represents either organically complexed dissolved Fe(III) or colloidal Fe(III) oxyhydroxides, we then also assumed that Fe(III) in pore water had the same isotopic composition as that of Fe(III) oxyhydroxides. This assumption is challenged by the potential isotopic fractionation during ligand-promoted dissolution of Fe(III) oxyhydroxides and Fe(III) complexation with organic matter. However, experimental studies suggest that complexation of Fe(III)–organic ligand yields <1‰ isotopic fractionation (Dideriksen et al., 2008; Morgan et al., 2011).

The Fe isotopic fractionation can be calculated from the difference between two Fe pools. In this case study, the isotopic fractionation between aqueous Fe(II) and Fe(III) oxyhydroxides (–2.6 ± 0.5‰) is broadly consistent with previous experimental laboratory work using iron-reducing bacteria and pure Fe minerals such as hematite and goethite (Crosby et al., 2005; Johnson et al., 2005; Wu et al., 2009). This is encouraging considering the complicated mineral compositions in Lake Tantaré sediment. The isotopic fractionation between aqueous Fe(II) and sorbed Fe(II) (–0.9 ± 0.4‰) was also calculated and is roughly consistent with previous work, which measured the Fe fractionation (0.2–0.9‰) between sorbed Fe(II) and aqueous Fe(II) in the laboratory setting (Crosby et al., 2007; Wu et al., 2009, 2010).

4.4 Conclusion

MC-ICP-MS is a powerful technique for studying stable metal isotopes, providing reliable and rapid isotopic analysis for samples from the natural environment. The application of Fe isotope signatures to biogeochemical Fe cycling is particularly insightful. Fe isotopes can be used to investigate constraints on redox conditions and Fe reaction pathways in modern oceans and ancient settings. The advantages of using MC-ICP-MS to analyze stable metal isotope signatures include its high ionization efficiency, low detection limits (ppb concentrations), high precision, and rapid analysis. However, MC-ICP-MS is expensive to operate due to the large volumes of pure argon gas that are consumed during measurements and the costs of routine instrument maintenance. Due to the variable composition of natural samples, one should bear in mind that a preliminary chemical separation is required to achieve a separation between interfering ions and the metal species of interest, and accompanying standards must be made in the same sample matrix. In addition to standalone MC-ICP-MS techniques, combined MC-ICP-MS with liquid chromatography and MC-ICP-MS with laser ablation are being increasingly explored as powerful analytical techniques for the *in situ* analysis of rock samples.

4.5 References

Albarède, F., Beard, B.L., 2004. Analytical methods for non-traditional isotopes, in Johnson, C.M., Beard, B.L., Albarede, F. (Eds.), *Geochemistry of Non-traditional Stable Isotopes*. The Mineralogical Society of America, Washington, DC, USA, pp. 113–152.

Anbar, A.D., 2004. Iron stable isotopes: Beyond biosignatures. *Earth and Planetary Science Letters* 217, 223–236.

Anbar, A.D., Roe, J.E., Barling, J., Nealson, K.H., 2000. Nonbiological fractionation of iron isotopes. *Science* 288, 126–128.

Azrieli-Tal, I., Matthews, A., Bar-Matthews, M., et al., 2014. Evidence from molybdenum and iron isotopes and molybdenum–uranium covariation for sulphidic bottom waters during Eastern Mediterranean sapropel S1 formation. *Earth and Planetary Science Letters* 393, 231–242.

Barling, J., Anbar, A.D., 2004. Molybdenum isotope fractionation during adsorption by manganese oxides. *Earth and Planetary Science Letters* 217, 315–329.

Beard, B.L., Johnson, C.M., 1999. High precision iron isotope measurements of terrestrial and lunar materials. *Geochimica et Cosmochimica Acta* 63, 1653–1660.

Beard, B.L., Johnson, C.M., 2004. Fe isotope variations in the modern and ancient earth and other planetary bodies. *Reviews in Mineralogy and Geochemistry* 55, 319–357.

Beard, B.L., Johnson, C.M., Cox, L., et al., 1999. Iron isotope biosignatures. *Science* 285, 1889–1892.

Beard, B.L., Johnson, C.M., Skulan, J.L., et al., 2003a. Application of Fe isotopes to tracing the geochemical and biological cycling of Fe. *Chemical Geology* 195, 87–117.

Beard, B.L., Johnson, C.M., Von Damm, K.L., Poulson, R.L., 2003b. Iron isotope constraints on Fe cycling and mass balance in oxygenated Earth oceans. *Geology* 31, 629–632.

Becker, J.S., 2007. *Inorganic Mass Spectrometry: Principles and Applications*. Wiley, Chichester, England.

Belshaw, N.S., Zhu, X., Guo, Y., O'Nions, R.K., 2000. High precision measurement of iron isotopes by plasma source mass spectrometry. *International Journal of Mass Spectrometry* 197, 191–195.

Bergquist, B.A., Boyle, E.A., 2006. Iron isotopes in the Amazon River system: Weathering and transport signatures. *Earth and Planetary Science Letters* 248, 54–68.

Borrok, D.M., Wanty, R.B., Ridley, W.I., et al., 2007. Separation of copper, iron, and zinc from complex aqueous solutions for isotopic measurement. *Chemical Geology* 242, 400–414.

Brantley, S.L., Liermann, L.J., Guynn, R.L., et al., 2004. Fe isotopic fractionation during mineral dissolution with and without bacteria. *Geochimica et Cosmochimica Acta* 68, 3189–3204.

Bufflap, S.E., Allen, H.E., 1995. Sediment pore water collection methods for trace metal analysis: A review. *Water Research* 29, 165–177.

Bullen, T.D., White, A.F., Childs, C.W., Vivit, D.V., Schulz, M.S., 2001. Demonstration of significant abiotic iron isotope fractionation in nature. *Geology* 29, 699.

Burdige, D.J., 2006. *Geochemistry of Marine Sediments*. Princeton University Press, Princeton, NJ, USA.

Burton, K.W., Vigier, N., 2011. Lithium isotopes as tracers in marine and terrestrial environments, in Baskaran, M. (Ed.), *Handbook of Environmental Isotope Geochemistry*. Springer, Berlin, Germany, pp. 41–59.

Busigny, V., Planavsky, N.J., Jézéquel, D., et al., 2014. Iron isotopes in an Archean ocean analogue. *Geochimica et Cosmochimica Acta* 133, 443–462.

Buss, H.L., Mathur, R., White, A.F., Brantley, S.L., 2010. Phosphorus and iron cycling in deep saprolite, Luquillo Mountains, Puerto Rico. *Chemical Geology* 269, 52–61.

Byrne, J.M., Klueglein, N., Pearce, C., et al., 2015. Redox cycling of Fe(II) and Fe(III) in magnetite by Fe-metabolizing bacteria. *Science* 347, 1473–1476.

Caiazza, N.C., Lies, D.P., Newman, D.K., 2007. Phototrophic Fe(II) oxidation promotes organic carbon acquisition by *Rhodobacter capsulatus* SB1003. *Applied and Environmental Microbiology* 73, 6150–6158.

Chen, J.-B., Busigny, V., Gaillardet, J., Louvat, P., Wang, Y.-N., 2014. Iron isotopes in the Seine River (France): Natural versus anthropogenic sources. *Geochimica et Cosmochimica Acta* 128, 128–143.

Chever, F., Rouxel, O.J., Croot, P.L., et al., 2015. Total dissolvable and dissolved iron isotopes in the water column of the Peru upwelling regime. *Geochimica et Cosmochimica Acta* 162, 66–82.

Cloquet, C., Carignan, J., Lehmann, M.F., Vanhaecke, F., 2008. Variation in the isotopic composition of zinc in the natural environment and the use of zinc isotopes in biogeosciences: A review. *Analytical and Bioanalytical Chemistry* 390, 451–463.

Collins, R.N., Waite, T.D., 2009. Isotopically exchangeable concentrations of elements having multiple oxidation states: The case of Fe(II)/Fe(III) isotope self-exchange in coastal lowland acid sulfate soils. *Environmental Science & Technology* 43, 5365–5370.

Couture, R.-M., Gobeil, C., Tessier, A., 2010a. Arsenic, iron and sulfur co-diagenesis in lake sediments. *Geochimica et Cosmochimica Acta* 74, 1238–1255.

Couture, R.-M., Shafei, B., Cappallen, P.V., Tessier, A., Gobeil, C., 2010b. Non-steady state modeling of arsenic diagenesis in lake sediments. *Environmental Science & Technology* 44, 197–203.
Crosby, H.A., Johnson, C.M., Roden, E.E., Beard, B.L., 2005. Coupled Fe(II)-Fe(III) electron and atom exchange as a mechanism for Fe isotope fractionation during dissimilatory iron oxide reduction. *Environmental Science & Technology* 39, 6698–6794.
Crosby, H.A., Roden, E.E., Johnson, C.M., Beard, B.L., 2007. The mechanisms of iron isotope fractionation produced during dissimilatory Fe(III) reduction by *Shewanella putrefaciens* and *Geobacter sulfurreducens*. *Geobiology* 5, 169–189.
Cullum, D.C., 1994. *Introduction to Surfactant Analysis*. Chapman & Hall, London, UK.
Dauphas, N., John, S.G., Rouxel, O., 2017. Iron isotope systematics. *Reviews in Mineralogy and Geochemistry* 82, 415–510.
Davison, W., 1993. Iron and manganese in lakes. *Earth-Science Reviews* 34, 119–163.
Dekov, V.M., Vanlierde, E., Billström, K., et al., 2014. Ferrihydrite precipitation in groundwater-fed river systems (Nete and Demer river basins, Belgium): Insights from a combined Fe-Zn-Sr-Nd-Pb-isotope study. *Chemical Geology* 386, 1–15.
Dideriksen, K., Baker, J.A., Stipp, S.L.S., 2008. Equilibrium Fe isotope fractionation between inorganic aqueous Fe(III) and the siderophore complex, Fe(III)-desferrioxamine B. *Earth and Planetary Science Letters* 269, 280–290.
Dodson, M.H., 1963. A theoretical study of the use of internal standards for precise isotopic analysis by the surface ionization technique: Part I – General first-order algebraic solutions. *Journal of Scientific Instruments* 40, 289–295.
Ehrenreich, A., Widdel, F., 1994. Anaerobic oxidation of ferrous iron by purple bacteria, a new type of phototrophic metabolism. *Applied and Environmental Microbiology* 60, 4517–4526.
Escoube, R., Rouxel, O.J., Sholkovitz, E., Donard, O.F.X., 2009. Iron isotope systematics in estuaries: The case of North River, Massachusetts (USA). *Geochimica et Cosmochimica Acta* 73, 4045–4059.
Fantle, M.S., Bullen, T.D., 2009. Essentials of iron, chromium and calcium isotope analysis of natural materials by thermal ionization mass spectrometry. *Chemical Geology* 258, 50–64.
Gauger, T., Byrne, J.M., Konhauser, K.O., et al., 2016. Influence of organics and silica on Fe(II) oxidation rates and cell–mineral aggregate formation by the green-sulfur Fe(II)-oxidizing bacterium *Chlorobium ferrooxidans* KoFox – Implications for Fe(II) oxidation in ancient oceans. *Earth and Planetary Science Letters* 443, 81–89.
Gelting, J., Breitbarth, E., Stolpe, B., Hassellöv, M., Ingri, J., 2010. Fractionation of iron species and iron isotopes in the Baltic Sea euphotic zone. *Biogeosciences* 7, 2489–2508.
Glass, J.B., Wolfe-Simon, F., Anbar, A.D., 2009. Coevolution of metal availability and nitrogen assimilation in cyanobacteria and algae. *Geobiology* 7, 100–123.
Guelke, M., von Blanckenburg, F., Schoenberg, R., Staubwasser, M., Stuetzel, H., 2010. Determining the stable Fe isotope signature of plant-available iron in soils. *Chemical Geology* 277, 269–280.
Guo, H., Liu, C., Lu, H., et al., 2013. Pathways of coupled arsenic and iron cycling in high arsenic groundwater of the Hetao basin, Inner Mongolia, China: An iron isotope approach. *Geochimica et Cosmochimica Acta* 112, 130–145.
Homoky, W.B., John, S.G., Conway, T.M., Mills, R.A., 2013. Distinct iron isotopic signatures and supply from marine sediment dissolution. *Nature Communications* 4, 2143.
Homoky, W.B., Severmann, S., Mills, R.A., Statham, P.J., Fones, G.R., 2009. Pore-fluid Fe isotopes reflect the extent of benthic Fe redox recycling: Evidence from continental shelf and deep-sea sediments. *Geology* 37, 751–754.
Hotz, K., Walczyk, T., 2013. Natural iron isotopic composition of blood is an indicator of dietary iron absorption efficiency in humans. *Journal of Biological Inorganic Chemistry* 18, 1–7.
Ilina, S.M., Poitrasson, F., Lapitskiy, S.A., et al., 2013. Extreme iron isotope fractionation between colloids and particles of boreal and temperate organic-rich waters. *Geochimica et Cosmochimica Acta* 101, 96–111.

Ingri, J., Malinovsky, D., Rodushkin, I., et al., 2006. Iron isotope fractionation in river colloidal matter. *Earth and Planetary Science Letters* 245, 792–798.
Jiao, Y., Kappler, A., Croal, L.R., Newman, D.K., 2005. Isolation and characterization of a genetically tractable photoautotrophic Fe(II)-oxidizing bacterium, *Rhodopseudomonas palustris* strain TIE-1. *Applied and Environmental Microbiology* 71, 4487–4496.
John, S.G., Mendez, J., Moffett, J., Adkins, J., 2012. The flux of iron and iron isotopes from San Pedro Basin sediments. *Geochimica et Cosmochimica Acta* 93, 14–29.
Johnson, C.M., Beard, B.L., 1999. Correction of instrumentally produced mass fractionation during isotopic analysis of Fe by thermal ionization mass spectrometry. *International Journal of Mass Spectrometry* 193, 87–99.
Johnson, C.M., Beard, B.L., Klein, C., Beukes, N.J., Roden, E.E., 2008. Iron isotopes constrain biologic and abiologic processes in banded iron formation genesis. *Geochimica et Cosmochimica Acta* 72, 151–169.
Johnson, C.M., Roden, E.E., Welch, S.A., Beard, B.L., 2005. Experimental constraints on Fe isotope fractionation during magnetite and Fe carbonate formation coupled to dissimilatory hydrous ferric oxide reduction. *Geochimica et Cosmochimica Acta* 69, 963–993.
Kappler, A., Pasquero, C., Konhauser, K.O., Newman, D.K., 2005. Deposition of banded iron formations by anoxygenic phototrophic Fe(II)-oxidizing bacteria. *Geology* 33, 865–868.
Kendall, B., Dahl, T.W., Anbar, A.D., 2017. The stable isotope geochemistry of molybdenum. *Reviews in Mineralogy and Geochemistry* 82, 683–732.
Labatut, M., Lacan, F., Pradoux, C., et al., 2014. Iron sources and dissolved-particulate interactions in the seawater of the Western Equatorial Pacific, iron isotope perspectives. *Global Biogeochemical Cycles* 28, 1044–1065.
Li, W., Beard, B.L., Johnson, C.M., 2015. Biologically recycled continental iron is a major component in banded iron formations. *Proceedings of the National Academy of Sciences of the United States of America* 112, 8193–8198.
Li, W.-Y., Teng, F.-Z., Ke, S., et al., 2010. Heterogeneous magnesium isotopic composition of the upper continental crust. *Geochimica et Cosmochimica Acta* 74, 6867–6884.
Liu, K., Wu, L., Couture, R.-M., Li, W., Van Cappellen, P., 2015. Iron isotope fractionation in sediments of an oligotrophic freshwater lake. *Earth and Planetary Science Letters* 423, 164–172.
Lovley, D.R., 1993. Dissimilatory metal reduction. *Annual Review of Microbiology* 47, 263–290.
Malinovsky, D., Rodyushkin, L.V., Shcherbakova, E.P., et al., 2005. Fractionation of Fe isotopes as a result of redox processes in a basin. *Geochemistry International* 43, 797–803.
Malinovsky, D., Stenberg, A., Rodushkin, I., et al., 2003. Performance of high resolution MC-ICP-MS for Fe isotope ratio measurements in sedimentary geological materials. *Journal of Analytical Atomic Spectrometry* 18, 687–695.
Mansfeldt, T., Schuth, S., Häusler, W., et al., 2011. Iron oxide mineralogy and stable iron isotope composition in a Gleysol with petrogleyic properties. *Journal of Soils and Sediments* 12, 97–114.
Marechal, C.N., Telouk, P., Albarede, F., 1999. Precise analysis of copper and zinc isotopic compositions by plasma-source mass spectrometry. *Chemical Geology* 156, 251–273.
Matthews, A., Zhu, X., O'Nions, R.K., 2001. Kinetic iron stable isotope fractionation between iron (-II) and (-III) complexes in solution. *Earth and Planetary Science Letters* 192, 81–92.
Meija, J., Yang, L., Mester, Z., Sturgeon, R.E., 2012. Correction of instrumental mass discrimination for isotope ratio determination with multi-collector inductively coupled plasma mass spectrometry, in Vanhaecke, F., Degryse, P. (Eds.), *Isotopic Analysis: Fundamentals and Applications Using ICP-MS*. Wiley-VCH, Weinheim, Germany, pp. 113–137.
Melton, E.D., Swanner, E.D., Behrens, S., Schmidt, C., Kappler, A., 2014. The interplay of microbially mediated and abiotic reactions in the biogeochemical Fe cycle. *Nature Reviews. Microbiology* 12, 797–808.
Molot, L.A., Watson, S.B., Creed, I.F., et al., 2014. A novel model for cyanobacteria bloom formation: The critical role of anoxia and ferrous iron. *Freshwater Biology* 59, 1323–1340.

Morgan, J.L.L., Wasylenki, L.E., Nuester, J., Anbar, A.D., 2011. Fe isotope fractionation during equilibration of Fe-organic complexes. *Environmental Science & Technology* 44, 6095–6101.
Moynier, F., Vance, D., Fujii, T., Savage, P., 2017. The isotope geochemistry of zinc and copper. *Reviews in Mineralogy and Geochemistry* 82, 543–600.
Nishizawa, M., Yamamoto, H., Ueno, Y., et al., 2010. Grain-scale iron isotopic distribution of pyrite from Precambrian shallow marine carbonate revealed by a femtosecond laser ablation multi-collector ICP-MS technique: Possible proxy for the redox state of ancient seawater. *Geochimica et Cosmochimica Acta* 74, 2760–2778.
Percak-Dennett, E.M., Beard, B.L., Xu, H., et al., 2011. Iron isotope fractionation during microbial dissimilatory iron oxide reduction in simulated Archaean seawater. *Geobiology* 9, 205–220.
Percak-Dennett, E.M., Loizeau, J.-L., Beard, B.L., Johnson, C.M., Roden, E.E., 2013. Iron isotope geochemistry of biogenic magnetite-bearing sediments from the Bay of Vidy, Lake Geneva. *Chemical Geology* 360–361, 32–40.
Pinheiro, G.M.S., Poitrasson, F., Sondag, F., Cochonneau, G., Vieiraa, L.C., 2014. Contrasting iron isotopic compositions in river suspended particulate matter: The Negro and the Amazon annual river cycles. *Earth and Planetary Science Letters* 394, 168–178.
Planavsky, N., Rouxel, O.J., Bekker, A., et al., 2012. Iron isotope composition of some Archean and Proterozoic iron formations. *Geochimica et Cosmochimica Acta* 80, 158–169.
Poitrasson, F., Viers, J., Martin, F., Braun, J.-J., 2008. Limited iron isotope variations in recent lateritic soils from Nsimi, Cameroon: Implications for the global Fe geochemical cycle. *Chemical Geology* 253, 54–63.
Radic, A., Lacan, F., Murray, J.W., 2011. Iron isotopes in the seawater of the equatorial Pacific Ocean: New constraints for the oceanic iron cycle. *Earth and Planetary Science Letters* 306, 1–10.
Rehkamper, M., Schonbachler, M., Andreasen, R., 2012. Application of multiple-collector inductively coupled plasma mass spectrometry to isotopic analysis in cosmochemistry, in Vanhaecke, F., Degryse, P. (Eds.), *Isotopic Analysis: Fundamentals and Applications Using ICP-MS*. Wiley-VCH, Weinheim, Germany, pp. 275–315.
Rouxel, O., Auro, M., 2010. Iron isotope variations in coastal seawater determined by multicollector ICP-MS. *Geostandards and Geoanalytical Research* 34, 135–144.
Savage, P., Moynier, F., Harvey, J., Burton, K., 2015. The behavior of copper isotopes during igneous processes. AGU conference, San Francisco 390, 451–463.
Schiff, S.L., Tsuji, J.M., Wu, L., et al., 2017. Millions of boreal shield lakes can be used to probe Archaean ocean biogeochemistry. *Scientific Reports* 7, 46708.
Scholz, F., Severmann, S., McManus, J., Hensen, C., 2014. Beyond the Black Sea paradigm: The sedimentary fingerprint of an open-marine iron shuttle. *Geochimica et Cosmochimica Acta* 127, 368–380.
Severmann, S., Anbar, A.D., 2009. Reconstructing paleoredox conditions through a multitracer approach: The key to the past is the present. *Elements* 5, 359–364.
Severmann, S., Lyons, T.W., Anbar, A., McManus, J., Gordon, G., 2008. Modern iron isotope perspective on the benthic iron shuttle and the redox evolution of ancient oceans. *Geology* 36, 487.
Severmann, S., McManus, J., Berelson, W.M., Hammond, D.E., 2010. The continental shelf benthic iron flux and its isotope composition. *Geochimica et Cosmochimica Acta* 74, 3984–4004.
Shi, B., Liu, K., Wu, L., et al., 2016. Iron isotope fractionations reveal a finite bioavailable Fe pool for structural Fe(III) reduction in nontronite. *Environmental Science & Technology* 50, 8661–8669.
Skulan, J.L., Beard, B.L., Johnson, C.M., 2002. Kinetic and equilibrium Fe isotope fractionation between aqueous Fe(III) and hematite. *Geochimica et Cosmochimica Acta* 66, 2995–3015.
Song, L., Liu, C.-Q., Wang, Z.-L., et al., 2011. Iron isotope fractionation during biogeochemical cycle: Information from suspended particulate matter (SPM) in Aha Lake and its tributaries, Guizhou, China. *Chemical Geology* 280, 170–179.
Sossi, P.A., Nebel, O., O'Neill, H.S.C., Moynier, F., 2018. Zinc isotope composition of the Earth and its behaviour during planetary accretion. *Chemical Geology* 477, 73–84.

Staubwasser, M., Blanckenburg, F., Schoenberg, R., 2006. Iron isotopes in the early marine diagenetic iron cycle. *Geology* 34, 629–632.

Stookey, L.L., 1970. Ferrozine – a new spectrophotometric reagent for iron. *Analytical Chemistry* 42, 779–781.

Strelow, F.W.E., 1980. Improved separation of iron from copper and other elements by anion-exchange chromatography on a 4% cross-linked resin with high concentrations of hydrochloric acid. *Talanta* 27, 727–732.

Tangalos, G.E., Beard, B.L., Johnson, C.M., et al., 2010. Microbial production of isotopically light iron(II) in a modern chemically precipitated sediment and implications for isotopic variations in ancient rocks. *Geobiology* 8, 197–208.

Teng, F.-Z., 2017. Magnesium isotope geochemistry. *Reviews in Mineralogy and Geochemistry* 82, 219–287.

Teng, F.-Z., Dauphas, N., Watkins, J.M., 2017. Non-traditional stable isotopes: retrospective and prospective. *Reviews in Mineralogy and Geochemistry* 82, 1–26.

Teng, F.-Z., Li, W.-Y., Rudnick, R.L., Gardner, L.R., 2010. Contrasting lithium and magnesium isotope fractionation during continental weathering. *Earth and Planetary Science Letters* 300, 63–71.

Teutsch, N., Schmid, M., Müller, B., et al., 2009. Large iron isotope fractionation at the oxic–anoxic boundary in Lake Nyos. *Earth and Planetary Science Letters* 285, 52–60.

Teutsch, N., von Gunten, U., Porcelli, D., Cirpka, O.A., Halliday, A.N., 2005. Adsorption as a cause for iron isotope fractionation in reduced groundwater. *Geochimica et Cosmochimica Acta* 69, 4175–4185.

Tsikos, H., Matthews, A., Erel, Y., Moore, J.M., 2010. Iron isotopes constrain biogeochemical redox cycling of iron and manganese in a Palaeoproterozoic stratified basin. *Earth and Planetary Science Letters* 298, 125–134.

Van Heghe, L., Delanghe, J., Van Vlierberghe, H., Vanhaecke, F., 2013. The relationship between the iron isotopic composition of human whole blood and iron status parameters. *Metallomics: Integrated Biometal Science* 5, 1503–1509.

Vanhaecke, F., 2012. Single-collector inductively coupled plasma mass spectrometry, in Vanhaecke, F., Degryse, P. (Eds.), *Isotopic Analysis: Fundamentals and Applications Using ICP-MS*. Wiley-VCH, Weinheim, Germany, pp. 31–75.

Vanhaecke, F., Degryse, P., 2012. *Isotopic Analysis: Fundamentals and Applications Using ICP-MS*. Wiley-VCH, Weinheim, Germany.

Vanhaecke, F., Kyser, K., 2012. The isotopic composition of the elements, in Vanhaecke, F., Degryse, P. (Eds.), *Isotopic Analysis: Fundamentals and Applications Using ICP-MS*. Wiley-VCH, Weinheim, Germany, pp. 1–29.

Walczyk, T., Blanckenburg, F., 2002. Natural iron isotope variations in human blood. *Science* 295, 2065–2066.

Wasylenki, L.E., Rolfe, B.A., Weeks, C.L., Spiro, T.G., Anbar, A.D., 2008. Experimental investigation of the effects of temperature and ionic strength on Mo isotope fractionation during adsorption to manganese oxides. *Geochimica et Cosmochimica Acta* 72, 5997–6005.

Wiederhold, J.G., Kraemer, S.M., Teutsch, N., et al., 2006. Iron isotope fractionation during proton-promoted, ligand-controlled, and reductive dissolution of goethite. *Environmental Science & Technology* 40, 3787–3793.

Wieser, M., Schwieters, J., Douthitt, C., 2012. Multi-collector inductively coupled plasma mass spectrometry, in Vanhaecke, F., Degryse, P. (Eds.), *Isotopic Analysis: Fundamentals and Applications Using ICP-MS*. Wiley-VCH, Weinheim, Germany, pp. 77–91.

Wombacher, F., Eisenhauer, A., Böhm, F., et al., 2011. Magnesium stable isotope fractionation in marine biogenic calcite and aragonite. *Geochimica et Cosmochimica Acta* 75, 5797–5818.

Wu, L., Beard, B.L., Roden, E.E., Johnson, C.M., 2009. Influence of pH and dissolved Si on Fe isotope fractionation during dissimilatory microbial reduction of hematite. *Geochimica et Cosmochimica Acta* 73, 5584–5599.

Wu, L., Beard, B.L., Roden, E.E., Kennedy, C.B., Johnson, C.M., 2010. Stable Fe isotope fractionations produced by aqueous Fe(II)-hematite surface interactions. *Geochimica et Cosmochimica Acta* 74, 4249–4265.

Wunder, B., Meixner, A., Romer, R.L., et al., 2007. Lithium isotope fractionation between Li-bearing staurolite, Li-mica and aqueous fluids: An experimental study. *Chemical Geology* 238, 277–290.

Xie, X., Johnson, T.M., Wang, Y., et al., 2014. Pathways of arsenic from sediments to groundwater in the hyporheic zone: Evidence from an iron isotope study. *Journal of Hydrology* 511, 509–517.

PART III

IMAGING TECHNIQUES

5 Scanning Probe Microscopy

ADAM F. WALLACE

Abstract

Scanning probe microscopy (SPM) is a suite of related imaging methods, in which variations in the interaction force between a probe and a sample surface are used to generate image contrast. These instruments are incredibly sensitive; they can measure forces on the order of those required to break physical and chemical bonds, and under the most optimal conditions, atomic-scale resolution can be achieved. Although SPM is still primarily used for imaging, it is increasingly being used to measure nanoscale properties and interaction forces. This chapter serves as an introduction to the fundamentals of SPM and to the most prevalent methods needed for the investigation of mineral–microbe interactions.

5.1 Introduction: A Brief History of Microscopy and the Emergence of SPM

According to archaeological evidence (Enoch and Lakshminarayanan, 2000), the first primitive lenses emerged during the Bronze Age, but they were most likely used for decorative purposes rather than magnification (Plantzos, 1997). However, during the Middle Ages, polished glass and rock crystal (i.e., quartz) "reading stones" became widely utilized throughout Europe (Quercioli, 2011). By the dawn of the seventeenth century, lens technology had advanced significantly, and the first compound optical microscopes and telescopes were being used for scientific investigations. Arguably, the microscopic age began in earnest with the publication of Robert Hooke's *Micrographia* (Hooke, 1665) in the winter of 1665. A little more than a decade after this seminal work, a Dutch fabric retailer turned lens maker (Antonie van Leeuwenhoek) perfected the manufacture of lenses capable of ~250× magnification, and in a series of letters to the Royal Society of London, reported what modern scholars consider to be the first documented observations of protista and bacteria (Gest, 2004; Porter, 1976). Based on these discoveries, many consider van Leeuwenhoek to have been the first microbiologist.

With the close of the nineteenth century came the recognition that the ultimate spatial resolution of traditional light microscopy was constrained by the diffraction limit, a distance equal to approximately half the wavelength of the incident radiation (~0.2 μm for visible light). However, the contemporaneous discovery of the electron by Sir J. J. Thomson (Thomson, 1897) and the subsequent understanding of the dual wave–particle nature of light and refinement of the theory of radiation by De Broglie, Einstein, Schrödinger, and others (Niaz, 2009) opened the door to the development of instruments capable of achieving superior resolution with higher-energy (and shorter-wavelength) sources of radiation than visible light. The first prototype electron microscope was unveiled by Knoll and Ruska (Knoll and Ruska, 1932) in 1932 based on the theories and electromagnetic lens designs asserted by Hans Busch (Busch, 1926). Modern instruments are still based on Ruska's design and have produced images with sub-Ångstrom resolution (Mohr et al., 2014; Zhu et al., 2009). However, somewhat lower resolution is typically attainable on biological samples, which must also be prepared specially to withstand the high-vacuum conditions inside the microscope.

In 1982, researchers at IBM Zürich (Gerd Binnig, Heinrich Rohrer, and others) invented a new type of nonoptical electron microscopy, scanning tunneling microscopy (Binnig et al., 1982a) (STM), based on the quantum tunneling phenomenon. Initially, Binnig and coworkers demonstrated the transfer of electrons between two conductive materials across a vacuum gap by applying a bias voltage between a sharpened tungsten probe (or tip) and a platinum plate (Binnig et al., 1982b). Subsequently, they recognized that the rate of charge transfer (i.e., the tunneling current) between the tip and the plate could be controlled by adjusting the size of the vacuum gap, and they devised a setup in which the tip–plate distance z could be rapidly adjusted to maintain a constant tunneling current as the position of the tip varied in the x–y plane above the sample. These data could then be used to construct topographic images of the surface with sub-Ångstrom resolution in the z direction.

The development of STM was revolutionary, although its utility was limited to conductive samples, primarily in vacuum. However, by 1986 Binnig had augmented the method to yield yet another type of SPM, atomic force microscopy (Binnig et al., 1986) (AFM), which was capable of imaging nonconductive surfaces. This advance was achieved by monitoring the physical interaction between a stylus-like probe and the sample in place of the tunneling current. Like STM, AFM was originally used to generate three-dimensional representations of nanoscale topographic features; however, users soon realized that the properties of the probe and sample could be tailored to explore an extremely wide variety of interaction forces. Modern commercially available scanning probe microscopes can be used for STM, and a variety of off-the-shelf and custom AFM imaging modes in vacuum, air, and fluid environments, making SPM one of the most versatile nonoptical imaging techniques available today.

This chapter is a basic primer for geoscientists with no prior SPM experience who are interested in using these techniques to investigate mineral–microbe interactions (Figure 5.1). Particular focus is placed upon AFM methods, which are most commonly used in studies of biological materials under physiologically relevant conditions (i.e., in fluid). More advanced readers are encouraged to explore the primary literature and advanced texts on SPM (Baró and Reifenberger, 2012; Bhushan and Fuchs, 2008; Morita, 2006). Manuals for commercially available microscopes are also good sources of background and instrument-specific information.

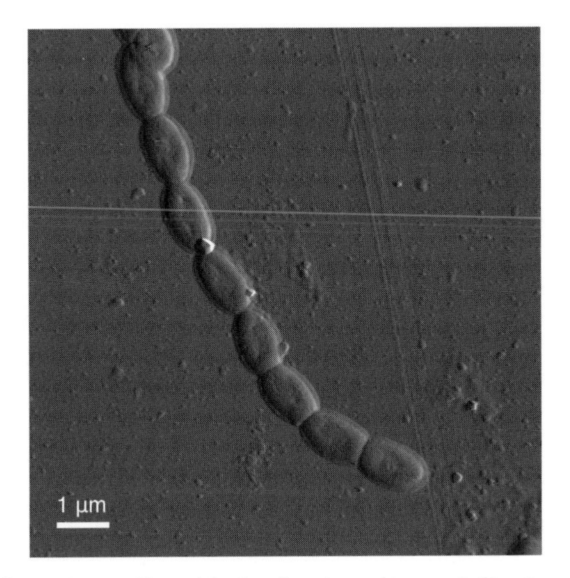

Figure 5.1 AFM contact mode deflection image of bacterial cells adhered to a thin organic film atop a gold sputter-coated silicon wafer. Mineral nanoparticles are also shown in association with the organic film.

5.2 Basic Operating Principles of SPM

Image contrast in SPM is derived from the interaction between a probe and a sample surface. Generally, an interaction parameter is set to a user-specified target value and the microscope works to maintain that setpoint as the probe is rastered over the surface. In some instruments, the sample is displaced relative to a fixed probe, while in others, the reverse is true. The relative sample/tip position is controlled by three scanners that control the displacement of the sample/tip in the x, y, and z directions. These scanners are composed of piezoelectric materials (i.e., a material that deforms mechanically in an electric field) that are capable of achieving extremely small displacements. SPMs utilize a feedback loop to determine the amount of displacement needed to maintain the interaction setpoint. That is, when the interaction parameter deviates from the target value, an electronic signal is sent through the microscope hardware to its control software, which determines how much voltage is applied to the piezoelectric scanners to return the interaction parameter to its initial value.

5.2.1 STM

In STM, an atomically sharp conductive probe is placed within a few Ångstroms of a conductive sample. A bias potential is applied between the tip and the sample to facilitate the flow of electrons, by quantum tunneling, from the tip to the sample. There are two basic modes of STM image acquisition that produce topographic information.

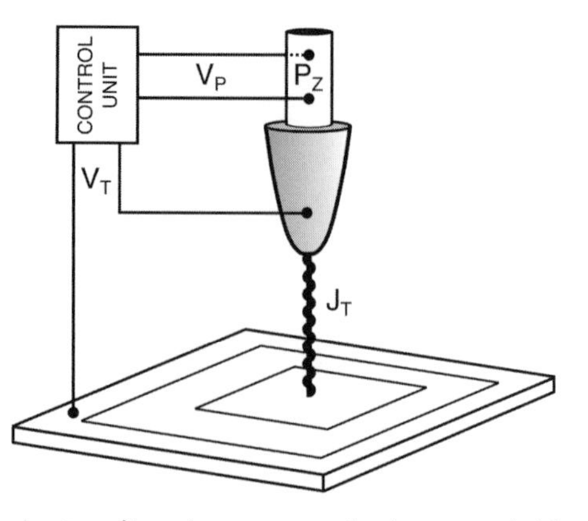

Figure 5.2 Schematic depiction of a scanning tunneling microscope operating in constant height mode (after Binnig et al., 1982a). The *x*- and *y*-piezo scanners (P_X and P_Y) raster the tip over the surface as the feedback loop (moderated by the control unit) determines the *z*-piezo (P_Z) displacement needed to hold the tunneling current fixed. V_P and V_T are the voltages supplied to the *z*-piezo and STM tip, respectively. J_T is the tunneling junction between the tip and the conductive sample.

In *constant current mode*, the real space surface topography is obtained by monitoring the displacement of the *z*-piezoelectric scanner required to maintain a constant tunneling current (Figure 5.2). Alternatively, in *constant height mode*, the tip–sample distance remains constant, and variations in the tunneling current reflect variations in the underlying topography. In this mode, the feedback loop can be disabled to facilitate faster scanning rates, but the surface topography is reported as a current, in contrast to a direct real space measurement, when the current is held constant. Disabling the feedback loop also limits the applicability of constant height mode to extremely flat surfaces, because the probe will run aground of the sample surface if the topographic roughness is greater than the tip–sample distance.

5.2.2 AFM

AFM- and STM-based imaging techniques differ primarily in the way their respective probes interact with sample surfaces. STM probes are rigid metal filaments that do not directly contact the sample. However, AFM probes are flexible cantilevers that bend within force range of the sample surface (Figure 5.3). As the interaction between the tip and the sample can persist over fairly large separation distances, it is possible to image with or without tip–sample contact (i.e., contact and noncontact imaging modes). At large separation distances, the tip does not directly contact the sample surface, and the cantilever deflection is influenced by long-range interaction forces. In noncontact modes, the tip–sample interaction force is typically attractive; however, in contact mode, imaging is performed exclusively in the repulsive regime.

The position of the cantilever is usually monitored by means of a laser beam deflection technique as shown schematically in Figure 5.4. Here, a laser is reflected off the top of the

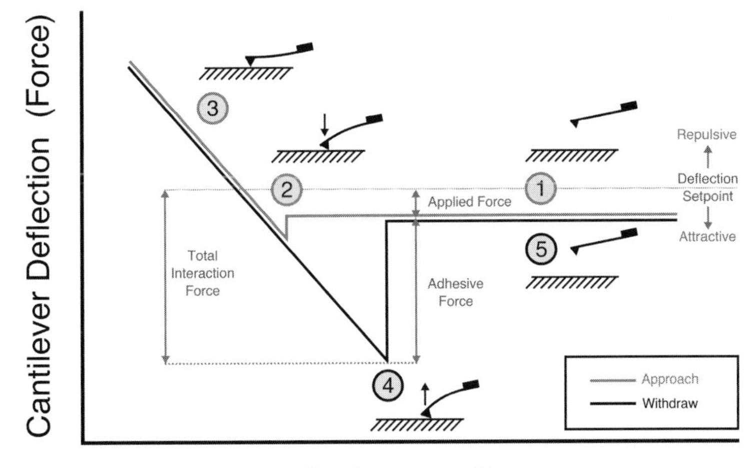

Figure 5.3 Characteristic force–distance curve annotated with a pictorial view of the AFM cantilever response during an approach and withdraw cycle. At point 1 the cantilever slowly approaches the surface. At point 2 the cantilever snaps into contact with the surface due to van der Waals forces. Upon continued advancement of the cantilever towards the sample, the tip–sample interaction transitions from attractive to repulsive (point 3). During withdrawal, strong adhesive forces keep the tip in contact with the surface at tip–sample distances beyond the point of initial contact identified during tip approach (point 4). Continued retraction of the cantilever induces the tip to disengage from the surface (point 5).

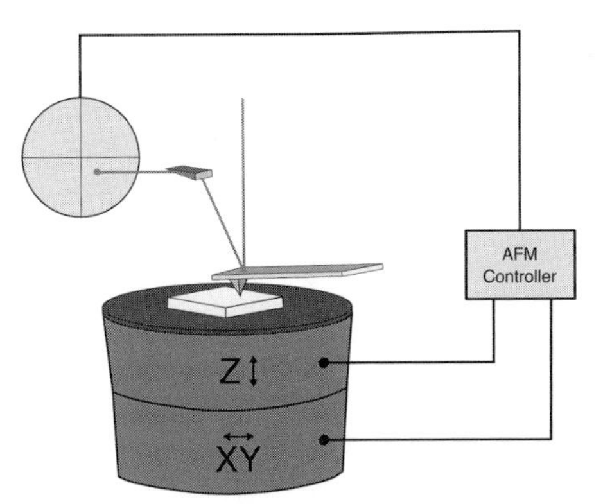

Figure 5.4 Laser beam deflection method. The tip–sample interaction is monitored by tracking the position of a cantilever reflected laser spot on a quadrant photodiode. The AFM feedback loop operates on the signal detected by the photodiode.

cantilever as the tip scans the surface. The laser beam is directed into a photodiode that converts the light intensity into a voltage to be used by the microscope's electronic feedback loop. The photodiode is usually divided into quadrants, which enables the simultaneous tracking of the total deflection as well as its vertical and lateral components.

5.3 AFM Imaging Modes

The most basic AFM imaging modes are *static*, meaning that the deflection of the cantilever arises only from the tip–sample interaction rather than from an external excitation source. Conversely, in *dynamic* imaging modes the cantilever is made to oscillate, usually at or near its resonant frequency.

5.3.1 Static Modes

Imaging with a static probe is done in either *contact mode* or *lift (noncontact) mode*. In contact mode the tip engages and scans the surface at a user-specified value of the total cantilever deflection. Images are generated from the instantaneous total deflection value as well as its vertical and lateral components. The vertical deflection signal contains topographic information and is used to construct a "height" image. The lateral deflection signal arises from the torsional displacement of the cantilever and is sensitive to topography as well as the frictional forces between the tip and the sample (Figure 5.5). Due to the dual sensitivity of the lateral force signal, the influence of the topography must be removed to obtain true "friction" contrast; typically, relatively flat samples are used for quantitative friction imaging.

In lift mode, the tip scans above the sample surface (i.e., in the attractive regime) without making contact, and the deflection of the cantilever is driven by variations in the noncontact forces acting between the tip and the sample. At small separation distances the interaction force is dominated by van der Waals (VDW) interactions, whose strength correlates with the underlying surface topography. However, as the separation distance increases, the contribution of VDW forces decreases and the cantilever deflection becomes more sensitive

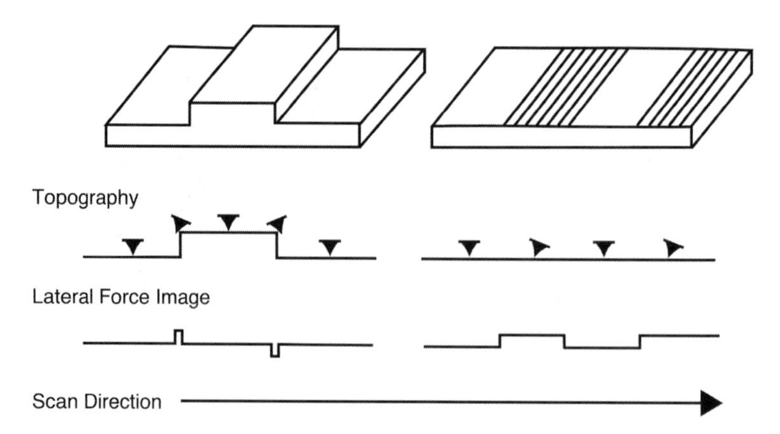

Figure 5.5 Torsional response of the cantilever as it scans from left to right over block models of a rough surface (left) and a flat but chemically heterogeneous surface (right). The lateral force image highlights how topography and friction can contribute to the lateral force signal.

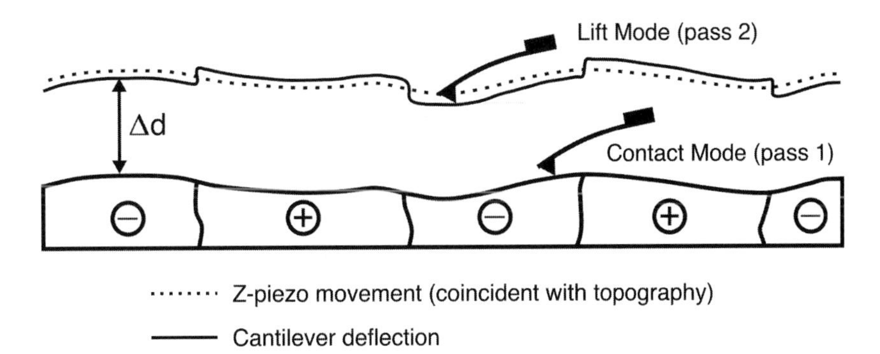

Figure 5.6 Principle of the lift mode. In a first scan line, the topography is measured (contact mode). In a second scan line, the topography is retraced with an offset Δd (dashed line). The deflection due to the long-range magnetic interaction is measured relative to this retraced height (solid line). Adapted from *Scanning Probe Microscopy.* NanoScience and Technology, Ch. 13: Static Atomic Force Microscopy, 2015, page 181, Bert Voigtländer, © Springer-Verlag Berlin Heidelberg 2015. With permission of Springer.

to longer-range forces. However, this transition is gradual and varies from sample to sample. Therefore, great care must be taken to remove the influence of surface topography when mapping long-range forces using noncontact mode. In electrostatic force microscopy (EFM) and magnetic force microscopy (MFM), a two-pass technique is generally used to ensure that the surface topography does not contaminate the long-range force signal (Girard, 2001; Passeri et al., 2014) (Figure 5.6). In the first pass, noncontact imaging is performed near the surface in order to obtain the sample topography or alternatively, the surface where the VDW forces between the tip and the sample are constant. The second pass occurs at a greater tip–sample distance but follows the topographic profile obtained from the first pass. In this way, the influence of the underlying topography on the long-range forces is essentially subtracted from the image obtained during the second pass. Note that in most commercial instruments EFM and MFM can also be performed in dynamic mode with an oscillating cantilever (Abelmann et al., 2005).

5.3.2 Dynamic Modes

To this point, we have only considered imaging modes in which the deflection of the cantilever occurs passively in response to the attractive or repulsive forces exerted upon it by the sample. However, many advanced applications use a dynamically actuated probe that senses the sample as it oscillates at high frequency (Figure 5.7A). This type of tip–sample interaction is far gentler than contact mode imaging, because lateral interaction forces are eliminated, making dynamic modes the standard choice for soft biological samples and high-resolution imaging in liquid.

The two most common dynamic imaging modes, amplitude modulation (Hölscher and Schwarz, 2007) (AM-AFM) and frequency modulation (Jarvis et al., 2008) (FM-AFM), are named for the quantity (i.e., amplitude or frequency) that the AFM feedback loop operates on (AM-AFM is commonly known as *tapping mode*). In these modes, the probe oscillates at or

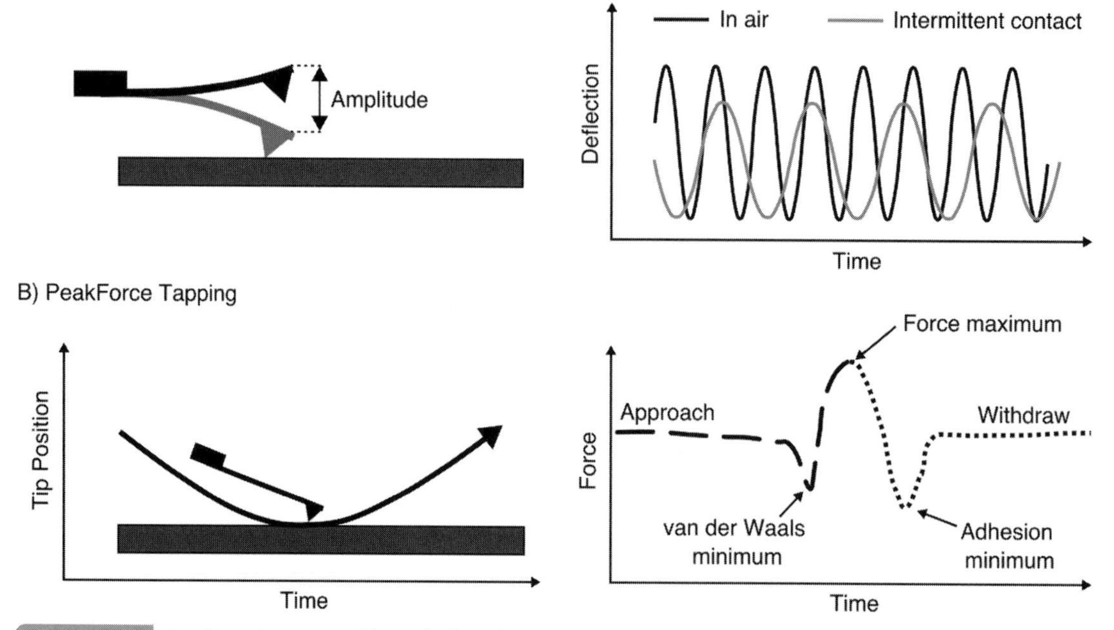

Figure 5.7 Cantilever dynamics in (A) standard amplitude modulation tapping (after Moreno-Herrero and Gomez-Herrero, 2012) and (B) PeakForce Tapping®. In AM-AFM, the cantilever oscillates at its resonant frequency and the amplitude and oscillation frequency decrease in contact with the sample. In PeakForce Tapping® mode, the cantilever is driven far below its resonant frequency in a trajectory akin to that shown in the lower left panel. The AFM feedback operates on the maximum tip–sample force. Panel B adapted from *Procedia Engineering*, 36, Chu, J.-Y., Hsu, W.-S., Liu, W.-R., Lin, H.-M., Cheng, H.-M., and Lin, L.-J. (2012) A novel inspection for deformation phenomenon of reduced-graphene oxide via quantitative nano-mechanical atomic force microscopy, pages 571–577, Copyright (2012), with permission from Elsevier.

near its resonant frequency in the attractive regime, and surface contact is intermittent. In AM-AFM, the amplitude of the cantilever oscillation is obtained far above the sample surface (this is called the free amplitude). As the tip–sample distance decreases, the tip and the sample begin to interact, inducing the amplitude to decrease by a factor that is roughly proportional to the tip–sample interaction force. Imaging is performed at a user-specified value of the amplitude shift (i.e., the difference between the free and engaged amplitudes). In FM-AFM, the cantilever amplitude is held constant and imaging is performed at a user-prescribed shift in the resonant frequency (i.e., the difference between the free and engaged resonances).

Bruker has recently introduced a new dynamic mode into its multimode instrument line that operates at much lower frequencies than the cantilever resonance (Schillers et al., 2016). In this proprietary mode (PeakForce Tapping® mode), the cantilever dynamics differ substantially from those employed in traditional on-resonance tapping modes (Figure 5.7B). As the cantilever is driven towards the surface, it enters a local force minimum defined by the balance of attractive VDW forces and short-range repulsive forces between the tip and the sample. Following contact, the interaction force first passes through

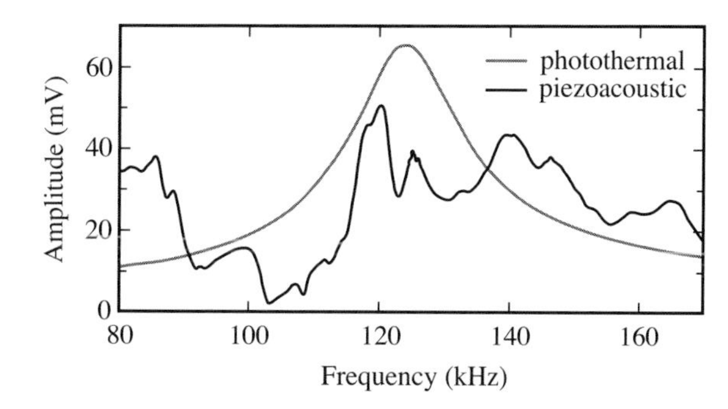

Figure 5.8 Comparison of the piezoacoustic and photothermal cantilever excitations methods in water. Reprinted from Kocun, M., Labuda, A., Gannepalli, A., and Proksch, R. (2015) Contact resonance atomic force microscopy imaging in air and water using photothermal excitation, Rev. Sci. Instrum. 86, 083706, with the permission of AIP Publishing.

a force maximum (or peak). As the cantilever drive transitions from approach to withdraw, the tip–sample interaction force enters another local minimum because the adhesive forces between the tip and the sample become dominant. In PeakForce Tapping® mode, the AFM feedback loop acts to hold the maximum tip–sample interaction force constant.

In most commercial instruments, the oscillation frequency is controlled by a piezoacoustic device that is mechanically coupled to the tip holder. This excitation technique is advantageous because it is inexpensive and can act upon any type of cantilever, regardless of its construction or composition. However, the piezoacoustic method does have one significant drawback, in that it is particularly problematic for fluid-phase imaging. Because the tip and the tip holder are mechanically coupled, the resonance spectrum is a convolution of signals from those sources and typically exhibits many resonance peaks (Labuda et al., 2011) (Figure 5.8). Even so, the task of finding the peak corresponding to the native response of the cantilever is not particularly challenging in air. However, when imaging is performed in liquid, the liquid becomes part of the mechanically coupled system, and the resonance spectrum becomes even more convoluted (Rogers et al., 2002). Several alternative cantilever excitation methods have been developed to specifically obtain the intrinsic resonance response of the probe in mechanical isolation. Several of these methods require specially made probes. Both magnetic (Florin et al., 1994) and electrostatic (Umeda et al., 2010) fields have been used to drive cantilever oscillation, but these methods can only operate on probe materials that are sensitive to those fields. As such, field-driven excitation methods place significant restrictions on the types of materials from which AFM probes can be manufactured. Cantilevers may also be coated with a thin piezoelectric film that can be specifically excited by electrical contact (Rogers et al., 2004); this method is quite effective but challenged by the difficulty of establishing proper electrical contact in fluid. In yet another method, the cantilever is replaced by a piezoelectric tuning fork (typically made of quartz) with a probe affixed to one prong (Edwards et al., 1997). In this setup, a voltage is applied to the fork to force the tip to oscillate in mechanical isolation, and the dampening of that signal is used to monitor the tip–sample distance during scanning. Due to the standard use of the laser beam deflection technique by commercial AFM manufacturers, tuning fork sensors of this type are not yet widely utilized and are only available from small suppliers. Within the past few years, a promising new technique has emerged, in which a secondary laser is used to drive the cantilever to oscillate; this photothermal excitation method (Labuda et al., 2014) is now available as an option for some commercial microscopes.

5.4 Practical Aspects of AFM Setup

As evident from the preceding discussion, AFM is distinctly different from traditional microscopy in a number of aspects. While the nature of a sample may influence the way it is prepared for optical or electron microscopy (scanning electron microscopy [SEM], Chapter 6, or transmission electron microscopy [TEM], Chapter 7), fundamentally, the investigative probe remains the same (i.e., electromagnetic radiation). However, how one sets up and operates an AFM depends considerably on the nature of the sample (e.g., hard vs. soft, hydroscopic vs. hydrophobic), the working environment (air vs. liquid), and the type of information desired (imaging or force measurement). Moreover, the physical characteristics of each probe that emerge from the microfabrication process are somewhat variable (even within the same production lot). Due to this, the behavior of an AFM can be variable from use to use. Here, we discuss the most basic procedures and considerations necessary for the standard use of an AFM (i.e., contact mode and dynamic mode imaging in air and liquid).

5.4.1 Probe Selection and Preparation

AFM probes are very often sold in relatively small numbers (5–20) and are shipped affixed to the surface of a silicone gel adhesive. When a probe is selected for use, some of the adhesive molecules can transfer to the portion of the probe that was in contact with the gel. Because the cantilever and tip do not contact the gel, this is less of a concern for operation in air. However, in solution, the bulk of the probe is submerged in fluid, and any adsorbed adhesive molecules can be unwittingly introduced into the sample cell. Therefore, as a general precaution, it is wise to remove this potential source of contamination by treating probes before use. This is usually done by briefly treating the probe in a plasma or UV/ozone cleaner. If this type of equipment is not available, probes can also be treated with highly oxidizing solutions (e.g., mixtures of sulfuric acid and hydrogen peroxide, called piranha solution) to degrade any adsorbed organic material.

Two types of cantilever geometries are generally used for imaging applications: rectangular and V-shaped. Both can be used for contact mode in air and liquid and for dynamic mode imaging in fluid. The rectangular geometry is used exclusively for dynamic imaging in air because these cantilevers are stiff enough to resist capillary forces that can cause the tip to snap into contact with the surface undesirably. Rectangular cantilevers are also preferred for friction imaging because they are considerably more sensitive to lateral deflection than V-shaped cantilevers. V-shaped cantilevers typically exhibit relatively low

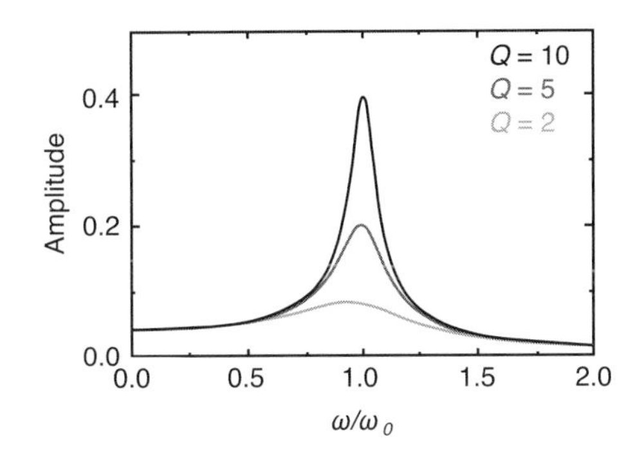

Figure 5.9 Amplitude of an oscillating cantilever, modeled as a simple harmonic oscillator with various values of the quality factor Q. Reproduced from Moreno-Herrero and Gomez-Herrero (2012) with permission. © 2012 Wiley-VCH Verlag & Co. KGaA.

force constants (<1 N/m) and resonant frequencies <50 kHz, whereas rectangular cantilevers span a much wider range (tens of N/m and tens to thousands of kHz).

For dynamic imaging, users must also be aware of the quality factor (Q factor), which is a measure of the dynamic response of the cantilever around its resonant frequency (Figure 5.9). Formally, the Q factor is the ratio of the energy stored in the oscillating cantilever to the amount of energy dissipated per oscillation. This means that the resonance response of a cantilever with a high Q factor is more sustained, or undamped, than that of a low-Q cantilever. The oscillation of a cantilever with a high Q factor is more stable and leads to better performance in dynamic mode imaging. However, when a cantilever is transferred from air to liquid, the resonant frequency shifts downward in response to the viscosity of the surrounding medium and the Q factor decreases substantially. Therefore, when using dynamic imaging modes in liquid, cantilevers with high Q factors must be selected to ensure that the probe resonates sufficiently in the liquid environment (Moreno-Herrero and Gomez-Herrero, 2012).

5.4.2 Laser Alignment

For some instruments, focusing the laser on the topside of the cantilever is now handled by the microscope's control software. However, for the vast majority of instruments currently on the market and in use in laboratories, laser alignment is a manual operation. A typical AFM setup features a low-power optical system that contains the tip, laser, and sample in its path. Provided that the laser spot is readily apparent in the optical scope's field of view, the x–y position of the tip and/or laser can be easily adjusted so that the laser reflects off the tip and onto the photodiode. Once the laser spot is focused on the detector, its position on the cantilever is usually fine-tuned to obtain the greatest response from the photodiode (this signal is called the "sum" signal). On some instruments, the laser beam reflects off the cantilever onto an adjustable mirror before reaching the detector; in this case, the position of the mirror must also be adjusted to maximize the sum signal. Once the sum signal is optimized, the laser alignment is completed by adjusting the position of the photodiode so that the laser spot rests near the common corner of the four sectors (i.e., at the center).

The initial placement of the laser spot on the cantilever is, unfortunately, not always trivial. Locating the laser can be difficult if it is initially focused at a significant distance from the tip, or when its current alignment allows the laser to interact with opaque components of the AFM tip holder. Furthermore, depending on the nature of the sample, the spot may sometimes appear rather diffuse, which can obscure its true position. In general, finding a misplaced laser spot is a bit easier if the illumination of the optical scope is turned down. However, skilled operators often use alternative strategies to place the laser spot on the cantilever that do not rely on the built-in optical system. Many users employ the “paper” method to ensure that the laser is focused on the cantilever before attempting to adjust the mirror. In this approach, the laser spot is initially placed on the chip portion of the probe, which has a much larger footprint than the cantilever that extends from its edge. Thereafter, the user moves the laser spot until it just barely disappears off the edge on the same side of the chip from which the cantilever protrudes. By simultaneously inserting a piece of white paper between the probe and the mirror and moving the laser laterally along the front end of the chip, the user can find the cantilever by watching for the appearance of the laser spot on the paper. This author prefers a variation of this strategy when using multimode instruments. In these microscopes, the AFM head, which contains the tip holder, laser, and mirror, can be detached from the scanner it rests upon without disconnecting the laser’s power source. When the head is held just a few centimeters over a table surface in a dim room, the shadow of the chip and cantilever can be easily recognized as the laser position is adjusted. The edges of the sample chamber are also apparent and can be used to rapidly locate the position of the laser relative to the tip.

5.4.3 Initial Approach to the Sample Surface

Modern AFMs have very precise motors that control the distance between the tip and the sample and automatically engage the surface at the interaction setpoint. However, these motors often step very slowly to avoid crashing the tip, and therefore, the operator usually needs to bring the tip into near contact with the surface manually before relying on the automatic engage motors. In the usual procedure, the operator focuses the optical microscope on the substrate surface with the shadow of the overlying cantilever in the field of view. The operator then lowers the probe by manually advancing the motor or in some cases by adjusting thumbscrews. As the cantilever descends, it gradually comes into focus on the sample. To ensure that the tip does not come into unwanted contact with the surface, it is important to stop advancing the motor before the cantilever (or its shadow) comes into full focus. When attempting to image the surface of a transparent material such as a crystal, additional care must be taken to ensure that optical scope is indeed focused on the surface rather the bottom of the sample or a planar feature within the material. This can be avoided by focusing the optical scope with the edge of the sample in view. Also, when operating in air, flexible cantilevers (i.e., those with a low force constant) can snap into contact with the surface due to capillary forces between the surface and tip-adsorbed water. Therefore, when imaging in dynamic modes in air, a relatively stiff cantilever should be used. In liquid, capillary forces are eliminated and less stiff cantilevers may be used for dynamic mode imaging (Malotky and Chaudhury, 2001).

5.4.4 Sample Preparation for the Liquid Environment

For AFM analysis, samples are usually mounted on a magnetic puck that adheres to the sample stage. In air, the method of affixing the sample to the puck is relatively unimportant, and a variety of adhesive materials may be used. However, because many adhesives can be readily solubilized in liquid, samples must be handled somewhat differently, depending on the nature of the fluid cell. Some instruments use an o-ring to contain the fluid, while others rely on the fluid surface tension to maintain the droplet shape and size. For cells that seal with an o-ring, the sample must be either large enough for the o-ring to sit atop it or small enough to fit inside the ring. If the o-ring rests on the sample, the choice of adhesive is arbitrary, because the interface between the sample and the puck will never come into direct contact with the fluid; however, this method is only feasible if the sample is flat enough for the o-ring to form an effective seal. If the sample is small enough to fit inside the o-ring, the magnetic puck is often covered with a glass coverslip that serves as a flat substrate for the o-ring to sit on. The sample is affixed to the glass coverslip using an adhesive. Because the adhesive will be exposed to the fluid, it should be chosen to be as resistant to the chemical environment as possible. For aqueous solutions, this author prefers to use specialized mounting waxes designed for electron microscopy applications that are resistant to acidic etching solutions (e.g., MWM100 Black Wax from Electron Microscopy Sciences: www.emsdiasum.com/microscopy/products/materials/adhesives.aspx).

5.5 AFM Imaging Artifacts

The AFM's sensitivity to exceptionally small variations in the tip–sample interaction is the root of its power; however, because these interactions are extremely delicate, they are prone to various types of interruption that can result in imaging artifacts. Various artifacts associated with the scanner and feedback loop can be abated by modifying various control settings. However, some issues arising from the tip itself may only be remedied by replacement. Here, we discuss the physical origins and manifestations of only the most common imaging artifacts; many additional artifacts have been identified, and users must always exercise caution when interpreting AFM images.

5.5.1 Tip Effects

The shape of an AFM tip (i.e., conical, pyramidal, etc.) can have a significant influence on the fidelity of the image through a mechanism known as "tip convolution." When a tip encounters a positive topographic feature on a surface, it is the tip sidewall rather than the apex that makes first contact with the feature. This interaction results in an apparent broadening of the feature in the lateral dimension (Allen et al., 1992; Shen et al., 2017).

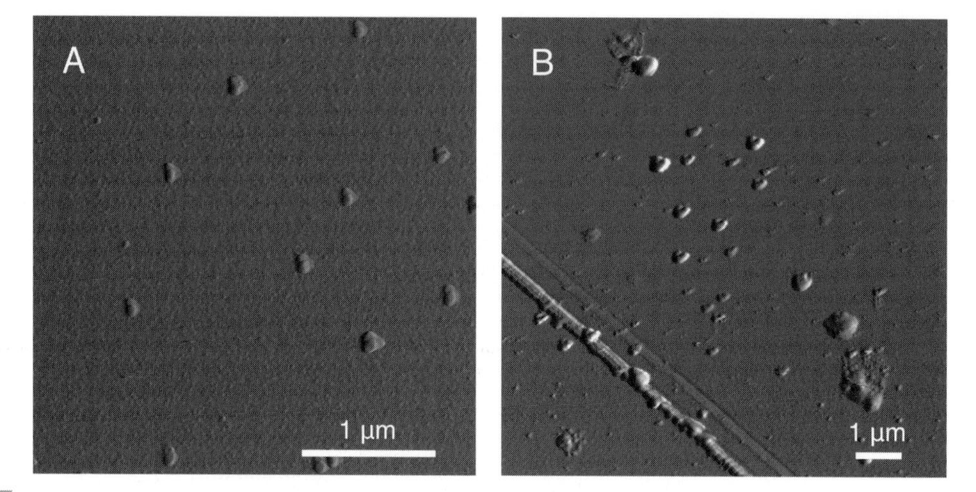

Figure 5.10 AFM contact mode deflection images showing tip convolution (A) and double tip (B) artifacts. In (A), all the particles exhibit the same general shape and orientation. Moreover, each particle has two discernible layers, with the second layer always in the southwest. The tip convolution persists in (B); however, a slight doubling effect is evident, particularly among the smaller features.

This effect can be pronounced if the feature is tall, or if the lateral dimensions of the feature are comparable to or smaller than the tip's radius of curvature. For instance, when spherical nanoparticles adsorbed to a surface are imaged, the apparent diameter of the particles must be considered an upper limit on the size, because the tip convolution effect always makes positive topographic features look larger. Conversely, tip convolution effects make negative topographic features, such as pits, appear smaller than they actually are. Returning to the nanoparticle example, if the probe shape is perfectly conical, the extent of the broadening would be the same in all directions regardless of the scan orientation. However, if a pyramidal probe is used, round particles might appear elongated, or even faceted in extreme cases. With a pyramidal tip, the extent of the broadening effect depends on the tip's orientation, and therefore, it is possible to determine the severity of the tip convolution effect by performing scans at multiple angles relative to the sample orientation. When nanoparticles are imaged, tip convolution often manifests as the particles all exhibiting the same general shape and orientation (Figure 5.10A). Notably, tip convolution effects only influence the fidelity of lateral features; the vertical dimension is unaffected.

AFM tips can also become chipped, blunted, or contaminated during imaging. A chipped or contaminated tip often exhibits multiple protrusions near the tip apex that may interact with the sample during scanning. When this occurs, a "double image" can result (Figure 5.10B). Tips may also become blunt with prolonged use. This increases the effective radius of curvature at the tip apex and apparent broadening of surface features. While a double image is a very obvious artifact, artifacts from contamination and blunt tips may not be. Therefore, when working with a new or unfamiliar surface, it is useful to first image a sample with known characteristics.

5.5.2 Scanner Effects

AFM relies upon piezoelectric materials that can rapidly displace the sample or tip with a high degree of accuracy. However, the performance of this hardware is only as reliable as the quality of its calibration. Therefore, scanners should be routinely checked with calibration standards and serviced when necessary. Piezoelectric devices expand and contract in response to an applied voltage; however, because this response is nonlinear, scanner movement is often somewhat parabolic. This typically introduces some degree of curvature in the background of an image that has to be compensated for. Additionally, there is contribution to the image background that arises from sample tilt (Canale et al., 2011). By default, most software packages automatically correct for curvature and tilt in real time as images are collected. However, the data are always saved in unprocessed form, and any offline image processing requires the user to apply some type of background subtraction or image-flattening procedure. Image processing can introduce artifacts and should be applied judiciously.

Piezoelectric scanners also respond differently when driven in opposing directions along the same path (Mokaberi and Requicha, 2008). This hysteresis can result in the appearance of a distinct set of parallel lines, or a distorted region along one of the image's edges (e.g., see Figure 2 in Bracco et al., 2014). If the hysteresis occurs in the z-direction, the edges of features in the image appear abnormally intense due to the artificial distortion of the true height differences between adjacent topographic features. The effects of scanner hysteresis can usually be minimized or eliminated by reducing the scan speed.

The piezo response time is also nonlinear. When the piezo control hardware (the actuator) acts to maintain the voltage applied to the scanner, the majority of the total response occurs almost instantly. However, to a very slight degree, the piezo displacement continues for some time after the initial rapid response. This results in a phenomenon called "piezo creep," in which the image appears to stretch or compress (Habibullah et al., 2013) (Figure 5.11). Creep artifacts occur following large displacements associated with changing the scan area or location and dissipate with time. Therefore, one way to avoid creep artifacts is to wait until the distortion abates. The effects of scanner hysteresis and creep are greatly reduced by the use of a "closed" feedback loop, which uses a sensor to monitor the piezo's actual position and applies a real-time correction to counteract its nonlinear behavior.

Samples may also appear to shift or move between subsequent scans. These distortions are most often due to "thermal drift," which occurs because the microscope's components have different coefficients of thermal expansion, and their dimensions change at different rates as the instrument approaches thermal equilibrium (Rahe et al., 2010). Thermal drift can be minimized by controlling the temperature of the ambient environment and by allowing the AFM's electronic components to warm up sufficiently prior to use.

5.5.3 Feedback Settings

The AFM feedback loop is sensitive to a number of user-defined parameters, including the scan rate and size, set-point, and gain settings. The gain setting controls the coupling between the input signal to the photodiode and the output signal from the z-piezo scanner.

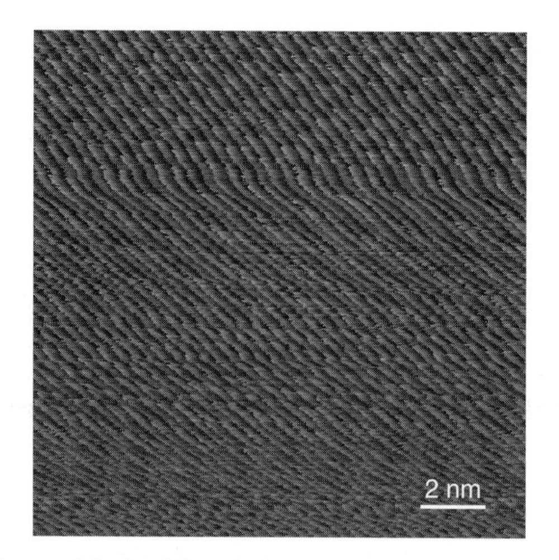

Figure 5.11 AFM contact mode height image of the (104) face of calcite. The irregular curvature of the features in the image is characteristic of piezo creep.

At higher gain values, the microscope's response to topographic changes on the sample is improved. Therefore, it is advantageous to set the gains (proportional and/or integral gain) as high as possible. If the gains are too low, the tip will not track the surface well, and its response to topographic features will be slow, resulting in the appearance of large streaks or shadows across the image (Figure 5.12A). However, if the gains are set too high, there is a certain amount of electronic noise related to oscillations in the feedback loop that can influence images. Interference related to the gain settings manifests in images as a grainy background texture (Figure 5.12C). The proper gain settings can be obtained easily by purposely increasing the gains until oscillations appear in the image and thereafter edging down the gains until the image noise dissipates (Figure 5.12B). This ensures that the feedback loop is as responsive as possible without introducing feedback-related artifacts into the image. Interference patterns can also appear in AFM images due to the scattering of laser light between the tip and the sample (Méndez-Vilas et al., 2002); these textures are not at all related to the feedback loop and can often be distinguished from electronic noise on the basis that the periodicity of the oscillations is insensitive to changes in imaging settings (i.e., scan rate or gain settings).

5.6 Nonimaging Applications

Although AFM is most widely employed as an imaging technique, it can also be used for a variety of nonimaging applications capable of yielding quantitative information about nanoscale material properties and interaction forces. For instance, the AFM can be set up as a nanoindentation device to measure local mechanical properties. While such methods are most often employed on hard materials, they are becoming increasingly utilized in cell and

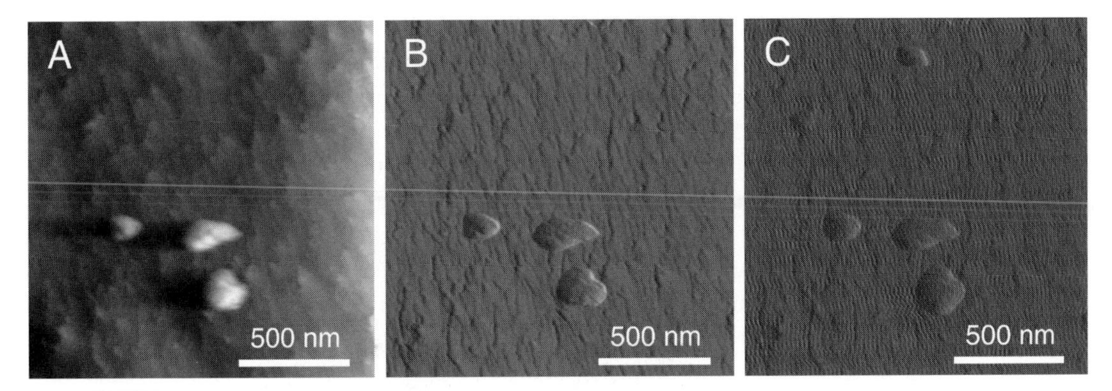

Figure 5.12 AFM contact mode deflection images of an etched calcite surface obtained for the same area at three different values of the integral gain. The sense of scanning is right to left (retrace) in all images. (A) Low gain. The tip responds to topographic changes sluggishly, leading to generally low resolution and large streaks in the wake of significant topographic features. (B) Sufficient gain. The feedback loop responds quickly and without generating electronic noise in the image. (C) High gain. The feedback loop responds quickly, but feedback oscillation features are apparent throughout the image.

tissue research, where variations in cell stiffness (elastic and viscoelastic response) can reflect changes in tissues related to cell differentiation (Engler et al., 2007) or in cell physiology, such as the onset of disease (Lekka, 2016). The AFM may also be used in force mode to quantify the interaction (or binding energy) between a substrate and a chemically modified AFM probe. Dynamic force measurements of this type are suitable for a wide array of geobiological applications, such as probing the adhesion of microbial cells and biofilm formation on minerals (Huang et al., 2015) and investigating the role of biopolymer matrices in directing the formation of biominerals (Hamm et al., 2014).

5.6.1 Cantilever Calibration

Whereas AFM imaging can be performed without knowing the properties of the cantilever with great precision, for quantitative force measurements the cantilever spring constant must be known accurately. Unfortunately, because force constants can vary considerably from their reported values, cantilevers must be calibrated prior to use. Calibration is greatly simplified by modern software tools; however, users who wish to make quantitative force measurements should become as familiar as possible with the primary literature.

One of the most widely utilized calibration techniques is the "thermal method" (Hutter and Bechhoefer, 1993; Lévy and Maaloum, 2002), in which it is assumed that the potential energy of the cantilever is equal to the ambient thermal energy ($k_B T$).

$$k = \frac{k_B T}{\langle d^2 \rangle} \qquad \text{(Eq.5.1)}$$

Here, k is the spring constant of the cantilever, k_B is Boltzmann's constant, T is temperature, and $\langle d^2 \rangle$ is the mean squared displacement of the cantilever. The mean squared displacement is obtained by monitoring the deflection signal of the cantilever over time. These data are then converted into the frequency domain (by Fourier transformation) to yield a power spectrum. Thereafter, the resonance peak is usually fitted to a Lorentzian function as in Eq. (5.3) and integrated to yield $\langle d^2 \rangle$.

However, the deflection in the preceding equation must be expressed in distance units rather than as a voltage. Therefore, before calibrating the cantilever, it is first necessary to determine how the vertical deflection signal, a voltage, varies with the z-position of the scanner (e.g., $\Delta V/\Delta z$). This is usually done by obtaining a force–distance curve (with the feedback loop disengaged) on a hard sample. In the hard contact region of the curve, the deflection signal varies linearly with the scanner position and the slope is equal to $\Delta V/\Delta z$. The inverse of this ratio is called the inverse optical lever sensitivity (InvOLS) and can be used to convert a measured deflection in volts to distance units (d = V ñ InvOLS). The InvOLS is sensitive to a number of factors, including the position of the laser on the cantilever. Therefore, if the laser spot is moved during the course of an experiment, the InvOLS must be reassessed. This can be very inconvenient if the sample under investigation is too soft. In addition, performing force measurements on hard substrates can damage the tip and prohibit further use. For these reasons, much work has been invested in developing approaches that do not require contact (Higgins et al., 2006).

The "Sader" method is a widely used method for obtaining the cantilever force constant without contact. Although it was first developed for V-shaped cantilevers (Sader, 1995; Sader et al., 1999), Sader's method has since been adapted for rectangular probes and extended to apply to cantilevers of arbitrary shape (Sader et al., 2012). The general Sader method relates the properties of the cantilever at resonance to the spring constant as follows:

$$k = \rho b^2 L \Lambda(\mathrm{Re}) \omega_R^2 Q \qquad \text{(Eq.5.2)}$$

where ρ is the density of the fluid environment (typically air), b and L are the cantilever width and length, Λ is a hydrodynamic function that depends on the shape of the cantilever and the Reynolds number (Re) of the fluid environment, ω_R^2 is the fundamental resonant frequency of the cantilever, and Q is the quality factor. The geometric dimensions of the cantilever are easily obtained by microscopy, and hydrodynamic functions have been tabulated for common cantilever types (e.g., rectangular and V-shaped). The resonant frequency and quality factor can be obtained by fitting the power spectrum of the cantilever response to a Lorentzian function as described earlier for the thermal method:

$$S(\omega) = P_{White} + \frac{P_{dc}\omega_R^4}{\left(\omega^2 - \omega_R^2\right)^2 - \frac{\omega^2 \omega_R^2}{Q^2}} \qquad \text{(Eq.5.3)}$$

Here, S(ω) is the power response function for a simple harmonic oscillator, P_{White} is a white noise baseline offset, and P_{dc} is the power response of the cantilever (obtained from the photodiode). Once this fit is complete, the force constant can be obtained directly from Eq. (5.2). Moreover, the InvOLS can be obtained analytically (Higgins et al., 2006):

$$\mathrm{InvOLS} = \sqrt{\frac{2k_B T}{\pi k \omega_R P_{dc} Q}} \qquad \text{(Eq.5.4)}$$

Using the noncontact method enables the cantilever to be rapidly recalibrated during the course of an experiment with minimal disruption. Lastly, note that the InvOLS values determined in contact and in noncontact are similar but distinct. The contact InvOLS describes the cantilever response under a static load, whereas the noncontact InvOLS describes the response of a free and dynamic cantilever. The two values differ by a constant value of ~9% ($InvOLS_{contact} = 1.09InvOLS_{noncontact}$).

5.6.2 Mechanical Properties (Reduced Elastic Modulus)

Once the cantilever has been calibrated, it is possible to obtain the actual interaction force between the tip and the sample from the vertical deflection signal. This information can be used to investigate the mechanical properties of the tip–sample system. The quantity that is most readily computed is the reduced elastic modulus E^*, which describes the combined deformation of the tip and sample under an applied force. E^* is distinct from the sample's elastic or Young's modulus E, which describes only the response of the sample to the applied force. Equation (5.5) defines the relationship between E^* and the Young's moduli and Poisson ratios ν_i of the tip and the sample.

$$\frac{1}{E^*} = \frac{1 - \nu_{samp}^2}{E_{samp}} + \frac{1 - \nu_{tip}^2}{E_{tip}} \qquad \text{(Eq.5.5)}$$

Therefore, for an unknown sample (i.e., one whose Poisson ratio is not known beforehand), the elastic modulus cannot be determined, whereas the reduced elastic modulus can.

The dependence of the interaction force on the tip–sample distance is usually modeled in the contact region with the Hertz model (Roa et al., 2011; Terán Arce et al., 2000). This model assumes that the interaction between a planar sample and a parabolic indenter (i.e., the tip) is negligible when the two are not in contact. The force F is described as follows:

$$F = \left(\frac{3}{4} E^* \sqrt{R_c}\right) h^{3/2} \qquad \text{(Eq.5.6)}$$

where R_c is the radius of curvature of the AFM tip and h is the depth of the tip's penetration into the sample. The reduced elastic modulus can be obtained by fitting this relationship to the contact region of the force–distance curve. This procedure is sensitive to the placement of the point on that curve that corresponds to the initial point of contact between the tip and the sample. For hard samples, the tip–sample contact point is well defined; however, it is typically less pronounced for soft samples (Figure 5.13). Software tools, such as the Matlab routine developed by Thomas et al. (2013), have been developed to make this analysis more routine and consistent.

Although there are many other models of the tip–sample interaction, the Hertz model has the advantage of being the simplest. However, to apply this or any other model, the radius of curvature of the tip must be assessed. Again, the manufacturer-specified value should not be used blindly. The shape of the tip can be assessed routinely using an AFM calibration grid with ultrasharp protrusions (Calabri et al., 2008) (e.g., grating TGT1 from NT-MDT Spectrum Instruments). An AFM height image taken on a calibration grid such as this

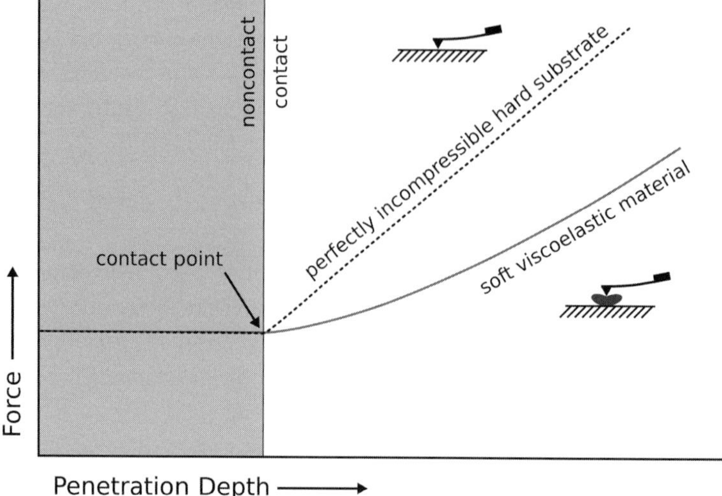

Figure 5.13 Schematic AFM force distance curves for indentation of a perfectly incompressible substrate and a soft viscoelastic material such as a single bacterium. The reduced modulus is obtained by fitting a model to such data. Due to its simplicity, the Hertz model is the most widely employed; however, more sophisticated models are being developed that provide superior fits for soft materials (Marsh and Waugh, 2013). Adapted from Fortier, H., Vanola, F., Wang, C., and Zou, S. (2016) AFM force indentation analysis on leukemia cells, Anal. Methods, 8, 4421–4431 – Published by the Royal Society of Chemistry. Content licensed under the Creative Commons Attribution 3.0 Unported License: https://creativecommons.org/licenses/by/3.0/.

displays features that reflect the actual shape of the AFM tip, and the radius of curvature can be obtained by analyzing topographic cross-sections.

5.6.3 Force Spectroscopy

Traditional spectroscopies investigate the response of a sample to some form of electromagnetic radiation. However, as the tip–sample interaction is entirely mechanical, the meaning of the term "force spectroscopy" (FS), which often appears in the AFM literature, is somewhat ambiguous. In fact, FS refers to any method in which AFM force curves are obtained as a function of some varied parameter. Due to this, FS has been used to probe a wide array of intermolecular forces, including VDW, electrostatic, solvation, hydrophobic, and steric forces. Very often, these measurements are performed with AFM tips that have been chemically modified to investigate specific interactions between a certain molecule and a substrate. In some studies, whole cells have been attached to AFM cantilevers. These types of measurements are of immense interest to geoscientists who wish to understand the nature of interactions between biomolecules and mineral surfaces.

The basic measurement in an FS experiment is the force–distance curve, which contains information about the adhesive forces acting between the tip and the sample. The work of adhesion (i.e., the adhesion energy) can also be obtained from the force–distance curve. Specifically, the work of adhesion can be obtained by integrating the area under the force–distance curve that is generated by pulling the tip off the surface (Friddle, 2014). If the tip–sample interaction is described as a bond that is coupled to a stiff spring (i.e., the AFM cantilever), the adhesion work is defined as

$$W = \frac{1}{2}\frac{f^2}{k} \quad \text{(Eq.5.7)}$$

where f is the bond rupture force and k is the cantilever spring constant. In this context, the rupture force is equivalent to the adhesion force as labeled in Figure 5.3. The rupture force that is observed in such an experiment depends upon the loading rate, that is, the rate at which force is applied to the spring to break its bond with the surface. Mathematically, the loading rate is the product of the cantilever spring constant and the tip retraction velocity.

The dependence of the bond rupture force on the loading rate forms the basis for dynamic force spectroscopy (DFS) techniques, which provide quantitative information about the kinetics and thermodynamics of tip–sample bonds. In a DFS experiment the tip is pulled off the surface over a range of loading rates. Many measurements are taken at each loading rate to fully characterize the distribution and obtain a statistically meaningful average value for the rupture force. At low loading rates, the dependence of the average rupture force on loading rate is relatively flat and asymptotically approaches the value for the bond–cantilever system at zero load. At high loading rates. the rupture force transitions to a kinetically controlled regime in which the rupture force increases proportionally to the natural log of the loading rate (Figure 5.14).

While there are a few models available to interpret DFS data (Bell, 1978; Dudko et al., 2006; Evans, 2001; Evans and Ritchie, 1997; Friddle et al., 2012), only the Friddle–De Yoreo model (Friddle et al., 2012) succeeds in describing the transition between the static and kinetically controlled regimes. In this model, the dependence of the average rupture force of a *single bond* versus loading rate is approximated as

$$\langle f \rangle \cong f_{eq} + f_\beta \ln\left(1 + \frac{re^{-\gamma}}{k_{off}(f_{eq})f_\beta}\right) \quad \text{(Eq.5.8)}$$

where f_{eq} is the unique force at which the unbinding and rebinding rates are equal, f_β is the thermal force scale, $k_{off}(f_{eq})$ is the unbinding rate, r is the loading rate, and $\gamma = 0.577$ is Euler's constant. A modified model is also available for the presence of multiple bonds. Fitting this function to a plot of the rupture force versus loading rate yields the equilibrium value of rupture force f_{eq}, which is related to the free energy difference between the bound and unbound states (i.e., the free energy of the unbound cantilever):

$$f_{eq} = \sqrt{2k\Delta G} \quad \text{(Eq.5.9)}$$

It is important to note that the free energy of the free cantilever is defined entirely by its force constant k. As such, the free energy difference between the bound and unbound states also depends upon the stiffness of the cantilever; see Eq. (5.9). The force constant is therefore a reference value or standard state for binding energies obtained by DFS.

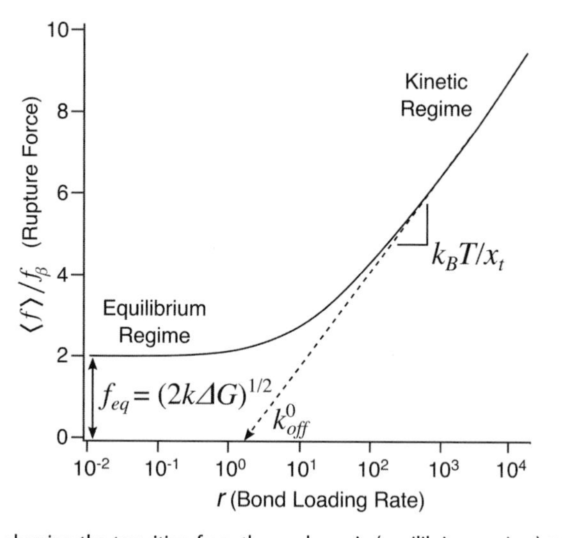

Figure 5.14 Idealized DFS spectrum showing the transition from thermodynamic (equilibrium regime) to kinetic control with increasing pulling rates. Adapted from Methods, 60, Noy, A. and Friddle, R., Practical single molecule force spectroscopy: How to determine fundamental thermodynamic parameters of intermolecular bonds with an atomic force microscope, pages 142–150, Copyright (2013), with permission from Elsevier.

5.7 Conclusions

The suite of techniques that fall under the umbrella of scanning probe microscopy is incredibly diverse and continues to grow. This chapter serves as an introduction to the core subset of those methods in order to provide a firm conceptual and practical foundation upon which uninitiated users can build their expertise. This point is critical, as developments in probe technology, advances in microscope hardware and software, and improved models of cantilever dynamics (both on and off resonance) continue to enable the development of new imaging modes and spatially resolved measurements. For instance, SPM is now being coupled with infrared (Dazzi and Prater, 2017) and Raman (Kumar et al., 2015) spectroscopy; fiber optic probes (Huckabay et al., 2013) are being used to break the far-field resolution limit of optical microscopy (scanning near-field optical microscopy [SNOM]), and high-speed AFM instruments have hit the commercial market that enable image acquisition at television frame rates.

In geochemical investigations, SPM has been used primarily as an imaging technique, and a large body of literature has evolved in which AFM is used to characterize mineral surfaces and measure growth and dissolution rates. Despite the great promise of these methods to provide insights into mineral–microbe interactions, a comparatively small body of this work has been concerned with the influence of biopolymers on mineral nucleation (Giuffre et al., 2013; Hamm et al., 2014; Tao et al., 2015; Wallace et al., 2009), growth/

dissolution (Elhadj et al., 2006; Friddle et al., 2010), and morphology (Chen et al., 2014; Orme et al., 2001). Therefore, there are still many opportunities to use SPM techniques to expand upon the current body of knowledge in this area.

5.8 References

Abelmann, L., van den Bos, A. and Lodder, C. (2005) 'Magnetic Force Microscopy – Towards Higher Resolution', in Hopster, H. and Oepen, H. P. (eds) *Magnetic Microscopy of Nanostructures*. Berlin, Heidelberg: Springer Berlin Heidelberg, pp. 253–283. doi:10.1007/3-540-26641-0_12.

Allen, M. J., Hud, N. V, Balooch, M., et al. (1992) 'Tip-radius-induced artifacts in AFM images of protamine-complexed DNA fibers', *Ultramicroscopy*, 42, pp. 1095–1100. doi:http://dx.doi.org/10.1016/0304-3991(92)90408-C.

Baró, A. M. and Reifenberger, R. G. (eds) (2012) *Atomic Force Microscopy in Liquid: Biological Applications*. Weinheim: Wiley-VCH Verlag GmbH & Co. KGaA.

Bell, G. I. (1978) 'Models for the Specific Adhesion of Cells to Cells', *Science*, 200(4342), pp. 618–627.

Bhushan, B. and Fuchs, H. (2008) *Applied Scanning Probe Methods XIII: Biomimetics and Industrial Applications*. Berlin, Heidelberg: Springer Berlin Heidelberg (NanoScience and Technology).

Binnig, G., Quate, C. F. and Gerber, C. (1986) 'Atomic force microscope', *Physical Review Letters*, 56(9), pp. 930–933.

Binnig, G., Rohrer, H., Gerber, C. and Weibel, E. (1982a) 'Surface studies by scanning tunneling microscopy', *Physical Review Letters*, 49(1), pp. 57–61.

Binnig, G., Rohrer, H., Gerber, C. and Weibel, E. (1982b) 'Tunneling through a controllable vacuum gap', *Applied Physics Letters*, 40(2), pp. 178–180. doi:10.1063/1.92999.

Bracco, J. N., Stack, A. G. and Higgins, S. R. (2014) 'Magnesite step growth rates as a function of the aqueous magnesium: carbonate ratio', *Crystal Growth & Design*, 14(11), pp. 6033–6040. doi:10.1021/cg501203g.

Busch, H. (1926) 'Berechnung der Bahn von Kathodenstrahlen im axialsymmetrischen elektromagnetischen Felde', *Annalen der Physik*, 386(25), pp. 974–993. doi:10.1002/andp.19263862507.

Calabri, L., Pugno, N., Menozzi, C. and Valeri, S. (2008) 'AFM nanoindentation: tip shape and tip radius of curvature effect on the hardness measurement', *Journal of Physics: Condensed Matter*, 20 (47), p. 474208.

Canale, C., Torre, B., Ricci, D. and Braga, P. C. (2011) 'Recognizing and Avoiding Artifacts in Atomic Force Microscopy Imaging', in Braga, P. C. and Ricci, D. (eds) *Atomic Force Microscopy in Biomedical Research: Methods and Protocols*. New York, Dordrecht, Heidelberg, London: Humana Press (Springer Science+Business Media), *Methods in Molecular Biology*, 736, pp. 31–43. doi:10.1007/978-1-61779-105-5_3.

Chen, C.-L., Qi, J., Tao, J., Zuckermann, R. N. and DeYoreo, J. J. (2014) 'Tuning calcite morphology and growth acceleration by a rational design of highly stable protein-mimetics', *Scientific Reports*, 4, p. 6266.

Dazzi, A. and Prater, C. B. (2017) 'AFM-IR: technology and applications in nanoscale infrared spectroscopy and chemical imaging', *Chemical Reviews*, 117(7), pp. 5146–5173. doi:10.1021/acs.chemrev.6b00448.

Dudko, O. K., Hummer, G. and Szabo, A. (2006) 'Intrinsic rates and activation free energies from single-molecule pulling experiments', *Physical Review Letters*, 96(10), p. 108101.

Edwards, H., Taylor, L., Duncan, W. and Melmed, A. J. (1997) 'Fast, high-resolution atomic force microscopy using a quartz tuning fork as actuator and sensor', *Journal of Applied Physics*, 82(3), pp. 980–984. doi:10.1063/1.365936.

Elhadj, S., De Yoreo, J. J., Hoyer, J. R. and Dove, P. M. (2006) 'Role of molecular charge and hydrophilicity in regulating the kinetics of crystal growth', *Proceedings of the National Academy of Sciences*, 103(51), pp. 19237–19242. doi:10.1073/pnas.0605748103.

Engler, A. J., Rehfeldt, F., Sen, S. and Discher, D. E. (2007) 'Microtissue Elasticity: Measurements by Atomic Force Microscopy and Its Influence on Cell Differentiation', in Wang, Y.-L. and Discher, D.E. (eds) Cell Mechanics. London, San Diego, CA: Academic Press, *Methods in Cell Biology*, 83, pp. 521–545. doi:https://doi.org/10.1016/S0091-679X(07)83022-6.

Enoch, J. M. and Lakshminarayanan, V. (2000) 'Duplication of unique optical effects of ancient Egyptian lenses from the IV/V Dynasties: lenses fabricated ca 2620–2400 BC or roughly 4600 years ago', *Ophthalmic and Physiological Optics*, 20(2), pp. 126–130. doi:10.1046/j.1475-1313.2000.00496.x.

Evans, E. (2001) 'Probing the relation between force–lifetime–and chemistry in single molecular bonds', *Annual Review of Biophysics and Biomolecular Structure*, 30(1), pp. 105–128. doi:10.1146/annurev.biophys.30.1.105.

Evans, E. and Ritchie, K. (1997) 'Dynamic strength of molecular adhesion bonds', *Biophysical Journal*, 72(4), pp. 1541–1555.

Florin, E., Radmacher, M., Fleck, B. and Gaub, H. E. (1994) 'Atomic force microscope with magnetic force modulation', *Review of Scientific Instruments*, 65(3), pp. 639–643. doi:10.1063/1.1145130.

Fortier, H., Vanola, F., Wang, C., and Zou, S. (2016) 'AFM force indentation analysis on leukemia cells', *Analytical Methods*, 8, 4421–4431. doi:10.1039/C6AY00131A.

Friddle, R. (2014) 'Direct Measurement of Interaction Forces and Energies with Proximal Probes', in Gower, L.B. and DiMasi, E. (eds) *Biomineralization Sourcebook: Characterization of Biominerals and Biomimetic Materials*. Boca Raton, FL: CRC Press, pp. 307–318. doi:10.1201/b16621-24.

Friddle, R. W., Noy, A. and De Yoreo, J. J. (2012) 'Interpreting the widespread nonlinear force spectra of intermolecular bonds', *Proceedings of the National Academy of Sciences*, 109(34), pp. 13573–13578. doi:10.1073/pnas.1202946109.

Friddle, R. W., Weaver, M. L., Qiu, S. R., et al. (2010) 'Subnanometer atomic force microscopy of peptide–mineral interactions links clustering and competition to acceleration and catastrophe', *Proceedings of the National Academy of Sciences*, 107(1), pp. 11–15. doi:10.1073/pnas.0908205107.

Gest, H. (2004) 'The discovery of microorganisms by Robert Hooke and Antoni van Leeuwenhoek, Fellows of The Royal Society', *Notes and Records of the Royal Society of London*, 58(2), p. 187 LP-201.

Girard, P. (2001) 'Electrostatic force microscopy: principles and some applications to semiconductors', *Nanotechnology*, 12(4), p. 485.

Giuffre, A. J., Hamm, L. M., Han, N., De Yoreo, J. J. and Dove, P. M. (2013) 'Polysaccharide chemistry regulates kinetics of calcite nucleation through competition of interfacial energies', *Proceedings of the National Academy of Sciences*, 110(23), pp. 9261–9266. doi:10.1073/pnas.1222162110.

Habibullah, Pota, H. R., Petersen, I. R. and Rana, M. S. (2013) 'Creep, hysteresis, and cross-coupling reduction in the high-precision positioning of the piezoelectric scanner stage of an atomic force microscope', *IEEE Transactions on Nanotechnology*, 12(6), pp. 1125–1134. doi:10.1109/TNANO.2013.2280793.

Hamm, L. M., Giuffre, A. J., Han, N., et al. (2014) 'Reconciling disparate views of template-directed nucleation through measurement of calcite nucleation kinetics and binding energies', *Proceedings of the National Academy of Sciences*, 111(4), pp. 1304–1309. doi:10.1073/pnas.1312369111.

Higgins, M. J., Proksch, R., Sader, J. E., et al. (2006) 'Noninvasive determination of optical lever sensitivity in atomic force microscopy', *Review of Scientific Instruments*, 77(1), p. 13701. doi:10.1063/1.2162455.

Hölscher, H. and Schwarz, U. D. (2007) 'Theory of amplitude modulation atomic force microscopy with and without Q-Control', *International Journal of Non-Linear Mechanics*, 42(4), pp. 608–625. doi:http://dx.doi.org/10.1016/j.ijnonlinmec.2007.01.018.

Hooke, R. (1665) *Micrographia, or, Some physiological descriptions of minute bodies made by magnifying glasses: with observations and inquiries thereupon*. London: Royal Society of London.

Huang, Q., Wu, H., Cai, P., Fein, J. B. and Chen, W. (2015) 'Atomic force microscopy measurements of bacterial adhesion and biofilm formation onto clay-sized particles', *Scientific Reports*, 5, p. 16857.

Huckabay, H. A., Armendariz, K. P., Newhart, W. H., Wildgen, S. M. and Dunn, R. C. (2013) 'Near-field scanning optical microscopy for high-resolution membrane studies', *Methods in Molecular Biology*, 950, pp. 373–394. doi:10.1007/978-1-62703-137-0_21.

Hutter, J. L. and Bechhoefer, J. (1993) 'Calibration of atomic-force microscope tips', *Review of Scientific Instruments*, 64(7), pp. 1868–1873. doi:10.1063/1.1143970.

Jarvis, S. P., Sader, J. E. and Fukuma, T. (2008) 'Frequency Modulation Atomic Force Microscopy in Liquids', in Bhushan, B., Fuchs, H., and Tomitori, M. (eds) *Applied Scanning Probe Methods VIII: Scanning Probe Microscopy Techniques*. Berlin, Heidelberg: Springer Berlin Heidelberg, pp. 315–350. doi:10.1007/978-3-540-74080-3_9.

Knoll, M. and Ruska, E. (1932) 'Das Elektronenmikroskop', *Zeitschrift für Physik*, 78(5), pp. 318–339. doi:10.1007/BF01342199.

Kocun, M., Labuda, A., Gannepalli, A. and Proksch, R. (2015) 'Contact resonance atomic force microscopy imaging in air and water using photothermal excitation', *Review of Scientific Instruments*, 86(8), p. 83706. doi:10.1063/1.4928105.

Kumar, N., Mignuzzi, S., Su, W. and Roy, D. (2015) 'Tip-enhanced Raman spectroscopy: principles and applications', *EPJ Techniques and Instrumentation*, 2(1), p. 9. doi:10.1140/epjti/s40485-015-0019-5.

Labuda, A., Cleveland, J., Geisse, N. A., et al. (2014) 'Photothermal excitation for improved cantilever drive performance in tapping mode atomic force microscopy', *Microscopy and Analysis*, 28(3), pp. S21–S25.

Labuda, A., Kobayashi, K., Kiracofe, D., et al. (2011) 'Comparison of photothermal and piezoacoustic excitation methods for frequency and phase modulation atomic force microscopy in liquid environments', *AIP Advances*, 1(2), p. 22136. doi:10.1063/1.3601872.

Lekka, M. (2016) 'Discrimination between normal and cancerous cells using AFM', *Bionanoscience*, 6, pp. 65–80. doi:10.1007/s12668-016-0191-3.

Lévy, R. and Maaloum, M. (2002) 'Measuring the spring constant of atomic force microscope cantilevers: thermal fluctuations and other methods', *Nanotechnology*, 13(1), p. 33.

Malotky, D. L. and Chaudhury, M. K. (2001) 'Investigation of capillary forces using atomic force microscopy', *Langmuir*, 17(25), pp. 7823–7829. doi:10.1021/la0107796.

Marsh, G. and Waugh, R. E. (2013) 'Quantifying the mechanical properties of the endothelial glycocalyx with atomic force microscopy', *Journal of Visualized Experiments*, (72), p. e50163. doi:doi:10.3791/50163.

Méndez-Vilas, A., González-Martín, M. L. and Nuevo, M. J. (2002) 'Optical interference artifacts in contact atomic force microscopy images', *Ultramicroscopy*, 92(3), pp. 243–250. doi:http://dx.doi.org/10.1016/S0304-3991(02)00140-7.

Mohr, P. J., Newell, D. B. and Taylor, B. N. (2014) 'CODATA recommended values of the fundamental physical constants: 2014', *Journal of Physical and Chemical Reference Data*, 45, p. 043102. doi:10.1063/1.4954402.

Mokaberi, B. and Requicha, A. A. G. (2008) 'Compensation of scanner creep and hysteresis for AFM nanomanipulation', *IEEE Transactions on Automation Science and Engineering*, 5(2), pp. 197–206. doi:10.1109/TASE.2007.895008.

Moreno-Herrero, F. and Gomez-Herrero, J. (2012) 'AFM: Basic Concepts', in Baró, A.M. and Reifenberger, R.G. (eds) *Atomic Force Microscopy in Liquid: Biological Applications*. Weinheim: Wiley-VCH Verlag GmbH & Co. KGaA, pp. 1–34. doi:10.1002/9783527649808.ch1.

Morita, S. (ed.) (2006) *Roadmap of Scanning Probe Microscopy*. Berlin, Heidelberg: Springer Berlin Heidelberg (NanoScience and Technology).

Niaz, M. (2009) 'Wave–Particle Duality: De Broglie, Einstein, and Schrödinger', in Critical Appraisal of Physical Science as a Human Enterprise: Dynamics of Scientific Progress. Dordrecht: Springer Netherlands, pp. 159–165. doi:10.1007/978-1-4020-9626-6_12.

Noy, A., and Friddle, R. (2013) 'Practical single molecule force spectroscopy: How to determine fundamental thermodynamic parameters of intermolecular bonds with an atomic force microscope', *Methods*, 60, pp. 142–150. doi:https://doi.org/10.1016/j.ymeth.2013.03.014.

Orme, C. A., Noy, A., Wierzbicki, A., et al. (2001) 'Formation of chiral morphologies through selective binding of amino acids to calcite surface steps', *Nature*, 411(6839), pp. 775–779.

Passeri, D., Dong, C., Reggente, M., et al. (2014) 'Magnetic force microscopy', *Biomatter*, 4(1), p. e29507. doi:10.4161/biom.29507.

Plantzos, D. (1997) 'Crystals and lenses in the Graeco-Roman world', *American Journal of Archaeology*, 101(3), pp. 451–464. doi:10.2307/507106.

Porter, J. R. (1976) 'Antony van Leeuwenhoek: tercentenary of his discovery of bacteria', *Bacteriological Reviews*, 40(2), pp. 260–269.

Quercioli, F. (2011) 'Fundamentals of Optical Microscopy', in Diaspro, A. (ed.) *Optical Fluorescence Microscopy: From the Spectral to the Nano Dimension*. Berlin, Heidelberg: Springer Berlin Heidelberg, pp. 1–36. doi:10.1007/978-3-642-15175-0_1.

Rahe, P., Bechstein, R. and Kühnle, A. (2010) 'Vertical and lateral drift corrections of scanning probe microscopy images', *Journal of Vacuum Science & Technology B, Nanotechnology and Microelectronics: Materials, Processing, Measurement, and Phenomena*, 28(3), pp. C4E31–C4E38. doi:10.1116/1.3360909.

Roa, J. J., Oncins, G., Diaz, J., Sanz, F. and Segarra, M. (2011) 'Calculation of Young's Modulus value by means of AFM', *Recent Patents on Nanotechnology*, pp. 27–36. doi:http://dx.doi.org/10.2174/187221011794474985.

Rogers, B., Manning, L., Sulchek, T. and Adams, J. D. (2004) 'Improving tapping mode atomic force microscopy with piezoelectric cantilevers', *Ultramicroscopy*, 100(3), pp. 267–276. doi:http://dx.doi.org/10.1016/j.ultramic.2004.01.016.

Rogers, B., York, D., Whisman, N., et al. (2002) 'Tapping mode atomic force microscopy in liquid with an insulated piezoelectric microactuator', *Review of Scientific Instruments*, 73(9), pp. 3242–3244. doi:10.1063/1.1499532.

Sader, J. E. (1995) 'Parallel beam approximation for V-shaped atomic force microscope cantilevers', *Review of Scientific Instruments*, 66(9), pp. 4583–4587. doi:10.1063/1.1145292.

Sader, J. E., Chon, J. W. M. and Mulvaney, P. (1999) 'Calibration of rectangular atomic force microscope cantilevers', *Review of Scientific Instruments*, 70(10), pp. 3967–3969. doi:10.1063/1.1150021.

Sader, J. E., Sanelli, J. A., Adamson, B. D., et al. (2012) 'Spring constant calibration of atomic force microscope cantilevers of arbitrary shape', *Review of Scientific Instruments*, 83(10), p. 103705. doi:10.1063/1.4757398.

Schillers, H., Medalsy, I., Hu, S., Slade, A. L. and Shaw, J. E. (2016) 'PeakForce Tapping resolves individual microvilli on living cells', *Journal of Molecular Recognition*, 29(2), pp. 95–101. doi:10.1002/jmr.2510.

Shen, J., Zhang, D., Zhang, F.-H. and Gan, Y. (2017) 'AFM tip-sample convolution effects for cylinder protrusions', *Applied Surface Science*, 422, pp. 482–491. doi:https://doi.org/10.1016/j.apsusc.2017.06.053.

Tao, J., Battle, K. C., Pan, H., et al. (2015) 'Energetic basis for the molecular-scale organization of bone', *Proceedings of the National Academy of Sciences*, 112(2), pp. 326–331. doi:10.1073/pnas.1404481112.

Terán Arce, P. F. M., Riera, G. A., Gorostiza, P. and Sanz, F. (2000) 'Atomic-layer expulsion in nanoindentations on an ionic single crystal', *Applied Physics Letters*, 77(6), pp. 839–841. doi:10.1063/1.1306909.

Thomas, G., Burnham, N. A., Camesano, T. A. and Wen, Q. (2013) 'Measuring the mechanical properties of living cells using atomic force microscopy', *Journal of Visualized Experiments*, (76), p. 50497. doi:10.3791/50497.

Thomson, J. J. (1897) 'XL. Cathode rays', *Philosophical Magazine Series 5*, 44(269), pp. 293–316. doi:10.1080/14786449708621070.

Umeda, K-i., Oyabu, N., Kobayashi, K., et al. (2010) 'High-resolution frequency-modulation atomic force microscopy in liquids using electrostatic excitation method', *Applied Physics Express*, 3(6), p. 65205.

Voigtländer, B. (2015) 'Static Atomic Force Microscopy', in *Scanning Probe Microscopy: Atomic Force Microscopy and Scanning Tunneling Microscopy*. Berlin, Heidelberg: Springer Berlin Heidelberg, pp. 177–186. doi:10.1007/978-3-662-45240-0_13.

Wallace, A. F., DeYoreo, J. J. and Dove, P. M. (2009) 'Kinetics of silica nucleation on carboxyl- and amine-terminated surfaces: insights for biomineralization', *Journal of the American Chemical Society*, 131(14), pp. 5244–5250. doi:10.1021/ja809486b.

Zhu, Y., Inada, H., Nakamura, K. and Wall, J. (2009) 'Imaging single atoms using secondary electrons with an aberration-corrected electron microscope', *Nature Materials*, 8(10), pp. 808–812.

6 Applications of Scanning Electron Microscopy in Geomicrobiology

JEREMIAH SHUSTER, GORDON SOUTHAM, AND FRANK REITH

Abstract

From mineralized biofilms of ancient or "extreme" environments to the nth replicate of laboratory-based biofilm experiments, geomicrobiological samples containing microbes associated with primary minerals or secondary (biogenic) mineral precipitates are highly diverse. The foremost advantage of scanning electron microscopy for geomicrobiology is that it provides high-resolution micrographs of cells and biofilms in association with minerals. These micrographs provide visual evidence of biogeochemical processes, which helps to explain phenomena that occur in the natural environment or in the laboratory. In addition to high-resolution secondary electron or backscatter electron modes of imaging, scanning electron microscopes can be equipped with a range of microanalytical tools, thereby extending the breadth of analytical capacities. Analytical techniques such as energy dispersive spectroscopy and electron backscatter diffraction analysis characterize the chemical composition and crystallography of biofilms and (bio)minerals, respectively, down to the micrometer scale. In addition, a focused ion beam can be used for nanomachining samples to provide a view beneath the outer surface of a sample or can be used as a technique for preparing samples for transmission electron microscopy. To provide the reader with an overview of tools and techniques, this chapter will explain a number of widely used preparation techniques, including whole-mounts, petrographic thin sections, polished blocks, and focused ion beam milling. By using these techniques, various types of geomicrobiological materials will be examined and used to guide and develop the necessary skills for interpreting biogeochemical processes from structural and chemical information obtained through secondary electron and backscatter electron micrographs and associated microanalyses.

6.1 Introduction

Characterizing interactions between microbes and minerals in samples obtained from either natural environments or laboratory-based experiments involves the explanation of structure and chemistry in relation to function (Beveridge, 1981). This can be achieved through electron microscopy (EM) and associated microanalytical tools. When given the first opportunity to use EM as an analytical technique, a microscopist will often be asked, "What do you hope to see?" While this question sounds rhetorical, it helps to plan a strategy for obtaining meaningful data. Therefore, what one hopes to see depends strongly on the fundamental knowledge about a given sample in order to determine: i) what constituents are of most interest, ii) which preparation techniques are required/practicable, iii) what analytical condition(s) are most suitable for characterization, and iv) how these data will help to tell a biogeochemical story. Addressing these questions is important, because scanning electron microscopy (SEM; also indicating a scanning electron microscope) is generally considered a starting point for the structural characterization of geomicrobiological materials. Furthermore, data created using these techniques can help determine whether higher-resolution analyses, such as transmission electron microscopy (TEM) or synchrotron-based analyses, will be beneficial.

An SEM has the capacity to resolve 10s of nanometer-scale features at high magnification, providing excellent two- and three-dimensional structural characterization of geomicrobiological materials. In this chapter, a number of the basic SEM principles, as well as associated preparatory and analytical techniques, are presented.

An SEM contains an electron gun that generates electrons and operates under low or high vacuum. Electrons are accelerated with the help of electromagnetic lenses producing a focused electron beam that typically ranges in energy between 1 and 40 kV. The focused beam acts as a fine probe that scans the surface of a sample and rasterizes a digital image, i.e., a micrograph, of the sample (Figure 6.1). Micrographs are generated by excited electrons that are ejected from the surface of a sample and collected by a detector. The resolution of the resulting micrograph is dependent on the size of the electron beam and how it interacts with the sample materials. Unlike TEM, samples for SEM can be up to centimeter-size structures, as sample holders and SEM chambers can accommodate larger samples. An electron beam can penetrate several micrometers in depth from the sample surface, producing an interaction volume that can be pear- to hemisphere-shaped. The shape and size of the interaction volume are dependent on the energy of the beam as well as the elemental composition of the sample material that it is penetrating. For example, higher-energy electron beams produce larger interaction volumes, whereas sample

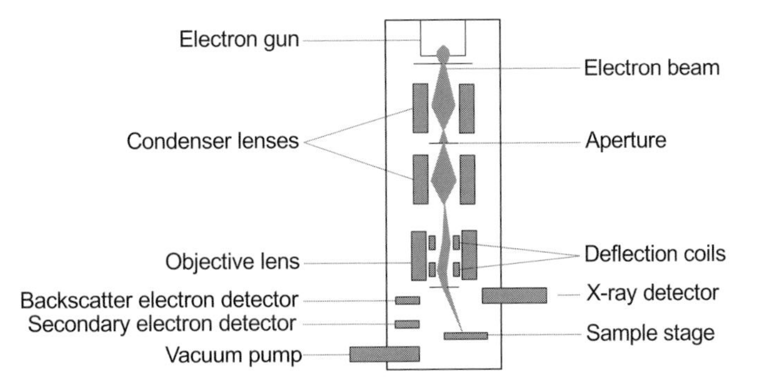

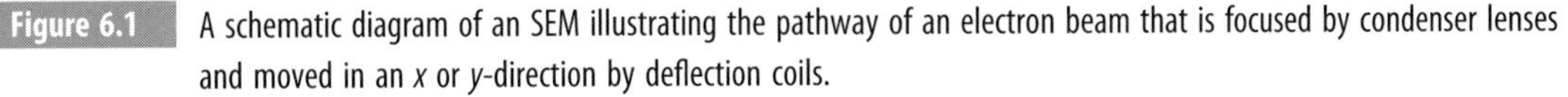
Figure 6.1 A schematic diagram of an SEM illustrating the pathway of an electron beam that is focused by condenser lenses and moved in an x or y-direction by deflection coils.

materials composed of higher–atomic mass elements will decrease interaction volumes. The interaction volume, and therefore, the energy of the electron beam, is important for developing secondary electron (SE) and backscatter electron (BSE) mode images. In general, 1 to 5 kV electron beams are best suited for SE imaging, since nanometer-scale surface textures can be resolved. Higher-energy electron beams, generally greater than 10 keV, are best suited for BSE mode imaging and obtaining microanalytical data, e.g., energy dispersive spectroscopy (EDS). It should be noted that there are no uniform analytical parameters, because different microscopes will have different optimal settings, and each sample is unique. Therefore, when using an SEM for the first time, it is recommended to try various parameters to determine which works best for a given sample.

In terms of sample preparation, a variety of advanced techniques, such as cryo-SEM and serial block-face SEM, are available for characterizing microbes in association with minerals (see Dohnalkova et al., 2011; Krember et al., 2015 and references therein; Walther and Müller, 1999). These techniques are more advanced and often require in-depth expertise. Therefore, this chapter will focus on "conventional" sample preparations that were originally designed for geological and biological samples and adapted to suit the analytical advantages of SEM. More importantly, this chapter will provide examples of structural and chemical characterization of various samples to demonstrate how interpretations of biogeochemical processes can be made.

6.2 Bacterial Cell and Biofilm Preservation

Geomicrobiological samples often contain both rigid structures, such as nanometer- to centimeter-scale minerals, and delicate structures, such as bacterial cells, biofilms, and extracellular polymeric substances (EPS). Because the composition of samples can vary, the preparation technique required for optimal characterization also needs to be varied. For EM, the preservation of cells via fixation is paramount for characterizing microbes in association with minerals. Fixation is necessary to prevent or limit alteration of or damage to cellular structures by further sample processing, e.g., ethanol/acetone dehydration, or irradiation by the electron beam within the microscope. Adding fixatives, such as glutaraldehyde ($C_5H_8O_2$), to samples preserves cells by crosslinking proteins, which results in the stabilization of cellular fine structure and makes cells osmotically unresponsive. Fixation helps retain the original morphology of bacteria so that after processing the cells appear as they occurred in their living state. Therefore, fixation is best performed immediately after collecting a sample in the field or from laboratory experiments to limit any structural alteration resulting from sample storage. Electron microscopy–grade glutaraldehyde, i.e., 50 or $70\%_{(aq)}$, is a high-purity solution that provides optimal fixation (Sabatini et al., 1962). It is important to note that glutaraldehyde is a carcinogenic chemical, and appropriate personal protective equipment should be worn at all times when working with this chemical. Furthermore, all sample processing involving chemicals should be performed in a fume hood. Although glutaraldehyde/formaldehyde mixtures are available (Karnovsky, 1965), a 2–2.5 wt.$\%_{(aq)}$ glutaraldehyde concentration is generally used for primary fixation, since greater concentrations can affect cell fine structures. For samples that are already in a solution, fixation can be achieved by adding EM-grade glutaraldehyde to obtain a final 2–2.5 wt.$\%_{(aq)}$ concentration. Conversely, samples can be placed in EM-grade glutaraldehyde that has been diluted with filter-sterilized (0.45 μm pore size) deionized water or a buffer such as cacodylate, phosphate buffered saline (PBS), or HEPES (2-[4-(2-hydroxyethyl) piperazineethane-sulfonic acid). These buffers can be used to maintain a circumneutral pH during fixation if necessary. While various concentrations and mixtures of buffers have been conventionally used for biological tissues, they should be used cautiously to avoid altering any mineral constituents within a geomicrobiological sample. See Rasmussen and Albrechtsen (1974) and Migneault et al. (2004) for details regarding glutaraldehyde reactivity in various aqueous solutions. The reaction rate of crosslinking proteins using glutaraldehyde is temperature dependent; therefore, fixation can occur within minutes at room temperature or within hours at refrigerator temperatures. Fixing samples in 2–2.5 wt.$\%_{(aq)}$ glutaraldehyde for 24 h at room temperature is a conventional method to ensure thorough fixation of bacterial cells, especially if the sample contains a high density of cells, e.g., biofilms. Depending on the microorganisms in the sample, the temperature during fixation should be similar to the growth conditions, so that drastic changes in temperature do not result in fine-structure damage. Since glutaraldehyde is an organic compound, it can be oxidized, and the rate of degradation increases with temperature. Once samples have been fixed, it is best to store them in sealed containers with little to no headspace and keep in a refrigerator until further processing. There is a wealth of scientific literature that describes various sample preparations for EM analysis; therefore, various methods can be adapted or modified to best suit a sample (see Hayat, 1959 and references therein). In the following examples, different types of geomicrobiological materials will be prepared using conventional techniques. The first example will focus on whole-mount sample preparation as a method to characterize three-dimensional structures of biofilms occurring on the surface of a mineral substrate. The second and the third examples demonstrate cross-sectional characterization of mineralized biofilms on petrographic thin sections and as polished blocks, respectively. Additionally, different microanalytical techniques for chemical and mineralogical characterization will be briefly discussed.

6.3 The Whole-mount Sample

The structural characterization of microorganisms growing on the surface of minerals can be achieved via the examination of whole-mount samples. Because the composition of geomicrobiological samples can be complex, techniques for whole-mount sample preparation involve the preservation of organic constituents while not altering the inorganic, i.e., mineral, constituents. The analysis of whole-mount samples can be used as a first step to characterizing the surface structure and chemical composition of samples. Micrographs of whole-mount samples can provide a three-dimensional view of the surface at high resolution. This perspective gives viewers a sense of depth based on the size of the primary object of interest, e.g., a bacterial cell or secondary mineral precipitate, relative to its surroundings.

6.3.1 Fixation and Ethanol Dehydration

Immediately after sampling, place the geomicrobiological material in 2.5 wt.$\%_{(aq)}$ glutaraldehyde under the conditions that are best suited for the sample, as previously discussed. In general, EM analysis requires a relatively small amount of material. In general, smaller-volume containers are best suited for sample preparation in the laboratory, because lower volumes of chemical solutions will be required and spills can be avoided. Therefore, sealable containers such as plastic cryo-vials or Eppendorf tubes (Figure 6.2A) are ideal, whereas scintillation vials can be used for biofilms, since these samples tend to be bigger and more mat-like. In addition, the fixation and dehydration procedures can be performed in the same container.

Allow the sample to settle at the bottom of the container. Mineral samples such as gold grains will easily sink to the bottom of a container. However, if a sample is comprised of

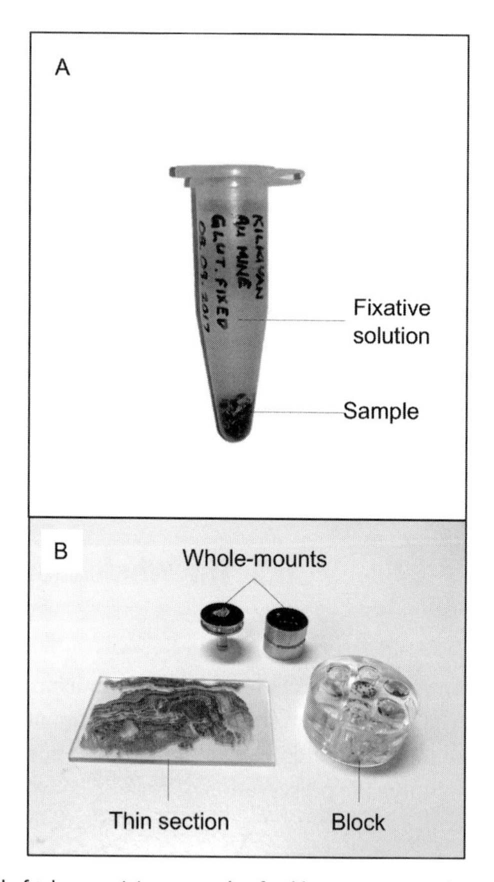

Figure 6.2 (A) A 1 mL-volume Eppendorf tube containing a sample of gold grains associated with black sediment within 2.5$\%_{(aq)}$ glutaraldehyde. Notice that there is no headspace within the Eppendorf tube, so that the sample will always be submersed within the fixative solution during transport and storage. (B) Whole-mounts (bacterially formed gold grains), a petrographic thin section (a terrace iron formation), and polished blocks (natural gold grains).

minerals that are easily suspended in solution or contains planktonic cells, then the sample may require centrifugation. A speed of 10 000 × *g* for ca. 1 minute is usually sufficient to pellet suspended minerals. Once the sample has settled or has been pelleted, remove and discard the supernatant containing the fixative in an appropriate waste container as specified by the respective laboratory protocol. Add $25\%_{(aq)}$ absolute-grade ethanol (i.e., diluted with sterile, distilled, deionized water), close the container, and mix by vortex or agitation to evenly expose the sample to ethanol. Incubate for 15 minutes and then discard the supernatant. Repeat the previous step with 50, $75\%_{(aq)}$, and three times 100% absolute-grade ethanol. During incubation, containers should be kept sealed to minimize the evaporation of ethanol, especially at higher concentrations. Furthermore, the length of time for incubations can be reduced by using a microwave processing system to accelerate equilibrium. The purpose of this dehydration procedure is to gradually substitute water with ethanol, an organic solvent, because hydrated samples cannot be placed in a standard SEM. Furthermore, water to air has a high surface tension and can cause considerable damage to cellular fine structure and cell morphology upon drying; therefore, an organic solvent is necessary. When it comes to preserving samples as three-dimensional structures, the use of absolute-grade ethanol is necessary for further sample processing procedures, i.e., sample drying. Acetone, another organic solvent, is used for sample preparations where samples are embedded in specific resins (see Section 6.5.1). It is recommended that the dehydration procedure be fully completed once the procedure has begun, since prolonged exposure to higher concentrations of ethanol can degrade cell membranes even if they have been thoroughly fixed.

6.3.2 Sample Drying

After the last dehydration step, transfer the samples to microporous specimen capsules, e.g., SPI® Supplies, suspend in 100% absolute-grade ethanol, and critical point (CP) dry. Procedures for CP drying may vary, since different CP dryer brands and models can be either manual or automated. The purpose of CP drying is to substitute ethanol with liquid carbon dioxide (CO_2) so that the cellular structures are retained under dry conditions. Ethanol is considered an intermediate fluid, since it is miscible with CO_2 at 31.1 °C and 1072 psi. At this "critical point," the transition from a liquid to a gaseous state is slow and surface tension is reduced, thereby eliminating damage to the sample. Since this technique involves the use of CO_2 gas in a highly pressurized chamber, it is extremely important to follow the training and safety procedures stipulated by the laboratory in which the CP dryer is housed. Prior to using any CP dryer, one should always confirm that the instrument has been properly installed/maintained and is functioning normally.

The limitation of CP drying is that samples can be exposed to vigorous agitation during this procedure, which could disrupt loosely consolidated mineral constituents within a sample. Therefore, CP drying may not always be an appropriate option, and a method using hexamethyldisilazane (HMDS) can be used as an alternative (Bray et al., 1994). HMDS is an organic solvent with low surface tension that rapidly evaporates at room temperature, when exposed to air. After the last step of dehydration, remove the supernatant and add a 1:1 volume ratio of $100\%_{(aq)}$ HMDS and $100\%_{(aq)}$ absolute-grade ethanol to the sample. Close the container to prevent the HMDS:ethanol solution from evaporation and incubate for 10 minutes. After incubation, remove and discard the HMDS:ethanol solution and repeat this step. Add 100% HMDS, close the container, and incubate for 10 minutes. After incubation, remove and discard the supernatant. Repeat this step involving 100% HMDS two more times. Once the final 100% HMDS supernatant has been removed and discarded, allow the samples to air dry overnight in a fume hood. See Bray et al. (1994) for comparisons between sample preparation using CP drying and HMDS.

6.3.3 Sample Mounting

Whole-mount samples are conventionally placed onto aluminum stubs using carbon adhesive tab. The size and type of aluminum stub to use is dependent on the stub holder of the SEM (Figure 6.2B). Once the sample is mounted onto a stub, the samples usually require a depositional coating such as carbon, iridium, gold, silver, platinum, or osmium to reducing charging effects during analysis. The type and amount of deposition depend on the chemical composition and topography of the sample. For example, the composition of the depositional coating should not be an element that naturally occurs within the sample, because this can interfere with analysis and quantification of the element. Furthermore, samples that act more like insulators or have greater variations in topography may require thicker depositions of a more conductive coating. Therefore, the choice of deposition is a judgment call that requires some preliminary knowledge about the sample. Depositions should ideally be kept to a minimum so that nanometer-scale structures are not "masked" and the identification of elements within the sample by EDS is not skewed.

6.3.4 Interpretation of Whole-mount Samples

In the following example, gold grains and a polymetallic sulfide ore sample were fixed with $2.5\%_{(aq)}$ glutaraldehyde, ethanol dehydrated, CP dried, mounted onto stubs, and coated with a 10 nm deposition of iridium. The samples were examined using a JEOL JSM-7100F field emission-scanning electron microscope (FE-SEM) operating at 3 kV.

When using an SEM as an analytical tool for characterizing whole-mount samples of geomicrobiological materials, SE detection mode is the routine method for generating detailed micrographs of the sample surface. Secondary electron mode uses a lower-energy electron beam that interacts with electrons from atoms within the outermost surface, i.e., less than 100 nm depth, of the sample. In the first micrograph, a surface of a gold grain is covered by a biofilm (Figure 6.3A). The variation in grey scale, as demonstrated in this image, is dependent on the number of secondary electrons emitted from the sample surface that reach the detector. Therefore, the edges of three-dimensional structures will appear brighter and give the appearance of depth. In this case, cells were easily identified as rod-shaped, since the edges appear whiter. High-resolution SE micrographs generally have a narrow depth of field whereby objects gradually become unfocused with increasing distance from the focal plane, giving a perception of three-dimensional space. It should be noted that EPS, in the living state, occurs as an amorphous gel surrounding cells. Since EPS does not have the same structural integrity as cells, CP drying or HMDS cannot

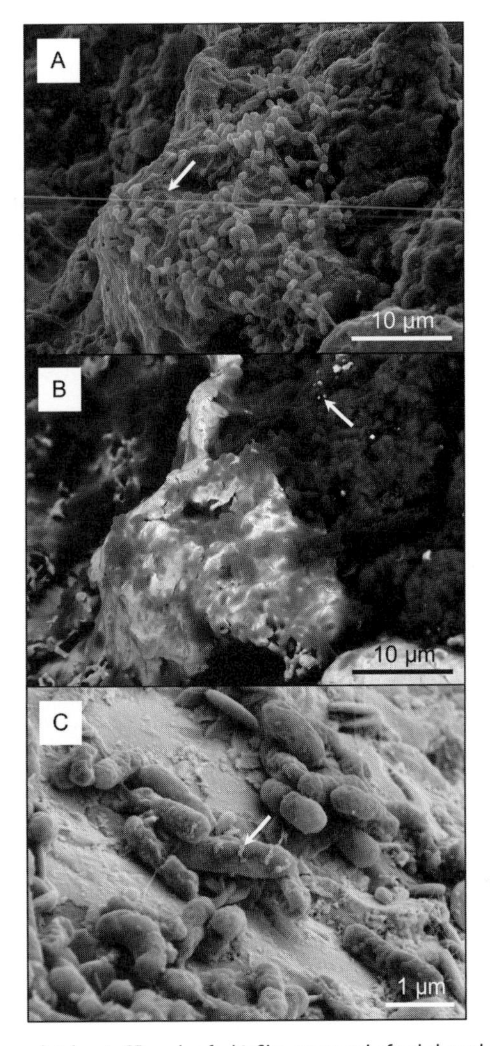

Figure 6.3 A high-resolution SEM micrograph taken in SE mode of a biofilm composed of rod-shaped cells on the surface of a gold grain from Cameroon (A). Note that EPS appears as string-like structures on the extracellular surface of cells connected to substrates and to other cells (arrow, A). A second micrograph of the same biofilm in BSE mode (B). Notice the gold nanoparticles (arrow, B), which were indistinguishable in the SE micrograph. A high-resolution SE-SEM micrograph of iron-oxidizing bacteria attached to the surface of a metal sulfide ore from Capillitas Mine, Argentina (C) (Shuster et al., 2016). Note the presence of nanometer-scale mineral precipitates on the extracellular surface (arrow, C).

preserve the original appearance. Therefore, EPS often appears as nanometer-scale stringy or web-like structures in SE micrographs (Figure 6.3A, arrow), which is an unavoidable effect of sample processing.

Another routine method for generating structural and chemical data from a sample is through BSE mode imaging. In this case, a higher-energy beam interacts with electrons from atoms deeper within the sample, i.e., up to several micrometers in depth, and emitted electrons are captured by a backscatter detector. The same biofilm on the surface of the gold grain was imaged using the same FE-SEM operating at 15 kV (Figure 6.3B). The variations in grey scale are dependent on the elemental composition of the sample; therefore, elements of high atomic mass will appear brighter, while materials composed of low–atomic mass elements will appear darker. From the micrograph, it is evident that the gold grain acts as substrate to which the biofilm is attached; more importantly, the compositional differences of minerals and biofilms are clearly demonstrated. Well-defined surface structures are reduced in BSE mode in comparison to SE mode. The benefit of BSE mode is that it provides an easier method for identifying nanophase particles with different composition. For example, clay minerals that contain nanophase gold particles (Figure 6.3B, arrow) that are nearly unidentifiable in SE mode (compare Figure 6.3A) are easily seen using BSE imaging.

Understanding the physical attributes of microorganisms, including size, shape, organization, and distribution, is important for recognizing these organisms within micrographs. Most chemolithotrophic bacteria, and arguably some spores, commonly occur between 0.5 and 2 μm in size, while secondary mineral precipitates on the extracellular surface of bacterial cells occur in the nanometer-scale range (Figure 6.3C). As with any analytical technique, the ability to recognize organic and inorganic structures within a sample and capture publishable-quality micrographs develops with greater experience. There are interactive online modules, such as MyScope™ (Australian Microscopy & Microanalysis Research Facility), that can assist in learning how to generate EM images. Since some whole-mount samples can be considered modern analogues of past biogeochemical environments, the ability to properly characterize microbes in relation to minerals becomes vital for interpreting more complex structures such as microfossils.

6.4 The Petrographic Thin-section Sample

6.4.1 Petrographic Thin-section Sample Preparation

Passive or active mineralization by bacteria can lead to the preservation of cells and biofilms as microfossils within the rock record (see Westall, 1999 and references therein). The advantage of analyzing geomicrobiological samples containing microfossils as petrographic thin sections or polished blocks is that cross-sections of mineralized cells and biofilms can be characterized. Micrographs of these structures allow viewers to distinguish similarities and differences between extracellular and intracellular mineralization. Furthermore, polished thin sections or blocks can present a large sample area in which the spatial organization of mineralized cells relative to other structures, i.e., pore spaces or millimeter-scale mineral grains, can be characterized.

For more rock-like materials, samples do not necessarily need to be fixed with glutaraldehyde, because any mineralized bacterial cells would occur as microfossils. These types of samples can be made into petrographic thin sections, which will be briefly described in this section, as there are different lapidary tools and equipment that use a variety of techniques. Using a rock saw, a sample is cut to be approximately 8–10 mm thick and to fit the dimensions of a glass slide. A glass slide is ground on one side to roughen the surface so that the sample can be attached with epoxy resin. Once the resin has cured, the sample is then polished using various polishing compounds to approximately 40 μm thickness. It should be noted that depending on the porosity or friability of the sample, impregnation with epoxy resin may be required to help hold the sample together prior to attachment to the glass slide and during polishing. Petrographic thin sections are often intermittently sonicated during polishing as well as after the final polish to remove any residual loose material. The surface of polished thin sections should be as clean as possible, since residual debris can affect the image quality of a micrograph. Plasma cleaners can be used to remove any oils or other contaminants from the surface of polished thin sections. During plasma cleaning of a sample, air, oxygen, hydrogen, argon, or nitrogen gases are ionized under low pressure in a chamber. Different gases can alter the surface properties of mineral surfaces, such as metal sulfides, which are sensitive to rapid oxidation. Therefore, basic knowledge of a sample's mineralogical composition is necessary before deciding which gas to use in order to avoid sample alteration. Once a polished thin section is clean, it can be coated with a conductive deposition, as previously described in the first example. A thinner deposition, i.e., generally less than 10 nm, of a conductive coating on the surface of thin sections is usually sufficient to reduce charging effects during analysis.

6.4.2 Interpretation of Polished Thin Sections

For this example, canga, a consolidated iron-duricrust from Brazil, was made into a polished petrographic thin section (see Levett et al., 2016 for details), coated with 8 nm-thick iridium deposition, and examined using a JEOL JSM-7100F FE-SEM operating at 15 kV in BSE mode.

Thin sections are relatively flat; therefore, using lower-energy beams and SE mode is usually only beneficial for confirming that the surface is flat and that structures observed in BSE mode are not artifacts of sample processing, e.g., air bubbles or debris flow in epoxy. Understanding bacterial cell structure in three-dimensional space is helpful for identifying microfossils from two-dimensional cross-sections. Compare the micrograph of a three-dimensional biofilm in Figure 6.3C with the two-dimensional mineralized biofilm in Figure 6.4A. In cross-section, coccoid cells will always appear circular, whereas rod-shaped and filamentous cells may appear as rods, filaments, circular, or oval, depending on the orientation of the cells. From the BSE micrograph, these microfossils are generally less than 3 μm in size (Figure 6.4A). At lower magnification, it is evident that these shapes are part of a larger, laminated structure containing layers of cells with different morphologies (Figure 6.4B). The size, shape, organization, and distribution of these structures can be interpreted as microfossils of a microbial consortium that once comprised a living biofilm. The lowest magnification provides an overall *in situ* context suggesting that the biofilm formed within the pore space between mineral grains, i.e., the rock–water interface (Figure 6.4C). Furthermore, the variations of the grey scale between the extracellular and intracellular regions suggest two modes of mineralization or cell preservation. This example highlights the importance of using a series of micrographs with different magnifications to fully justify evidence-based characterization for the interpretation of past biogeochemical processes from the rock record.

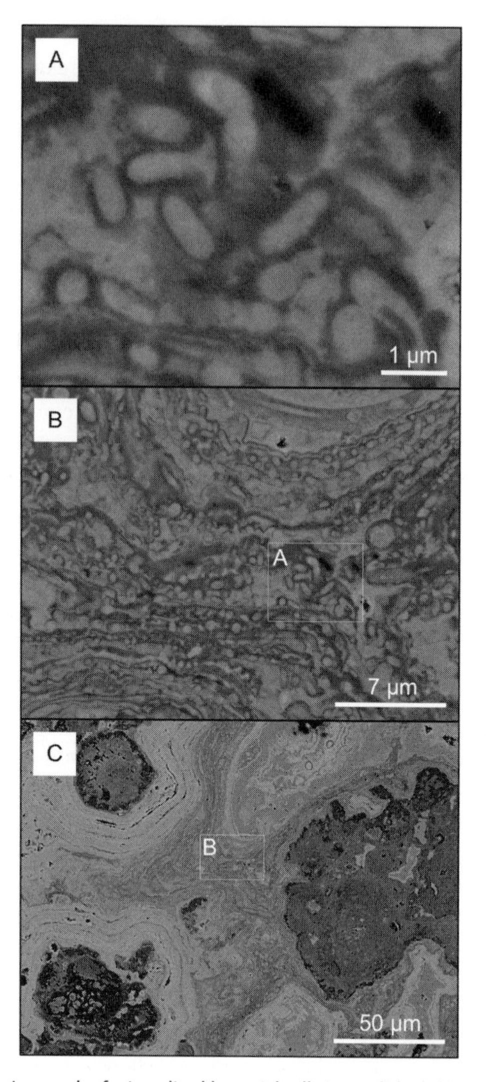

Figure 6.4 A high-resolution BSE-SEM micrograph of mineralized bacterial cells in a polished thin section of canga from Carajás mineral province, Brazil (A). These microfossils represented one type of cell morphology within the biofilm (B). A low-resolution BSE-SEM micrograph demonstrating that the once living biofilms occurred within the pore space between grains of canga (C). See Levett et al. (2016).

6.5 The Polished Block Sample

6.5.1 Polished Block Sample Preparation

The difference between polished blocks and petrographic thin sections is that the former have greater volume. This may pose challenges, because not all SEMs come equipped with holders that can accommodate samples that are prepared as large blocks. Therefore, it is best to know the dimensions of a sample holder and to determine whether processing samples by this method is practicable. The advantage of polished blocks is that geomicrobiological samples containing a complex network of both mineralized and nonmineralized organic material can be characterized.

Once the sample has been fixed, water in the sample must be replaced with an organic solvent. Ethanol is used if the sample will be embedded in London Resin (LR) White resin, whereas acetone is used for embedding in Epon epoxy resin. The dehydration with the organic solvent is performed by a similar method to that previously described: samples are placed in sequential 25, 50, 75 $\%_{(aq)}$, and three times 100% organic solvent for 15 minutes at each concentration. Remove and discard the organic solvent, add a mixture of organic solvent:resin with a ratio of 3:1 to the sample, and incubate for 30 minutes in a sealed container. Repeat this step using 1:1 and 1:3 ratios of organic solvent:resin. Place the sample in 100% resin and incubate for 1 hour. Repeat this step three times to completely replace any residual organic solvent with resin. The substitution of the organic solvent with resin is generally performed in this manner because resin is viscous and the gradual substitution enables even impregnation into the sample. Note that reduced incubation times are possible when using a microwave processing system to accelerate equilibrium. See Giammara (1993) for details regarding microwave embedding of samples in various types of resin. Finally, transfer the sample into a mold and ensure that the sample is completely immersed in resin. Cure the resin-embedded sample according to the conditions specified by the manufacturer of the resin. Once the resin has completely cured, the block can be cut and polished. It should also be mentioned that samples prepared as polished blocks can be further processed into petrographic thin sections if desired.

6.5.2 Interpretation of Polished Blocks

In this example, a thrombolite was embedded in epoxy resin, polished, coated with a 5 nm-thick deposition of iridium, and examined using a JEOL JSM-7100F FE-SEM operating at 10 kV in BSE mode. Since blocks are fairly large volumes of material, charging effects during analysis may be a problem. Therefore, in addition to the routine depositional coating, conductors such as aluminum or copper tape or silver paint can be applied to the side of the block to provide greater electron conduction and grounding.

For characterization, structures within polished blocks, as with thin sections, are best viewed in BSE mode, because SE mode will only show the flat surface. From the polished block of the thrombolite sample, it is evident that there are two different mineral phases: an abundant mineral (darker grey; Figure 6.5A) and a less abundant mineral (light grey; Figure 6.5A). The more abundant mineral phase, identified as stevensite ($(Ca_{0.5},Na)_{0.33}(Mg,Fe)_3Si_4O_{10}(OH)_2 \bullet n(H_2O)$; see Burne et al., 2014), primarily occurs as micrometer-size spherical minerals on the extracellular surface and presumably within the cell wall, resulting in the preserved structure of filamentous bacteria (Figure 6.5B). Interestingly, finer-grained stevensite appears to preserve smaller cells, interpreted as being associated heterotrophic bacteria (Figure 6.5B). In this example, it is important to note that any nonmineralized cells were "lost" in the resin, since both materials are carbon-

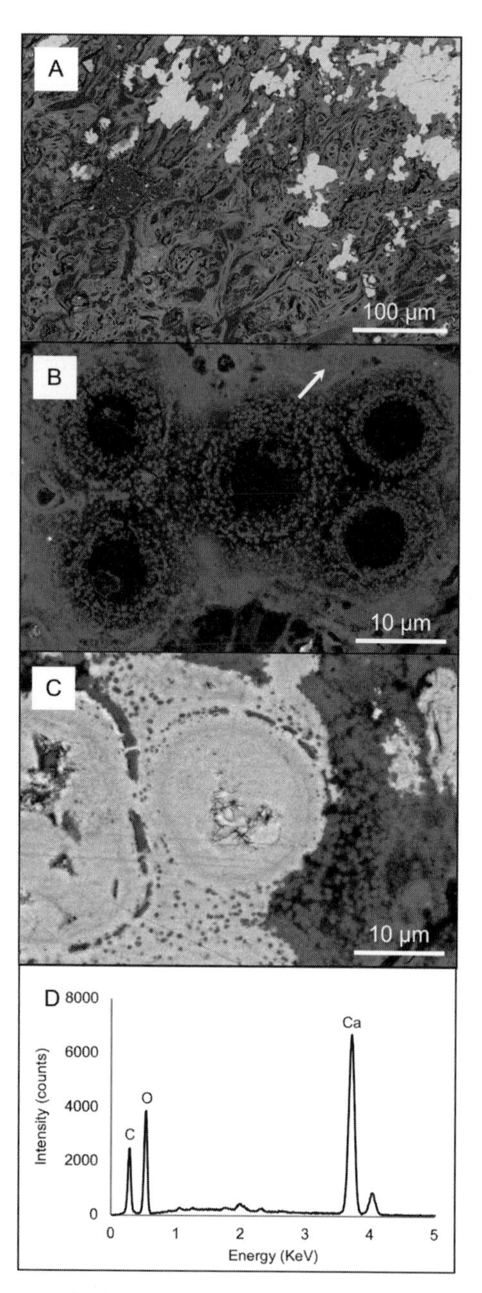

Figure 6.5 A low-resolution BSE SEM micrograph of a polished block of thrombolite from Lake Clifton, Australia (A). Filamentous bacteria were extracellularly mineralized in nanometer-scale colloids of stevensite (B). Note that the precipitation of finer "stevensite" appeared to also mineralize associated heterotrophs and presumably EPS, thereby preserving their structure (arrow, B). A secondary episode of mineralization appears to "overprint" the stevensite mineralization and evidence of microfossils (C). The second mineral is composed primarily of Ca, C, and O based on EDS analysis (D).

based and appear the same shade of grey in BSE mode. The second mineral phase appears lighter, indicating that it is likely composed of higher–atomic mass elements relative to stevensite. Furthermore, this mineral appears to overprint the preservation of mineralized cells, suggesting that it likely occurred later in time (Figure 6.5C). Therefore, by characterizing different phases and comparing them with each other, it is possible to reconstruct the likely environmental conditions that resulted in the formation of these structures.

Micrographs provide a wealth of information in regard to microbe and mineral structure; however, this is only a part of the story. EDS is a microanalytical technique that identifies the elements present within a sample. This chemical analysis is often obtained as either spot/line scans or maps and can be quantitative or semi-quantitative. Most importantly, this analysis complements the structural characterization of minerals. In general, when electrons from the outer shell fill the hole left by an excited electron from the inner shell, energy is released in the form of X-rays. The energy of these X-rays is characteristic of the atomic structure of the respective element. As these X-rays are detected, a spectrum of elements present within a sample is obtained. In this example, the EDS spot analysis detected carbon, oxygen, and calcium from the less abundant mineral that appeared to overprint stevensite. This semi-quantitative analysis suggests that this material is a carbonate mineral (Figure 6.5D).

6.6 Focused Ion Beam Milling

6.6.1 Basics of Controlled Etching

In some instances, geomicrobiological samples require microanalyses that conventional preparatory techniques, such as the ones discussed earlier, cannot provide. For example, ultrafine and controlled etching using a focused ion beam (FIB) can produce polished surfaces with topographical variations less than few nanometers. With this sort of precise sample preparation, microbe–mineral interfaces and mineral grain boundaries and orientations can be characterized in detail at the nanometer scale. Furthermore, rock-like samples can be made into ultrathin sections for TEM analysis, since rocks cannot be sectioned using an ultramicrotome (see Section 7.5).

Some SEMs can be equipped with a focused (gallium) ion beam. Unlike electron beams, ion beams bombard the surface of a sample and thereby eject atoms. With this capability, a FIB can be used as a nanomachining tool to mill nanometer- to micrometer-scale sections with high precision. The exact parameters of the ion beam will vary depending on the elemental composition of the sample, since materials will sputter differently. For example, samples with a relatively homogeneous composition, such as gold grains, are relatively easy to FIB mill in comparison to samples with heterogeneous compositions, such as polymetallic sulfide ores. Whether the surface of a sample is flat or articulated, e.g., petrographic thin section vs. whole-mount sample, a localized coating of platinum is often deposited *in situ* prior to milling. This coating acts as a protective layer of the targeted surface and allows high-accuracy milling. The required thickness of the deposition is dependent on the surface evenness and composition of the sample. The dimensions of the deposition are dependent on the intended width of the milled section. It is important to note that sputtered material can be redeposited onto the sample surface proximal to the milled section; therefore, it is important to be aware of the potential for artifacts from this technique.

6.6.2 Applications of FIB Milling

The following examples will briefly highlight the advantages of using an FIB to characterize biogeochemically transformed precious metal samples. In the first example, an FIB was used to mill the surface of a gold grain that was initially prepared as a whole-mount sample. The advantage of this technique is that milled "troughs" provide a cross-sectional view of the sample that is tens of micrometers below the outer surface. Relative to cross-sections obtained from resin-embedded and polished samples, this technique is much more precise, since exact locations can be selected for milling. The disadvantage of this technique is that FIB milling is practicably restricted to making micrometer-size troughs. From the gold grain, fine structures of the interstitial spaces within the polymorphic layer and its association with the grain boundary are visible (Figure 6.6A). See Reith et al. (2010) for details regarding FIB-milled sections. It should be mentioned that FIB milling removes roughly 10 to 20 nm-thick "slices" of material, depending on the size of the ion beam. Therefore, it is possible to FIB mill surfaces as sequential slices that can then be imaged and stitched together to reconstruct a three-dimensional model.

FIB milling can also be used to ultrafine-polish samples. A platinum grain was initially prepared as a polished block. FIB polishing of this sample revealed nanometer-scale grain boundaries of secondary platinum crystals at the outer edge of the grain (Figure 6.6B). This sample was FIB polished because an ultraflat surface was necessary for electron backscatter diffraction (EBSD) analysis – an auxiliary crystallographic microanalysis that provides crystal orientation of mineral composites. Using FIB milling and EBSD, it was determined that the dissolution of platinum–iron alloys and the reprecipitation of secondary platinum as nanophase crystals were analogous to biogeochemical processes contributing to secondary gold enrichment on gold–silver alloy grains (see Campbell et al., 2015 for details).

Since most rock material cannot be easily made into ultrathin sections for TEM analysis, FIB milling can also be used to target and to make "lift-outs" that can be mounted onto TEM grids. Lift-outs are small slices that are generally a few micrometers in length and can be as thin as a few hundred of nanometers thick. The exact dimensions of the lift-out are dependent on the analytical needs using the TEM. In this example, a lift-out was milled from a sample of Gunflint chert that was originally prepared as a polished thin section. This sample contained bacterial microfossils occurring as pyritized filamentous bacteria within a chert matrix (Figure 6.6C). The lift-out was made as a transect through the mineralized cell envelope and what once was the cytoplasm (Figure 6.6D). Once lift-outs are made, they are then "welded" onto a TEM grid. Materials such as chert cannot be made into conventional ultrathin sections using an ultramicrotome and would severely damage a diamond knife; however, preparation of these types of geomicrobiological samples as lift-outs is now possible.

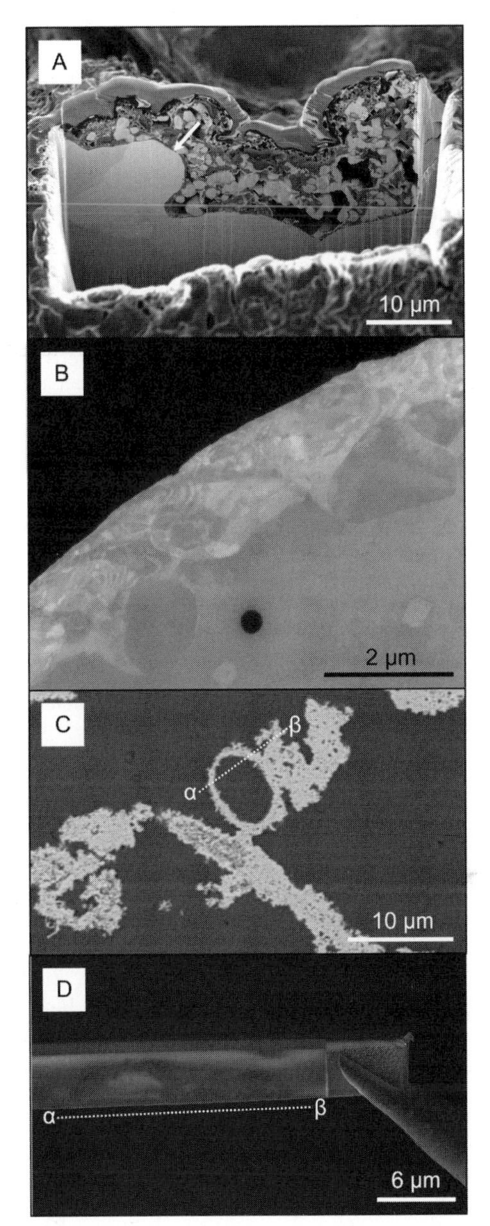

Figure 6.6 A high-resolution BSE-SEM micrograph of FIB trough, cross-section through a polymorphic layer (A) (Reith et al., 2010) in contact with the consolidated surface of a crystalline gold grain from Queensland, Australia (arrow, A). A high-resolution BSE-SEM micrograph of an FIB-polished platinum grain from New South Wales, Australia, demonstrating a microcrystalline rim composed of secondary platinum (B) (Campbell et al., 2015). A high-resolution BSE-SEM micrograph of a filamentous microfossil from the Gunflint chert, Canada (C). A high-resolution SE-SEM micrograph of an FIB section lift-out. The dashed lines illustrate where the FIB section lift-out was taken from the original polished thin-section sample (D).

6.7 Conclusion

SEM provides the magnification and resolution required for characterizing the structure and chemistry of geomicrobiological samples. For geomicrobiologists, this technique can be the first line of visual evidence for interpreting biogeochemical processes. As highlighted throughout this chapter, there are various methods by which samples can be prepared for the characterization of different structures. Overall, this analytical technique requires the preservation of both biological and mineralogical constituents of a sample, which can be challenging for fine-grained or hydrated materials. This chapter briefly discussed sample preparation techniques for whole-mounts, polished thin sections/blocks, and the application of FIB milling. With the increasing sophistication of SEMs, the creativity in sample preparation and the analytical capacity of SEMs are endless. While the methods described in this chapter are conventional, they should only be considered as a basic starting point. Any method can be modified or adapted to enable the highest-quality characterization of any geomicrobiological material. It should be noted, however, that the greater the number of treatments applied to a sample, the higher the probability of altering or losing any ultrafine structures. The characterization of structure in association with chemistry serves as a guide to help interpret the association of microbes with minerals. SEM as an analytical technique enables geomicrobiologists to visualize structures down to the nanometer scale and elucidate a "bacterial view" of the biogeochemical world.

Acknowledgments

We thank T. Simpson for FIB sectioning assistance at the Nanofabrication Laboratory at the Western University, Canada, A. Levett at the Centre for Microscopy and Microanalysis at the University of Queensland, Australia, and A. Basak at Adelaide Microscopy, University of Adelaide, Australia for SEM imaging assistance.

6.8 References

Australian Microscopy & Microanalysis Research Facility (AMMRF), MyScope™ http://ammrf.org.au/myscope/

Beveridge, T.J., 1981. Ultrastructure, chemistry, and function of the bacterial wall. *International Review of Cytology*. 72, 229–317.

Bray, D.F., Bagu, J. and Koegler, P., 1994, Comparison of hexamethyldisilazane (HMDS), Peldri II, and critical point drying methods for scanning electron microscopy of biological specimens. *Microscopy Research and Technique*. 26, 489–495.

Burne, R.V., Moore, L.S., Christy, A.G., et al., 2014, Stevensite in the modern thrombolites of Lake Clifton, Western Australia: A missing link in microbialite mineralisation? *Geology*. 42, 575–587.

Campbell, S.G., Reith, F., Etschmann, B., et al., 2015, Surface transformation of platinum grains from Fifield, New South Wales, Australia. *American Mineralogist*. 100, 1236–1243.

Dohnalkova, A.C., Marshall, M.J., Arey, B.W., et al., 2011, Imaging hydrated microbial extracellular polymers: Comparative analysis by electron microscopy. *Applied and Environmental Microbiology*. 77, 1254–1262.

Giammara, B.L. 1993. Microwave embedment for light and electron microscopy using epoxy resins, LR White and other polymers. *Scanning*. 15, 83–87.

Hayat, M.A. 1959. *Principles and Techniques of Electron Microscopy: Biological Applications*. New York, Van Nostrand Reinhold Company.

Karnovsky, M.J., 1965. A formaldehyde-glutaraldehyde fixative of high osmolality for use in electron microscopy. *Journal of Cell Biology*. 27, 137A–138A.

Krember, A., Lippens, S., Bartunkova, S., et al., 2015, Developing 3D SEM in a broad biological context. *Journal of Microscopy*. 259, 80–96.

Levett, A., Gagen, E., Shuster, J., et al., 2016, Evidence of biogeochemical processes in iron duricrust formation. *Journal of South American Earth Sciences*. 71, 131–142.

Migneault, I., Dartiguenave, C., Bertrand, M.J. and Waldron, K.C., 2004, Glutaraldehyde: Behaviour in aqueous solution, reaction with proteins, and application to enzyme crosslinking. *BioTechniques*. 37, 790–802.

Rasmussen, K.E. and Albrechtsen, J., 1974, Glutaraldehyde: The influence of pH, temperature and buffering on the polymerisation rate. *Histochemistry*. 38, 19–26.

Reith, F., Fairbrother, L., Nolze, G., et al., 2010, Nanoparticle factories: Biofilms hold the key to gold dispersion and nugget formation. *Geology*. 38, 843–846.

Sabatini, D.D., Bensch, K. and Barrnett, R.J., 1962, Cytochemistry and electron microscopy: The preservation of cellular ultrastructure and enzymatic activity by aldehyde fixation. *Journal of Cell Biology*. 17, 19–58.

Shuster, J., Lengke, M., Márquez-Zavalía, M.F. and Southam, G., 2016, Floating gold grains and nanophase particles produced from the biogeochemical weathering of a gold-bearing ore. *Economic Geology*. 111, 1485–1494.

Walther, P. and Müller, M., 1999, Biological ultrastructure as revealed by high resolution cryo-SEM of block faces after cryo-sectioning. *Journal of Microscopy*. 196, 279–287.

Westall, F., 1999, The nature of fossil bacteria: A guide to the search for extra-terrestrial life. *Journal of Geophysical Research*. 104, 16437–16451.

7 Applications of Transmission Electron Microscopy in Geomicrobiology

JEREMIAH SHUSTER, FRANK REITH, AND GORDON SOUTHAM

Abstract

Geomicrobiological samples obtained from the natural environment or from laboratory-based experiments often contain microbial cells, primary minerals, and/or (biogenic) secondary minerals as extracellular precipitates on cell surfaces. The advantage of transmission electron microscopy is that it provides high-resolution imaging of ultrafine structures of cells and minerals. Transmission electron microscopes can be equipped with a range of microanalytical tools, such as selected area electron diffraction and energy dispersive spectroscopy, that complement the imaging capacities of the electron microscope. These two analytical techniques are used to characterize the crystallography and chemical composition of (secondary) minerals, respectively. In addition, lift-outs that have been focus ion beam milled from rock-like materials are analogous to ultrathin sections and can be characterized in a similar manner using transmission electron microscopy. This chapter will demonstrate conventional techniques for preparing samples as whole-mounts and ultrathin sections, which provide a three-dimensional and two-dimensional perspective of samples, respectively. Whole-mounts and ultrathin sections are the most fundamental sample preparation techniques that can be used to characterize geomicrobiological materials. Therefore, in geomicrobiological studies, transmission electron microscopy is an ideal technique to demonstrate and interpret the structure–function relationship of how microbes contribute to the biogeochemistry of a given system.

7.1 Introduction

Transmission electron microscopy (TEM; also indicating a transmission electron microscope) is a microanalytical technique that can help geomicrobiologists to characterize the structure and chemistry of microorganisms as well as the association of microbes with minerals. A TEM, much like a scanning electron microscope (see Chapter 6), uses an electron gun to generate an electron beam under high vacuum. A TEM typically operates between the energies of 60 and 200 kV and can magnify samples up to 10^7 times, depending on the model of the TEM. A higher-energy beam will provide greater resolution, thereby attaining more detail of fine structures at greater magnification. Various lenses are used to adjust the electron beam so that it passes, i.e., transmits, through a sample to produce an image onto a fluorescent screen (Figure 7.1). Micrographs were originally made by exposing film to the electron beam and developing the film afterwards;

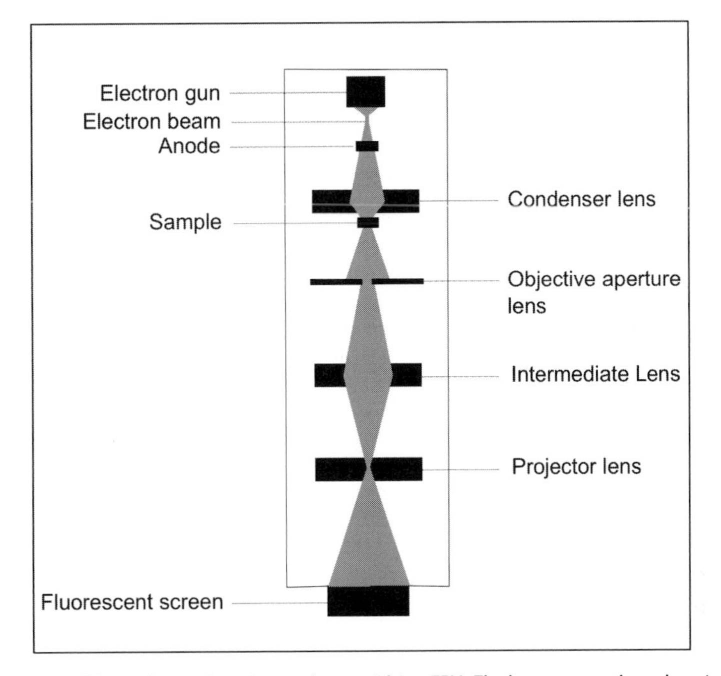

Figure 7.1 A schematic diagram of the pathway of an electron beam within a TEM. The beam passes through various lenses and through the sample, thereby generating a transmitted image on a fluorescent screen.

however, TEMs are now equipped with a digital camera that captures the transmitted image.

A TEM can provide high-resolution micrographs of the unconstrained growth of secondary minerals nucleating from the extracellular surface of microorganisms. The successful imaging of these materials is often dependent on sample preparation. Therefore, it is important to note that the art of the science is not simply taking an aesthetic micrograph but also involves careful sample preparation. Basic knowledge about any given sample is required before starting; this can often be obtained through light microscopy or scanning electron microscopy. A microscopist needs to consider critical factors of a sample, e.g., single cells versus biofilm, because this will determine the method of sample preparation and the analytical conditions best suited to characterize these materials. Careful planning will provide true and meaningful data, which can be used to interpret biogeochemical processes. Sample preparation is arguably one of the most important aspects of electron microscopy. In regard to TEM sample preparation, there is no "one way" to prepare a sample. Numerous techniques for sample preparation have been described and assessed in detail throughout the literature (see Hayat, 1959). Microscopists often prepare samples using different techniques or using one technique with slight modifications. The purpose of this is to obtain the highest-quality micrographs showing the structures which they are interested in; hence, it is not uncommon to spend more time preparing a sample than actually obtaining the TEM micrographs. Sample preparation is dependent on the nature of the sample and the type of information an investigator wants to obtain. Furthermore, TEM sample preparation requires dexterity, which often comes with practice. This chapter focuses on the conventional techniques for making whole-mount and ultrathin sections and provides a basic understanding of how to interpret micrographs when using these techniques. It is extremely important to realize that sample preparation for electron microscopy involves hazardous chemicals, such as fixatives or heavy metal solutions for staining. Some of these chemicals can be very toxic, and caution must be used at all times. Refer to all respective material safety data sheets prior to handling any chemical.

7.2 Preservation of Organic Structures

The preservation of organic structures is necessary when using a TEM for analyzing the functional role of bacteria within biogeochemical systems. In general, the purpose of fixation is to preserve the physical and chemical structure of microorganisms by arresting cells from carrying out any further biochemical reactions. Glutaraldehyde ($C_5H_8O_2$) is a noncoagulant fixative with low molecular weight. During the process of fixation using glutaraldehyde, amino-functional groups of proteins are crosslinked, resulting in increased membrane rigidity and thereby preserving cell morphology. Electron microscopy–grade glutaraldehyde is a conventional fixative because it rapidly penetrates specimens due to its low viscosity. For details regarding fixation procedures, see Section 6.1. It is important to remember that fixation (where possible) and any procedure involving fixed samples should be performed in a fume hood.

7.3 TEM Grids

To analyze a sample using a TEM, samples are placed on grids. These grids are roughly 3 mm in diameter, 10–25 μm in thickness, composed of copper, palladium, nickel, or gold, and contain a mesh comprised of a varying number of square-, hexagon-, or rectangular-shaped gaps. These gaps can also come in different sizes, which are large enough to see the sample. Grids should only be handled on the outer perimeter using an anti-capillary tweezer or (reverse) forceps, because the mesh is very delicate and can easily be damaged. Some grids come with a Formvar film on one side, which is typically carbon coated (Figure 7.2A). Carbon-coated Formvar film dissipates charging, which provides additional stability of microbial cells or nanophase minerals. This is especially useful when using a higher-energy beam at higher magnification. When samples are prepared as (ultra)thin sections, TEM grids without Formvar film or carbon coating are sometimes preferred, since minor imperfections in the Formvar film could distort an image. Therefore, the type of grid is discretional and often an individual preference. When a grid is chosen, however, the elemental composition of any minerals within a sample should be different from the metal composition of the grid to avoid skewing any Energy Dispersive Spectroscopy (EDS) analysis; e.g., a noncopper TEM grid would be best for copper-bearing minerals of a sample.

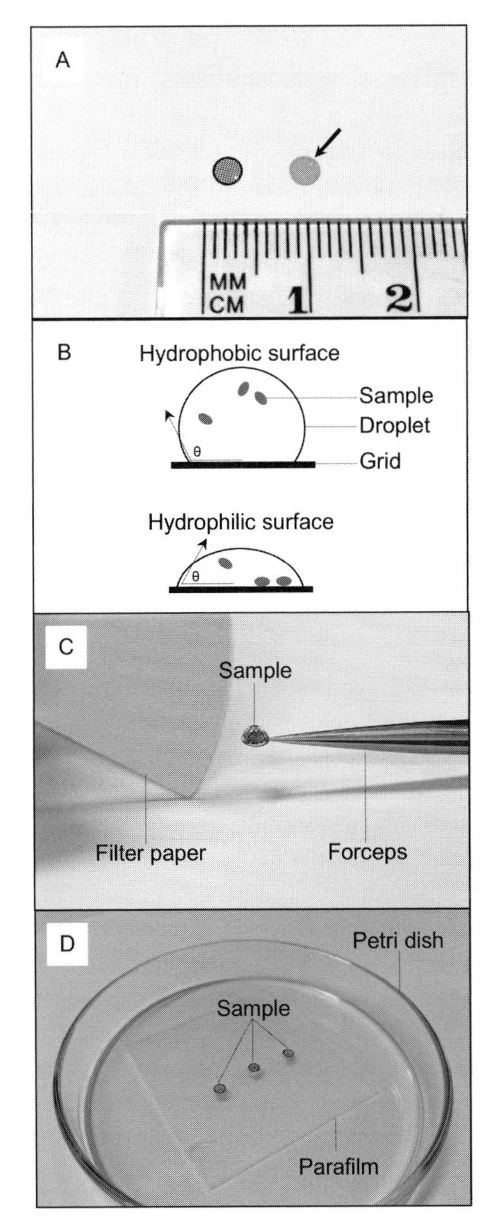

Figure 7.2 A photograph of 100 square-mesh copper grid (A). The side with the carbon-coated Formvar film appears matte and is the surface on which samples are placed (arrow, A). A schematic diagram illustrating the shape of a droplet when the grid surface is hydrophobic and hydrophilic. Theta (θ) represents the angle between the plane of the grid and the outer edge of the droplet that is in contact with the grid. Larger θ angles indicate that the grid surface is hydrophobic, whereas lower θ angles indicate a hydrophobic surface (B). A photograph of a sample droplet placed on a 100 square-mesh copper grid. A quartered piece of filter paper is touched to the outer edge of the grid to wick off the droplet (C). A photograph of multiple whole mounts being prepared at once using the floating grids onto droplets of samples (D).

7.4 The Whole-mount Sample

Whole-mount samples are the simplest technique for preparing geomicrobiological samples for TEM analysis. This technique can be used as a starting point for viewing a sample and to determine whether alternative preparations, e.g., ultrathin sections for cross-sectional views, are necessary and will provide additional information. When the sample is viewed on the fluorescent screen and in micrographs, microbial cells and minerals will have a slightly three-dimensional appearance. This "overhead" perspective is analogous to viewing samples with a traditional phase-contrast light microscope, albeit at much higher resolution. As previously mentioned, a small amount of sample is required for TEM analysis. Individual cells and minerals less than hundreds of micrometers in size are best suited to this technique. Larger cells, e.g., filamentous cyanobacteria, biofilms, and large minerals can be more difficult to image. Therefore, the primary limitation of this technique is the size of the sample.

7.4.1 Single-drop Whole Mount

To prepare a sample as a whole-mount, ensure that any microbial constituents within the sample have been given enough time to be thoroughly fixed, i.e., immersed in $2.5\%_{(aq)}$ glutaraldehyde for 24 h, and that microbial and mineral constituents are homogenized in the fluid phase. To make whole-mounts, use anti-capillary tweezers or reverse forceps to hold a grid that contains a Formvar–carbon film. Place a droplet of homogenized sample onto the Formvar film and leave it for 2–3 minutes. During this time, cells and minerals within the droplet will adsorb with their greatest surface area in contact with the Formvar film. While this length of time is generally sufficient, more or less time may be required, because materials sorb at different rates depending on hydrophobicity and charge character. If the droplet appears more spherical than dome-shaped, then the grid surface is hydrophobic (Figure 7.2B). A glow discharge system, e.g., EMS GloQube™ or PELCO easiGlow™, is an instrument that uses a plasma current to make TEM grids hydrophilic. If a glow discharge system is not available, a simple trick is to expose the Formvar film to a UV light for 5–10 minutes. Once a droplet has been on the grid surface for the necessary length of time, gently touch a piece of dry Whatman® filter paper to the edge of the grid. The filter paper will wick, i.e., draw off by capillary absorption, the majority of the droplet (Figure 7.2C). As it does this, a thin layer of liquid will be drawn over the grid surface. Allow the thin, liquid layer to air dry. If the fixed sample solution contains a buffer or if the sample naturally has a high salt content, e.g., halophilic microbial samples, mineral precipitates can form once the thin, liquid layer has dried. See Section 6.1. These precipitates are considered artifacts of sample preparation and can be rinsed away using filter-sterilized, deionized water. Using a sterile syringe attached to a 0.45 μm pore-size filter unit, a gentle stream of water droplets can be passed across the surface of the grid when held at a slight angle. Another method for rinsing is adding a droplet of water on the surface of the grid while simultaneously drawing it off using a piece of filter paper. After rinsing, allow the grid to air dry. Prior to putting a sample into a TEM, grids can be placed on a sterile glass slide, and a light microscope can be used to check that the sample is on the grid. Note: if the concentration of cells or minerals within a sample solution is dilute, then sequential droplets can be placed on a grid and wicked off prior to rinsing and drying. Alternatively, a sample can be concentrated by centrifugation, e.g., 12 000 × *g* for 30 seconds, which will create a pellet of solid material. By removing a fraction of the supernatant and resuspending the pelleted material by vortex, it is possible to make the sample more concentrated. Keep in mind that too much material on a grid can "clog" the mesh, thereby making imaging difficult or impossible.

7.4.2 Floating Grid Whole-Mount

An alternative method for preparing whole-mount samples is by floating grids onto a droplet of sample. Start by placing a piece of Parafilm® inside a sterile, glass Petri dish and placing a droplet of fixed and homogenized sample onto the Parafilm®. Using forceps, place the carbon-coated Formvar side of a grid on the droplet and allow the grid to float for 2–3 minutes (Figure 7.2D). Avoid plunging the grid into the droplet, as this may get the sample on both sides of the grid, making it difficult to image. As with the single-drop method, the length of time required to allow cells or minerals to adsorb to the Formvar–carbon film can vary. Once a grid has been floating for a desired amount of time, remove the grid using forceps. As previously described, wick the grid using a piece of filter paper, allow the thin layer of liquid to air dry, rinse with sterile, deionized water, and redry. The benefit of this alternative method is that multiple samples can be prepared at once, because anti-capillary tweezer or reverse forceps are not required to hold the grid in place.

7.4.3 Interpretation of Whole-mount Samples

Like scanning electron microscopy (SEM) micrographs, the variation in grey scale is the most recognizable visual feature of TEM micrographs. This grey scale is attributed to differences in the atomic mass of elements comprising a sample. Unlike SEM micrographs taken in backscatter mode (see Section 6.3.4), materials composed of greater–atomic mass elements will appear darker; i.e., they are electron dense. Cells and extracellular polymeric substances will generally appear blurry, with rounded and soft edges outlining their shape, because they are composed of amorphous carbon-based compounds. On the contrary, minerals will appear darker, with distinct edges defined by their crystal boundaries, because they are generally composed of greater–atomic mass elements with an ordered structure. Therefore, a basic understanding of cellular morphology and mineralogy will help with characterizing samples and making interpretations of biogeochemical processes. In the following examples, samples of nanophase gold occurring as octahedral platelets and colloids were prepared using the single-drop whole-mount method. These samples were not fixed using glutaraldehyde because they did not contain microbial cells. In the first example, the sample was examined using a Philips CM-10 TEM operating at 80 kV (Figure 7.3A).

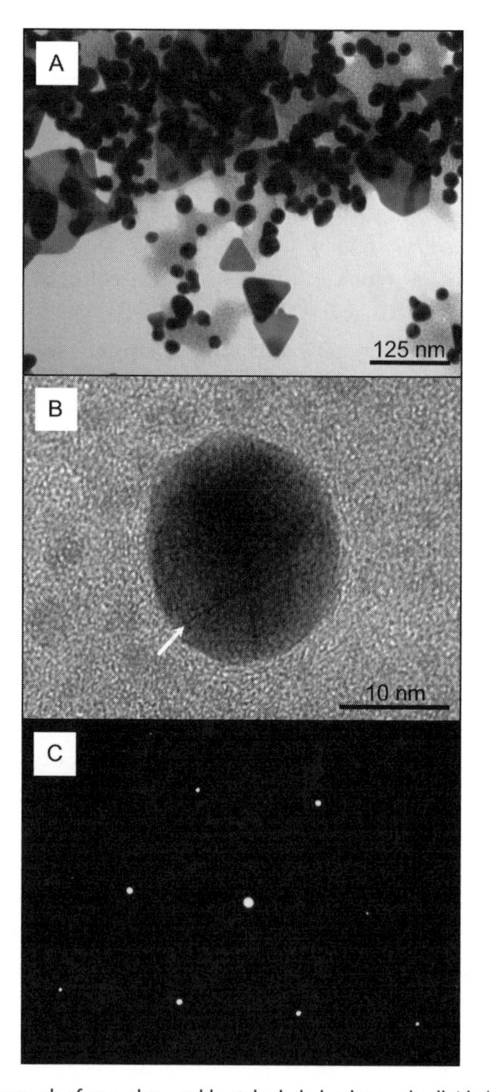

Figure 7.3 A transmission electron micrograph of nanophase gold octahedral platelets and colloids (A). The nanoparticles were produced by exposure of delftibactin to gold(III) chloride (see Johnston et al., 2013 for details). A micrograph of a gold colloid removed from the surface of gold grains (B; see Shuster et al., 2017a). A representative SAED pattern confirming the identity of elemental gold from an octahedral platelet (C) as seen in the previous micrograph.

Octahedral platelets preferentially adsorbed with the greatest surface area in contact with the Formvar film; hence, these varying-size crystals appear as pseudo-triangular shapes. Variation in electron density of platelets is attributed to overlapping crystals or folding/bending of an individual crystal. It is important to note that both octahedral platelets and colloids are composed of elemental gold. Colloids appear darker than the octahedral platelets because there are a greater number of gold atoms that are closely arranged. In the second example, colloidal gold was characterized using a JEOL JEM-

2100 TEM operating at 200 kV (Figure 7.3B). At high magnification and energy, gold atoms making up the crystal lattice structure of the colloid are visible. These atoms are more visible at the outer edges of the colloid compared with the centers, which are more electron dense. By comparing these two examples, it is evident how different magnifications and energy levels can provide different structural characterization.

Selected area electron diffraction (SAED) is a quantitative analytical technique that can be used to identify specific minerals and is ideal for whole-mount samples because crystals often exhibit a preferential, planar orientation on the Formvar. In general, the electron beam passes through a mineral and produces a series of spots that correspond to the diffraction conditions of a crystal's structure. The principle of this technique is analogous to X-ray diffraction in the context of crystal structure analysis; however, the resolution of SAED analysis is on the submicrometer scale; see Bozzola and Russel (1992) and references therein for more detail on SAED.

7.5 The Ultrathin Section

Ultrathin sections are 50 to 100 nm thick slices of a sample. This technique was originally developed for biological specimens but has been adapted for geomicrobiological samples. The procedure of making ultrathin sections is laborious relative to whole-mount samples. In general, the aqueous component of a sample needs to be replaced by a solid medium, i.e., resin, to support organic and inorganic structures so that ultrathin sections can be made. The primary benefit of ultrathin sections is that extracellular versus intracellular regions of a cell can be differentiated. Viewing both sides of the cellular envelope may be important for characterizing microbe–mineral interactions such as modes of passive or active biomineralization. Another benefit of this technique is that larger cells, such as filamentous bacteria, and regions of biofilms can be made into ultrathin sections. The limitation of this technique is that highly mineralized samples such as microbialites are difficult (sometimes impossible) to cut into an ultrathin section using the diamond knife during ultramicrotoming. Extensively mineralized samples may also damage a diamond knife, an expensive and specialized tool that is used for cutting ultrathin sections from a resin-embedded sample.

7.5.1 Enrobing

Ultrathin sections require only a small amount of sample material. Therefore, it is more practical to work with samples in 1 to 2 mL-volume Eppendorf tubes, since fixation, staining (an optional step described later), and enrobing can be performed in the same tube. Start by making a pellet of a fixed sample by centrifugation. A speed of 12 000 × *g* for 30–60 seconds is usually sufficient to form a pellet that is not too compacted (Figure 7.4A and B). Note that centrifugation at higher speeds or for prolonged durations can distort the shape or completely disrupt cells, especially if minerals are present. Set aside the pelleted sample to make the enrobing medium. When setting a pelleted sample aside, it is best to leave all or some of the supernatant to prevent the pellet from dehydrating.

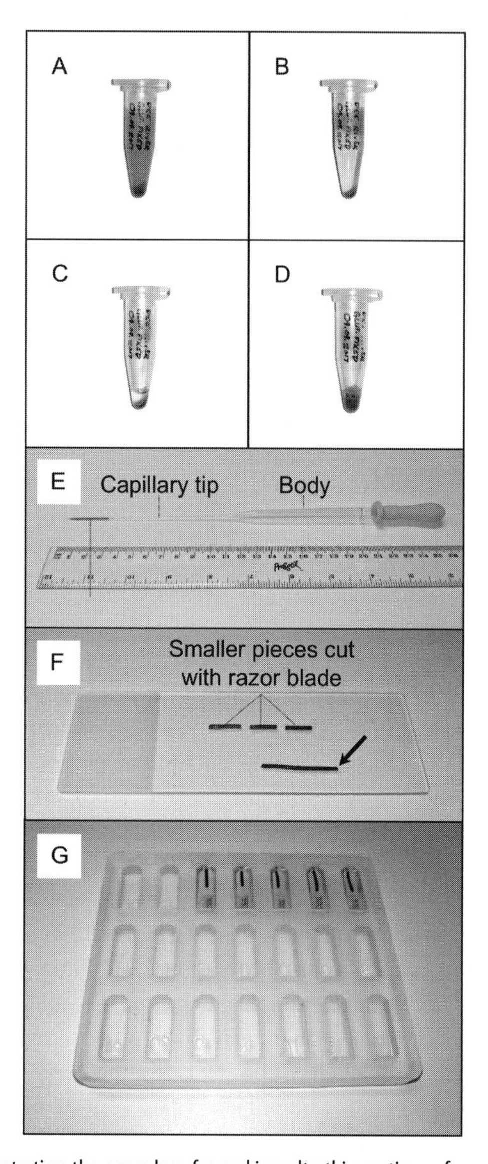

Figure 7.4 A series of photographs illustrating the procedure for making ultrathin sections of a sample containing both microbes and minerals, i.e., an iron-oxidizing bacterial consortium, in a 1.5 mL-volume Eppendorf tube. Glutaraldehyde was added to the sample for a final $2.5\%_{(aq)}$ concentration, and the sample was incubated for 24 hours. Note that there is little to no headspace, ensuring that all constituents of the sample were immersed in the fixative solution (A). A pellet of bacteria and minerals after centrifugation (B). Approximately 50 μL of dissolved $2.5\%_{(w/v)}$ Noble agar added to a pellet (C) and mixed using the tip of a transfer pipette (D). For making the "noodle shape," the enrobed sample is only drawn up and kept at the capillary tip of the Pasteur pipette (E). As it cools and begins to solidify, it is pushed out onto a clean glass slide (arrow, F) and cut into smaller pieces using a razor blade (F). Enrobed and dehydrated samples are placed at one end of the wells of a mold. Each well is filled with resin and cured. Note that the labels, written in pencil, are placed at the opposite end of the well (G).

Noble agar is the solidifying agent that is conventionally used as the enrobing medium. The purpose of enrobing samples is to provide additional support for cells and emplace the minerals. This is necessary because it is often difficult to embed cells directly into resin due to its viscosity. Enrobing also concentrates a sample into a defined volume, which makes it more practical for making ultrathin sections using an ultramicrotome. To make the enrobing medium, start by preparing a $2\%_{(w/v)}$ Noble agar solution using filter-sterilized, deionized water in a sterile and heat-resistant glass container. Cover the container with a piece of aluminum foil to prevent contamination. The $2\%_{(w/v)}$ Noble agar solution will initially look slightly opaque. To completely dissolve the Noble agar, bring the solution to a gentle boil until it appears transparent. Bunsen burners, microwaves, heating blocks, or autoclaves can be used to boil the enrobing medium. Keep in mind that this equipment should only be used as specified by the laboratory; furthermore, personal protective equipment is necessary when handling hot solutions in glass containers. As a general rule, the volume of enrobing medium should not exceed more than 30% of the container's capacity to avoid the agar boiling over. It is possible to use $2\%_{(w/v)}$ agarose as the enrobing medium instead of Noble agar. A difference between these two agents is the rate at which the enrobing mediums begin to solidify. At room temperature, agarose generally polymerizes more quickly than Noble agar. The benefit of slower polymerization, however, is that it allows more time to perform the enrobing procedure, which requires some dexterity, especially for beginners. The type of enrobing medium is an individual choice.

To enrobe a sample, remove and discard the supernatant from the Eppendorf, leaving only the pelleted sample. The supernatant contains residual fixative and should be discarded appropriately. When the enrobing medium has cooled to a temperature at which the glass can be handled with bare hands, it is ready for use. If the enrobing medium is still hot, it can distort or destroy the fine structure of the cells. Use a glass Pasteur pipette to add approximately 50 μL of enrobing medium to the pellet. Using the tip of the same pipette, stir the sample so that the pellet is homogenized within the enrobing medium (Figure 7.4C and D). Long glass Pasteur pipettes, i.e., greater than 10 cm in length, are often easier because the tip acts as a better stirring rod and can also be used as a mold (described later). Once the pellet is thoroughly mixed in the Noble agar solution, allow the enrobed sample to polymerize at the bottom of the Eppendorf tube. Once polymerized, the enrobed sample will have a cone-like shape. There is an option, however, to make an enrobed sample into a noodle-like shape, which is achieved by using the glass Pasteur pipette. The benefit of making enrobed samples into noodle shapes is that the sample is concentrated into a straight line, which makes it more convenient for (re)sectioning with the ultramicrotome. To make the noodle shape, draw up the enrobed sample up the capillary tip of the glass Pasteur pipette (Figure 7.4E). During this step, the enrobed sample should still be warm and fluid. As the enrobed sample cools, slowly squeeze it out of the pipette and onto a clean glass slide. As the enrobed sample cools and is pushed out of the pipette, it will form a noodle shape (Figure 7.4F). This is a temperature- and time-sensitive procedure. It is important to understand that heat from the enrobed sample will transfer to the tip of the pipette, which may increase the rate of cooling and polymerization. A way to check whether the enrobed sample is starting to become more solid is by moving it no more than 10 mm up and down the length of the capillary tip. This is done by gradually applying and releasing pressure on the pipette bulb. As the enrobed sample begins to polymerize, it will have more traction, and its movement within the pipette will reduce. Do not draw the enrobed sample too far up the pipette; it will either get stuck in the "body" of the pipette or completely solidify before it can be pushed out of the opening of the capillary tip. This optional step of shaping the enrobed sample requires precision and timeliness. Therefore, it is highly recommended to practice this technique on any extra samples to become familiar with the tactility before applying it to a precious sample intended for analysis. Using a new razor blade or scalpel, cut the enrobed samples into approximately 5 mm-long segments (Figure 7.4F). For cone-like samples that have solidified inside the Eppendorf tube, use a clean probe to remove the sample. If the sample is stuck, add a couple of drops of filter-sterilized water to loosen the sample or cut the tube to make it easier to fish out. Place the enrobed sample on a clean glass slide and cut into desired pieces with a razor blade or scalpel: rectangular blocks are usually easiest to cut. The pieces should be small enough to fit into a block mold. Note that new razor blades or scalpels make cleaner cuts and reduce the chances of contamination. When working with sharp tools, exercise caution and dispose of used blades and scalpels accordingly. Furthermore, cutting enrobed samples on glass slides is a preference; however, any clean smooth surface can be used.

7.5.2 Dehydration

In general, the purpose of dehydrating an enrobed sample is to gradually replace water with an organic solvent, because hydrated samples cannot be analyzed inside a conventional TEM. For TEM ultrathin sections, the viscosity of resin must be reduced using an organic solvent so that it infiltrates the entire sample (see Section 7.5.3). The organic solvent used for the dehydration is dependent on the type of resin that is used for embedding. For example, ethanol is used with London Resin (LR) White (Electron Microscopy Sciences©), a type of resin that can polymerize rapidly with a chemical accelerator. This type of resin is suitable for geomicrobiological samples containing oxygen-sensitive minerals such as nanophase metal sulfides. Procure 812 resin (Electron Microscopy Sciences©) is a slow-polymerizing resin that requires acetone (or propylene oxide) to reduce its viscosity. There are a variety of other resins that are commercially available, and the choice of one over another is dependent on the nature of the sample. It is a good idea, therefore, to plan ahead and ask a microscopist's advice on which resin and corresponding organic solvents are best suited to a given sample. See Glauert (1974) or Mollenhauer (1963) and references therein for more information about different types of resin embedding. For this chapter, this section will focus on the dehydration of enrobed sample intended for embedding in Procure 812 resin.

Once the enrobed sample has been cut using a clean razor or scalpel, place the enrobed sample into a clean container that has a sealable cap, such as a 7 mL glass scintillation vial with a screw cap. Other types of containers can be used as long as they are sealable and are resistant to acetone. Add enough $25\%_{(aq)}$ acetone (diluted with distilled deionized water) that the enrobed sample is suspended in the solution. Seal the container to prevent the acetone from evaporating, gently agitate the container to mix the samples in the solution, and incubate for 15 minutes. Note that as the enrobed samples dehydrate, they will eventually sink to the bottom of the scintillation vial and will also become harder. After incubation, remove and discard the fluid phase in the appropriate manner. Repeat this step using 50, $75\%_{(aq)}$, and 100% acetone. Repeat the last step involving 100% acetone two more times to ensure that all water has been replaced.

As previously mentioned, Procure 812 resin can be mixed with acetone or propylene oxide to make it less viscous. It is easiest to perform the dehydration with acetone as it requires fewer steps and causes less extraction of cell constituents (Grahams and Beveridge, 1990). If acetone is not available, a dehydration series using absolute ethanol can be performed. However, additional steps to substitute ethanol with propylene oxide are required after the ethanol dehydration, since Procure 812 can be mixed with propylene oxide. For this added procedure, mix 100% absolute ethanol with propylene oxide in a 3:1 ratio. Remove and discard the last 100% absolute ethanol and add the 3:1 (ethanol: propylene oxide) mixture. Seal the container to prevent evaporation and allow the sample to incubate for 20 minutes at room temperature. After incubation, remove and discard the solution using a sterile pipette and repeat this procedure using 1:1 and 1:3 (ethanol: propylene oxide) and 100% propylene oxide. Repeat the last step using 100% propylene oxide to ensure that all the ethanol has been removed. It is important to note that propylene oxide is a highly volatile organic solvent and should also be used in a fume hood.

7.5.3 Resin Embedding

Once all the water has been replaced with an organic solvent, the next step is to gradually substitute the organic solvent with resin. This resin-embedding process can be performed in the same containers that were used for the dehydration. Start by preparing a stock of Procure 812 resin as indicated by the manufacturer's instructions. Note that this type of resin is comprised of different constituents that when mixed in varying proportions will polymerize as a soft, medium, or hard resin. It is recommended to prepare a resin with medium strength, because it holds organic constituents in place and still has enough flexibility to prevent minerals from falling out of the resin while using the ultramicrotome. Hard resins tend to be more brittle and will not section evenly; furthermore, hard resins will reduce the sharpness of ultramicrotome knives, thereby reducing their longevity. In separate scintillation vials or other acetone-resistant containers, prepare aliquots of 100% acetone (or propylene oxide if used instead) with 100% Procure resin in ratios of 3:1, 1:1, and 1:3. Remove and discard the 100% acetone dehydration solution from the container with the enrobed samples. Add the 3:1 acetone:resin mixture to the enrobed sample and seal the container to prevent evaporation of the acetone. Gently agitate the container to mix the samples within the solution and incubate for 30 minutes at room temperature. Mixing can be performed by tapping the container or placing the containers in an automated end-over-end or circular rotator operating at a slow speed. After incubation, remove and discard the 3:1 acetone:resin mixture in the appropriate manner. Repeat this procedure with the remaining 1:1 and 1:3 acetone:resin solutions and three times with 100% Procure 812 resin. The last resin exchanges should be incubated for 1 hour per exchange. It is important to understand that substituting an organic solvent with resin is performed gradually and with increasing resin concentrations because this reduces the potential of causing any distortion to ultrafine cellular structures. If a microwave is available, this instrument can be used to reduce the time required for incubation by accelerating the rate of organic solvent and resin penetration into the sample (Giammara, 1993). Once all the acetone has been replaced by resin, transfer the sample pieces into wells of an embedding mold with the sample at one end. If the wells do not have embossed numbers to identify each block, small pieces of paper with sample identifications can be placed in the molds with the sample. Only use pencil for writing sample names on the paper, since ink will diffuse into the resin; furthermore, keep the paper label clear from the sample, so that it does not interfere during ultramicrotome sectioning (Figure 7.4G). Completely fill each well containing a sample with 100% Procure 812 resin and ensure that the resin does not have any air bubbles. Any air bubbles that do form can be removed by popping them with a clean probe, e.g., a needle or toothpick, but be careful not to move the samples within the well. Once all the samples have been arranged within the well and the resin has been added, place the entire mold in a 60 °C incubation oven for 48 hours to polymerize the resin.

7.5.4 Block Trimming and Ultramicrotome Sectioning

Once the resin has polymerized, the embedded samples (now commonly referred to as blocks) are ready for trimming and ultramicrotome sectioning. There are different approaches by which an embedded sample can be trimmed; furthermore, the type of ultramicrotome can vary according to brand or model. Therefore, this section will highlight the basic concepts and principles of trimming and sectioning to provide an understanding of what to anticipate when it comes to performing these procedures. Trimming a block involves cutting a trapezoid shape at the end of the block where the sample is located. There are three reasons for trimming the trapezoid shape. First, it reduces the amount of resin that the knife must cut through during sectioning. Second, the trapezoid shape gives the strongest support to the embedded sample, thereby reducing the potential for sample distortion by the compression of the knife during sectioning. Third, the base of the trapezoid is the longest dimension (≤2 mm in length) that first comes into contact with the knife. This orientation allows ultrathin sections to slice off the block as a continuous "ribbon," which is easier to collect onto a grid (Figure 7.5). New razor blades are best for trimming the trapezoid on the block because they are sharper and make cleaner cuts. Trimming the sides of the blocks should be done with thin and gradual slices. Deep and jagged cuts should be avoided, as these could cause fractures within the resin. The trimmed edges should be approximately 50° to the trapezoid face containing the sample, creating a pyramidal shape. Once the block has been trimmed, a glass knife is generally used to cut a "mirror surface" on the trapezoid-face. In general, an ultramicrotome and a diamond knife, e.g., Diatome, are used to cut ultrathin sections. The diamond knife is often attached to a well that is filled with filter-sterilized water. As the ultrathin sections are cut, they come off the block as a ribbon and float on the surface of the water. Ultrathin sections ranging between 60 and 90 nm in thickness are ideal for imaging, because semi-thin sections (>100 nm) contain a greater volume of sample, which is more material for the electron beam to transmit through. As the diamond knife is sectioning the block, the thickness of the floating sections can be determined based on the color of the section when it reflects light. Ultrathin sections of

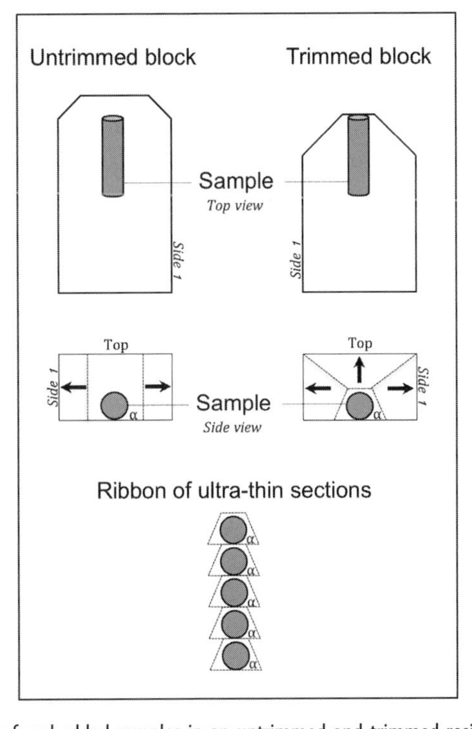

Figure 7.5 A schematic representation of embedded samples in an untrimmed and trimmed resin block. In top view, the sample is a noodle shape and its length is parallel to the longest length of the well (also see Figure 7.4G). Arrows indicate the downward-sloping sides due to the shape of the well of the mold. In the side view, the sample-bearing face is indicated by α. The block is trimmed with a razor blade so that the sample-bearing face (α) is in the shape of a trapezoid and forms an overall pyramid-like end. Downward-sloping sides (arrows) should be ca. 50° to the trapezoid face. When the trapezoid face is cut with a diamond knife, it will form a ribbon of serial ultrathin sections.

70 nm thickness will appear silver, whereas 100 nm semi-thin sections will appear gold (see Peachy, 1958 for details). Once a desired number of ultrathin sections have been cut, they are collected onto a grid and allowed to air dry before being analyzed.

7.5.5 Interpretation of Ultrathin Section Samples

Unlike the perspective of a whole-mount sample (Figure 7.3A and B), material within an ultrathin section will appear two-dimensional. Keep in mind that an ultrathin section represents a very thin "slice" through any three-dimensional object. Therefore, nanophase particles smaller than the thickness of the ultrathin section can be seen in their entirety, whereas a fraction of larger-size minerals will occur within an ultrathin section. In the following example, a sample of naturally occurring nanophase gold colloids was observed in a clay substrate immobilized as a whole mount (Figure 7.6A) as well as an ultrathin section (Figure 7.6B). Both samples were characterized using a JEOL 1010 TEM operating at 100 kV. When the sample is prepared as a whole mount, the gold colloids appear darker in

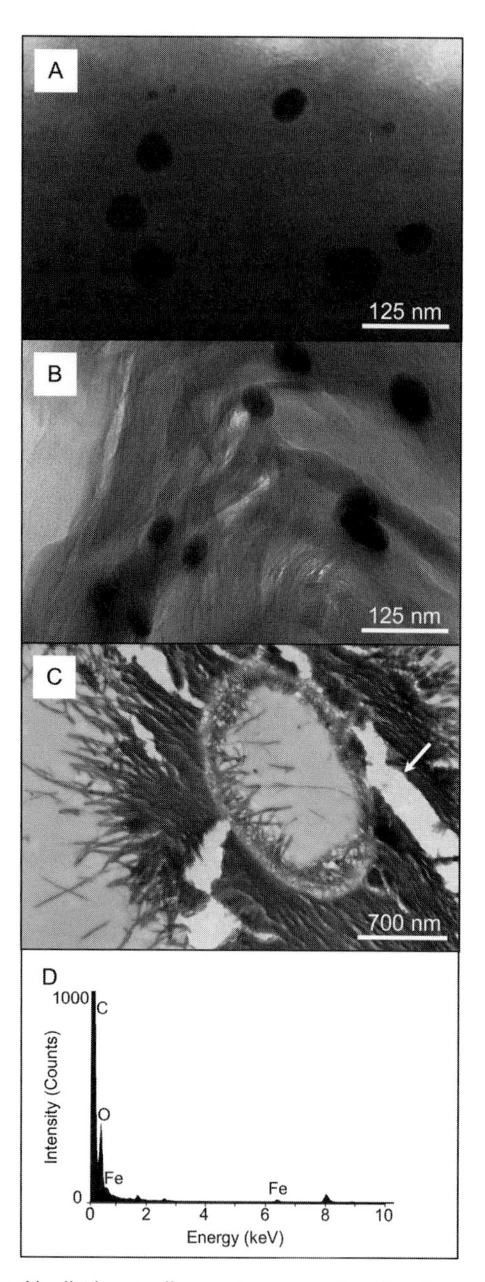

Figure 7.6 A micrograph of nanophase gold colloids naturally occurring among clay minerals prepared as a whole-mount (A). A micrograph of the same sample represented in the previous micrograph but prepared as an ultrathin section (B; see Shuster et al., 2017a). A micrograph of an ultrathin section of mineralized microbial cell enriched from an acid mine drainage environment (A; see Shuster et al., 2017b). An EDS spectrum of the semi-acicular minerals illustrated in the previous micrograph (D).

comparison to the clay matrix, which is attributed to the difference in electron density of the materials. The clay minerals appear somewhat amorphous because the sheets are stacked with the greatest surface area in contact with the grid. In the ultrathin section, however, the structure and orientation of the sheets are visible.

For microbial samples, only a fraction of a cell will be represented in ultrathin sections. Therefore, the morphology and orientation of cells will determine the cross-sectional or tangential-section shape. Cocci will appear circular; rod-shaped and spirillum cells can produce rod, circular, or oval shapes; filamentous cells can look filamentous, circular, or oval. Depending on the sample, ultrathin sections can help characterize different stages of biomineralization, from nonmineralized to extensively mineralized cells.

Microfossils are examples of highly mineralized cells and are relicts of microorganisms. In its simplest representation within an ultrathin section, a microfossil often consists of a mineralized cell envelope that preserves the original morphology of the microbial cell. Based on the need for a template to produce a hollow structure, the interpretation of these materials as microfossils can be arguably justified when a broad range of biomineralization stages are preserved and represented in a sample. In the next example, a microbial consortium was enriched from an acid mine drainage environment. The consortium was fixed with glutaraldehyde, prepared as an ultrathin section, and characterized using a Philips CM10 TEM operating at 80 kV (Figure 7.6C). In this sample, hollow structures were interpreted as microfossils because they were the size and shape of microbial cells. Although the original cells and organic materials were not observed, the extracellular surfaces of some cells were extensively mineralized with semi-acicular iron oxyhydroxide minerals, thereby preserving the overall structure of the cell and the cell wall. Since there was a lower degree of intercellular mineralization, it is reasonable to suggest that the microbial cell could have acted as a site for mineral nucleation even after death. It is possible that pieces of minerals can fall out of an ultrathin section when the plastic matrix does not provide enough structural support (Figure 7.6C, arrow). EDS is a semi-quantitative analytical technique that can identify the elemental composition of a mineral assemblage within a sample. The mineralized sample was characterized with a Philips EM400T equipped with a LINK X-ray EDS detector operating at 100 kV. At the nanometer scale, the acicular minerals that mineralized the extracellular surface of cells were composed of iron, sulfur, and oxygen (Figure 7.6D).

7.6 Staining – an Optional Procedure

Because bacteria, Formvar film, and resin are primarily composed of carbon, it can be difficult to differentiate these materials, especially in ultrathin sections. Soluble heavy metals can be used to stain various components of a cell to increase its electron density, i.e., darkness, in order to achieve greater contrast and resolution compared with the resin or Formvar background. An added benefit of staining is that the heavy metal can act as a conductor and reduce the potential for cellular damage by the electron beam.

The effectiveness of a stain to enable greater structural characterization is dependent on the concentration of the cellular component and the molecular density of the stain. In general, staining procedures can be performed either during fixation or after embedding. Since stains are heavy metal solutions, it should be noted that they can potentially affect the structure and chemistry of mineral constituents within a sample. Therefore, staining is purely an optional procedure. If staining is employed as a treatment, it should be performed with vigilance, because heavy metals are highly toxic substances. Therefore, staining solutions must be prepared using personal protective equipment including double nitrile gloves, goggles, facemask, and laboratory coat. All stain preparations and staining procedures should be conducted in a certified fume hood under the guidance of an authorized individual. Different heavy metal solutions can be used for staining, and a variety of techniques can be found in the literature (see Hayat, 1959 and references therein; Mc Cutcheon and Southam, 2018). Therefore, this section will briefly highlight the use of osmium tetroxide and uranyl acetate as stains after glutaraldehyde fixation, and uranyl acetate and lead citrate staining of ultrathin sections, respectively.

It is possible that some metals could precipitate out of solution; therefore, centrifugation (10 000 × g for 3 minutes) or filtration (0.1 μm pore-size filter) may be required to remove any precipitates prior to use. Metal precipitates in the staining solution will create unevenness in the staining and potentially incorrect mineral identification. The benefit of osmium tetroxide is that it acts both as a stain for lipids and as a fixative. The process of staining as well as fixing with osmium tetroxide is called "osmication" and is considered a post-fixation procedure because it is performed after the primary fixation step (see Section 7.1). Once a sample has been fixed and pelleted, remove and discard the supernatant. Add approximately 1 mL of $2\%_{(aq)}$ osmium tetroxide solution to the pelleted sample, seal the container, and incubate at room temperature for 30 to 60 minutes. It should be noted that $2\%_{(aq)}$ metal solutions are conventionally used, since higher metal concentrations can overstain samples. Furthermore, the time required to stain a sample with any metal solution is dependent on the cell density and the concentration of the staining solution. In general, a higher density of cells and biofilm may require more time, i.e., 60 minutes, especially if the staining solution is more dilute than 1%. After incubation, remove and discard the osmium tetroxide solution in the appropriate manner and rinse the stained pellet three times with filter-sterilized, distilled water. At this stage, the sample can be prepared as a whole-mount sample (Figure 7.7A) or processed further for ultrathin sections.

Post-staining is the process of staining samples that have already been prepared as whole-mounts (negative stain) or ultrathin sections (positive staining) and have been placed on grids. This procedure can be performed whether or not osmication has been performed, since the use of additional heavy metals such as uranyl acetate and lead citrate for ultrathin sections will increase the contrast intensity. Like osmium tetroxide solution, uranyl acetate and lead citrate solutions should be centrifuged or filtered prior to use to remove any precipitates. To post-stain a whole-mount, the immobilized sample is reacted with stain for 1 to 2 seconds, and then the stain is removed using filter paper, leaving a thin film of stain on the grid. For ultrathin sections, staining with uranyl acetate is generally easier and uses the floating grid method as described in Section 7.2.3. Within a sterile glass Petri dish, place

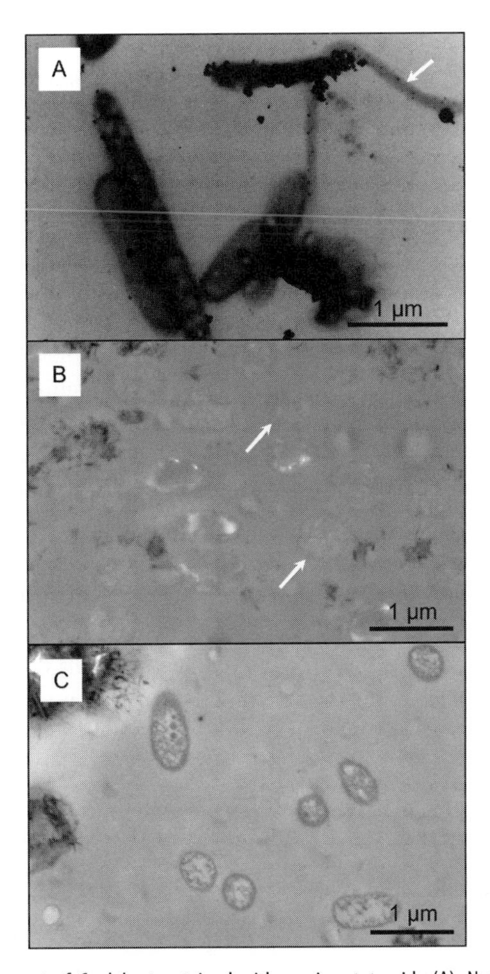

Figure 7.7 A micrograph of a whole-mount of *Caulobacter* stained with osmium tetroxide (A). Note that nanophase particles ranging from 10 to 100 nm in size are associated minerals from the natural environment. More importantly, the osmium stain provided a shade gradation, providing a subtle indication of the height of the cell and stalk (arrow, A). A micrograph of an iron/sulfur-oxidizing microbial consortium enriched from Rio Tinto made into an ultrathin section without staining. Nonmineralized cells are difficult to differentiate from the resin background (arrows, B). A micrograph of a different region of the same ultrathin section seen in Figure 7.7B and post-stained with uranyl acetate (C). The structure of the cells is more prominent relative to the resin background. From this micrograph, the microbial cells can be interpreted to be rod-shaped cells.

a 10 μL drop of $2\%_{(aq)}$ uranyl acetate solution onto a piece of Parafilm®. Float the grid containing a sample on a droplet of uranyl acetate solution for 5 to 10 minutes. Remove the grid and rinse with filter-sterilized, distilled water. Grids can be rinsed using the procedure described in Section 7.2.2 or by gently dipping the entire grid into a beaker of filter-sterilized, distilled water. Wick off any residual water using a piece of filter paper and allow to air dry. Cells that originally appeared "invisible" within the ultrathin section will be more pronounced (compare Figure 7.7B and C). For staining with 2% lead citrate, the grid is floated on a drop of 2%$_{(aq)}$ lead citrate adjacent to a NaOH pellet (to capture CO_2 from the atmosphere and prevent $PbCO_3$ artifacts from forming on the section) for 5–10 minutes. The grid is then rinsed with a few drops of 0.05 N NaOH and then with deionized water and air dried.

7.7 Characterization of Focused Ion Beam–milled Sections

In some cases, geomicrobiology samples such as thrombolites or canga, a consolidated iron-rich rock, can be partially or completely mineralized, thereby making conventional embedding techniques impracticable. Focused ion beam (FIB) milling is a technique commonly used in scanning electron microscopy analysis (see Section 6.6.2). This technique is ideal for preparing TEM "lift-outs" of lithified samples. In general, a lift-out is a precisely cut section of a sample. Using the FIB, a lift-out is polished to only a few nanometers in thickness. Once the lift-out has been attached to a grid, it can be characterized using a TEM in the same way as ultrathin sections are analyzed. The advantage of lift-outs is that microfossils and nanophase minerals contributing to the mineralization of microbial cells (Figure 7.8) in lithified materials that cannot be ultrathin sectioned can be analyzed and compared with modern analogues, as demonstrated in the interpretation of ultrathin sections.

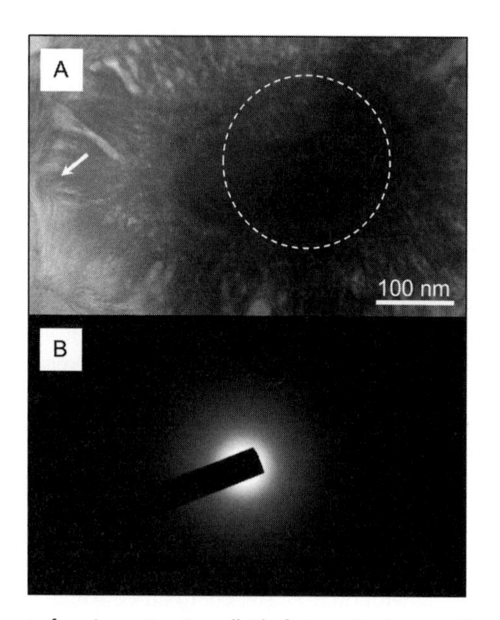

Figure 7.8 A micrograph of an FIB lift-out of a micrometer-size colloid of stevensite that contributed to the mineralization of filamentous cells in a thrombolite (A; see also Figure 6.5B). The center of the colloid appears amorphous (dashed circle, A) whereas the outer edge of the colloid appears to have nanometer-scale "laminated" structure (arrow, A). SAED had no diffraction pattern, suggesting that the center of the colloid was amorphous (B).

7.8 Conclusion

TEM enables the characterization of ultrafine microbial structures and nanophase particles at high magnification and resolution. As demonstrated throughout this chapter, the preservation of biological and mineralogical constituents of a sample is essential for adequate characterization. There are a variety of techniques for sample preparation that have their own respective advantages and can be reasonably adapted to any given sample. Furthermore, with the increasing sophistication of TEMs, the possibilities for sample preparation can be limitless and are themselves an art. This chapter provided a basic outline with examples of how to prepare various types of geomicrobiological materials as whole-mount and ultrathin section samples. While these techniques are conventional, they provide a fundamental basis for understanding the importance of sample preparation when it comes to the analysis of natural materials and the associated structure–function relationship of microbe–mineral interactions. It is important to realize that these techniques should be considered as a starting point that can be modified to achieve the best possible characterization of a sample. Therefore, TEM analysis provides micrographs, which represent snapshots of microbe–mineral interactions, while quantitative SAED and semi-quantitative EDS provide chemical analysis of geomicrobiological materials. Together, the structural and chemical characterization can help make interpretations of biogeochemical processes from samples derived from laboratory-based experiments or from the natural environment.

7.9 References

Bozzola, J.J. and Russel, L.D. 1992. *Electron microscopy: Principles and techniques for biologists*. Boston, Jones and Bartlett Publishers.

Giammara, B.L. 1993. Microwave embedment for light and electron microscopy using epoxy resins, LR white and other polymers. *Scanning*, 15, 82–87.

Glauert, A. 1974. *Fixation, dehydration and embedding of biological specimens*. Amsterdam, North-Holland Publishing Company.

Graham, L.L. and Beveridge, T.J. 1990. Evaluation of freeze-substitution and conventional embedding protocols for routine electron microscopic processing of Eubacteria. *Journal of Bacteriology*, 172, 2141–2149.

Hayat, M.A. 1959. *Principles and techniques of electron microscopy: Biological applications*. New York, Van Nostrand Reinhold Company.

Johnston, C.W., Wyatt, M.A., Ibrahim, A., et al. 2013. Gold biomineralisation by a metallophore from a gold-associated microbe. *Nature Chemical Biology*, 9, 241–243.

McCutcheon, J. and Southam, G. 2018. Advanced biofilm staining techniques for TEM and SEM in geomicrobiology: Implications for visualizing EPS architecture, mineral nucleation, and microfossil generation. *Chemical Geology*, 498, 115–127.

Mollenhauer, H.H. 1963. Plastic embedding mixtures for use in electron microscopy. *Stain Technology*, 39, 111–114.

Peachy, L.D. 1958. Thin sections: I. A study of section thickness and physical distortion produced during microtomy. *Journal of Biophysical and Biochemical Cytology*, 4, 233–242.

Shuster, J., Reith, F., Cornelis, G., et al. 2017a. Secondary gold structures: Relics of past biogeochemical transformations and implications for colloidal gold dispersion in subtropical environments. *Chemical Geology*, 450, 154–164.

Shuster, J., Reith, F., Izawa, M.R.M., et al. 2017b. Biogeochemical cycling of silver in acidic weathering environments. *Minerals*, 7(11), 218.

8 Whole Cell Identification of Microorganisms in Their Natural Environment with Fluorescence *in situ* Hybridization (FISH)

NATUSCHKA M. LEE

Abstract

One of the main goals in biogeochemistry is to explore the global relationships between organisms and chemical elements in different ecosystems. A diversity of analytical techniques based on chemical-physical or molecular biological procedures are available to explore different organisms and their abiotic and biotic interactions in a variety of ecosystems. Even though many of these modern analytical techniques are irreplaceable in today´s research, most of them can only provide indirect results because they are built on a "black-box" approach, where the biological species in an ecosystem or a geological environment are disrupted for extraction of nucleic acids, proteins, etc. Essential biological information, such as the morphology of specific species, their location, distribution, association with other organisms in their natural environment, and individual activities and functions, is therefore lost. Fluorescence *in situ* hybridization (FISH) helps retrieve this information without either cultivation or extraction of cell components, and can therefore provide a quick and useful complement to different "black-box"-based approaches. FISH is based on fluorescently labeled gene probes with a unique nucleotide composition designed to match specific genes in different cellular species. Thus, different biological species can be identified simultaneously with different gene probes labeled with different fluorochromes in their natural environment. The technique has undergone extensive development with around 30 variations for different applications. FISH is evaluated either by microscopy (e.g., fluorescence microscopy, Raman micro spectroscopy, Nano-SIMS), or by nonmicroscope-based methods, such as flow cytometry, microarray technology, or molecular biological methods such as proteomics. This chapter will serve as a guide for sample preparation, selection of appropriate FISH protocols, evaluation and design of gene probes, and evaluation of FISH experiments.

8.1 Introduction

Since the pioneering studies by Lourens Baas Becking in the 1930s, one of the main research goals in biogeochemistry has been to explore "the relationships and interactions between organisms and the Earth" (Becking, 2015). An assortment of techniques are available to study these interactions, including cultivation, microscopy, chemical-physical analytical methods, and molecular biological methods, each with its benefits and disadvantages. Today, advanced molecular biology based on extracted cell components, such as nucleic acids for gene sequencing, is the most commonly used technique for the identification of organisms in their environment. However, while these methods offer several benefits, such as the possibility to obtain cultivation-independent and, at least, semi-quantitative results from large quantities of environmental samples, other types of information are lost with these cell-disruptive methods. Thus, essential information for a larger understanding of individual organisms within their environment is lost, such as their morphologies, locations, distributions, *in situ* physiology, and association patterns of different cellular species both within their biological community as well as within their environmental context. Standard microscopic methods, even when combined with specific stains to visualize specific chemical or biological structures, are of limited use, as they cannot provide a detailed identification of microorganisms due to their small size and limited morphological variety. However, the fluorescence *in situ* hybridization (FISH) technique can overcome many of these limitations, since it is based on the use of fluorescently labeled gene probes with a unique nucleotide composition, which can be made to specifically match (hybridize) with the corresponding unique gene fragment within an intact organism and can then be visualized by fluorescence microscopy or evaluated by other means. Thus, compared with detection techniques based on extracted nucleic acids, results retrieved using FISH are based on whole, intact cells in their natural environment (Amann and Fuchs, 2008).

The first FISH procedure (oligonucleotide FISH) in the microbiological field was developed in the late 1980s based on the FISH procedures developed previously in the medical field for exploration of chromosome structure and function (Volpi and Bridger, 2008). Since then, oligonucleotide FISH for microorganisms has been applied with great success in many different research fields. FISH became an essential part of the novel ribosomal RNA approach for a cultivation-independent identification of microorganisms in different types of environment (Amann and Fuchs, 2008). However, the oligonucleotide FISH had two main drawbacks: i) it could not provide sufficient information about the activity or the function of the identified organisms; and ii) certain organisms, in particular in oligotrophic ecosystems, could not be explored due to weak gene probe signals and/or a confounding background. The most common explanations for this are low cellular activity, due to lack of sufficient amounts of substrate, or a large amount of either autofluorescing mineral particles or biochemical compounds, such as chloroplasts in phototrophs or the cofactor F420 in methanogens. Thus, during the last two decades, several modifications and expansions of the original oligonucleotide FISH procedure have been made. This chapter will give a brief overview of different FISH techniques and how they can complement other analytical techniques (Section 8.2), sample preparation (Sections 8.3 and 8.4), the most commonly used FISH techniques (oligonucleotide FISH and catalyzed reporter deposition FISH [CARD-FISH]; Sections 8.5 and 8.6), existing gene probes for FISH and how new probes can be designed for specific purposes (Section 8.7), and how to critically evaluate FISH results (Section 8.7). These surveys are accompanied by several protocols provided in the online supplementary materials (1 to 8).

8.2 Application Fields for FISH and Overview of Different FISH Procedures

FISH can be applied to archaea, bacteria, eukarya, some viruses, including bacteriophages (e.g., Boas et al., 2016, Chou et al., 2013, Dang and Sullivan, 2014), and even to some extent to spores (e.g., Weerasekara et al., 2013). Thus, FISH can be applied to samples from all kinds of ecosystems – from microbial cultures in the laboratory to complex environmental systems (terrestrial, aquatic, geological objects such as rocks, caves, subsurface, and macroorganisms). Furthermore, FISH can also be applied to a variety of anthropogenic systems, from the human body to biotechnological processes such as food production and biological wastewater treatment, or items subject to microbial degradation, such as pipelines, art or historical objects, and the surfaces of furniture, vehicles, or buildings. FISH experiments are mostly evaluated by different types of microscopy, such as fluorescence microscopy (in particular confocal laser scanning microscopy; Amann and Fuchs, 2008), electron microscopy (e.g., Gérard et al., 2005), nanoscale secondary ion mass spectroscopy (Dekas et al., 2016; Eichorst et al., 2015), or Raman microscopy (Wang et al., 2016). Quantitative FISH results can be obtained from microscopic images via digital image software (Daims, 2009). FISH results can also be evaluated by indirect methods, such as flow cytometry (Neuenschwander et al., 2015), microarray technology (Nikolaki and Tsiamis, 2013), or spectral imaging (Valm et al., 2013). Depending on the cell type, sample type, and research goals, different types of FISH can be employed. Today, around 30 different FISH procedures exist for different types of application, depending on the biological species, ecosystem, and research goals. Each is based on different strategies, in order to i) overcome methodological limits of previously developed FISH procedures, ii) retrieve additional specific information, such as *in situ* physiology or gene expression, or iii) enrich specific cell species for further single-cell analysis, such as gene/genome sequencing or omics.

The various FISH procedures differ from each other by employing different strategies, such as alternative amplification strategies of probe signals (e.g., polynucleotide FISH, CARD-FISH, double labeling of oligonucleotide probes [DOPE]-FISH, and multilabeled oligonucleotides [MiL]-FISH); targeting alternative genes (e.g., functional genes instead of ribosomal genes, which are the main gene target for most FISH applications); or alternative probe chemistries (e.g., peptide nucleic acid [PNA]-FISH and locked nucleic acid [LNA]-FISH), or by combining with other analytical procedures, instruments, and FISH protocols (e.g., microautoradiography [MAR-FISH], stable isotope detection, cell-sorting, magneto-FISH, spectral imaging, quantum dot FISH, combinatorial labeling and spectral imaging [CLASI]-FISH, Raman spectroscopy, NanoSIMS, *in situ* RCA, gene FISH, phage FISH, and bioorthogonal noncanonical amino acid tagging [BONCAT] FISH). A brief survey of these and a guideline on their suitability for different types of samples and research purposes are presented in the supplementary information and, e.g., in Lee et al. (2011), Schimak et al. (2016), and Wagner and Haider (2012).

Compared with molecular methods, such as next-generation sequencing methods, the procedure for FISH is generally simple, inexpensive, and fast (usually ~0.5–3 days), consisting of five principal steps (which apply to most FISH techniques):

i) Sampling (cells from laboratory cultures or complex samples with cells from an arbitrary environment), followed by appropriate fixation and storage of samples (Section 8.3, online supplementary materials 8.2 and 8.3).
ii) Selection of an appropriate FISH procedure, gene probes, and controls (gene probes as well as reference samples (Sections 8.4 and 8.5, Table 8.3, online supplementary materials 8.1).
iii) Performance of the FISH procedure, exemplified by the oligonucleotide FISH procedure and the CARD-FISH procedure (Sections 8.5 and 8.6, online supplementary material 8.7).
iv) Evaluation of FISH results and comparison of FISH results with other methods for an optimal and critical evaluation (Section 8.8, online supplementary files 8.8).

8.3 Sampling and Fixation

Irrespective of the chosen FISH procedure and combinations with other methods, it is essential to employ optimal sample preparations in order to avoid false positive or false negative results. Depending on the organism and the ecosystem, different precautions must be considered. There follow some examples of guidelines on sampling and fixation of samples prior to FISH.

8.3.1 Sampling Strategy

The optimal cell concentration for FISH studies is around 10^7 cells/mL. Samples with cell numbers below or above this may have to be diluted, concentrated (usually by centrifugation or filtration), fractionized (e.g., by filters with different pore sizes), homogenized, or sectioned. Depending on the nature of the sample, it may have to be either washed or pretreated to remove disturbing parameters (e.g., debris, precipitates). A confounding background caused by mineral precipitates such as calcium carbonates can be reduced by decalcification through a chemical treatment (e.g., Shiraishi et al., 2008). Alternatively, cells can be extracted from their environment by chemicals (e.g., using a weak detergent to dissolve aggregates) and/or by density gradient centrifugation (e.g., Caracciolo et al., 2005; see also online supplementary material 8.2).

Cells should be sampled and fixed as soon as possible from their natural environment to avoid shifts in activity level (ribosome content) and/or population structure. If possible, avoid repeated freezing/thawing of samples prior to the fixation. Cells should preferably be collected during their exponential growth state to achieve a strong gene probe signal, since the intensity of the gene probe signal is dependent on the number of ribosomes, which in turn often reflects the level of cell activity (e.g., Rossetti et al., 2007). However, this may not be the case for certain organisms and must therefore be explored individually for different species (see, e.g., Wagner et al., 1995).

Sampling can be done from all kinds of sample types (liquid, solid, or atmospheric). Solid samples must be dissolved in an appropriate buffer (usually a neutral phosphate saline buffer [e.g., 1× PBS; see Section 8.5], but depending on the sample type, other buffers may also be used). Biofilm samples on rocks, buildings, etc. can be cut or removed with special swabs or forceps or even with adhesive tape (e.g., Cutler et al., 2012). Samples from diluted aquatic systems or from the atmosphere must be concentrated onto a membrane filter with a pump. If the structure of the sample must remain intact, special modes can be used to enclose your sample in, e.g., filter paper, agarose or agar (e.g., 0.1–1%, depending on the sample type), or specially designed trays with filters (e.g., Foissner, 1992).

8.3.2 Fixation of Samples prior to FISH

The purpose of the fixation is to i) preserve the morphology of the cells, ii) denature proteins and enzymes, and iii) prevent degradation of the target RNA or DNA. In this way, the shape and nucleic acid content of the cells can be maintained during the hybridization process, since this consists of several chemical-physical stress moments (e.g., increased salt concentration and temperature, and exposure to detergents). In some rare cases, FISH may also work on nonfixed samples and frozen samples (Yilmaz et al., 2010); however it is strongly recommended to explore the suitability of such conditions for unknown samples in order to guarantee reliable results. Due to the complexity and diversity of cell types (size, cell wall type, phylogeny, etc.), various fixation procedures must be employed for different species. The two most common fixation reagents for prokaryotes are paraformaldehyde (PFA, Section 8.3.3) and ethanol (EtOH, Section 8.3.4) (Amann, 1995). The PFA fixation works well for many species from all three domains, but in particular for the Gram-negative species of bacteria. The EtOH fixation (Section 8.3) is usually recommended for Gram-positive species of bacteria, but may also work well on other species within the archaea, bacteria, and eukarya (e.g., protozoa and fungi). Many Gram-negative bacteria species may still produce good FISH signals with the EtOH fixation protocol but may possibly lyse after a longer storage time (weeks or longer, depending on the involved species). For samples with an unknown composition of biological species, it may be wise to employ at least two different types of fixation procedure in parallel (e.g., PFA as well as EtOH). Eukaryotes and viruses may demand other fixation protocols (see the online supplementary material 8.3). A few studies have also employed more sophisticated fixation protocols based on other chemical reagents, such as formalin, methanol, acetone, chloroform, glutaraldehyde, osmium tetroxide, mercury chloride, and picric acid. However, for practical reasons, it is recommended to first start with the two abovementioned fixation procedures (PFA and/or EtOH) and modify them as far as possible before other procedures are considered for further optimization. The standard fixation time is about 3 h but can vary depending on the biological species. Successful FISH results have also been obtained from shorter fixation times down to 10 minutes (e.g., for certain clinical strains for which commercial protocols of PNA-FISH [AdvandDX] or quantum dot FISH [miacom diagnostics] have been developed). Certain species or sample types (e.g., hypersaline prokaryotes) may, however, demand a longer fixation time, up to 12 h or even longer (e.g., Ánton et al., 1999). It is important to avoid over-fixation of samples, as this will decrease the permeabilization efficiency of the gene probes through the cell walls. After fixation, samples are normally stored in a 50% ethanol–buffer solution (to avoid freezing of cells) at –20 °C for longer time periods (even for years). Storage at +4 °C is acceptable for shorter periods (a few days).

8.3.3 PFA Fixation

This protocol is generally recommended for Gram-negative prokaryotes, but it may also apply for other species (e.g., certain Gram-positive prokaryotes, archaea, and eukarya). The protocol below explains A) how to prepare the PFA solution and B) how to fix samples with the PFA solution.

A. **Preparation of the PFA solution** (modified from Amann, 1995)

Chemicals

- Sterile Na_xPO_4 stock solution (phosphate buffer, pH 7.2):
  A) Na_2HPO_4 200 mM = 35.6 g/L (prepare 1 liter)
  B) NaH_2PO_4 200 mM = 27.6 g/L (prepare 250 mL)
  Adjust pH of solution A to 7.2–7.4 by using solution B
- Sterile 1× phosphate buffered saline solution (1× PBS, pH 7.2):
  130 mM sodium chloride
  10 mM sodium phosphate buffer Na_xPO_4
  To prepare 1× PBS: NaCl 130 mM = 7.6 g/L
  Na_xPO_4 10 mM = 50 mL Na_xPO_4 stock solution/L
  Distilled water to 1 000 mL
- Sterile 3× phosphate buffered saline solution (3× PBS, pH 7.2):
  390 mM sodium chloride
  30 mM sodium phosphate buffer Na_xPO_4
  To prepare 3× PBS: NaCl 390 mM = 22.8 g/L
  Na_xPO_4 30 mM = 150 mL Na_xPO_4 stock solution/L
  Distilled water to 1000 mL
- Paraformaldehyde solution (PFA) usually within the concentration range 4–8%: Use high-grade, fresh paraformaldehyde (powder), e.g., from Sigma (# P-6148) and store under proper conditions, in the dark, at +4 °C (or under argon/liquid nitrogen).
- 2 M NaOH and 2 M HCl (for pH adjustments).

Procedure for preparation of PFA (Note: Use protective mask or eye glasses and gloves, and perform all work in a fume hood):

1) For preparation of, e.g., a 100 mL fixation solution: heat 65 mL distilled and deionized H_2O to 60 °C.
2) For preparation of a 4% PFA solution, add 4 g PFA. (Comment: The solution becomes "milky.")

3) To dissolve the milky appearance, add one or a few drops (slowly, drop by drop – avoid overdosage) of 2 M NaOH solution and stir rapidly (use a magnetic stirrer at a medium speed) until the solution has nearly clarified (takes 1–2 min).
4) Remove from the heat source and add 33 mL of 3× PBS. Check that the pH is neutral by dropping a few drops of the solution on a pH indicator strip.
5) Filtrate solution slowly through a 0.2 μm filter into a new tube (e.g., a 15–50 mL Falcon tube). If the filter gets clogged, use a new filter. Collect all PFA waste in specially designed PFA waste bins stored in the fume hood. Follow the regulations for how PFA waste should be handled at your institute.
6) Quickly cool down the solution to 4 °C (e.g., place in an ice-bath) and store it at this temperature up to 24 h (maximum 1 week; however, it is better to aliquot the fixation solution and store at –20 °C for longer periods [at least for months, possibly even for years]). It is also possible to store PFA under liquid nitrogen. Thaw frozen aliquots just before usage and use within 24 hours. Avoid too frequent freezing and thawing.

B. **Standard procedure for PFA fixation** (Note: modifications may be necessary, depending on the nature of the sample and sampling technology):

1) Ensure that you have taken all precautions for optimal sampling as described in Section 8.4 and that your sample is well mixed before you add the fixative.
2) Add three volumes of 4% PFA fixative to one volume of sample (end concentration 1% PFA).* For example, for a 2 mL tube, mix 0.5 mL sample with 1.5 mL 4% PFA.
   * Ratio of fixative versus biomass depends on the PFA concentration. You can use, e.g., a stronger PFA solution to reduce the total amount of volume (less to wash in the following steps).
3) Fix for a minimum of 0.5 h up to 3 h (general recommended time, but exceptions may be relevant for certain organisms that may need even longer fixation times) at 4 °C. If necessary or unavoidable, samples can also be fixed for longer periods (e.g., overnight). However, it is recommended to explore the optimal fixation time for unknown samples. Avoid too weak fixation or too strong fixation of your samples. Furthermore, for systematic comparison of different samples, the same fixation time should be used.
4) Pellet sample by centrifugation (approx. 1–2 min, 5 000 rpm; modify harvest conditions as appropriate; the main goal is simply to produce a pellet of your fixed sample) and remove the fixative (supernatant).
5) Wash cells in 1× PBS (or other suitable buffer) via centrifugation as described earlier, and then resuspend in one volume of 1× PBS.
6) If necessary, repeat steps 2–3. Residues of PFA may produce autofluorescing particles, so it is important to remove all traces of the PFA reagent.
7) For end-fixation (storage at longer periods), add one volume (better ≥1.1 in relation to the volume used for resuspension) of ice-cold 96–99% EtOH and mix (final concentration of EtOH should be 50%). Before you add the EtOH, ensure that your sample is well mixed (use a pipette tip or tooth pick to carefully stir up the pellet).

8) For short-term storage, store at room temperature or at +4 °C. For long-term storage, store fixed samples at –20 °C.

8.3.4 EtOH Fixation

This protocol is generally recommended for Gram-positive prokaryotes but may also apply for other organisms (certain Gram-negative prokaryotes, archaea, and eukarya).

Chemicals: Ice-cold 96–99% EtOH, molecular grade.

Procedure (modified from Amann, 1995):

1) Harvest your sample (where appropriate wash your sample via centrifugation with a suitable buffer) and dissolve in a suitable buffer (e.g., 1× PBS; see supplementary file 2).
2) Before you add EtOH, ensure that your sample is well mixed (use a pipette tip or tooth pick to carefully stir up the pellet).
3) Add one volume (better ≥1.1) of ice-cold 96–100% EtOH (molecular grade) to one volume of sample and mix (final concentration of EtOH should be minimum 50%).
4) For short-term storage, store at room temperature or at +4 °C. For long-term storage, store at –20 °C.

8.4 General Preparations for FISH

Once cells have been fixed (see Section 8.3 and the online supplementary material 8.3), they can be used for FISH analysis. Prior to the FISH procedure, it is recommended to pay attention to the following:

a) Select the appropriate FISH protocol (e.g., oligonucleotide FISH [Section 8.5] or CARD-FISH [Section 8.6 and the online supplementary materials 8.7] or any other protocol as outlined in the online supplementary material 8.1), gene probes, and controls for the gene probes as well as for the samples (see Section 8.7 and Table 8.3).
b) Consider whether a post-treatment of the fixed cells, prior to FISH, is necessary. This is usually the case for certain Gram-positive prokaryotes and for other taxa such as certain archaea species with rigid or unusual cell wall chemistry (e.g., Meier et al., 1999; Nakamura et al., 2006; Weerasekara et al. 2013). Such cell walls must be slightly denatured with enzymes or by some other chemical-physical parameter (e.g., heat or acid treatment) to increase the permeability of the gene probes through the cell wall (see online supplementary material 8.4).
c) Consider the type of support on or in which the FISH experiment can be performed. The most common support is the microscope slide, onto which fixed cells can be easily pipetted and dried. After this, the microscope slide with the dried cells can then be used directly for the hybridization procedure and thereafter evaluated by

a microscope. Different kinds of microscope slides can be used – with or without coating. The most common type of microscope slides used for FISH are Teflon coated with numbered reaction wells (usually around 10). Such microscope slides allow simultaneous analyses of several samples on the same microscope slide. Depending on the adhesive nature of the sample, coating of the microscope slide may be necessary to enhance the attachment of the cells to the slide. Coated slides can either be purchased or made manually with different coating agents, such as gelatin or poly-L-lysine, depending on the chemical-physical nature of the cell wall package (see online supplementary material 8.5).

The FISH procedure can also be performed on other types of support, such as embedded aggregates, sections from larger embedded samples (online supplementary material 8.6), or membrane filters (the last are commonly used for filtration of aquatic or atmospheric samples). Depending on the diameter of the membrane filter, it may be necessary to cut the membrane filter into smaller pieces with a razor blade (advice: cut out a defined number of slices so that the area used for the FISH experiment can be estimated). The FISH procedure can also be performed in liquid, e.g., in an Eppendorf tube. In this case, all procedures (hybridization and washing), are done in a centrifuge. The hybridized sample can then be used directly for flow cytometry. However, prior to microscopy, the hybridized cells must then be pipetted onto a microscope slide and allowed to air dry in the dark. The benefit of performing the FISH procedure in solution is that unwanted detachment of cells (which may apply to certain types of sample) from microscope slides can be reduced. However, some typical drawbacks include a more tedious and time-consuming procedure, disruption of the original sample structure due to repeated centrifugation, and higher costs, since a larger volume of the reagents is needed.

After these considerations, the FISH procedure is quite simple and consists of six steps, which are rather similar among most FISH techniques:

1) Dehydration of the fixed cells in a series of different EtOH solutions (50%, 80%, 99%).
2) Preparation of a hybridization solution containing the gene probe with the appropriate concentration of formamide to allow the gene probes to target their goal as specifically as possible (for a deeper explanation, see Amann and Schleifer, 2001).
3) Application of the hybridization solution to the fixed cells on a suitable support.
4) Hybridization in a moist hybridization chamber at an appropriate temperature, normally at 46 °C, usually for about 90 min, though certain FISH procedures may demand up to 96 h (Yilmaz et al., 2006).
5) Removal of excess/unbound nucleic acid probe with a washing buffer (Section 8.5), normally at 48 °C for about 10–20 min. After this, the sample is carefully dried with pressurized air.
6) Mounting with a debleaching agent (Section 8.8) and microscopy of the hybridized samples. If it is not possible to evaluate the FISH experiment immediately, hybridized samples can be stored cold (between +4 and −20 °C) and in the dark for a shorter period (hours to a few days, depending on the sample and the intention of the experiment).

8.5 The Oligonucleotide FISH Procedure

8.5.1 General Background

Oligonucleotide FISH is the oldest and most commonly used FISH protocol, used for identifying microorganisms based on the universal phylogenetic marker ribosomal genes (depending on the domain, the small [SSU] or the large subunit [LSU]; Section 8.7). In principle, the protocol is simple and fast (~90 min or longer), and it consists of the following steps:

1) Ensure that your samples have been properly fixed and stored prior to the FISH experiment (see Section 8.3.2).
2) Pipette fixed cells onto microscope slides, which may be coated or not, and allow to dry. Once the cells are dry, dehydrate them in an incremental ethanol series and then allow the slides to dry.
3) Turn on an incubator oven to 46 °C and a water bath to 48 °C.
4) Prepare hybridization buffer with gene probes in sterile 2 mL Eppendorf tubes and washing buffer in a sterile 50 mL Falcon tube. Keep the hybridization buffer on ice in the dark. Prewarm the washing buffer in the water bath.
5) Apply the hybridization buffer with the gene probes on the microscope slide with the dehydrated cells. Place the slide in a moisture chamber (e.g., a 50 mL Falcon tube with filter soaked in the hybridization buffer) and incubate in the dark for at least 90 min at 46 °C.
6) After incubation, rinse the microscope slide with the prewarmed washing buffer and place the microscope slide into the tube with the washing buffer. Incubate in the water bath at 48 °C for at least 10 min up to 30 min.
7) Rinse the microscope slide at least once with cold distilled water. Dry with pressurized air.
8) Mount for microscopy (or store in the refrigerator or freezer).

The standard oligonucleotide FISH protocol works optimally on ecosystems with at least moderate access to nutrients and turnover rate, so that the cells may contain sufficiently high numbers of ribosomes and thus produce strong gene probe signals. However, some cellular species, such as those in certain biofilms, may still produce weak gene probe signals although the ribosome content may be high. This is usually attributed to i) rigid cell walls (e.g., in Gram-positive bacteria), which may reduce the permeabilization rate of the gene probes through the cell walls, ii) gene target inaccessibility problems, or iii) repellent charged components in the extracellular matrix. The gene probe signals can usually be improved by a post-treatment of the cells prior to the FISH procedure (online supplementary material 8.4), by HELPER FISH, based on probes that bind adjacent to the weak probe target site to open it up (Fuchs et al., 2000), or by employing another probe chemistry based on uncharged gene probes (e.g., PNA-FISH and LNA-FISH; see online supplementary material 8.1).

Further developments of the oligonucleotide FISH procedure have been performed to increase weak FISH probe signal intensity in cells, such as DOPE-FISH and MiL-FISH

(Behnman et al., 2012; Schimak et al., 2016; Stoecker et al., 2010). The FISH probe signal intensity can also be increased by other FISH techniques, such as polynucleotide FISH (see online supplementary material 8.1) and, more commonly, CARD-FISH (Section 8.6, online supplementary material 8.7).

8.5.2 Detailed Step-by-step Procedure

The oligonucleotide FISH procedure described here works for most prokaryotes (archaea, bacteria) and to some extent also for certain eukaryotic species (modified from Amann, 1995). Other FISH procedures must be used for other types of eukaryotic species – please consult the literature.

Chemicals (Advice: Prepare 0.5 to 1 L autoclaved solutions of all; these can be stored for longer periods. Use aliquots in, e.g., 50 mL Falcon tubes for daily work):

- 5 M NaCl
- 1 M Tris/HCl pH 8.0
- Sterile distilled and deionized H_2O or MQ (ultrapure water)
- Formamide of high quality (e.g., Merck, # 1.09684.1000), properly stored in the refrigerator in the dark. Do not use old solutions. Commercial formaldehyde solutions may also be used, but be aware of the quality
- 10 % (w/v) sodium dodecyl sulfate (SDS) (do not autoclave, but dissolve SDS in an autoclaved bottle with sterile water)
- 0.5 M ethylene diamine tetraacetic acid (EDTA) pH 8.0
- Fixed sample (see Section 8.3)
- Controls (probes and appropriate reference cells; see Section 8.7)
- Oligonucleotide probes (singly labeled or DOPE labeled; see, e.g., Stoecker et al., 2010) can be provided from, e.g., Biomers (Ulm, Germany). Store at −20 °C, in the dark, as aliquoted working solutions; see concentrations in Section 8.7.

Procedure:

1) Plan your FISH experiment (select appropriate samples, probes, and controls).
2) Check list to get started:
   a. Turn on the hybridization oven (46 °C) and the water bath (48 °C). Check that the temperatures are stable and correct.
   b. Check that the EtOH series (50, 80, ≥96%) solutions, FISH reagents, probes with suitable fluorochromes, and if appropriate, DNA stains are available.
   c. Use sterile pipette tips and sterile reagent tubes. Use new sterile Falcon tubes for the washing buffer. Used/dishwashed Falcon tubes can be used for the moisture chamber.
   d. Other materials: Teflon-coated microscope slides (with or without additional coating), racks (2 mL, 50 mL), paper tissue.
   e. Prepare a special waste bin to collect liquid waste with formamide and solid waste contaminated with formamide (e.g., pipette tips, tissue).

Table 8.1 Hybridization buffer for oligonucleotide FISH

%Formamide (v/v)	Formamide (μL)	MQ (μL)
0	0	1 600
5	100	1 500
10	200	1 400
15	300	1 300
20	400	1 200
25	500	1 100
30	600	1 000
35	700	900
40	800	800
45	900	700
50	1 000	600
60	1 100	500
65	1 200	400
70	1 300	300

Pipette into a 2 mL Eppendorf reaction tube:

a) 360 μL 5 M NaCl
b) 40 μL 1 M Tris/HCl pH 8.0
c) Add sterile distilled deionized H_2O or MQ (ultrapure water) and formamide as listed in Table 8.1, depending on the stringency of the probes used
d) Add 10% (w/v) SDS 4 μL (SDS is important because it denatures native ribosome structures by removing ribosomal proteins, thus increasing the accessibility of the target sites).

3) Pipette fixed cells (1–20 μL, depending on the cell concentration and/or whether the sample is to be quantified. If the cell concentration of your sample is low, you can increase the cell concentration in the well by simply adding more cells after drying) on a Teflon-coated microscope slide. Allow to air dry overnight (recommended; some samples do show better adhesion if more time is allowed for the drying) or dry quickly (~20 min) between 45 and 60 °C (e.g., in your hybridization oven). If you are doing FISH in solutions in tubes, pipette your cells into tubes. Centrifuge and remove supernatant.
4) In the meantime, prepare the hybridization buffer (Table 8.1) and store it at room temperature (use within next few hours). Use the safety hood and gloves when pipetting the formamide.

This should be used within the next few hours (if formamide was included). The buffer with formamide should always be freshly prepared for a FISH experiment. The formamide solution should be of a high quality and stored in the dark (e.g., wrapped in aluminum foil) in the fridge. When you add the formamide, work in the safety hood and use gloves. If you purchase a larger volume (e.g., 1 liter from Merck, #1.09684.1000) it is recommended that you aliquot this into, e.g., 15–50 mL Falcon tubes and store these at −20 °C prior to use. (Note 1: Formamide is sensitive to degradation, and degraded formamide can generate unspecific FISH signals; Note 2: If probes with different formamide concentrations for optimal stringencies are to be used, then FISH must be performed in several subsequent steps, starting with the probe with the highest formamide concentration.)

5) Dehydrate the fixed, dried cells on the microscope slide in an increasing ethanol series (1–5 min each in 50, 80, and ≥96% ethanol (denatured EtOH diluted with distilled water). If you are doing FISH in solutions in tubes, pipette EtOH into tubes. Incubate and centrifuge; remove supernatant. Proceed with next washing step. Remove the supernatant after the last washing procedure.
6) In the meantime, thaw the fluorochrome-labeled oligonucleotide probes (diluted to appropriate working solutions with appropriate pH, generally ~30 ng/µL for Cy3- and Cy5-labeled probes or ~50–100 ng/µL for FLUOS-labeled probes). Protect them from light with, e.g., aluminum foil. Vortex and spin down the probe solution.
7) Pipette 8–10 µL of the well-mixed hybridization buffer onto the wells without scratching the surface. If you are doing FISH in solutions in tubes, pipette the hybridization buffer into the tube.
8) Add 1 µL of each working solution of probe without scratching the surface. (Note: If two or more probes are applied simultaneously, consider whether it might be necessary to adjust probe concentrations; i.e., use a higher probe concentration of the working solutions.) If you are doing FISH in solutions in tubes, pipette the probes into the tube.

Comment on steps 7–8: If you are preparing a large number of slides using the same probes, you may simplify your work by preparing a *master mix* (of your hybridization reagents and probes) in one tube and then applying the master mix on each well on the microscope slide. It is recommended to include "one extra" sample for your master mix. For example, if you wish to do FISH on six samples distributed on six wells, pipette into a small tube [6 + 1] × 8 µL hybridization buffer and [6 + 1] × 1 µL working solution of probe; mix and spin down; distribute 7 (from 6 + 1) µL of master mix on each well.

9) Prepare a 50 mL hybridization moisture chamber (e.g., a 50 mL Falcon tube) by folding a piece of tissue into the tube and pouring the rest of the hybridization buffer onto the tissue. If you are doing FISH in solutions in tubes, simply place the tube with the hybridization buffer and probes into the hybridization oven (Figure 8.1).
10) Immediately transfer the microscope slide into the hybridization tube, place the tube horizontally in a rack, and then place the rack in the oven (46 °C).
11) Incubate the hybridization chamber with the microscope slide in the hybridization oven (46 °C) for a minimum of 60 min, up to 96 h. The standard average hybridization time is ~90 min.
12) In the meantime, prepare the washing buffer (Table 8.2) and preheat this buffer at 48 °C (water bath).

Comment: Prepare a bottle of cooled (+4 °C) sterile distilled, deionized H_2O or MQ (ultrapure water) for the final washing step.

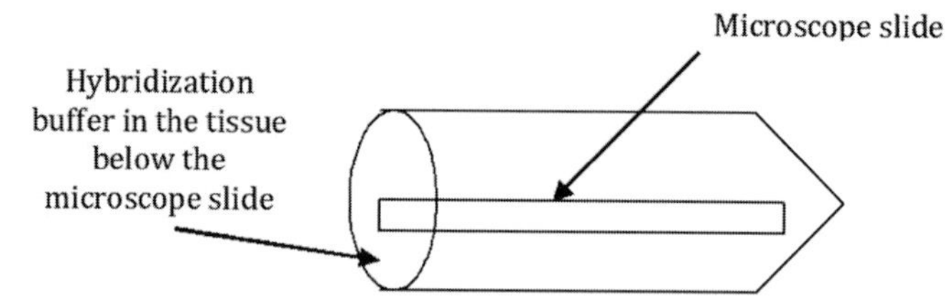

Figure 8.1 Simple layout of the moisture chamber, using a Falcon tube in a horizontal position. The dry tissue is first placed evenly on the bottom of the Falcon tube; then it is made moist by distributing the hybridization buffer onto the tissue with a pipette. After this, the microscope slide is placed carefully into the Falcon tube. The Falcon tube is closed tightly with a lid and the moisture chamber is then placed horizontally in a tube rack, which is placed into the hybridization oven.

13) Rinse the microscope slide with the washing buffer and incubate the microscope slide in the washing buffer for 10–20 min in a preheated water bath (48 °C). If cells are to be quantified, the washing procedure should be standardized for all processed samples. If you are doing FISH in solutions in tubes, wash via centrifugation. Act quickly to avoid too much cooling.
14) Remove the washing buffer with cold, distilled water (avoid cell detachment), carefully remove excess water with tissue paper, and dry the microscope slide as soon as possible carefully with compressed air. If you are doing FISH in solutions in tubes, remove the water via centrifugation. Allow to dry.
15) If a counterstain such as a universal DNA stain is to be used (e.g., DAPI or SYBR Green I [Note: Select the DNA stain so that it will not overlap with the fluorochromes on the probes!]), pipette the DNA stain onto the sample (e.g., 10–20 μL of DAPI 1 μg/mL or SYBR Green I diluted with distilled water 1:10 000), incubate in the dark for ~10–20 min at room temperature. Wash the microscope slide with cold, distilled water and dry the microscope slide as soon as possible carefully with compressed air. Note: DNA stains are toxic compounds, so use a protected space (e.g., a safety hood) clearly indicated by safety labels. Collect all waste in a special tray and treat the waste according to the instructions of your institute. If you are doing FISH in solutions in tubes, add DNA stains into the tube and remove and wash via centrifugation.
16) Before microscopy, drop single drops of embedding solution on the microscope slide and place a coverslip on the microscope slide. If you are doing FISH in solutions in tubes, pipette your cells onto a microscope slide and allow to dry in the dark, drop single drops of embedding solution on the microscope slide, and place a coverslip on the microscope slide. Press carefully on the cover slip to allow the mounting medium to get soaked up fully by all wells with samples. Keep the slides in the dark prior to microscopy. Comment: If you do not have time to microscope the sample directly after the washing procedure, you may store it in a Falcon tube (wrapped in aluminum foil) in the fridge, or for longer periods (days) at –20 °C.
17) Keep embedded microscope slides in a dark box until you can microscope them. For evaluation, you can use evaluation protocols like those suggested in the online supplementary material 8.8.

Table 8.2 Washing buffer for oligonucleotide FISH

% Formamide in the hybridization buffer	NaCl (μL)
0	9 000
5	6 300
10	4 500
15	3 180
20	2 150
25	1 490
30	1 020
35	700
40	460
45	300
50	180
55	100
60	40
70	no NaCl, only 350 μL EDTA

Mix in a 50 mL Falcon tube:

1) 1 mL 1 M Tris/HCl pH 8.0
2) 0.5 mL 0.5 M EDA (Note: Add this only if the formamide concentration is >20% in the hybridization buffer)
3) Add NaCl corresponding to the table.
4) Add sterile distilled, deionized H_2O or MQ (ultrapure water) up to 50 mL.
5) Add 50 μL 10% (w/v) SDS.
6) Preheat the washing buffer at 48 °C prior to use.

18) After microscopy:
    a. If you do not wish to save the microscope slides for further microscopy or keep them for a record, throw away the Teflon-coated slides in a special dedicated waste tray according to the regulations of your institute.
    b. If you wish to reuse your slides for further microscopy or for a record, remove the coverslip, rinse the slide carefully with distilled water, and dry carefully with pressurized air. Store in the dark at −20 °C.

8.6 Catalyzed Reporter Deposition-FISH (CARD-FISH)

CARD-FISH was developed in the late 1990s to increase the weak probe signal intensities observed with oligonucleotide FISH when it was applied to cells with a low ribosome content, which is often the case in extreme, oligotrophic ecosystems and geological environments (e.g., Amann and Fuchs, 2008). In CARD-FISH, horseradish peroxidase (HRP)-labeled gene probes are used during the hybridization. After the hybridization, fluorescently labeled (e.g., A488 [green] or A555 [red]) tyramine molecules are added to the sample. This will initiate the catalyzed reporter deposition so that tyramine molecules will activate the HRP on the gene probes and thus produce highly reactive intermediates, which will bind to moieties of proteins. In this way, numerous fluorescent molecules will be introduced into the vicinity of the probe target site, resulting in a higher gene probe signal intensity and sensitivity. CARD-FISH can be used to target both multicopy (ribosomal) and low-copy genes (e.g., housekeeping genes or functional genes), even for expressed genes such as mRNA (Pernthaler and Amann, 2004). Furthermore, CARD-FISH can be combined with several other FISH procedures and analytical methods (see examples in online supplementary material 8.1). Some of the major drawbacks of CARD-FISH compared with oligonucleotide FISH are that it is more tedious and time consuming (1.5 days or longer per probe), more expensive, and less flexible, as only one HRP-labeled probe can be employed per hybridization.

CARD-FISH can be performed manually (online supplementary material 8.7) or at least partly with kits (e.g., from Molecular Probes, Tyramide Signal Amplification Kits, MP 20911). Several CARD-FISH protocols are available online for different types of ecosystems:

a) General overview (e.g., Kubota, 2013).
b) For aquatic samples and sediments: Wendeberg (2010).
c) For soil: Schmidt et al. (2012).

Brief overview of procedure (adapted from Teira et al., 2004): Prior to the CARD-FISH procedure, the samples must be fixed (Section 8.3, online supplementary materials 8.3), placed on a suitable supporter (e.g., microscope slide, membrane filter; see Section 8.4), and where appropriate, embedded, e.g., with a protective layer of a 0.1% agarose solution. The CARD-FISH procedure consists principally of six steps, spanning a period of 1.5 to 4 days, or even longer, depending on the number of desirable gene probe combinations.

Day 0:

a) Ensure that the cells have been properly fixed and stored as described in Section 8.3.2. Distribute fixed cells on a suitable support material, e.g., a microscope slide or membrane filter, or in an Eppendorf tube.
b) Synthesize (or purchase an appropriate kit) tyramide, the fluorescent agent that activates the HRP-labeled probe once it has been hybridized to the gene target in the cell. Multiple aliquots can be made and stored at −20 °C for a longer time period (at least months).

Day 1:

c) Permeabilize the cells on your support material by pipetting enzymes onto the fixed cells.

   Choose the enzyme (e.g., lysozyme, proteinase K) depending on gene probe and sample type (compare with former studies).

d) Inactivate endogenous peroxidases in your sample by incubating it with H_2O_2 (see, e.g., Pavlekovic et al., 2009).
e) Start the hybridization reaction by mixing the HRP-labeled gene probe with your prepared sample.

Day 2:
f) Start the catalyzed reporter deposition by adding the tyramide to the sample that has been hybridized with the HRP-labeled gene probe.
g) Wash your sample.
h) Evaluate by microscope.

Unfortunately, the CARD-FISH protocol applies only for one HRP gene probe per hybridization experiment. If more HRP gene probes are desired in order to allow the simultaneous detection of several species within your sample, then several CARD-FISH experiments must be performed subsequently (one experiment per HRP gene probe). After each experiment, the HRP from the previous hybridization must be inactivated; however, the probe signals from the former hybridization experiment will usually remain (multi-color CARD-FISH; see Wendeberg, 2010),

8.7 Gene Probes for FISH

8.7.1 Targets for Gene Probes

Common gene targets for most FISH applications are the ribosomal genes, since these are conservative, ubiquitous genes present in every biological species and thus suitable for a global identification on a phylogenetic evolutionary basis. The ribosomal genes employed for FISH can be divided into three categories:

- The small subunit (SSU): 16S rRNA (length around 1 542 nucleotides) and 18S rRNA (length around 1 869 nucleotides)
- The large subunit (LSU): 23S rRNA (length around 2 906 nucleotides) and 28S rRNA (length around 5 070 nucleotides)
- The intergenic spacer region (ISR) or the internal transcribed spacer (ITS): length between 150 and 1 500 nucleotides.

FISH procedures have been developed for all these gene target categories. The basic principle is that the length and conservation degree of the gene target determine the options for specific probe designs; thus, the longer the gene target (the LSU), the more target options there are (see, e.g., Manz et al., 1992). However, the most common gene target for FISH is currently the shorter SSU, since the SSU gene database is approximately six times larger than the LSU gene database (based on the Silva database version 132 from December 2017, the SSU database consists of 6 073 181 gene sequences, whereas the LSU gene database consists of 907 382 gene sequences; Glöckner et al., 2017). For ISR/

ITS gene sequences, even fewer gene sequence data have been retrieved, and no quality-based gene databases are available. FISH procedures have also been developed against nonribosomal genes, such as housekeeping or functional genes, employing other FISH procedures, such as mRNA CARD-FISH (e.g., Pernthaler and Amann, 2004), recognition of individual genes (RING)-FISH (Zwirglmaier, 2005), gene FISH (Moraru et al., 2010), and *in situ* RCA (see, e.g., review by Wagner and Haider, 2012). Although these protocols show much promising potential, they are time consuming, tedious, and often expensive and are therefore not yet used in routine analyses.

8.7.2 Databases for Gene Probes

So far, nearly 3 000 gene probes targeting ribosomal genes have been developed for oligonucleotide FISH. Many of them are summarized in the online database "probeBase" (Greuter et al., 2016). Unfortunately, there are no online databases for other types of gene probes used in other FISH protocols such as PNA-FISH or CARD-FISH. Since many of these probes were designed one or two decades ago and our knowledge about the unknown microbial diversity is expanding regularly, it is crucial to explore whether published gene probes are still up to date. This can be done by the probe match tools provided by online databases like the RDP database (Cole et al., 2014) and the Silva database (Glöckner et al., 2017) or in one's own database (e.g., using the bioinformatics software package ARB; Ludwig et al., 2004), based on one's own sequences as well as the online reference Silva database (Glöckner et al., 2017).

8.7.3 Design of Gene Probes

Gene probes for FISH can be designed from gene sequence databases with appropriate bioinformatics software. For an optimal probe design, it is essential that the gene sequence database meets certain quality criteria with respect to sequence quality, length, alignment, number of sequences, and phylogenetic diversity. General databases like NCBI contain all updated relevant gene sequence data, but without any quality-based evaluation and alignment. Quality-based evaluated and aligned ribosomal gene sequence databases can be retrieved from the RDP database (Cole et al., 2014) and the Silva database (Glöckner et al., 2017). The RDP database focuses on 16S rRNA gene sequence data for archaea and bacteria, and to some extent, 28S rRNA gene sequence data for fungi. The Silva database focuses on both the SSU and the LSU gene sequence data from all three domains. Appropriate gene sequence data for different organism groups can be downloaded from these online databases for further phylogenetic analysis and biomarker design. Both of these databases also contain other useful functions, such as an online gene probe evaluation for quick check-ups of the target spectrum of gene probes. Downloaded gene sequence data from the RPD and Silva databases, as well as one's own generated gene sequence data, can be evaluated by bioinformatics software packages such as ARB (Ludwig et al., 2004). The ARB software is a graphically oriented package for Unix platforms, comprising various tools for gene sequence database handling (any kinds of gene sequences as well as protein-coding genes), data analysis

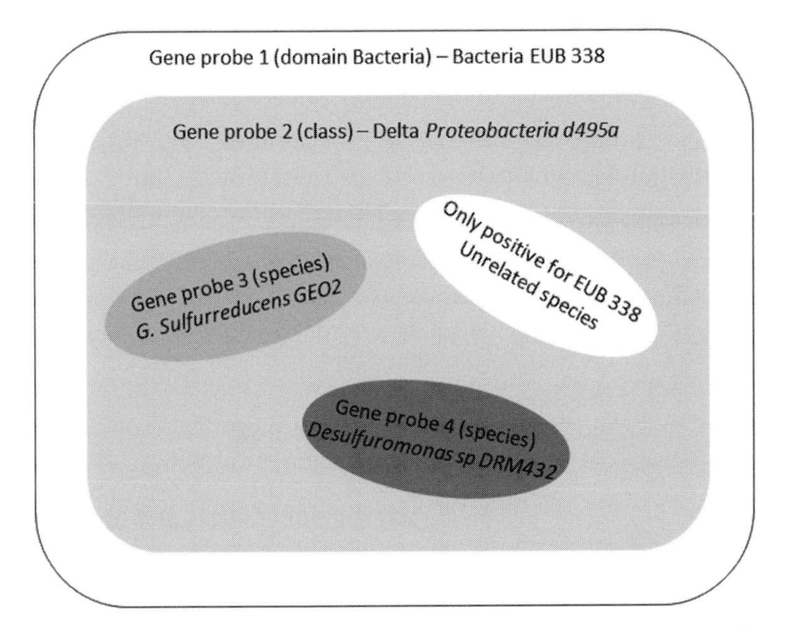

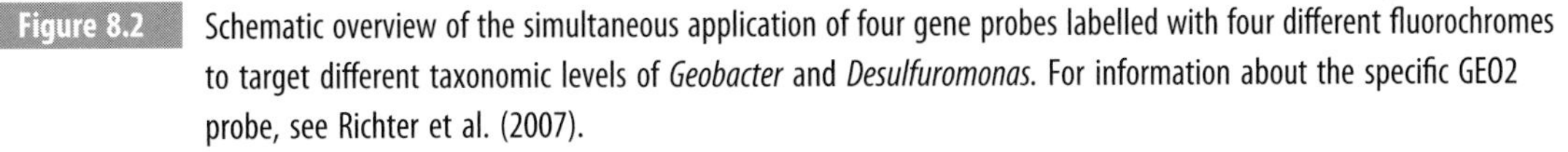
Figure 8.2 Schematic overview of the simultaneous application of four gene probes labelled with four different fluorochromes to target different taxonomic levels of *Geobacter* and *Desulfuromonas*. For information about the specific GEO2 probe, see Richter et al. (2007).

(including evaluation and design of gene probes and polymerase chain reaction [PCR] primers), and visualization of gene sequences (secondary structure with probe matches, etc.). A central database of processed (aligned) sequences and any type of additional data linked to the respective gene sequence entries is structured according to phylogeny or other user-defined criteria.

For an efficient gene probe design, it is essential to align the gene sequences properly and to reconstruct phylogenetic trees, so that appropriate target groups can be identified and selected for the probe design based on an evolutionary phylogenetic understanding (Figure 8.2). After this, probe design is easily done via a special probe design menu. After the design, the probes must be further evaluated, theoretically as well as practically. For the theoretical *in silico* evaluation of the novel probes, various types of online programs or databases can be employed, such as i) MATHFISH, a web tool for the computational evaluation of RNA-targeted FISH probes using mathematical models (Yilmaz et al., 2010) and ii) LOOP OUT, a program that allows potential FISH probes to be screened for nontarget binding at off-target sites containing single nucleotide insertions or deletions. Probe design can also be performed with other software, such as PRIMROSE (Asheldford et al., 2002). Once the optimal conditions for the *in silico* designed probes have been performed, it is essential to evaluate the novel probes in laboratory experiments in order to check whether they produce appropriate gene probe signal intensities and to explore the conditions (the formamide concentration) for optimal identification (stringency). This is usually done with a formamide series (e.g., from 0 to 60%; Amann and Schleifer, 2001).

8.7.4 Nomenclature of Gene Probes

FISH gene probes can be named in three different ways (exemplified with the domain probe EUB 338 against bacteria):

- ***Short name***: Abbreviation of target group and starting position on the *Escherichia coli* 16S rRNA gene sequence. Example: EUB 338-I (Eubacteria).
- ***Comprehensive nomenclature***: Precise description of the main features of the probe: target gene, target group, target group level, 3′ end of the probe, and probe length (Alm et al., 1996). Example: EUB 338: S-D-Bact-0338-a-A-18.
- ***Probe accession number*** (Greuter et al., 2016): Identity number defined by the online probe base. Example: EUB 338: pB-00159.

8.7.5 Selection of Controls and Gene Probes

For optimal FISH experiments, it is critical to include both positive and negative sample references, as well as positive and negative gene probes. The purpose of these controls is to generate positive and negative controls – both are essential to ensure a reliable evaluation of the FISH experiment and thereby avoid false positive and false negative results. Positive controls should be represented by fixed cells, which are expected to react positively with the selected gene probes, and vice versa for the negative controls. It is also essential to include two other types of control: i) exclusion of gene probes in order to estimate the intensity of the natural background fluorescence; and ii) employment of a "nonsense" gene probe (a fluorochrome-labeled gene probe with a nucleotide sequence that will not bind to the desired species or even any biological species) in order to estimate the intensity of the unspecific attachment of the fluorochrome of the gene probe. A survey of these controls is presented in Table 8.3.

Employ optimal and stringent hybridization conditions, using an optimal formamide concentration, which has been experimentally determined by a formamide series (Amann and Schleifer, 2001).

Employ an intelligent combination of gene probes, preferentially based on the hierarchic probe concept (see Figure 8.1). For example, employ a combination of at least three different gene probes targeting different taxonomical levels (e.g., the phylum, the genus, and the species level, respectively). If each gene probe is labeled with a different fluorochrome, different overlap color patterns will reveal different phylogenetic relationships. This will enable a more reliable and precise identification of both known as well as potentially novel biological species (Amann and Schleifer, 2001).

8.7.6 Selection of Suitable Fluorochromes for Gene Probes

Choose fluorescent dyes based on the equipment (filters, laser beam type) of your fluorescence microscope. Gene probes can be labeled in at least three different ways,

Table 8.3 Overview of different control samples and probes for a reliable FISH experiment

Category	Biological	Technical
Positive controls	Expected target group[a]	Expected probes
	Other method for additional proof (e.g., PCR, sequencing)	e.g., Counterstain
Negative controls	Target group not related to the desired target group	Nonsense probe, NON-EUB, etc.
	Manipulated group (e.g., heat killed or inhibited)	No probe

[a] Pure culture, complex culture, clone, or environmental sample if pure cultures are not available.

with i) radioactive compounds (e.g., ^{35}S; Giovannoni et al., 1988); ii) nonfluorescent compounds, such as HRP, biotin, and digoxigenin (Schönhuber et al., 1997); and iii) fluorochromes (single, double, or even more), such as Cy-3, Cy-5, FLUOS, or Alexa derivates. Of these three alternatives, fluorochrome labeling presents the least complicated alternative and is thus most commonly used. Labeling of nucleotide probes can be performed manually in the laboratory (Amann, 1995), but the easiest way is to purchase labeled oligonucleotide probes from biotechnological companies (e.g., Biomers, Molecular Probes/Invitrogen, MWG Biotech, Operon, Sigma, and Thermo Hybaid). Single labeling is normally done at the 5′-terminus of the oligonucleotide; however, multiple labeling of probes is also possible (e.g., polynucleotide FISH, DOPE-FISH, and Mil-FISH; see online supplementary material 8.1).

8.7.7 Handling of Purchased Probes

Probes are generally, but not always, shipped in a lyophilized state. Upon arrival, the lyophilized probe should be dissolved in either sterile water or an appropriate buffer (depending on the fluorochrome or other labels, such as enzymes). The lyophilized probes are often, based on the instructions from the company, dissolved to 100 picomoles/L ("the stock solution"). For standard FISH/DOPE-FISH, gene probes labelled with Cy3 or Cy5, 30 ng/μL working solutions are normally prepared, while a higher concentration (50–100 ng/μL) is employed for certain types of fluorescein-labeled or HRP gene probes (for CARD-FISH). In this way, the stock solution normally lasts for at least 100–1 000 (depending on the applied concentrations, size of sample, etc.) hybridization experiments. In most cases, the stock solution should be stored at −20 °C in the dark, but other gene probes, such as the HRP gene probes for CARD-FISH, should be stored at +4 °C. For longer storage times, relyophilization of the probes might be more appropriate. To minimize frequent thawing and freezing of the stock solution (and thus degradation of the nucleic acid probes), aliquots of working solutions should be prepared.

8.8 Evaluation of FISH Results

FISH experiments can be evaluated in different ways: by microscopy, or by combination with other methods such as flow cytometry, microarray technology, spectral imaging, Raman spectroscopy, or NanoSIMS (online supplementary material 8.1). All these methods have their specific requirements. Here, only the handling of samples for fluorescence microscopy and confocal laser scanning microscopy will be addressed.

i. ***Mounting with an anti-debleaching medium***: Shortly prior to microscopy, distribute a few drops of a suitable anti-debleaching mounting medium (e.g., Citifluor, or others such as VectaShield, Moviol, or Molecular Probes) on the microscope slide and place a coverslip on the microscope slide. Press carefully on the coverslip to allow the mounting medium to soak up evenly on the microscope slide. Keep the slides in the dark at +4 °C prior to microscopy.
ii. ***Optimize the usage of the laser light during microscopy***: Avoid exposing your sample to light or laser for long periods to minimize rapid bleaching of the fluorochromes. If possible, start viewing samples with probes labeled with fluorochromes that demand lower laser energies (e.g., if you are using the fluorochromes Cy3, Cy5, and fluorescein-labeled probes, start with Cy3, then proceed with fluorescein-labeled probes, and then finally, proceed to Cy5).
iii. ***Evaluation of microscope observations***: The results observed by microscopy can be evaluated by manual counting or by digital image analysis software. Different types of digital image analysis software exist depending on the purpose and are provided either by the microscope company or by other research institutions. The most commonly used software is daime, which has been developed to meet the needs of biologists who explore microorganisms *in situ* in environmental and medical samples (Daims, 2009). The software is available both for Linux and for Microsoft Windows and is free of charge.
iv. ***Storage of FISH experiments***: The microscope slide with the FISH experiments can be stored for later re-evaluations (after removal of the coverslip, rinsing in cold, distilled water, and drying with pressurized air) in the dark at +4 °C (for short-term storage, usually a couple of days) or at −20 °C (for long-term storage, for weeks, months, or even longer).
v. ***Comparison of FISH results with other methods***: In many cases, it is sufficient to use only FISH to identify the species in a sample (on condition that appropriate controls, gene probes, and replicate experiments have been selected). However, in certain cases, especially when dealing with extreme or complex ecosystems, it is necessary to employ other methods (e.g., methods targeting the activity level or the function, amplification of specific genes by PCR/quantitative PCR [qPCR], or sequencing of genes/genomes) to ensure the identification or to retrieve more detailed results. One of the main drawbacks of FISH versus nucleic acid–based methods, such as PCR and gene sequencing, is that FISH has a higher detection limit than PCR. Thus, negative results obtained with FISH do not immediately suggest that the targeted species is not present; alternative explanations include that the cell concentration lies below the detection level of FISH,

or that the species of interest is present but inactive. Thus, employing a combination of studies based on FISH, gene amplification, gene sequencing, or other methods can expand the interpretation possibilities of the state of a microbial community (see, e.g., Juretschko et al., 2002).

8.9 Conclusions

FISH is a useful, quick, and simple cultivation-independent method for identification of intact organisms (prokaryotes, eukaryotes, and to some extent also viruses and spores). FISH can therefore provide valuable complementary information to today´s "black-box" analytical methods, which rely on disruption of organisms (e.g., extracted nucleic acids). In contrast to these, FISH can visualize, identify, and quantify organisms, demonstrate how they are located, distributed, and associated with other species, and reveal their *in situ* physiology and gene expression state on a single-cell level. Since the 1980s, several different types of FISH procedures have been developed to solve methodological obstacles observed with the first FISH procedure (oligonucleotide FISH) and to meet the sophisticated demands of contemporary research goals in microbial ecology and geomicrobiology. The most successful FISH applications are based on the detection of ribosomal genes, such as the 16S and the 18S rRNA genes, used by the two most commonly applied FISH procedures, oligonucleotide FISH and CARD-FISH. However, new, impressive, and sophisticated FISH procedures are emerging, which also allow further analysis on a single-cell level of other gene targets as well as other parameters (e.g., mRNA CARD-FISH, magneto-FISH, *in situ* rolling circle amplification (RCA) FISH, gene FISH, and BONCAT FISH). FISH can also be combined with a multitude of other analytical techniques, such as flow cytometry, electron microscopy, microautoradiography, stable isotope analysis, Raman-microspectroscopy, Nano-SIMS, and advanced molecular biological techniques, such as proteomics. Such combinations will expand the possibilities of detailed cultivation-independent observations on a single-cell level within a complex community in their natural environment, and thus provide novel research possibilities for whole-system bio-geochemical processes on Earth.

8.10 References

Alm E, Oerther D, Larsen N, Stahl D, Raskin L (1996). The Oligonucleotide Probe Database. *Appl. Environ. Microbiol.* 62: 3557–3559.

Amann RI (1995). In situ identification of microorganisms by whole cell hybridization with rRNA targeted nucleic acid probes, pp. 331–345. In Molecular Microbial *Ecology Manual.*

Amann RI, Fuchs BM (2008). Single-cell identification in microbial communities by improved fluorescence in situ hybridization techniques. *Nat Rev Microbiol.* 6(5): 339–348. doi:10.1038/nrmicro1888.

Amann R, Schleifer KH (2001). Nucleic acid probes and their application in environmental microbiology, pp. 67–82. In Garrity GM (ed.), *Bergey's Manual of Systematic Bacteriology*, 2nd Edition.

Antón J, Llobet-Brossa E, Rodríguez-Valera F, Amann R (1999) Fluorescence in situ hybridization analysis of the prokaryotic community inhabiting crystallizer ponds. *Environ Microbiol.* 1(6): 517–523.

Asheldford KE, Weightman AJ, Fry JC (2002). PRIMROSE: a computer program for generating and estimating the phylogenetic range of 16 rRNA oligonucleotide probes and primers in conjunction with the RDP-II database. *Nucleic Acids Res.* 30: 3481–3489.

Becking LB (2015). *Baas Becking's Geobiology*. Canfield DE (ed.). Wiley-Blackwell. 152 pp. ISBN: 978–0-470–67381-2, http://eu.wiley.com/WileyCDA/WileyTitle/productCd-0470673818,subjectCd-EN64.html#

Behnam F, Vilcinskas A, Wagner M, Stoecker K (2012). A straightforward DOPE-FISH method for simultaneous multicolor detection of six microbial populations. *Appl. Environ. Microbiol.* 78: 5138–5142.

Boas, V, Almeida C, Sillankorva S, et al. (2016). Discrimination of bacteriophage infected cells using locked nucleic acid fluorescent in situ hybridization (LNA-FISH). *Biofouling*. 32(2): 179–190. doi:10.1080/08927014.2015.1131821.

Caracciolo AB, Grenni P, Cupo C, Rossetti S (2005). In situ analysis of native microbial communities in complex samples with high particulate loads. *FEMS Microbiol. Lett.* 253: 55–58.

Chou YY, Heaton NS, Gao Q, et al. (2013). Colocalization of different influenza viral RNA segments in the cytoplasm before viral budding as shown by single-molecule sensitivity FISH analysis. *PLoS Pathog*. 9(5): e1003358. doi:10.1371/journal.ppat.1003358.

Cole JR, Wang Q, Fish JA, et al. (2014). Ribosomal Database Project: data and tools for high throughput rRNA analysis. *Nucleic Acids Res*. 42(Database issue): D633–D642.

Cutler NA, Oliver AO, Viles HA, Whiteley AS (2012). Non-destructive sampling of rock-dwelling microbial communities using sterile adhesive tape. *J. Microbiol Methods*. 91: 391–398.

Daims H (2009). Use of fluorescence in situ hybridization and the daime image analysis program for the cultivation-independent quantification of microorganisms in environmental and medical samples. Cold Spring Harbor Protocols website, http://cshprotocols.cshlp.org/content/2009/7/pdb.prot5253.full.

Dang VT, Sullivan MB (2014). Emerging methods to study bacteriophage infection at the single-cell level. *Front. Microbiol.* 2014; 5: 724.

Dekas AE, Connon SA, Chadwick GL, Trembath-Reichert E, Orphan VJ (2016). Activity and interactions of methane seep microorganisms assessed by parallel transcription and FISH-NanoSIMS analyses. *ISME J.* 10(3): 678–692.

Eichorst SA, Strasser F, Woyke T, et al. (2015). Advancements in the application of NanoSIMS and Raman microspectroscopy to investigate the activity of microbial cells in soils. *FEMS Microbiol. Ecol.* 91(10) pii: fiv106.

Foissner W (1992). Preparation of samples for scanning electron microscopy. In Lee JJ and Soldo AT (eds), *Protocols for Protozoology*. Society of Protozoologists, Lawrence, Kansas, USA. Section C. Fixation, staining, light and electron microscopical techniques, chapter 20.

Fuchs BM, Glöckner FO, Wulf J, Amann R (2000). Unlabelled helper oligonucleotides increase the in situ accessibility to 16S rRNA of fluorescently labelled oligonucleotide probes. *Appl. Environ. Microbiol.* 66: 3603–3607.

Gérard E, Guyot F, Philippot P, López-García P (2005). Fluorescence in situ hybridisation coupled to ultra small immunogold detection to identify prokaryotic cells using transmission and scanning electron microscopy. *J. Microbiol. Methods*. 63(1): 20–28.

Giovannoni SJ, Delong E, Olsen GJ, Pace NR (1988). Phylogenetic group-specific oligodeoxynucleotide probes for in situ microbial identification. *J. Bacteriol.* 170: 720.

Glöckner FO, Yilmaz P, Quast C, et al. (2017). 25 years of serving the community with ribosomal RNA gene reference databases and tools. *J. Biotechnol.* 261: 169–176.

Greuter D, Loy A, Horn M, Rattei T. (2016). probeBase – an online resource for rRNA-targeted oligonucleotide probes and primers: new features 2016. *Nucleic Acids Res.* 44(D1): D586–589.

Juretschko S, Loy A, Lehner A, Wagner M (2002). The microbial community composition of a nitrifying-denitrifying activated sludge from an industrial sewage treatment plant analyzed by the full-cycle rRNA approach. *Syst. Appl. Microbiol.* 25(1): 84–99.

Kubota K (2013). CARD-FISH for environmental microorganisms: technical advancement and future applications. *Microbes Environ.* 28: 3–12.

Lee NM, Meisinger DB, Schmid M, Rothballer M, Löffler FE (2011). Fluorescence in situ hybridization in geomicrobiology, pp. 854–880. In Reitner HJ and Thiel V (eds), *Encyclopedia in Geobiology*. Springer Verlag.

Ludwig L, Strunk O, Westram R, et al. (2004). ARB: a software environment for sequence data. *Nucleic Acids Res*. 32: 1363–1371.

Manz W, Amann R, Ludwig W, Wagner M, Schleifer KH (1992). Phylogenetic oligodeoxynucleotide probes for the major subclasses of proteobacteria: problems and solutions. *Syst. Appl. Microbiol.* 15: 593–600.

Meier H, Amann R, Ludwig W, Schleifer KH (1999). Specific oligonucleotide probes for in situ detection of a major group of gram-positive bacteria with low DNA G+C content. *Syst. Appl. Microbiol.* 22: 186–196.

Moraru C, Lam P, Fuchs BM, Kuypers MM, Amann R (2010). GeneFISH – an in situ technique for linking gene presence and cell identity in environmental microorganisms. *Environ. Microbiol.* 12(11): 3057–3073. doi:10.1111/j.1462-2920.2010.02281.x.

Nakamura K, Terada T, Sekiguchi Y, et al. (2006). Application of pseudomurein endoisopeptidase to fluorescence in situ hybridization of methanogens within the family Methanobacteriaceae. *Appl. Environ. Microbiol.* 72: 6907–6913.

Neuenschwander SM, Salcher MM, Pernthaler J (2015). Fluorescence in situ hybridization and sequential catalyzed reporter deposition (2C-FISH) for the flow cytometric sorting of freshwater ultramicrobacteria. *Front. Microbiol.* 6(247): 1–8.

Nikolaki S, Tsiamis G (2013). Microbial diversity in the era of omic technologies. *Biomed. Res. Int.* 2013; 2013: 958719. doi:10.1155/2013/958719. Epub October 24, 2013.

Pavlekovic M, Schmid MC, Schmider-Poignee N, et al. (2009). Optimization of three FISH procedures for in situ detection of anaerobic ammonium oxidizing bacteria in biological wastewater treatment. *J. Microbiol. Methods*. 78: 119–126.

Pernthaler A, Amann R (2004). Simultaneous fluorescence in situ hybridization of mRNA and rRNA in environmental bacteria. *Appl. Environ. Microbiol.* 70: 5426–5433.

Richter H, Lanthier M, Nevin KP, Lovley DR (2007). Lack of electricity production by *Pelobacter carbinolicus* indicates that the capacity for Fe(III) oxide reduction does not necessarily confer electron transfer ability to fuel cell anodes. *Appl. Environ. Microbiol.* 73: 5347–5353.

Rossetti S, Tomei MC, Blackall LL, Tandoi V (2007). Bacterial growth kinetics estimation by fluorescence *in situ* hybridization and spectrofluorometric quantification. *Lett. Appl. Microbiol.* 44: 643–648.

Schimak MP, Kleiner M, Wetzel S, et al. (2016). MiL-FISH: multilabelled oligonucleotides for fluorescence in situ hybridization improve visualization of bacterial cells. *Appl. Environ. Microbiol.* 82: 62–70.

Schmidt S, Eickhorst T, Tippkötter R (2012). Evaluation of tyramide solutions for an improved detection and enumeration of single microbial cells in soil by CARD-FISH. *J. Microbiol. Methods.* 91: 399.

Schönhuber W, Fuchs B, Juretschko S, Amann R (1997). Improved sensitivity of whole-cell hybridization by the combination of horseradish peroxidase-labelled oligonucleotides and tyramide signal amplification. *Appl. Environ Microbiol.* 63: 3268.

Shiraishi F, Zippel B, Neu TR, Arp G (2008). *In situ* detection of bacteria in calcified biofilms using FISH and CARD-FISH. *J. Microbiol Methods.* 75:103–108.

Stoecker K, Dorninger C, Daims H, Wagner M (2010). Double labelling of oligonucleotide probes for fluorescence in situ hybridization (DOPE-FISH) improves signal intensity and increases rRNA accessibility. *Appl. Environ Microbiol.* 76(3): 922.

Teira E, Reinthaler T, Pernthaler A, Pernthaler J, Herndl GJ (2004). Combining catalyzed reporter deposition-fluorescence in situ hybridization and microautoradiography to detect substrate utilization by bacteria and Archaea in the deep ocean. *Appl. Environ. Microbiol.* 70(7): 4411.

Valm AM, Welch JLM, Borisy GG (2013). CLASI-FISH: principles of combinatorial labeling and spectral imaging. *Syst. Appl. Microbiol.* 35: 496–502.

Volpi EV, Bridger JM (2008). FISH glossary: an overview of the fluorescence in situ hybridization technique. *BioTechniques* 45: 385–409.

Wagner M, Haider S (2012). New trends in fluorescence in situ hybridization for identification and functional analyses of microbes. *Curr. Opin. Biotechnol.* 23: 96–102.

Wagner M, Rath G, Amann R, Koops H-P, Schleifer KH (1995). In situ identification of ammonia-oxidizing bacteria. *Syst. Appl. Microbiol.* 18: 251–264.

Wang YY, Huang WE, Cui L, Wagner M (2016). Single cell stable isotope probing in microbiology using Raman microspectroscopy. *Curr. Opin. Biotechnol.* 41: 34–42.

Weerasekara ML, Ryuda N, Miyamoto H, et al. (2013). Double-color fluorescence in situ hybridization (FISH) for the detection of *Bacillus anthracis* spores in environmental samples with a novel permeabilization protocol. *J. Microbiol Methods*. 93(3): 177–184. doi:10.1016/j.mimet.2013.03.007.

Wendeberg A (2010). Fluorescence in situ hybridization for the identification of environmental microbes. Cold Spring Harbor Protocols, http://cshprotocols.cshlp.org/content/2010/1/pdb.prot5366.full

Yilmaz LS, Okten HE, Noguera DR (2006). Making all parts of the 16S rRNA of *Escherichia coli* accessible in situ to single DNA oligonucleotides. *Appl. Environ. Microbiol.* 72: 733–744.

Yilmaz LS, Parnekar S, Noguera DR (2010). mathFISH, a web tool that uses thermodynamics-based mathematical models for in silico evaluation of oligonucleotide probes for fluorescence in situ hybridization. *Appl. Environ. Microbiol.* 77: 1118–1122.

Yilmaz S, Haroon M, Rabkin B, Tyson G, Hugenholtz P (2010) Fixation-free fluorescence in situ hybridization for targeted enrichment of microbial populations. *ISME J.* 4(10): 1352–1356.

Zwirglmaier K (2005). Fluorescence in situ hybridisation (FISH) – the next generation. *FEMS Microbiol. Lett.* 246(2): 151.

PART IV

SPECTROSCOPY

9 X-ray Diffraction Techniques

DANIEL K. UNRUH AND TORI Z. FORBES

Abstract

X-ray diffraction techniques provide information regarding the formation and alteration of mineral phases that is critical for assessing geomicrobial processes. Of particular interest is the use of powder X-ray diffraction (pXRD) to identify unknown solid-state materials, determine the particle size of nanoscale mineral phases, and refine structure characteristics, such as unit cell parameters and atomic positions. The goal of this chapter is to provide practical knowledge for the successful preparation of solid mineral samples, optimal data collection strategies, and analysis of diffractograms collected from pXRD experiments. Specific uses of pXRD techniques in geomicrobiology are discussed to demonstrate the importance of diffraction in advancing our understanding of microbial communities in geologic systems.

9.1 Introduction: General Use and Fundamentals of Powder X-ray Diffraction

The interaction of X-rays with solid-state materials is an essential physical phenomenon that can be used to provide atomistic information about minerals and has been widely applied to biotic and abiotic systems (Benzerara et al., 2008; Bertel et al., 2012; Brantner et al., 2014; Das et al., 2014; Fandeur et al., 2009; Maillot et al., 2013; O'Reilly and Hochella, 2003). Powder X-ray diffraction (pXRD) is the most common technique used to evaluate many natural samples, because low-temperature geologic environments often produce polycrystalline mineral phases. Analysis of powders and poorly crystalline phases is particularly useful for geomicrobial studies, because most materials formed or utilized by microbial communities typically fall within this regime. The most important information that pXRD typically provides to the investigator is the identity of solid-state material produced through biomineralization processes. However, additional structural information can be gained through analysis by X-ray diffraction, including unit cell parameters, particle size, atomic position, and quantitative assessment of heterogeneous samples.

The current chapter provides practical information regarding the use of pXRD to understand geomicrobial processes. First, the basics of X-ray diffraction are discussed, followed by best practices for optimal data quality, and finally, a sampling of studies from the literature is provided to illustrate the use of pXRD in geomicrobiological investigations. Our intended audience is investigators who have a basic understanding of solid-state materials and are interested in utilizing diffraction techniques in their current and future research. Crystallographic nomenclature (lattice, d-spacing, *hkl* planes) is used throughout the chapter, and fundamental information on these topics can be found in basic mineralogy textbooks (Klein and Dutrow, 2007; Nesse, 2012; Wenk and Bulakh, 2004). For more advanced users, additional resources are also available, including several excellent books edited by Bish and Post (1989), Dinnebier and Billinge (2008), Pecharsky and Zavalij (2005), and Young (2002).

9.1.1 Fundamentals of Diffraction – X-ray Interaction with Matter

Diffraction of electromagnetic radiation or particles by matter is the basic underpinning of X-ray diffraction techniques, and utilizing this physical phenomenon can provide fundamental structural information for solid materials. Whether your sample is a beautiful single crystal or a polycrystalline array, diffraction data can provide fundamental knowledge of the structural nature of your sample. This section will provide the reader with a general overview of X-ray diffraction that will lay a foundation for understanding the technique. In addition, the basic experimental setup is described to provide introductory knowledge before entering the laboratory to perform the diffraction experiments.

X-rays are a subset of the electromagnetic spectrum that have wavelengths from 0.1 to 100 Å (10^{-10} m). When X-ray photons propagate through a substance, they can be scattered or absorbed (ionization or photoelectric effect), but it is the analysis of the scattered X-rays that provides the information on the nature of the crystallographic lattice. The wavelength of scattering for the X-ray photon can remain constant (coherent) or increase (incoherent), and it is the coherent interactions that result in diffraction. For coherent interactions to take place, the X-ray beam will interact with the electrons surrounding the nucleus, creating an oscillation with the same frequency as the electric field component of the electromagnetic wave. This results in radiation emitted in all directions that has the same wavelength as the incoming wave. When atoms are arranged in a periodic array, such as a crystalline lattice, the X-ray beam can interact with a multitude of electrons, and the coherently scattered waves can interact either destructively (canceling out the amplitude of the wave) or constructively (increasing the magnitude of the wave). This constructive interaction of the X-ray beam produces wave fronts that are in phase, and this cooperative scattering effect is known as diffraction.

9.1.2 Bragg's Law

Crystal habit is one distinguishing feature of mineral phases and is related to the arrangement of atoms within the material. In fact, the faces that appear on highly crystalline specimens are generally parallel to the planes of atoms that have the greatest density of lattice nodes. When an X-ray beam penetrates a crystal, the diffraction can occur as a "reflection," whereby the incident beam reflects back out of the crystal from these parallel planes. However, for the diffraction effect to be of measurable intensity, reflections from the individual planes must be in phase with each other to lead to constructive enhancements.

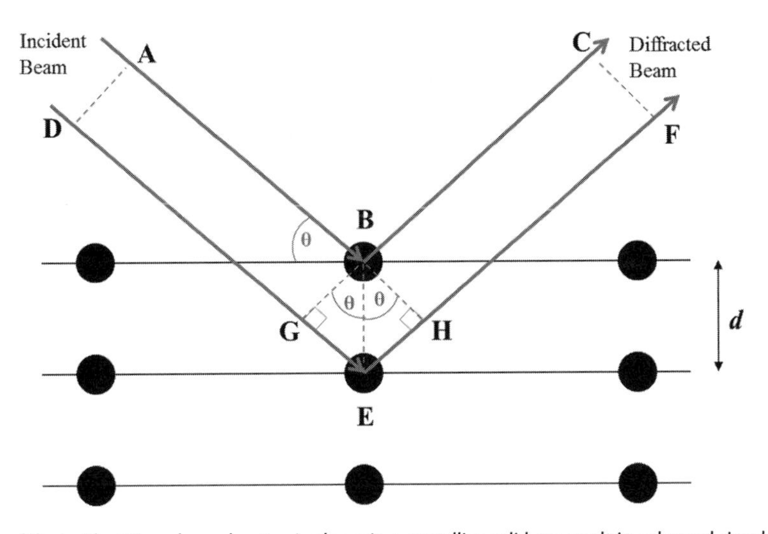

Figure 9.1 Diffraction of the incident X-ray beam by atomic planes in a crystalline solid can result in enhanced signals if all the waves are in phase. This can be accomplished if there is an integral wave value, as determined by Bragg's law.

The necessary conditions for in-phase diffraction or "reflection" by the parallel atomic planes can be expressed using Bragg's equation: Eq. (9.1). W.L. Bragg, who is generally considered one of the fathers of diffraction, determined that an integer value of the wavelength is related to the distance between the parallel atomic planes and the angle of the diffracted beam. Within the crystalline lattice, a set of parallel atomic planes can be described that are separated by a distance (d) (Figure 9.1). X-rays striking the first plane are reflected at the incident angle, θ, and must be reinforced by in-phase reflections (integral values) of the other atomic planes (e.g., second, third, fourth, etc.) for diffraction to be measurable. In other words, the wave path along the first atomic plane (ABC) is shorter than for the second plane (DEF), and this value must be a whole number of the wavelength ($n\lambda$) for diffraction conditions to be satisfied. In Figure 9.1, lines BG and BH are drawn perpendicular to AB and BC, so that AB = DG and BC = HF. To satisfy the in-phase condition, GE + EH = $n\lambda$. The line drawn between B and E is perpendicular to the atomic planes and is equal to the interplanar spacing (*d*). This creates two right triangles where $d\sin\theta$ = GE and $d\sin\theta$ = EH. Thus, Bragg's equation states:

$$2d\sin\theta = n\lambda \qquad \text{(Eq.9.1)}$$

9.1.3 Powder X-ray Diffraction

Atoms arranged in a long-range periodic array, as observed in crystalline mineral specimens, are all that is necessary to cause coherent X-ray scattering and satisfy Bragg's law. High-quality single crystals with dimensions between 50 and 300 μm can be utilized for single-crystal X-ray diffraction and provide the optimal means to understand atomic positions and compositions. Naturally occurring minerals, particularly those formed in low-temperature geologic environments or by microbial communities, rarely exist with these dimensions. Instead, polycrystalline minerals are much more common; thus, X-ray diffraction techniques for use on powder samples are widely employed in geology. To utilize Bragg's law in polycrystalline or powder samples, we must move beyond thinking about one individual set of lattice planes to understanding the sample as a whole.

Basic conditions that satisfy the in-phase "reflection" for a single lattice plane still hold true for powder X-ray diffraction. The major difference is that polycrystalline materials are randomly oriented, so that all atomic planes and related diffraction conditions can be accessed simultaneously. If the sample is prepared such that the crystallite orientation is truly random, for each family of atomic planes with characteristic interplanar distance (d) there are numerous particles whose orientation creates the proper θ angle with the incident beam to satisfy Bragg's law. Random orientation of the polycrystalline sample results in the diffraction maxima from a given set of *hkl* planes forming a series of nested cones with the incident beam as the axis. If a two-dimensional (2-D) detector is placed perpendicular to the incident X-ray beam, then a series of concentric circles, or the Debye–Scherrer rings, corresponding to the cones of diffraction can be observed (Figure 9.2a). Taking a slice through these rings is what is represented in a typical powder X-ray diffraction pattern, and the intensity of the diffraction rings is represented by the peak height in the diffractogram (Figure 9.2b). The specific positioning and intensity of the rings correlate to the structural positioning and identity of the atoms, resulting in a specific "fingerprint" for each mineral species.

9.2 Modern Powder Diffraction (pXRD) Instruments

All X-ray diffraction techniques rely on three major components: 1) the X-ray source; 2) the sample holder/stage; and 3) the detector (Figure 9.3). Historically, Debye–Scherrer, Gandolfi, and Guinier cameras were utilized for powder diffraction measurements, and photographic film was necessary to detect the cones of diffraction (Bish and Post, 1989). While modern instruments have moved beyond photographic film and towards the use of scintillating materials for detection, the underpinnings of pXRD instrumentation remain the same as when the technique was originally developed.

The X-ray source for most laboratory instruments consists of an X-ray tube that contains a tungsten filament as a cathode to provide the source of electrons. This tube is sealed under vacuum, and the electrons emitted from the cathode are accelerated onto a metal target that acts as the anode to the produced X-rays. A characteristic spectrum is produced from the metal target as a result of electronic transitions from outer shells to inner-shell vacancies that are produced by the bombardment of the target material. $K\alpha_1$ radiation is the dominate line spectrum used for diffraction techniques and the result of an L- to K-shell transition. Most powder X-ray diffractometers are equipped with a copper (Cu) or cobalt (Co) target and a monochromator that result in $K\alpha_1$ wavelengths of 1.5418 or 1.7902 Å, respectively. For a user, it is important to know the wavelength of the X-ray source, because it will impact the peak placement on the diffractogram and the resulting data analysis. In addition, the

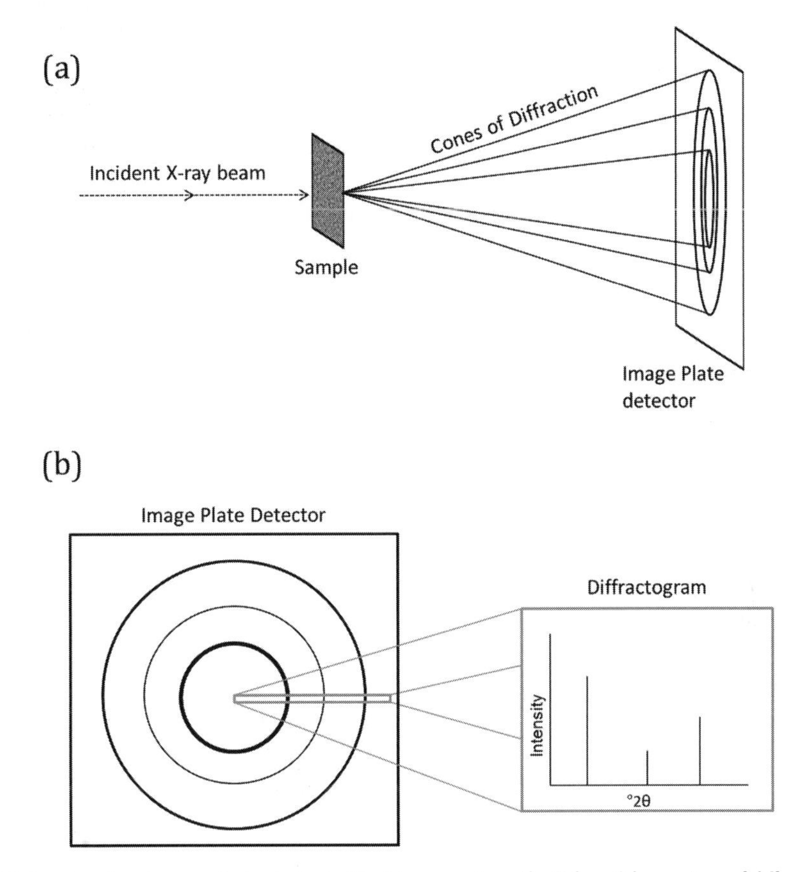

Figure 9.2 (a) A 2-D detector placed perpendicular to the incident beam captures the Debye–Scherrer rings of diffraction that result from a polycrystalline sample. (b) A slice through those concentric rings results in the typical diffractogram, with intensity on the y-axis and °2θ on the x-axis.

wavelength of the incident X-ray beam can also cause fluorescence of the sample and contribute to high background and low signal to noise ratios. This is particularly problematic when analyzing iron mineral phases with an instrument that has a Cu target. However, this can be avoided by switching to an instrument with Co radiation.

A detector is necessary to convert the individual X-ray photons into voltage pulses that can be transformed into information regarding the intensity and position of the diffracted beam. To accomplish this task, the detector is created from material that will produce ions upon exposure to X-ray photons and can be manufactured to cover different areas of the resulting cones of diffraction (Figure 9.4). Most instruments will have a point (0-D) detector, which requires that the detector scan through the range of reflected angles in order to locate all of the diffraction planes within the sample (Figure 9.4a). Initially, these detectors were typically scintillation detectors, in which the incoming X-ray beam would strike an NaI phosphor. The NaI material would convert the X-ray photons into light with a wavelength of approximately 410 nm and then transform the electromagnetic pulse into an electronic signal using a photomultiplier (Bish and Post, 1989). These detectors have very little noise, dead

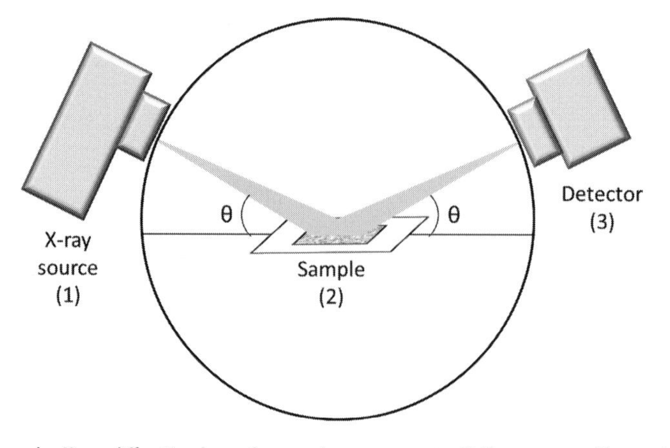

Figure 9.3 Instruments for powder X-ray diffraction have three major components: 1) X-ray source; 2) sample holder; and 3) detector, which are generally arranged on a focusing circle with a Bragg–Brentano geometry.

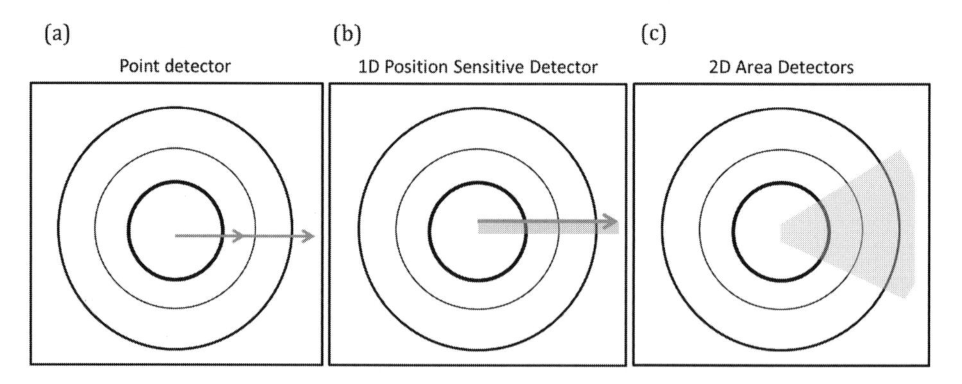

Figure 9.4 Comparison of detectors used for pXRD studies. (a) The point detector is scanned through 2θ space to determine the position of the diffraction cones. (b) A larger area is probed with the position sensitive detector and (c) the area detector provides the large range of 2θ values in one exposure. (Adapted from Bob He et al., 2000.)

time, or energy resolution but require significant counting times, resulting in longer measurement times. Many of the current detectors on the market are solid-state detectors that are made from semiconducting materials, such as Ge and Si(Li) (Bish and Chipera, 1989). Si(Li) is a p-type semiconductor that only requires Peltier cooling and has higher quantum counting efficiencies than scintillation detectors, leading to shorter data collection times. These solid-state materials can be used in point detectors but are also found in one-dimensional (1-D) linear or position-sensitive detectors (PSD). The advantage of the PSD detectors is that they provide a larger range of 2θ values that can be measured simultaneously, resulting in increased intensities and shorter measurement times (Figure 9.4b). Two-dimensional charge-coupled device (CCD) detectors are also available, which can provide additional positional information and decrease the overall measurement time to just a few minutes in well-crystallized materials. A 2-D detector can also measure the entire 2θ range and orientation of the sample simultaneously, providing very short measurement times for sensitive samples (Figure 9.4c). The tradeoff for this type of detector is lower resolution than the 0-D and 1-D varieties, but advanced technologies, point-beam X-ray optics, and larger computing power continue to provide enhancements in data quality (Bob He et al., 2000; Rudolf and Landes, 1994; Sulyanov et al., 1994). For each type of detector there are variations in quantum counting efficiency, linearity, and proportionality, but most detectors that are in use today provide excellent data for standard diffraction studies.

The sample holder is located between the X-ray source and the detector, and many instruments utilize Bragg–Brentano parafocusing geometry to provide accurate peak position with moderately fast measurement times. Data collection can be accomplished using θ-2θ mode, in which the X-ray source remains stationary, the sample is tilted at θ, and the detector rotates around the sample at twice this angle. A more commonly used mode is the Bragg–Brentano θ-θ geometry, which allows the sample to remain at a fixed angle, but both the X-ray source and the detector move at angle θ to produce the 2θ angle. In both cases, the sample must be properly aligned so that accurate positional and intensity information is obtained. It is important that sample loading and unloading are performed carefully so that the optimal instrument alignment is maintained. In addition, it is also valuable to prepare a calibration curve of observed versus theoretical 2θ for a well-characterized material, such as quartz, after alignment of the instrument is complete. This will provide a means of assessing the magnitude of the potential positional error and if performed on a regular basis, will be a valuable tool to monitor instrument alignment and the viability of the X-ray tube.

Typically, reflection mode is the most commonly used sample collection strategy, with the polycrystalline sample placed on a flat plate or a disc (Figure 9.5a). Sufficient material must be present on the sample holder to ensure that the incident beam is diffracted off the surface of the sample. With minimal material, the beam can also scatter off the sample holder, leading to a larger signal to noise ratio and potentially spurious peaks in the diffractogram. For Bragg–Brentano geometry, a flat sample is necessary because the X-ray beam is typically divergent. For rough surfaces, reflection mode can still be used, but different optics must be utilized to create a parallel beam. In transmission mode, the incoming beam is orthogonal to the sample and ideally focused using a collimator (Figure 9.5b). A small quantity of polycrystalline sample is placed inside a capillary tube, and a PSD or area detector is used to provide positional information. One benefit of transmission mode is that samples can be analyzed wet, which means that data can be collected for materials that dehydrate easily or oxidize under normal laboratory conditions.

9.3 Sample Preparation

The most common problems encountered during analysis via pXRD techniques are the direct result of improper preparation of samples. Taking time to create high-quality powder mounts will result in optimal data, save instrument time (and related costs), and prevent the need to recollect data on precious geologic samples. In this section, we will provide

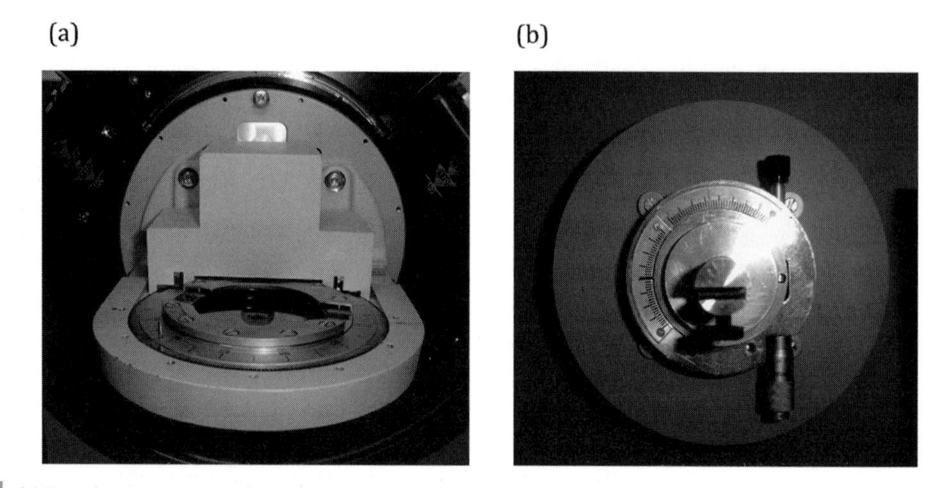

Figure 9.5 (a) Typical instrument stage for samples run in reflection mode compared with a stage (b) that can be used in either reflection or transmission mode depending on the orientation of the sample holder.

practical advice for collecting publishable diffractograms, particularly focusing on the most common sample preparation techniques.

High-quality samples for pXRD studies contain a large number of crystallites of a uniform size (<10 μm) that are randomly oriented with respect to the incident X-ray beam, resulting in homogeneous rings of scattered X-rays for each Bragg angle. For typical Bragg–Brentano geometry, the surface must be relatively flat, because additional roughness will impact the intensity of the reflected beam. In geology, samples typically do not start out fitting all, if any, of the criteria for collecting high-quality pXRD data. Thus, samples must 1) be ground to a homogeneous crystallite size; 2) have a flat surface; and 3) maintain random orientation of the crystallites. Other aspects, such as small sample quantities, difficult preparations, and use of internal standards, will also be discussed to provide a practical guide for the best-quality pXRD samples.

9.3.1 Crystallite Size

To ensure that the sample is of the proper particle size, grinding is generally the first step in the preparation process. Whether a simple agate mortar and pestle or a more sophisticated mechanical mill setup is used for sample preparation, the ultimate goal is to produce a homogeneous powder with particle size ranging from 1 to 10 μm. Manual grinding is preferred for softer materials, because overgrinding materials in a mill can result in strain- and particle-related broadening of the diffraction peaks and may even result in polymorphic phase changes for some mineral and solid-state species (Alam et al., 2010; Altheimer et al., 2013; Pesenti et al., 2008). To grind minerals by hand, a small amount of material (approximately 100–200 mg) is typically placed into an agate mortar with a solvent (e.g., acetone, methanol, or water) to help minimize structural damage to the sample. Adding a liquid to the sample is very important, as grinding dry samples results in the formation of cracks and defect sites in the crystalline lattice (Jackson, 1969). For harder materials, manual grinding is difficult, and a ball mill or shatter box can be used to prepare the powder. Again, caution must be used with a mechanical mill, because overgrinding can lead to significant issues with data analysis. Once the sample has been ground, the powder can be sieved to produce a more uniform sample, but be aware that this process could potentially introduce inhomogeneity due to density differences within the sample.

9.3.2 Creating a Powder Mount

A typical sample holder consists of a plastic plate or disc that contains a small cavity to accommodate the powdered sample. These holders can be specially ordered from the company that manufactured the diffractometer, but a good machinist can also make similar plates for a much lower cost. The most important consideration for manufacturing the sample holder is that that the sample height is correct. If there is an offset in height, the sample will be located slightly below or above the focusing circle, and positional error will be introduced into the diffractogram.

Top-loading is the most commonly used method for sample mounts due to ease of preparation. To create a top-loaded mount, the material is transferred into the sample holder, and the cavity is completely filled with the powdered sample. The exact mass of the sample needed to fill the cavity typically depends on the size of the cavity and the density of the material, but typically 50–200 mg is enough powder to create a high-quality sample for analysis by pXRD. To ensure the correct sample height and eliminate roughness, excess powder is removed by levelling off the cavity with either a glass slide or a razor blade (Figure 9.6). Do not pack or compress your powdered sample down into the cavity, as this may lead to nonrandom orientation of the crystallites within the top layer of your sample. Other specially designed sample mounts are also available, including back- and side-loading sample holders, but are less commonly used than the top-loading method. These types of holders are typically utilized to maximize the random orientation of the sample.

If 50–100 mg of powdered sample is unattainable, small sample sizes (5–20 mg) can still be analyzed by altering the available sample mounts. The major issue with small sample sizes is the inability to fill the entire cavity with material, leading to a decrease in the sample height and related positional errors. To alleviate this problem, an insert can be fashioned to fill most of the available volume of the cavity (Figure 9.7a). The cheapest option is to cut a glass microscope slide to the correct volume of the cavity, but often this will result in higher background due to scattering of the X-rays by the disordered atoms within the glass matrix. Zero-background surfaces (specialty-cut single crystals of silicon or quartz) provide diffractograms of much better quality but can be quite expensive to purchase. In either case, the surface must be coated with a layer of the powder that is thick enough, so that the X-rays are scattering off the sample and not the holder.

The best option for creating an ideal thick film on the surface is by preparing a slurry mount. This type of top-loading mount is created by adding a small amount of a solvent (usually acetone or ethanol) to the powder and mixing to create a suspension. A glass pipet is used to remove a small amount of the slurry, and then it is placed directly on the insert. By gently shaking the insert, the suspension can be spread on the surface of the insert to create a thick homogeneous surface, and the solvent will quickly evaporate, leaving a dry

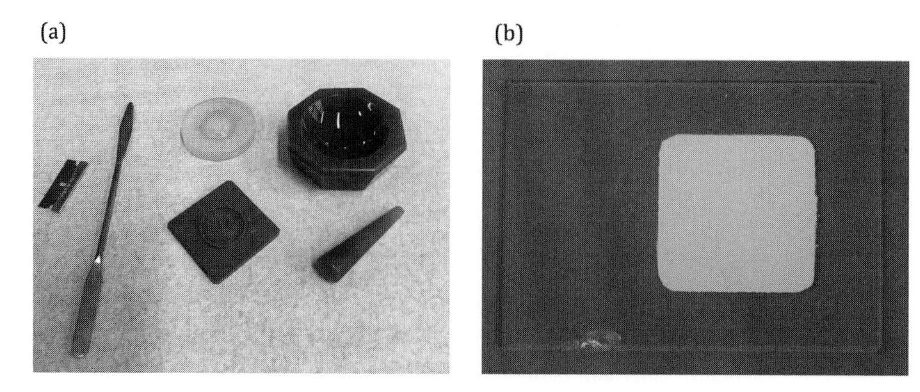

(a) Tools necessary to create a top-loaded sample include a razor blade, a spatula, sample holders, and a mortar and pestle. (b) The cavity of the sample holder should be completely filled with the solid sample (1–10 μm in size) and the surface should be flat and even with the top of the cavity.

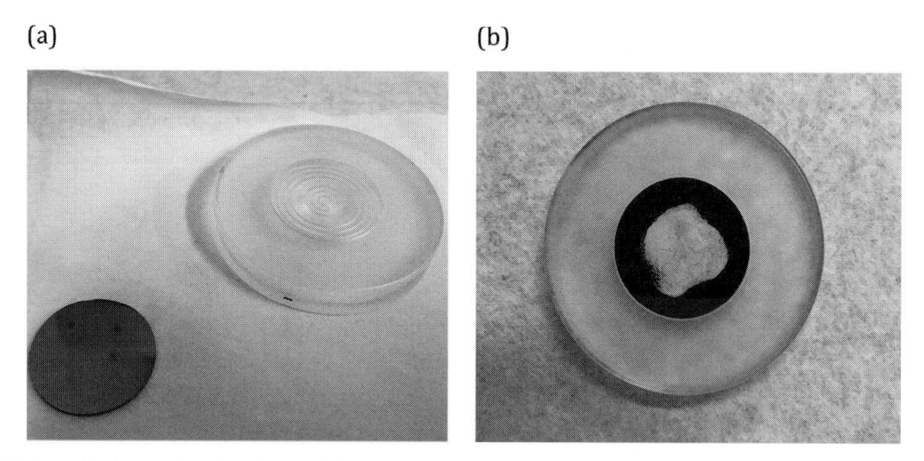

(a) A zero-background insert can be used if insufficient amount of powdered sample is not available for analysis. (b) A suspension composed of the sample and a volatile solvent (acetone, ethanol) can be used to create a film of the sample on the surface of the insert.

film on the surface of the insert. The benefit of this method is that the material can easily be scraped off and either reused for other characterization techniques or remade into a second powder mount if the first attempt was unsuccessful (Figure 9.7b). One important point to note is that the choice of solvent is important for a high-quality slurry mount. The sample must not dissolve in the solvent and quickly evaporate to ensure a random orientation of crystallites.

9.3.3 Random Orientation

The last component of a good sample mount is the random orientation of the crystallites within the cavity or film. If the mineral has a blocky or prismatic habit, then random orientation of the crystallites within the powder mount is easily achievable using either the top-loading or the slurry mount approach. However, many samples contain habits that are less likely to orient in random ways, preferring instead to stack parallel in preferred orientations (typically acicular, bladed, or tabular habits). To understand why this preferred orientation occurs, think about how the majority of the cards in a deck will fall either face-up or face-down when thrown onto the floor and very few (if any) will stand on a side. This is a similar situation to what can occur when creating a slurry mount when the crystallites in the powder are tabular or bladed in habit.

Preferred orientation will become apparent when analyzing the data, because specific peak intensities will be skewed either higher or lower than predicted, depending on the orientation of various diffraction planes to the incoming X-ray beam. To combat this issue, different sample preparation techniques can be utilized to force the crystals into more random orientations. The first approach is to use a frosted glass slide during the top-loading method to create a level surface. While compressing the material in a top-loading mount is usually not ideal, in this case, the frosted glass will provide a slightly rough, but level, surface, disrupting the orientation of the crystallites and minimizing some of the preferred orientation effects. Amorphous fillers (powdered glass, starch, gelatin, or boron) can also be added to disrupt the preferred orientation, but as stated before, will introduce some amorphous character into the pattern. Slurry mounts are not recommended with significant preferred orientation; an alternative method is to use grease to mount the sample. To create this mount, spread a layer of grease (vacuum grease or 3-in-1 oil is usually a good available option) on the surface and sprinkle the powdered sample to coat. Usually, this provides a surface that can be relatively rough, thereby minimizing the preferred orientation, but can decrease the intensity of the diffracted beam. The presence of the grease will contaminate the sample, preventing additional chemical characterization of the material, and will contribute to a higher background signal for the diffractogram. However, it remains a viable alternative to the slurry mount when preferred orientation is a problem. It is helpful to collect a diffractogram of the grease or amorphous fillers to ensure that any broad amorphous features can be accounted for during the data analysis step.

Clay minerals are particularly susceptible to preferred orientation because of the laminar mineral habit, and the methods described so far are generally ineffective at achieving ideal random mounts. While there are specialized methods to minimize preferred orientation (tubular aerosol suspension, liquid-phase spherical agglomerization, and spray drying), oriented mineral mounts can also be used as a diagnostic tool to identify the species (Jackson, 1969). The plate-like clay minerals have relatively similar dimensions along the aluminosilicate sheets, but the distances between the sheets (the 001 basal plane) change based upon the identity of the material. Therefore, preferred orientation is required to create oriented clay mounts so that the 001 plane is parallel to the sample surface. Advanced oriented clay mounts are beyond the scope of this chapter but can be found in Jackson (1969).

9.3.4 Internal and External Standards

Another aspect to consider when preparing the sample for analysis is the use of an internal reference material. For phase identification, an internal standard is usually not necessary unless you are concerned about the alignment of the instrument. However, if determination of percent crystallinity, phase percentages in a heterogeneous sample, or lattice parameters in a doped material is of interest, having an internal standard homogeneously mixed with your sample will help ensure the production of an accurate analysis. The choice of reference material depends on the identity of the sample, but the optimal internal standard is one that contains a few strong reflections at peak positions that do not overlap with those associated with the sample. If you are unsure of the peak location of your sample, a quick scan can provide the 2θ regions that would be optimal for the internal standard. Two of the most popular internal standards for geologic materials include quartz and corundum, but there is a suite of standards that can be purchased from the National Institute of Standards and Technology for use in pXRD experiments. The addition of an internal standard will also prevent the reuse of the sample in other chemical characterization methods. If this is not an acceptable scenario, then an external standard can also be used. In this case, a standard is chosen with discrete, high-intensity peaks but is analyzed separately from the sample of interest. It is important to note that extreme care must be taken to prepare and analyze both samples in a similar fashion so as not to produce systematic errors between the sample and the standard. External standards are generally used to check for instrument alignment or as a calibration curve for quantitative measurements; they are not useful for correcting specimen displacement error or transparency.

9.4 Data Collection and Analysis

Each diffractometer has slight variations in the physical configuration of the instrument and utilizes different software for data collection and analysis. While this section will not cover the specifics associated with these individual variations, we will discuss in more general terms the important factors to consider when establishing an experimental procedure. While a wide range of information can be gathered from the collected data, we will focus on the most widely used analytical methods that are of particular interest to geomicrobial investigations: phase identification, unit cell determination, and particle size measurement. Again, more detailed information on data collection can be found in Bish and Post (1989) or in the software manuals available with your X-ray diffraction instrumentation.

9.4.1 Data Collection Strategies

After optimal preparation of the sample, the next step is to determine the best instrument configuration, while also being conscious of the cost and collection time associated with the data collection. Modern pXRD instruments come with a variety of bells and whistles, but at the core, there are really a handful of basic components that the general user will need to consider. These include type of optics, collimating slits, and the data collection strategy.

In some pXRD facilities, there exists the possibility to quickly change between two types of incident beam optics: Bragg–Brentano and parallel beam. While Bragg–Brentano produces a larger flux of X-rays than parallel beam, the X-rays are not in phase with one another, and there is greater sensitivity to sample displacement. In comparison, parallel beam optics pass the incident X-ray beam through a set of focusing mirrors and slits, producing a coherent beam of X-rays. In general, Bragg–Brentano optics are ideal for both qualitative and quantitative analysis of powder mounts, whereas parallel beam optics can be utilized to study textures, stress, and thin-film materials (Verman and Kim, 2004). Parallel beam optics can also be very useful for mineral specimens, because if a relatively flat surface on a rock sample is available, no additional grinding or preparation is necessary. In addition, parallel beam optics can be focused to a smaller spot size using a collimator, which allows point detection on mineral surfaces or very small powder mounts. The downside to parallel beam optics is the lower X-ray flux, which requires longer count times and results in decreased peak intensities in the diffractogram (Verman and Kim, 2004).

Most diffractometers are equipped with collimating slits on the X-ray tube to truncate the incident beam and also secondary slits on the detector that eliminate the divergent reflected beam. These slits can be varied, either manually or through the automated software program. Smaller slit widths result in better resolution of the diffractogram but at the cost of signal intensity and longer scan times. Automated programs also have the capability of variable slit widths that change over the course of the scan and can be useful for divergent beams (Bragg–Brentano optics). The software-controlled variable slits open as the 2θ angle increases, so that the amount of sample in the area of the incident beam that hits the sample stays the same, leading to more accurate peak intensities throughout the entire data collection.

A wide range of software programs are available, depending on the company that manufactures the pXRD instrument, but most include data collection strategies that allow the user to determine 2θ range, step size, and count time. The 2θ range is rather arbitrary but should be chosen to capture the major peaks in the diffractogram. A typical 2θ range for sample collection is 5 to 90°, but that may need to be adjusted for materials with larger pore spaces (one will need to start at lower 2θ values). The X-ray tube and detector will move through the 2θ range in a series of small steps that need to be defined in the software. A step size of 0.05° in 2θ space will be adequate for phase identification, but when attempting particle size analysis, whole pattern fitting, and lattice constants, a smaller (0.02–0.005°) step size will be necessary. Count times typically depend on the scattering ability of the solid-state sample, the initial X-ray flux, the amount of material on the sample holder, and the type of detector. Typically, a count time per step of 0.05 to 1 s is reasonable for most materials, but a quick scan of the actual sample may provide more information regarding the proper amount of time for the optimal data collection strategy.

9.4.2 Initial Background Subtraction and Corrections

After the data are collected, a background correction is necessary before additional processing. This is of particular importance for methods that involve the use of amorphous binders or inserts. Background in pXRD originates from a variety of sources, including inelastic scattering; scattering from the air, sample holder, and surface; X-ray fluorescence; and detector electronic noise (Pecharsky and Zavalij, 2005). Higher backgrounds can also be attributed to the presence of biomass, which can be a problem for samples with substantial biological components (Brayner et al., 2009). Manual background subtraction can be performed by most software by initially running the binder or empty holder and then subtracting from the sample pattern. This will remove the large broad signals that are often observed when amorphous materials are present within the sample material. While performing manual background subtraction, care must be taken to ensure that both samples were collected in the same manner and that the peak intensities are large enough that they will not be masked by the amorphous signal. Other strategies for background subtraction include automatic fitting defined by the software, which is the preferred method for whole pattern fitting analysis. In many cases, the background is fitted to a low-order polynomial and can be refined along the angular range. Higher-order polynomials will increase the degree of curvature and can also be used to remove some of the noise associated with the presence of amorphous material. However, the user must be cautious with polynomial fits within a short 2θ range, because a higher-order polynomial is flexible enough to describe the peaks as well as the background.

In addition to the background subtraction, positional corrections can be made based on either the external calibration or an internal standard. An external calibration will identify the extent of instrumental error, and then the user can shift the x-axis of the sample diffractogram accordingly. Typically for well-aligned instruments this shift will be small (0.001–0.005 °2θ), but it may be an important consideration, depending on the sensitivity of the analysis. The use of an internal standard allows the positional correction to be sample specific and also helps identify errors in sample height. If the sample is located above the focusing circle, then the positional error for all peaks in the diffractogram will be shifted to a higher 2θ value. The opposite is true for a mount located slightly below the circle, as the positional error will be negative. Typically, the positional error due to height is relatively small for top-loaded mounts but can be up to 0.01–0.05 °2θ for slurry mounts.

9.4.3 Identifying Unknown Mineral Specimens

Just as fingerprints are unique to each individual person, the specific location and intensity of the peaks are unique for each mineral phase. The peak positions are based on the planes of diffraction, which in turn is related to the placement of the atoms in the crystalline lattice. As individual minerals are defined based upon their chemical composition and highly ordered atomic arrangement, the location of the peaks serves as a means to identify unknown materials.

The diffractogram itself does not provide enough information about an unknown sample to instantly classify the species, but comparing the collected pattern with a database of known phases will typically result in a positive identification. This is comparable to a fingerprint at a crime scene, because the perpetrator can only be identified through a match in the database. Many X-ray facilities have access to a powder diffract file (PDF) database through data analysis software, but a license can be purchased separately for research groups. The International Centre for Diffraction Data (ICDD) has the most comprehensive and searchable database through PDF-4+ products (www.icdd.com/products/pdf4.htm), including specific subsets for mineral phases (with data for nearly 97% of all known mineral types). Freeware and open access databases on the internet are also available, including the American Mineralogist Crystal Structure Database, which was established by Downs and Hall-Wallace (2003) (http://rruff.geo.arizona.edu/AMS/amcsd

.php). In all cases, the pattern can be matched based upon mineral name, chemical composition, or major diffraction peaks, providing a means to easily search a database containing thousands of mineral specimens to find a proper match. An example of the accurate identification of corundum (Al_2O_3) is given in Figure 9.8, with the diffraction peaks matched to those of PDF-97–016-4617.

During the search-match process for an unknown mineral specimen, there are several points to consider that will ensure the correct identification of the solid-state phase. First, all peak positions in the experimental pattern must completely match the peak positions of a pattern in the database. Additional peaks in the experimental pattern are most likely the result of the sample containing a secondary phase or contaminant, and a sample with additional peaks in the diffractogram cannot be characterized as homogeneous. It is also important that positional corrections have been made prior to utilizing the search-match database so that these errors are not incorporated into the search criteria. Commonly, peaks associated with an unknown sample may match the position of a known phase present in the database, but the reported intensities may vary from the sample. In many cases, this deviation in peak intensities is caused by sample preparation, particularly for materials that are prone to preferred orientation, so careful analysis of the data in this case is important to ensure accurate identification. Second, several patterns are typically available in the database for the same mineral phase, and it is important to assess the quality of the pattern before applying it to the experimental data. Various quality marks are provided in the database to give the user an idea of how the data were collected for the available patterns. In the ICDD database, an S (Star) and G (Good) represent reliable experimental data, but caution must be used for patterns with O (Low-precision) and M (Minimal acceptable) (ICDD, 2012). Other symbols, such as R (Rietveld), I (Indexed), and C (Calculated), represent data that include additional information regarding the relationship between the observed diffraction peaks and the *hkl* planes within the crystalline lattice, and are also acceptable for use in search-match functions. Last, minerals often contain chemical variability that cannot be accurately determined using pXRD. Therefore, it is important to couple the pXRD data with additional chemical characterization techniques, such as scanning electron microscopy-energy dispersive spectroscopy (SEM-EDS) (see Chapter 6), inductively coupled plasma mass spectrometry (ICP-MS), or electron microprobe analysis, to provide definitive proof of the exact composition of the unknown phase.

Confirming the identity of a mineral phase is the major use of pXRD for geologic studies. Of specific interest to the geomicrobiology community is the identification of solids formed from natural biomineralization processes, and laboratory-based microbial communities (Achal et al., 2012; Baskar et al., 2014; Cosmidis et al., 2014; Kukkadapu et al., 2001; Obst et al., 2009; Ona-Nguema et al., 2004; Peng et al., 2013; Piepenbrock et al., 2011; Roden et al., 2002; Sinha et al., 2014; Tazaki et al., 2003; Thorpe et al., 2014; Warren et al., 2001; Wu et al., 2011; Zachara et al., 2002; Zegeye et al., 2005). One example of phase identification of samples taken from a natural system is Frierdich et al. (2011), in which the deposition of mineral phases in a karst system was investigated by a several analytical techniques. The pXRD data revealed the presence of poorly crystalline Fe oxide deposits intermixed with quartz (SiO_2) and calcite ($CaCO_3$) grains in the solid samples. The authors were also able to subtract the

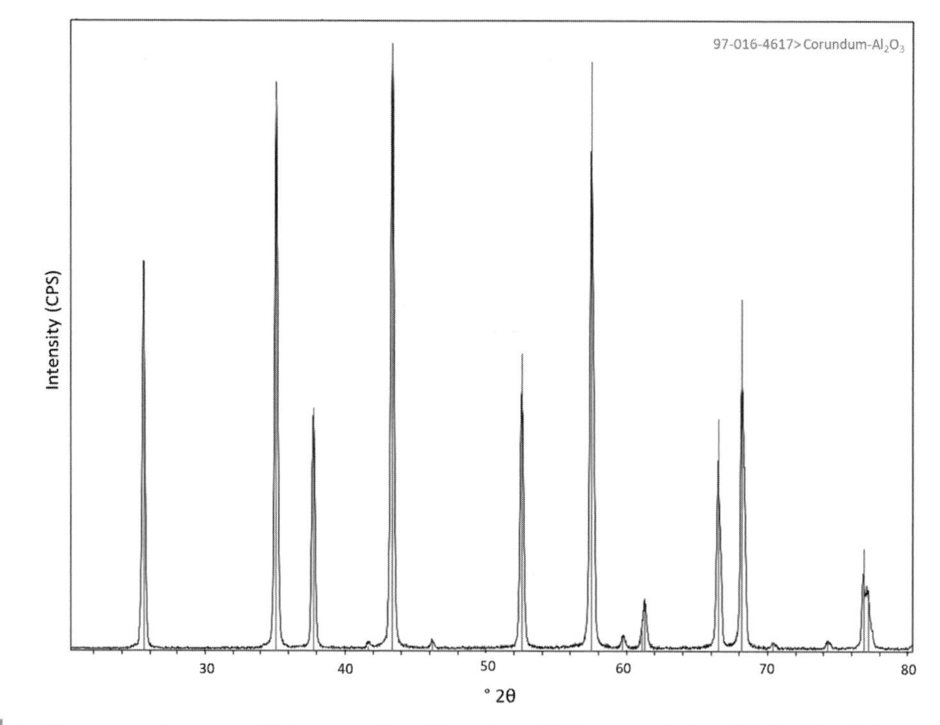

Figure 9.8 A diffractogram of corundum (Al_2O_3) was collected from 20 to 80 °2θ and matched to PDF-97–016-4617.

peaks associated with the bulk minerals from the diffractogram to provide more resolution of the poorly crystalline phases. Based upon the broad peaks near 36 and 63 °2θ, the mineral phase ferrihydrite ($Fe^{III}{}_2O_3{\cdot}0.5H_2O$) was suggested as the major iron oxide present in this sample. Biomineralization of ferrihydrite was then confirmed by the complementary SEM images, which found complex mixtures of fibrous and helical morphologies, indicative of the presence of specific organisms (bacteria from the genus *Leptothrix* and *Gallionella*). Brayner et al. (2009) investigated the biomineralization processes by which the cyanobacteria *Anabaena* and *Calothrix* produce akageneite along with an amorphous material attributed to the presence of biomass. In a more recent example, Ronholm et al. (2014), studied calcium carbonate phases precipitated from heterotrophic bacteria collected from cryophilic polar regions, revealing the presence of several mineral polymorphs. Of particular note was the formation of vaterite ($CaCO_3$), the rare calcium carbonate polymorph that has been linked to bacterial entombment in certain phylotypes.

9.4.4 Unit Cell Determination

The unit cell within a crystalline lattice is the smallest repeatable building block that captures all of the symmetry within the material. It is defined by both unit cell parameters (edge lengths and angles), which are again specific to individual mineral phases.

Perturbation in the unit cell (subtle variations in unit cell lengths and/or angles) can be caused by elemental substitutions or formation of defects within the crystalline lattice. X-ray diffraction techniques are very sensitive to these lattice changes, and pXRD provides an excellent avenue to explore subtle crystallographic transformations to understand geomicrobial processes.

Unit cell dimensions can be calculated manually based upon the position of the diffraction peaks within the diffractogram and is relatively straightforward when the material can be identified in the database and is highly symmetric. The basic procedure begins with generating a list of all possible combinations of *hkl* planes within the sample, using that list to calculate interplanar distances, assigning the *hkl* planes to the observed peaks, and refining the unit cell dimensions using the observed diffraction peaks. The relationship among interplanar *d*-spacings, *hkl* planes, and the unit cell lengths (a,b,c) is given in Eq. (9.2):

$$\frac{1}{d^2} = \frac{h^2}{a^2} + \frac{k^2}{b^2} + \frac{l^2}{c^2} \qquad \text{(Eq.9.2)}$$

The relationship can be simplified when the material crystallizes in a cubic unit cell, because all of the unit cell lengths are the same:

$$\frac{1}{d^2} = \frac{(h^2 + k^2 + l^2)}{a^2} \qquad \text{(Eq.9.3)}$$

It can also be extended to include the information provided in the diffractogram (namely, the peak position in °2θ) through Bragg's law:

$$\sin^2\theta = \frac{(h^2 + k^2 + l^2)\,\lambda^2}{4a^2} \qquad \text{(Eq.9.4)}$$

Based upon this relationship, the unit cell parameter (a) can be determined from the compiled pXRD data. An example of this procedure in practice is highlighted by relating the peak positions for uraninite (UO_2; space group Fm-3 m) to the indexed PDF. From the experimental data, the *d*-spacings were calculated from Bragg's law, and a table relating all of the positional information was constructed to demonstrate the relationship between observed °2θ, intensity, calculated *d*-spaces, *hkl* values, and the calculated unit cell parameter (Table 9.1). The equation that describes the relationship between the *d*-spacing and the unit cell parameters must be expanded for lower-symmetry materials, and additional information on manually calculating these values can be found in Pecharsky and Zavalij (2005).

Whole pattern fitting can also provide unit cell information but requires high-quality experimental data for modeling the diffractogram. In this type of experiment, it is recommended to carefully prepare the sample mount to minimize the preferred orientation and provide accurate peak intensities. The data collection strategy should also be programmed with a small step size and longer count times so that accurate analysis can be provided for not only peak location but also height and shape. In most cases, this type of analysis will require the use of a specialized program, such as Topas (Coelho, 2007) or Jade (Materials Data, 2002), but freeware programs such as FullProf (Rodriguez-Carvajal, 2001), GSAS (Larson and Von Dreele, 2004), and MAUD (Lutterotti et al., 1999) are also available.

Table 9.1 Example of calculating unit cell dimensions from pXRD data for uraninite (UO_2) collected using Cu Ka radiation (Downs and Hall-Wallace, 2003; Wyckoff, 1963).

Observed °2θ	Intensity (I/I_o)	Calculated *d*-spacing	*h*	*k*	*l*	Calculated unit cell parameter (a)
28.27	100	3.1571	1	1	1	5.4683
32.75	36.19	2.7341	2	0	0	5.4682
47.00	44.24	1.9333	2	2	0	5.4682
55.76	38.97	1.6487	3	1	1	5.4681
58.47	8.52	1.5785	2	2	2	5.4681
68.66	5.84	1.3671	4	0	0	5.4684
75.84	13.47	1.2545	3	3	1	5.4682
78.17	9.77	1.2227	4	2	0	5.4681
87.37	10.59	1.1162	4	2	2	5.4682

The advantage of the commercial software is the user-friendly interface, but the freeware programs are highly capable of whole pattern fitting and unit cell refinements. Peak fitting requires modeling of the background, peak shape, and intensities; therefore, the user should think critically about the values provided by the software, because instrument and sample errors can cause significant deviations in the data. Again, it is important to have secondary techniques to judge chemical compositions and provide high-quality, complementary evidence to support the diffraction studies.

Determination of unit cell parameters by pXRD has been applied to a range of geomicrobial and related systems. Pokroy et al. (2004, 2006) investigated anisotropic lattice distortion in biogenic calcite and aragonite by establishing a calibration curve for a range of materials with the composition $Mg_xCa_{1-x}CO_3$. This was applied to both the *a* and *c* lattice parameters, but the lattice distortion along the *c* axis was found to be larger, suggesting that there was anisotropic swelling of the biogenic calcite due to the incorporation of organic macromolecules. Villalobos et al. (2006) utilized whole pattern fitting and XRD simulation techniques to determine the unit cell and atomic positions for a biogenic manganese oxide produced by *Pseudomonas putida*. Fischer et al. (2011) investigated the formation of magnetite ($Fe^{II}Fe^{III}{}_2O_3$) nanoparticles by magnetotactic bacteria (*Magnetospirillum gryphiswaldense* and *M. magneticum*) and observed that significant differences in the lattice parameters for the iron oxide phase magnetite occur for culture bacteria when compared with isolated forms. These variations were then traced back to the oxidation state and stoichiometry of the magnetite that forms during magnetosome biomineralization (Fischer et al., 2011).

9.4.5 Calculating Particle Size

Reactive nanomineral phases have been identified as major products or drivers in many geochemical and geomicrobial processes, and pXRD is one of the tools that can be used to determine the average particle size in solid materials. Typical crystallite size for powder measurements on highly crystalline materials is 1–10 μm because discrete, homogeneous rings of diffraction occur within this size regime. As particle size decreases, the rings become more diffuse and the related peaks in the diffractogram broaden in response to smaller domains of coherent X-ray scattering. For nanominerals (typically defined between 1 and 100 nm), the broadening is significant, and the full-width half maximum of the peaks (FWHM) can be used to determine particle size based upon the Scherrer equation:

$$\tau = \frac{K\lambda}{\beta \cos\theta} \qquad \text{(Eq.9.5)}$$

where τ is the mean size of the crystalline domains, K is a dimensionless shape factor, λ is the wavelength of the X-ray radiation used in the experiment, β is the line broadening at FWHM, and θ is the Bragg angle of the particular peak of interest (Langsford and Wilson, 1978; Scherrer, 1918).

When using this equation to determine the average particle size of the experimental sample, there are several aspects that must be considered. First, K will typically be given a value of 0.9, which corresponds to a spherical or isotropic (blocky) particle, but for plate or needle-like crystallites the value can range from 0.8 to 2.0 (Monshi et al., 2012). Second, the Scherrer equation should be used on multiple peaks within the diffractogram to provide information on the error associated with the particle size. Third, it is very important to note that the Scherrer equation provides an average, lower boundary for the size of the crystalline domains in the sample, because peak broadening can also be influenced by multiple factors (e.g., lattice strain, dislocations, stacking faults, twinning, and even extensive grinding during the preparation stage). Therefore, pXRD should always be used as a complement to other techniques, such as transmission electron microscopy (TEM) and Brunauer–Emmett–Teller (BET) analysis, to ensure a proper analysis of the particle size of your material. Variations in particle size commonly occur between pXRD and other techniques, and it is very important to remember that pXRD is providing information about the size of the coherent domain of scattering, or the size of the periodic crystalline lattice. This can be different from the surface area measured by BET or what can be determined visually by TEM when aggregation of periodic crystallites occurs. Aggregation will lower the available surface area for BET measurements, suggesting a larger particle size than predicted by pXRD; therefore, diffraction provides a starting point for understanding the crystallite size, and the extent of aggregation can be further assessed using BET and electron microscopy.

The use of the Scherrer equation has been reported for a range of geochemical studies, most often paired with other techniques to confirm the particle size of the solid-phase material. Work by Till et al. (2014) describes the alteration of lepidocrocite (γ-FeO(OH)) by abiogenic and biogenic (*Shewanella putrefaciens* ATCC 8071) means. Reduction of the lepidocrocite to magnetite was observed in both experiments, and pXRD confirmed that the alteration product was cation-excess magnetite ($Fe_{3+x}O_4$). The particle size was determined be approximately 5–9 nm using the Scherrer equation, and these values were in good agreement with TEM observations. Another example of a study utilizing the Scherrer equation was performed by Ghashghaei and Emtiazi (2013). This work characterized biomineralization of calcite nanocrystals by chemoorganotrophic microorganisms. Particle size was determined by pXRD and confirmed by other analytical techniques. The observed nanoparticles ranged between 35 and 63 nm, but SEM analysis indicated aggregated particles of larger dimensions. The authors hypothesized that the aggregation was likely due to the bacteria acting as nucleation sites during the precipitation reaction, creating complex calcite architectures.

9.5 Conclusions

The goal of this chapter was to provide fundamental knowledge on X-ray diffraction and practical advice for collecting high-quality pXRD diffractograms that could be utilized in geomicrobial systems. Diffraction techniques can be used with different experimental designs, but due to the polycrystalline nature of most geologic samples, pXRD provides the best means of identifying unknown phases and providing structural information on the solid-state sample. One of the most important points to take away from this chapter is that time spent creating the optimal sample mount is a wise investment because it will lead to high quality data and a range of structural information. Methods for matching unknown phases to the PDF database, analyzing unit cell information, and determining particle sizes are the major uses of pXRD in geomicrobial studies. Advanced pXRD methods, particularly Rietveld and quantitative assessment of solid-state mixtures, were not specifically covered in this chapter but can also be applied to these systems (Young, 2002).

Geomicrobial samples also push the limits of diffraction techniques that can be performed by laboratory-based instruments due to small sample sizes, complex matrices, and heterogeneous samples. The application of pXRD to synchrotron sources is continuing to advance the numbers and types of samples that can be analyzed using diffraction techniques, opening this methodology to a wide range of applications (Beazley et al., 2009; Bell et al., 2007; Dick et al., 2009; Toner et al., 2009; Yee et al., 2006). The structural information provided by these advanced light sources is vital for improving our understanding of biomineralization processes and biogeochemical cycling, and a mechanistic understanding of the role of microbial communities in the natural world.

9.6 References

Achal, V, Pan, X, Fu, Q and Zhang, D 2012, 'Biomineralization based remediation of As(III) contaminated soil by *Sporosarcina ginsengisoli*' *Journal of Hazardous Materials*, vol. 201–202, pp. 178–184.

Alam, S, Patel, S and Bansal, AK 2010, 'Effect of sample preparation method on quantification of polymorphs using PXRD' *Pharmaceutical Development and Technology*, vol. 15, no. 5, pp. 452–459.

Altheimer, BD, Pagola, S, Zeller, M and Mehta, MA 2013, 'Mechanochemical conversions between crystalline polymorphs of a complex organic solid' *Crystal Growth & Design*, vol. 13, no. 8, pp. 3447–3453.

Baskar, S, Baskar, R, and Routh, J 2014, 'Speleothems from Sahastradhara Caves in Siwalik Himalaya, India: Possible biogenic inputs' *Geomicrobiology Journal*, vol. 31, no. 8, pp. 664–681.

Beazley, MJ, Martinez, RJ, Sobecky, PA, Webb, SM and Taillefert, M 2009, 'Uranium biomineralization as a result of bacterial phosphatase activity: Insights from bacterial isolates from a contaminated subsurface' *Geomicrobiology Journal*, vol. 26, pp. 431–441.

Bell, ATM, Coker, VS, Pearce, CI, et al. 2007, 'Time-resolved synchrotron X-ray powder diffraction study of biogenic nanomagnetite' *Zeitschrift fuer Kristallographie. Supplement Issues*, vol. 26, no. 2, pp. 423–428.

Benzerara, K, Morin, G, Yoon, TH, et al. 2008, 'Nanoscale study of As biomineralization in an acid mine drainage system' *Geochimica et Cosmochimica Acta*, vol. 72, no. 16, pp. 3949–3963.

Bertel, D, Peck, J, Quick, JQ and Senko, JM 2012, 'Iron transformations induced by an acid-tolerant *Desulfosporosinus* species' *Applied and Environmental Microbiology*, vol. 78, pp. 81–88.

Bish, DL and Chipera, SJ 1989, 'Comparison of a solid state Si detector to a conventional scintillation detector-monochromator system in X-ray powder diffraction analysis' *Powder Diffraction*, vol. 4, pp. 137–143.

Bish, DL and Post, JE (eds.) 1989, *Modern Powder Diffraction*, The Mineralogical Society of America, Washington, DC.

Bob He, B, Preckwinkel, U and Smith, KL 2000, 'Fundamentals of two-dimensional X-ray diffraction (XRD)2' *Advances in X-ray Analysis*, vol. 43, pp. 273–280.

Brantner, J, Haake, ZJ, Burwick, JE, et al. 2014, 'Depth-dependent geochemical and microbiological gradients in Fe(III) deposits resulting from coal mine-derived acid mine drainage' *Frontiers in Microbiology*, vol. 5, pp. 1–15.

Brayner, R, Yepremain, C, Djediat, C, et al. 2009, 'Photosynthetic microorganism-mediated synthesis of akaganeite (β-FeOOH) nanorods' *Langmuir*, vol. 25, no. 17, pp. 10062–10067.

Coelho, A 2007, Topas Academic V4.1, Coelho Software, Brisbane.

Cosmidis, J, Benzerara, K, Morin, G, et al. 2014, ' Biomineralization of iron-phosphates in the water column of Lake Pavin (Massif Central, France)' *Geochimica et Cosmochimica Acta*, vol. 126, pp. 78–96.

Das, R, Ali, ME and Abd Hamid, SB 2014, 'Current applications of X-ray powder diffraction – a review' *Reviews on Advanced Materials Science*, vol. 38, pp. 95–109.

Dick, GJ, Clement, BG, Webb, SM, et al. 2009, 'Enzymatic microbial Mn(II) oxidation and Mn biooxide production in the Guaymas Basin deep-sea hydrothermal plume' *Geochimica et Cosmochimica Acta*, vol. 73, no. 21, pp. 6517–6530.

Dinnebier, RE and Billinge, SJL (eds.) 2008, *Powder Diffraction Theory and Practice*, Royal Society of Chemistry, Cambridge.

Downs, RT and Hall-Wallace, M 2003, 'The American Mineralogist crystal structure database' *American Mineralogist*, vol. 88, no. 1, pp. 247–250.

Fandeur, D, Juillot, F, Morin, G, et al. 2009, 'XANES evidence for oxidation of Cr(III) to Cr(VI) by Mn-oxides in a lateritic regolith developed on serpentinized ultramafic rocks of New Caledonia' *Environmental Science & Technology*, vol. 43, no. 19, pp. 7384–7390.

Fischer, A, Schmitz, M, Aichmayer, B, Fratzl, P and Faivre, D 2011, 'Structural purity of magnetite nanoparticles in magnetotactic bacteria' *Journal of the Royal Society Interface*, vol. 8, pp. 1011–1018.

Frierdich, AJ, Hasenmueller, EA and Catalano, JG 2011, 'Composition and structure of nanocrystalline Fe and Mn oxide cave deposits: Implications for trace element mobility in karst systems' *Chemical Geology*, vol. 284, no. 1, pp. 82–96.

Ghashghaei, S and Emtiazi, G 2013, 'Production of calcite nanocrystal by a urease-positive strain of *Enterobacter ludwigii* and study of its structure by SEM' *Current Microbiology*, vol. 67, no. 4, pp. 406–413.

ICDD (2012), PDF-4+ 2012(Database), edited by Dr. Soorya Kabekkodu, International Centre for Diffraction Data, Newtown Square, PA.

Jackson, ML 1969 Soil Chemical Analysis – Advanced Course, Self-published, Madison.

Klein, C and Dutrow, B 2007 *Manual of Mineral Science*, 23rd edn., John Wiley & Sons, Inc., Hoboken.

Kukkadapu, RK, Zachara, JM, Smith, SC, Fredrickson JK and Liu, CX, 2001, 'Dissimilatory bacterial reduction of Al-substituted goethite in subsurface sediments' *Geochimica et Cosmochimica Acta*, vol. 65, pp. 2913–2924.

Langsford, JI and Wilson, AJC 1978, 'Scherrer after sixty years: A survey and some new results in the determination of crystallite size' *Journal of Applied Crystallography*, vol. 11, pp. 102–133.

Larson, AC and Von Dreele, VB 2004, 'General Structure Analysis System (GSAS)' *Los Alamos National Laboratory Report LAUR 86–748*.

Lutterotti, L, Matthies, S and Wenk, HR 1999, 'MAUD: A friendly Java program for material analysis using diffraction' *IUCr: Newsletter of the CPD*, vol. 21, pp. 14–15.

Maillot, F, Morin, G, Juillot, F, et al. 2013, 'Structure and reactivity of As(III)- and As(V)-rich schwertmannites and amorphous ferric arsenate sulfate from the Carnoulès Acid Mine Drainage, France: Comparison with biotic and abiotic model compounds and implications for As remediation' *Geochimica et Cosmochimica Acta*, vol. 104, pp. 310–329.

Materials Data 2002, Jade, Materials Data, Inc., Livermore, CA.

Monshi, A, Foroughi, MR and Monshi, MR 2012, 'Modified Scherrer equation to estimate more accurately nano-crystallite size using XRD' *World Journal of Nano Science and Engineering*, vol. 2, no. 3, pp. 154–160.

Nesse, WD 2012, *Introduction to Mineralogy*, Oxford University Press, Oxford.

Obst, M, Dynes, JJ, Lawrence, JR, et al. 2009, 'Precipitation of amorphous $CaCO_3$ (aragonite-like) by cyanobacteria: A STXM study of the influence of EPS on the nucleation process' *Geochimica et Cosmochimica Acta*, vol. 73, no. 14, pp. 4180–4198.

Ona-Nguema, G, Carteret, C, Benali, O, et al. 2004, 'Competitive formation of hydroxycarbonate green rust I vs hydroxysulphate green rust II in *Shewanella putrefaciens* cultures' *Geomicrobiology Journal*, vol. 21, pp. 79–90.

O'Reilly, SE and Hochella, MF 2003, 'Lead sorption efficiencies of natural and synthetic Mn and Fe-oxides' *Geochimica et Cosmochimica Acta*, vol. 67, no. 23, pp. 4471–4487.

Pecharsky, VK and Zavalij, PY (eds.) 2005, *Fundamentals of Powder Diffraction and Structural Characterization of Materials*, Springer Science and Business Media, Inc., New York.

Peng, X, Chen, S and Xu, H 2013, 'Formation of biogenic sheath-like Fe oxyhydroxides in a near-neutral pH hot spring: Implications for the origin of microfossils in high-temperature, Fe-rich environments' *Journal of Geophysical Research: Biogeosciences*, vol. 118, pp. 1397–1413.

Pesenti, H, Leoni, M, and Scardi, P 2008, 'XRD line profile analysis of calcite powders produced by high energy milling' *Zeitschrift fur Kristallographie*, vol. 27, no. 27, pp. 143–150.

Piepenbrock, A, Dippon, U, Porsch, K, Appel, E and Kappler, A 2011, 'Dependence of microbial magnetite formation on humic substance and ferrihydrite concentrations' *Geochimica et Cosmochimica Acta*, vol. 75, pp. 6844–6858.

Pokroy, B, Quintana, JP, Caspi, EAN, Berner, A and Zolotoyabko, E 2004, 'Anisotropic lattice distortions in biogenic aragonite' *Nature Materials*, vol. 3, no. 12, pp. 900–902.

Pokroy, B, Fitch, AN, Marin, F, Kapon, M, Adir N and Zolotoyabko, E 2006, 'Anisotropic lattice distortions in biogenic calcite induced by intra-crystalline organic molecules' *Journal of Structural Biology*, vol. 155, pp. 96–103.

Roden, EE, Leonardo, MR, and Ferris, FG 2002, 'Immobilization of strontium during iron biomineralization coupled to dissimilatory hydrous ferric oxide reduction' *Geochimica et Cosmochimica Acta*. vol. 66, pp. 2823–2839.

Rodriguez-Carvajal, J 2001, 'Recent developments of the program FULLPROF' *Commission on Powder Diffraction (IUCr) Newsletter*, vol. 26, pp. 12–19.

Ronholm, J, Schumann, D, Sapers, HM, et al. 2014, 'A mineralogical characterization of biogenic calcium carbonates precipitated by heterotrophic bacteria isolated from cryophilic polar regions' *Geobiology*, vol. 12, no. 6, pp. 542–556.

Rudolf, PR and Landes, BG 1994, 'Two-dimensional X-ray diffraction and scattering of microcrystalline and polymeric materials' *Spectroscopy*, vol. 9, no. 6, pp. 22–33.

Scherrer, P 1918, 'Bestimmung der Grosse und der Inneren Struktur von Kolloidteilchen Mittels Rontgenstrahlen' *Nachrichten von der Gesellschaft der Wissenschaften*, vol. 26, pp. 98–100.

Sinha, A, Singh, A, Kumar, S, Khare, SK and Ramanan, A 2014, 'Microbial mineralization of struvite: A promising process to overcome phosphate sequestering crisis' *Water Research*, vol. 54, pp. 33–43.

Sulyanov, SN, Popov, AN and Kheiker, DM 1994, 'Using a two-dimensional detector for X-ray powder diffractometry' *Journal of Applied Crystallography*, vol. 27, pp. 934–942.

Tazaki, K, Rafiqul, IABM, Nagai, K, and Kurihara, T, 2003, '$FeAs_2$ biomineralization on encrusted bacteria in hot springs: An ecological role of symbiotic bacteria' *Canadian Journal of Earth Sciences*, vol. 40, no. 11, pp. 1725–1738.

Thorpe, CL, Boothman, C, Lloyd, JR, et al. 2014, 'The interactions of strontium and technetium with Fe(II) bearing biominerals: Implications for bioremediation of radioactively contaminated land' *Applied Geochemistry*, vol. 40, pp. 135–143.

Till, JL, Guyodo, Y, Lagroix, F, Ona-Guema G, and Brest, J 2014, 'Magnetic comparison of abiogenic and biogenic alteration products of lepidocrocite' *Earth and Planetary Science Letters*, vol. 395, pp. 149–158.

Toner, BM, Santelli, CM, Marcus, MA, et al. 2009, 'Biogenic iron oxyhydroxide formation at mid-ocean ridge hydrothermal vents: Juan de Fuca Ridge' *Geochimica et Cosmochimica Acta*, vol. 73, no. 2, pp. 388–403.

Verman, B and Kim, B 2004, 'Analytical comparison of parallel beam and Bragg-Brentano diffractometer performances' *Materials Science Forum*, vol. 443–444, pp. 167–170.

Villalobos, M, Lanson, B, Manceau, A, Toner, B and Sposito, G 2006, 'Structural model for the biogenic Mn oxide produced by *Pseudomonas putida*' *American Mineralogist*, vol. 91, no. 4, pp. 489–502.

Warren, LA, Maurice, PA, Parmar N and Ferris, FG 2001, 'Microbially mediated calcium carbonate precipitation: Implications for interpreting calcite precipitation and for solid-phase capture of inorganic contaminants' *Geomicrobiology Journal*, vol. 18, no. 1, pp. 93–115.

Wenk, HR and Bulakh, A 2004, *Minerals: Their Constitution and Origin*, Cambridge University Press, Cambridge.

Wu, W, Li, B, Hu, J, et al. 2011, 'Iron reduction and magnetite biomineralization mediated by a deep-sea iron-reducing bacterium *Shewanella piezotolerans* WP3' *Journal of Geophysical Research: Biogeosciences*, vol. 116, no. G4.

Wyckoff, RWG 1963, *Crystal Structures – Volume 1*, Interscience Publishers, New York.

Yee, N, Shaw, S, Benning, LG and Nguyen, TH 2006, 'The rate of ferrihydrite transformation to goethite via the Fe(II) pathway' *American Mineralogist*, vol. 91, pp. 92–96.

Young, RA (ed.) 2002, *The Rietveld Method*, Oxford University Press, Oxford.

Zachara, JM, Kukkadapu, RK, Fredrickson, JK, Gorby YA and Smith, SC 2002, 'Biomineralization of poorly crystalline Fe(III) oxides by dissimilatory metal reducing bacteria (DMRB)' *Geomicrobiology Journal*, vol. 19, pp. 179–207.

Zegeye, A, Ona-Nguema, G, Carteret, C, et al. 2005, 'Formation of hydroxysulphate green rust 2 as a single iron (II-III) mineral in microbial culture' *Geomicrobiology Journal*, vol. 22, no. 7–8, pp. 389–399.

10 Application of Synchrotron X-ray Absorption Spectroscopy and Microscopy Techniques to the Study of Biogeochemical Processes

MAXIM I. BOYANOV AND KENNETH M. KEMNER

Abstract

The need for increasingly detailed characterization of geomicrobiological or geochemical systems has necessitated the use of increasingly sophisticated analytical techniques that enable direct measurements of the atomic-scale parameters used in mechanistic models. From the characterization techniques available today, those utilizing penetrating X-rays have the distinct advantage of being able to probe environmental samples in their natural hydrated state, without the need for complicated sample preparation that could potentially alter the system and without the limitation of surface-only sensitivity that is typical of particle and optical probes. In addition, biogeochemical samples of interest are often noncrystalline and consist of many chemical components, which complicates their structural characterization by conventional techniques. We will discuss how the methods of synchrotron X-ray absorption spectroscopy (XANES and EXAFS) overcome many of these limitations to provide the chemically specific, molecular-scale information needed to unravel the mechanisms controlling biogeochemical processes. Current X-ray focusing technology is capable of providing hard X-ray beams in the nanometer size range, which enables the application of X-ray fluorescence (XRF) spectroscopy techniques at spatial resolutions comparable to those of scanning electron microscopy, but with approximately 1000-fold higher elemental sensitivity. These methods can be applied to a diverse range of environmental systems, such as adsorption of heavy metals to minerals, plants, or bacterial cells, redox transformations of metals and radionuclides induced by microbial activity, biomineralization, nanomineral nucleation and precipitation, metal uptake, transport, and distribution within biomass, and many other areas of biogeochemistry.

10.1 Introduction

The energy-dependent absorption of X-rays by materials has been known since the 1920s (Bergengren, 1920; Stelling, 1925); however, the development of X-ray spectroscopy techniques with applications to various problems in chemistry, physics, and biology progressed rapidly only after dedicated synchrotron X-ray facilities began to be built in the 1970s. The publication of the seminal papers by Stern, Lytle, and Sayers in the 1970s (Lytle et al., 1975; Sayers et al., 1971; Stern, 1974; Stern et al., 1975) may be considered the beginning of quantitative X-ray absorption near-edge structure (XANES) and extended

X-ray absorption fine structure (EXAFS) spectroscopy analysis. Although measurements and qualitative analyses of X-ray absorption spectra were available earlier, this series of papers presented a theoretical basis for the EXAFS effect and provided a connection between the molecular coordination environment around an element of interest and its macroscopic X-ray absorption coefficient. This, in turn, enabled the use of X-ray absorption spectroscopy as a structural characterization tool, providing information such as interatomic distances and coordination numbers in disordered materials. Initial studies were limited to highly concentrated samples (e.g., metal foils) and were primarily concerned with method development (i.e., testing and refining the methods and theory of EXAFS by characterizing materials with known structure). Increased access to synchrotron X-ray beamlines (initially operating in parasitic mode at high-energy physics accelerators) resulted in broader application of EXAFS to studies of the structure of catalysts, doped materials, and disordered glasses, which were inaccessible at the time by other techniques and in which the analyte was present at the percentage level or higher. Technological advances in dedicated second- and third-generation storage ring X-ray sources enabled an increase in X-ray intensity by several orders of magnitude, which made viable the measurement of XANES and EXAFS spectra from dilute samples that are of interest to environmental scientists (down to tens and hundreds of ppm). The increased brilliance of third-generation X-ray sources also enabled X-ray imaging of biogeochemical systems with <100 nm spatial resolution.

One of the early applications of X-ray absorption spectroscopy to biological systems dates back to the 1980s, presenting EXAFS analyses of Fe and Ni in bacterial reaction centers and hydrogenases (Bunker et al., 1982; Lindahl et al., 1984). The unique advantages of EXAFS in probing metal–ligand coordination in disordered systems were recognized early on, and initial work was largely driven by an interest in understanding the coordination of metals within proteins (Scott, 1985). Geomicrobiologists began applying XANES and EXAFS when interest in the mechanistic details of the interaction between bacteria and dissolved contaminants from the perspective of bioremediation expanded (Buchanan et al., 1995; Francis et al., 1994). Later work applied X-ray absorption spectroscopy in studies of amorphous biomineral precipitation (Waldo et al., 1995; Watson et al., 2000), in understanding and modeling the adsorption of metals to minerals or to bacterial cell walls (Boyanov et al., 2003; Charlet and Manceau, 1992; Dent et al., 1992; Fendorf et al., 1997; Kelly et al., 2001; O'Day et al., 1994), in understanding bacterial respiration (Charnock et al., 2000; Polette et al., 2000; Williams et al., 1999), and in understanding contaminant transformations in soils (Foster et al., 1998). These studies brought XANES and EXAFS to the intersection of the biological and mineralogical domains, with X-ray techniques being able to distinguish between dissolved metal ions attaching to a biofilm and attaching to a mineral surface without the need for separating the components in the system (Kemner et al., 2004; Templeton et al., 2003). A more recent, significant impetus for using X-ray spectroscopy in environmental systems has been the understanding of radionuclide transformations from the perspective of spent nuclear fuel storage or bioremediation (Bargar et al., 2000a; Bernhard et al., 2001; Beyenal et al., 2004; Boyanov et al., 2011; Brooks et al., 2003; Catalano and Brown, 2005; Denecke et al., 2003; Fletcher et al., 2010; Francis et al., 2004; Hattori et al., 2009; Ikeda et al., 2007; Ilton et al., 2004; Latta et al.,

2014; Merroun et al., 2005; O'Loughlin et al., 2003; Scheinost et al., 2008; Singer et al., 2009; Sylwester et al., 2000; Veeramani et al., 2011; Waite et al., 1994; Watson and Ellwood, 2003; Zachara et al., 2007). In many of these studies, XANES or EXAFS was able to provide valuable information that was not accessible by other methods, highlighting the role of X-ray spectroscopy as an indispensable tool in the toolbox of biogeochemists.

Several excellent books on the general aspects of X-ray absorption spectroscopy exist (Bunker, 2010; Koningsberger and Prins, 1988; Penner-Hahn, 2003; Teo, 1986). They detail the theoretical basis, the synchrotron measurement, and the data analysis. More specialized tutorials and reviews on the application of X-ray absorption spectroscopy to metal speciation have also been written (Brown, 1990; Kelly et al., 2008; Kemner and Kelly, 2007; Schulze and Bertsch, 1995; Templeton and Knowles, 2009). The current chapter does not aim to reach the level of detail or the breadth of topics covered in these references. Instead, we will extract and present concisely a subset of this knowledge base that is applicable to the specific systems of interest to the geomicrobiologist or geochemist. The literature references are supplemented with our hands-on experience from two decades of studies in the area. The hope is to provide a facile starting point and a roadmap for scientists wishing to apply the synchrotron X-ray techniques of XANES, EXAFS, and XRF (micro)spectroscopy to their studies of biogeochemical processes.

10.2 Synchrotron X-ray Spectroscopy: Advantages, Applicability, Accessibility

Most synchrotrons are national user facilities that can be accessed through a competitive proposal program (more details to follow). Given the relative difficulty of including synchrotron techniques in one's research (e.g., the oversubscription rates and the corresponding delays in gaining access to a beamline, the steep learning curve in obtaining high-quality and artifact-free EXAFS spectra from geomicrobiological samples with low concentration of the analyte, and the nontrivial data analysis), it is important to plan experiments with a good understanding of the advantages and limitations of the different X-ray absorption techniques. In the following, we present their most important advantages and limitations, with advice on experimental design considerations. Despite the fact that significant time and effort may be involved in obtaining beamtime and carrying out an experiment, a single synchrotron measurement could answer directly a question that might otherwise take a series of indirect measurements, which might be susceptible to artifacts (e.g., possible interferences in colorimetric assays or sample alterations resulting from laboratory extraction procedures; Kim et al., 2003; Kwon et al., 2014).

1. ***Elemental specificity***: one of the great advantages that X-ray absorption spectroscopy offers is the ability to selectively probe an element of interest within a sample that consists of many other elements. Every element has a characteristic binding energy for the electrons in its structure (K-, L-, M-, N-shells, etc.). When the energy of an incident photon exceeds the binding energy of an electron in a shell, the photon is absorbed and

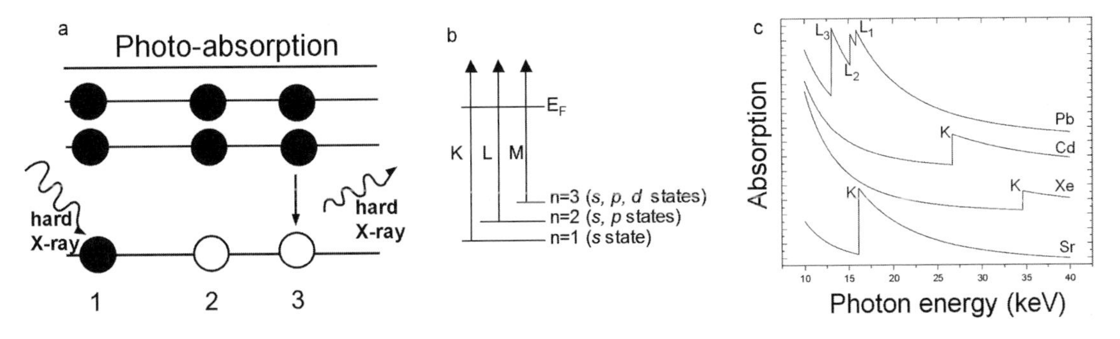

Figure 10.1 Mechanism of the photo-absorption process. (a) A photon is absorbed by an inner-shell electron (1), which ejects the electron and leaves an unoccupied electron state (2). Electrons from higher energy states fill the vacancy and a fluorescence X-ray photon is emitted (3). (b) Depending on the energy level (shell) from which the electron is ejected, there are K, L, M, etc. absorption edges of the elements. (c) A simulation of the isolated-atom X-ray absorption of several elements, which includes the absorption edge but excludes the fine structure (XANES and EXAFS) resulting from the arrangement of the atoms in a material.

a photoelectron is ejected (Figure 10.1). Thus, by scanning the energy of an incident X-ray beam across the binding energy of an electron shell, the atoms of interest can be *selectively* ionized. The X-ray absorption coefficient experiences a step-wise increase when photoionization occurs, and the energy values of these steps are termed "absorption edges," labeled according to the electron shell from which the photoelectron is ejected. For instance, spectroscopy at the energy of excitation of the electrons from the K (or 1s) shell of Fe is called Fe K-edge X-ray absorption spectroscopy (XAS), and spectroscopy at the excitation energy of the U L_{III} shell electrons is termed U L_{III}-edge XAS (XAS is a concise reference to both XANES and EXAFS spectroscopy of a given element). X-ray absorption by the electrons of interest occurs simultaneously with absorption by electrons with lower binding energies and with various photon-scattering processes, so the experimental absorption edge is always overlaid on a smoothly varying background (Figure 10.1c).

The energy range used by most hard X-ray synchrotron beamlines (~5 keV to ~30 keV) covers the K or L absorption edges of all elements with $Z > 22$ (Ti, V, Cr, etc.). Within this range, the absorption edges for the different elements are reasonably well separated and will not interfere with each other. However, overlap between edges can occur within the same sample, and care should be taken that an interfering element will not be present in more than a few molar percent of the element of interest. For elements with $Z > 55$ (i.e., Cs), the L absorption edges begin to overlap with the K absorption edges of elements $22 < Z < 55$ (e.g., for Sr and Pb on Figure 10.1c).

2. ***Broad applicability***: Many readers will be familiar with sample preparation for electron microscopy of biological materials, such as fixation and drying of the sample, flatness and small thickness requirements, the need for coating of an insulating sample to promote conductivity, and the need for a vacuum environment during the measurement. In contrast to these restrictive and potentially sample-altering requirements, X-ray

spectroscopy approaches allow measurements of the samples in their natural state. The large penetration depth of X-rays in most materials (including water) allows samples to be interrogated in their hydrated state under an appropriate environment (e.g., inert or oxidizing/reducing gases, dry or humid atmosphere, ambient or low/high pressure). Unlike diffraction, X-ray absorption techniques do not require long-range atomic order in the sample, so measurements of the molecular structure can be made in any physical state – single crystals, polycrystalline powders, amorphous solids and glasses, melts, liquid or frozen solutions, gases, or plasmas. In most cases, the interaction of the X-ray probe with the sample does not alter the sample, which allows measurements to be made *in situ* during a chemical or biological process. A redox process or the transformation of an element could be tracked while it happens. Unlike particle probes, X-rays can reach surface and inside atoms alike, allowing the interrogation of the entire sample (however, it is also possible to use surface-enhanced or spatially resolved XAS approaches). Together with the relatively low concentrations of the analyte needed (current detection levels are a few tens of ppm for XANES), these features make X-ray absorption spectroscopy a convenient and powerful probe.

3. ***Valence state and speciation (XANES):*** X-ray absorption near-edge spectroscopy (XANES) refers to the region of the spectrum from 10–20 eV below the absorption edge up to 150–200 eV above the absorption edge (Figure 10.2). This part of the spectrum carries information predominantly about the valence state of the element of interest. Because the binding energy of the electrons depends on the overall charge of the atom, the position of the adsorption edge is related to the valence (oxidation state) of the studied atoms. In general, the higher the valence, the greater the photon energy at which the absorption edge occurs. Figure 10.2a illustrates this dependence for the K edge of Fe, demonstrating that as the formal valence state changes between 0, +2, +2.33, +2.66, and +3, the edge position (defined sometimes as the energy position at half of the normalized absorption and sometimes as the position of the inflection point on the absorption edge) changes from 7 112 eV to 7 125 eV. Researchers have established linear dependencies

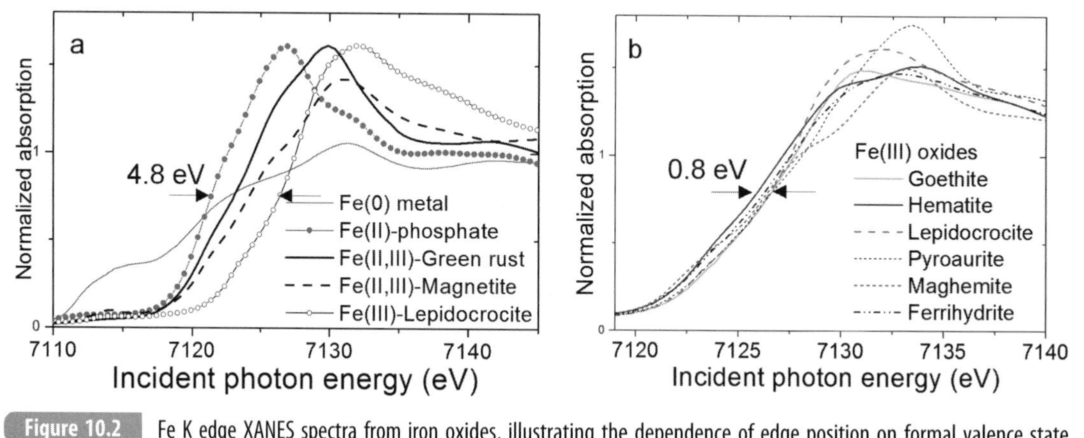

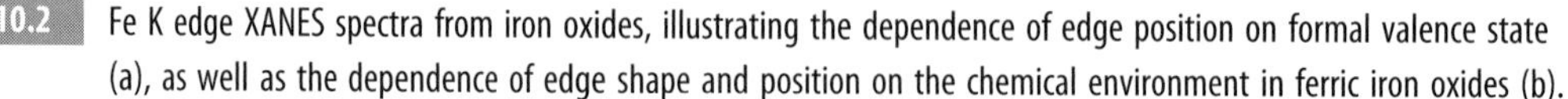
Figure 10.2 Fe K edge XANES spectra from iron oxides, illustrating the dependence of edge position on formal valence state (a), as well as the dependence of edge shape and position on the chemical environment in ferric iron oxides (b).

between the edge position and valence state for some elements, e.g., Delaney et al. (1998); Ressler et al. (2002). While these general trends are useful in determining a formal valence state in a mixed-species sample, it is important to remember that the edge position and shape also depend on the molecular structure around the absorber atom. This is illustrated in Figure 10.2b with several ferric iron minerals, in which Fe atoms of the same +3 valence are present in structures with different O-coordinated environments, resulting in measurable differences in the XANES spectra. This introduces a "structural" uncertainty in valence state determination by XANES, which, together with the energy resolution limitations of the measurement, results in valence state accuracy of about 10–15%. However, when the endmember species are known, the structural uncertainty can be eliminated and a case can be made for accuracy of about 5% in valence state determination (e.g., Latta et al., 2016; Rui et al., 2013). On the other hand, the dependence of XANES on atomic structure can be useful for species identification. The concentrations needed to collect XANES spectra are 10–100 times lower than the concentrations needed for EXAFS, so the sensitivity of XANES to structure can be particularly useful for low-concentration samples or for samples probed with a micro- or nanobeam (X-ray focusing elements can limit the scannable energy range and thus the ability to obtain EXAFS).

From a practical biogeochemistry perspective, the sensitivity of XANES to valence state may be the feature that is of most interest to the beginning nonspecialist. It allows the study of the redox transformations of major and minor elements in environmental samples, providing valuable insight into the metabolic pathways of bacteria or the bioavailability of elements (Bargar et al., 2000b; Francis et al., 1994; Vairavamurthy et al., 1993). The solubility or the toxicity of contaminants is often strongly dependent on redox state, so a simple valence state determination can provide information about the contaminant's mobility and risk (Jardine et al., 1999; Terzano et al., 2010; Vodyanitskii, 2013; Zachara et al., 2004). The relatively facile data analysis and the greater sensitivity to low concentrations of the analyte make XANES analysis an ideal entry point for the biogeochemist into X-ray spectroscopy.

4. ***Average local environment (EXAFS):*** The region of the X-ray absorption spectrum between 30 eV and 1 000 eV above the absorption edge is termed the extended X-ray absorption fine structure (EXAFS) spectrum (Figure 10.3b). This part of the spectrum carries information about the atomic surroundings of the element of interest. The EXAFS consists of tiny oscillations (fine structure) in the absorption coefficient that are barely visible until the smoothly varying post-edge background is subtracted (Figure 10.3c). The physical origin and the mathematical treatment of the phenomena leading to EXAFS are complicated and beyond the scope of this chapter – the interested physics-inclined reader is referred to the monographs (Bunker, 2010; Koningsberger and Prins, 1988; Teo, 1986). To provide a foundation for understanding the key features of an EXAFS spectrum, we include in the following a simplified explanation of the effect.

When the incident X-ray photon (black wavy arrow, Figure 10.3a) is absorbed by the atom of interest (solid circle), the ejected photoelectron travels as a quantum mechanical

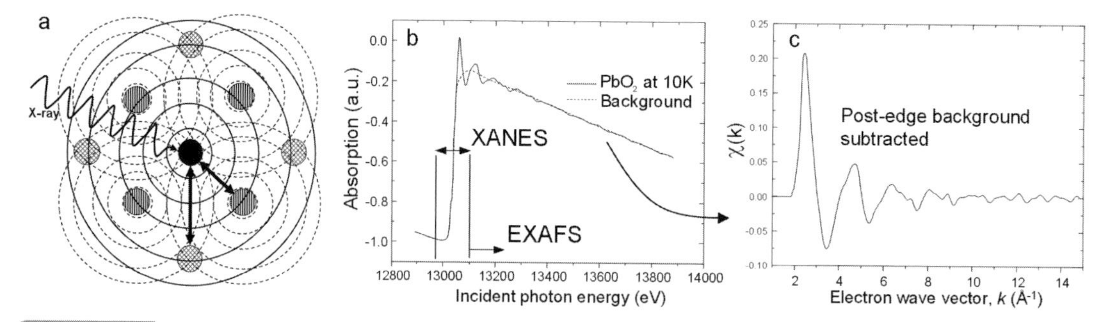

Figure 10.3 An illustration of the EXAFS effect. (a) Propagation of the outgoing electron wave (solid line circles) and its scatter from the first coordination shell (broken line circles). The interference between these electron waves creates the fine-structure modifications in the post-edge absorption coefficient (b). The oscillations are visible after a background curve is subtracted from the data (c).

wave within the material (solid, liquid, or gas), and its interaction with the surrounding atoms is analogous (in simple terms) to the scattering of a wave in a pond by nearby obstacles. The outgoing (solid circles) and the scattered (dashed circles) electron waves interfere with each other, and the rigorous treatment of this interference pattern shows that the probability for absorption of an X-ray photon is modified depending on whether the pattern has a maximum or a minimum at the location of the absorbing atom(solid circle). As the incident photon energy is scanned across the EXAFS range, the kinetic energy of the ejected electron and therefore, the wave vector k of its quantum mechanical wave are continuously increased, which in turn modifies the relative phase of the waves and their interference pattern. The end result is that by scanning the incident photon energy, the interference maxima and minima continuously alternate at the location of the absorbing atom, which results in a modification of the absorption coefficient that is periodic with respect to the wave vector k of the photoelectron (Figure 10.3c). The rigorous mathematical treatment reveals that the frequency of these oscillations is related to the length of the path traveled by the electron; i.e., for single backscattering from a neighboring atom, the frequency of oscillation in the absorption coefficient is proportional to the distance between the absorbing and the scattering atoms (arrows in Figure 10.3) or, in other words, the radius of the coordination shell, R_i. For a multiple-scattering (MS) path (e.g., a closed path along solid–striped–checked–solid atoms; Figure 10.3a), the resulting frequency of oscillation is the sum of the distances between the scattering atoms along the path of the photoelectron wave divided by two. Thus, the EXAFS from each analyte atom in a sample is a superposition of oscillations in k-space with frequencies corresponding to the interatomic distances between this atom and the surrounding atoms. Closer atoms are represented by lower frequencies in the EXAFS, whereas more distant atomic shells and MS paths participate with higher-frequency oscillations. With some approximations, the EXAFS spectrum $\chi(k)$ can be calculated in closed form for any atomic configuration using the following equation (Koningsberger and Prins, 1988):

$$\chi(k) = \frac{m}{2\pi\hbar^2} \sum_i \frac{t_i(2k)}{(kr_i)^2} \exp(-2r_i/\lambda) \exp(-2k\sigma_i^2) \sin(2kr_i + \delta_i(k)) \tag{10.1}$$

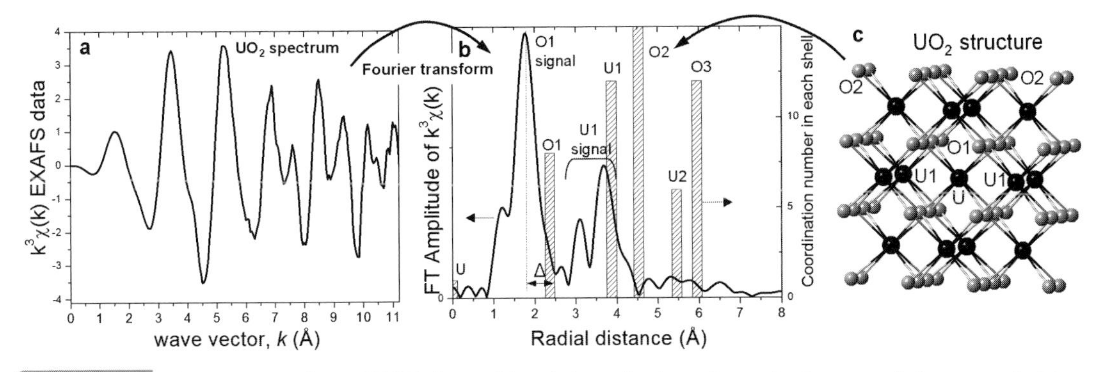

Figure 10.4 The Fourier transform of EXAFS data and its relation to local atomic coordination; example for nanoparticulate uraninite. (a) Measured and extracted $\chi(k)$ EXAFS data, weighed by k^3 to emphasize the diminishing amplitude at larger k. (b) Fourier transform of the data in (a). The vertical bars indicate the number of atoms in each shell (right axis) at the corresponding radial distance in (c) the crystal structure of UO_2.

Here, the summation is over all atoms around the central atom, $k = \sqrt{2m(E - E_0)}/\hbar$ is the wave vector of the photoelectron, where E is the photon energy and E_0 is the edge energy, m is the mass of the electron, r_i is the radial distance to the central atom, σ_i is a static and thermal disorder parameter within a coordination shell called the Debye–Waller factor, λ is a dampening parameter related to the electron's mean free path, $t_i(2k)$ is the backscattered amplitude of the electron wave, and $\delta_i(k)$ is a phase shift resulting from the interaction of the electron wave with the electron clouds of the central and scattering atoms. Because of the dampening effect of the exponential factor, spectra are typically presented as $k^n\chi(k)$ (n = 1,2,3) to emphasize the regions of the spectra where the signal becomes low (e.g., Figure 10.4a). Note that if the probed atoms in a sample are present in two or more coordination environments, the experimental EXAFS will be the molar fraction weighted average of the spectrum from each distinct species.

As seen in Eq. (10.1), the measured $\chi(k)$ data is a superposition of sine waves of different frequencies, similar to the sound wave created by simultaneously ringing several different bells. It is convenient to use signal processing methods to deconvolute the contribution of the individual frequencies in the overall signal. Applying a Fourier transform (FT) to the experimental data separates and visualizes the contribution of the frequency components in the EXAFS signal (Figure 10.4b). From Eq. (10.1) it can be seen that the multiplier of the free variable k is the interatomic distance R, so that the FT peak (or frequency) corresponding to a certain shell will be centered near its radial distance from the central atom. Thus, the FT of EXAFS data resembles and is often mistaken for a radial distribution function around the atom of interest. However, it should be kept in mind that the FT only visualizes the frequency content of the EXAFS data and that the structural information should be obtained from fits of the data using expressions similar to Eq. (10.1). A comparison of the radial distribution around U in UO_2 (bars) and the FT of its EXAFS data (line) is shown in Figure 10.4b. The central atom (marked "U") in the structure is at $R = 0$ Å in the FT. Some of the factors complicating the interpretation of FT EXAFS data as a radial distribution function include 1) the nearly linear phase shift $\delta(k)$ in the oscillating term of Eq. (10.1), resulting in a shift Δ of -0.2 to -0.6 Å in the FT relative to the true interatomic distance R; 2) the generally nonmonotonous shape of $t_i(2k)$ in Eq. (10.1), which can result in FT peaks that are not symmetric or are "doublets" and could be mistaken for more than one atomic shell, particularly for heavier atoms (the doublet peak for the U1 shell can be seen in Figure 10.4b); 3) the contributions from MS paths of the photoelectron in the EXAFS data, which are generally not directly related to a particular atomic shell (seen as various "shoulders" and small peaks in Figure 10.4b); 4) the limited range of the EXAFS signal, resulting in broadening of the FT peaks and overlap between the signals from neighboring shells, as well as interferences from truncation effects in the FT; and 5) the possibility of constructive or destructive interference between single- and/or multiple-scattering signals of similar frequency, which could then be misinterpreted as coordination shells.

This discussion provides a basis for the qualitative understanding of EXAFS data. From a practical perspective, the sensitivity of EXAFS to atomic structure provides a valuable tool to probe the speciation or the transformation of elements during a process of interest, particularly when such information is inaccessible by other experimental techniques. The analysis of EXAFS data could be as simple as direct comparison with a set of standards (fingerprinting). Transformations from one species to another during a process (e.g., particle ripening or redox transformations) can be monitored and quantified by linear combination (LC) analysis with spectral endmembers, which is also a relatively straightforward analysis to be carried out by a nonspecialist. At a more advanced level, processes such as the incorporation of contaminants into the structures of minerals or the determination of surface complexation mechanisms during adsorption can be studied by extracting the structural information from EXAFS data and looking for specific interatomic distances or the presence of signals from specific atomic shells. The latter is achieved by fitting the experimental data, which is discussed more in Section 10.4.

5. ***Chemical analysis and mapping (XRF analysis):*** Intense X-rays also provide a fast, sensitive, and nondestructive way to determine the elemental content of a biogeochemical sample. The principle is the same as in the energy dispersive X-ray analysis (EDX) done in electron microscopy, except that X-rays are used instead of electrons to excite the atoms in the sample. Probing with X-rays provides several advantages relative to electrons, such as 1000-fold higher sensitivity to most elements (down to ppm levels), greater penetration depth in the sample, less scatter within the sample resulting in less damage from irradiation, and no requirement for vacuum. Current focusing technologies based on Kirkpatrick–Baez mirrors can produce monochromatic X-ray beams that are about 1 μm in size (Eng et al., 1995), while those based on Fresnel optics can produce monochromatic X-ray beams less than 100 nm in size (Lai et al., 1992). This allows elemental content and speciation maps to be determined for very small domains in a sample.

Elemental mapping is based on the analysis of the characteristic X-ray fluorescence photons emitted by the excited atoms in a sample. When an incident X-ray photon is absorbed and an electron is ejected from an atom, the electrons from higher energy levels fill the unoccupied state, and an X-ray fluorescence (XRF) photon with energy equal to the difference between the two levels is emitted (Figure 10.1). Similarly to absorption edges, the differences between the energy levels in an atom are characteristic of each element.

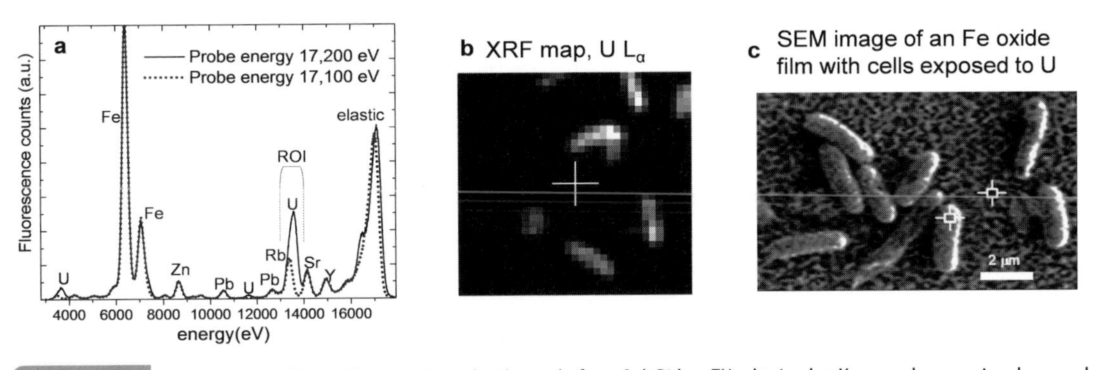

Figure 10.5 (a) XRF spectra from a U-contaminated soil sample from Oak Ridge, TN, obtained at X-ray probe energies above and below the U L_{III} edge. By integrating the peaks in regions of interest (ROI) in an XRF spectrum and raster scanning a sample with a small beam, a false-color elemental map (b) can be obtained even from single bacterial cells adhering to an iron oxide (c).

The energy position of the X-ray emission peaks can therefore be used for chemical identification. The amplitude of each peak is proportional to the number of emitting atoms, so quantitative elemental analysis is possible. Emitted X-rays are collected by energy dispersive Si or Ge detectors, which use pulse-height analysis to separate and count the photons of different energies to produce an XRF spectrum similar to the one shown in Figure 10.5a. The XRF peaks are identified by comparison with standards or tabulated values. The use of synchrotron X-rays allows tuning of the incident X-ray energy above and below an absorption edge so that the fluorescence from this element can be turned "on" and "off" to verify its presence, which is particularly important when the XRF peak of one element overlaps that of another (e.g., Rb and U in Figure 10.5a). A region of interest (ROI) window can be set up around a peak and the number of counts integrated. The sample is then raster-scanned in front of the beam and the integrated XRF intensity at each position is used to create elemental concentration maps, such as the one shown in Figure 10.5b for single bacterial cells. More sophisticated approaches record the XRF spectrum at each point and then use curve fitting to deconvolute overlapping XRF peaks or to remove interfering background (Pushie et al., 2014; Vogt et al., 2003).

The higher elemental sensitivity of synchrotron XRF analysis relative to charged particle probes, the high spatial resolution, and the ability to probe samples under ambient pressure and in their natural state provide a biogeochemist with a tool to study the elemental content and chemistry of dilute and sensitive samples such as biological tissues and bacterial cells: systems that are typically not amenable to EDX analysis with an electron probe. Synchrotron XRF microscopy has been recently applied to metal uptake by plants, roots, and biological tissues (Paunesku et al., 2006; Pushie et al., 2014), as well as to bacteria, whereby elemental maps of single, unfixed bacterial cells or subcellular organelles have been obtained (Kashiv et al., 2016; Kemner et al., 2004). XRF microscopy can also be combined with X-ray spectroscopy to obtain spatially resolved valence state or speciation information (Brinza et al., 2014; Etschmann et al., 2014; Glasauer et al., 2007; Hettiarachchi et al., 2006; Pushie et al., 2014).

10.3 Planning a Synchrotron Experiment

10.3.1 Choice of a Synchrotron

Nearly all synchrotrons have a user program, allowing researchers worldwide to submit a proposal for beamtime. If the proposal passes review, the research team is given free access to an appropriate beamline over a period of 2–5 days. Many factors are taken into consideration during the review – the common link between them is the likelihood of success of the experiment in the form of a publication, report, or other means of knowledge dissemination (beamlines and synchrotrons are evaluated on the impact of the science conducted there). Thus, it is helpful to argue in the proposal the importance of the synchrotron results in answering a pressing scientific question. It is also helpful to argue the competency of the research team in carrying out a successful measurement and analysis of the data. The staff operating the beamline will help in setting up the beamline and meeting one's requirements, but they typically will not help in designing these requirements or in the quality control of the collected data. Beamline staff typically will not analyze users' data – the latter activities usually elevate the staff's involvement to collaboration and coauthorship. The proposal will also need to justify why a particular beamline is suitable for the experiment. Most beamlines receive two to five times more proposals than are awarded beamtime, so it is best to apply to beamlines that meet but do not exceed the requirements of one's experiment. Beamlines that have greater capabilities (e.g., higher flux or smaller beams) are heavily oversubscribed, making it more difficult to obtain beamtime. It may be worthwhile redesigning the experiment to use a beamline with lower capabilities (e.g., by increasing the concentration of the analyte or by preparing a sample that does not require a small beam). Accessibility is an important consideration, as several beamruns may be needed before obtaining data suitable for publication.

10.3.2 Sample Preparation

The large penetration depth of hard X-rays allows the probe photons to reach buried atoms, and the emitted fluorescence can escape through the sample matrix and the ambient environment to reach the detector. Samples can be hydrated or dried solids, solutions, or gases. The atoms of interest can, in principle, be distributed in any way throughout the sample, including clustering in precipitates. However, due to intensity inhomogeneity across the beam probe as well as possible beam motion during the measurement, it is best to have as uniform an area concentration as possible on the length scale of the size of the beam. Thus, if the sample contains 100 μm-sized precipitates separated by 100 μm on the average, it is better to probe the system with a 1 000 μm beam (or greater) than with a 100 μm beam, so that many precipitates are always present in the beam and any particle size variation or beam motion does not result in large changes in the number of fluorescing atoms. For the same reason, variations in the thickness of the sample should also be small. Ideally, the part of the sample that is in the beam should be a parallel slab of material with a uniform area density (concentration) of the element of interest and a uniform background matrix. The optimal concentration depends on the element, on the mode of measurement (transmission for high concentration or fluorescence for low concentration), and on the X-ray absorption of the other elements in the sample (Bunker, 2010; Kelly et al., 2008; Koningsberger and Prins, 1988). In the next paragraph, we suggest several sample mounting procedures that attempt to meet these requirements for the different samples that may be encountered by a geomicrobiologist.

Solid standards: XANES and EXAFS analysis is based on comparison to polycrystalline standards. These could include metal oxides, hydroxides, carbonates, phosphates, sulfides, etc., in which the element of interest is present at high concentration in a known crystal structure. As explained in more detail in Section 10.3.3, an appropriate measurement approach for concentrated samples is transmission mode, in which the X-ray intensity before and after the sample is used to obtain the EXAFS. The alternative fluorescence mode measures the intensity of the fluorescence emitted from the sample to obtain the EXAFS and is suitable for lower-concentration samples. To produce a uniform sample, polycrystalline standards should be ground and sieved to particles that are less than several tens of microns in size. The rule-of-thumb requirement for particle size is "less than one absorption length," which means that when a transmission spectrum is measured on a sample of this thickness, the difference between the absorption coefficients below and above the edge would be 1. Some analysis programs can calculate the absorption length for arbitrary materials (e.g., Hephaestus; Ravel and Newville, 2005). Once the particle size requirement is met as closely as possible, the powdered material can be mounted in one of two ways. The powder can be diluted in a light material, such as boron nitride or graphite, and then pressed into tablets (Stern et al., 1980). The optimal edge step for a transmission measurement is 0.5–1.0. Again, a program can estimate the dilution needed so that this absorption length is obtained for the physical thickness of the tablet. Alternatively, the powdered material can be attached to the adhesive side of Kapton tape (Stern et al., 1980). Kapton is a polyimide film developed by DuPont. Its high mechanical and thermal stability and high transmittance to X-rays make it the preferred sample containment material for X-ray absorption spectroscopy. When preparing samples in this way, a piece of tape is secured on a table with the adhesive layer facing up, and the powdered material is gently moved without pressing along the tape to achieve uniform coverage (if uniform transparency against light is achieved, that is usually sufficient). After several passages, the smaller particles are retained on the tape, whereas the larger particles fall off. The tape can then be folded at the beamline as many times as necessary to achieve the optimal X-ray thickness. While both of these methods can produce suitable samples, the latter method has the advantages of simplicity, the preferential mounting of smaller particles, and the flexibility to adjust the sample thickness at the beamline. However, the former method will be preferable for samples that may react with the adhesive. For some studies, a suitable standard may be the metallic form of an element. Most spectroscopy beamlines carry metallic foils of appropriate thickness as energy calibration standards.

Suspensions: Most biogeochemists will be interested in the characterization of hydrated solids resulting from the interactions between bacteria, minerals, and a solution matrix.

Depending on the uptake of the analyte in the solids and the absorption of the X-rays in the background matrix, the material can be mounted in several different ways. If the concentration is low (e.g., between 20 and 3 000 ppm) and the background matrix is relatively transparent at the energy of interest, then a thick sample (2–4 mm) may be preferable in order to increase the number of atoms sampled by the beam and thus, the efficiency of the measurement. A convenient sample holder is a 2–4 mm-thick plastic slide in which a hole has been drilled. The hole is closed on one side with Kapton film and tape; the centrifuged or filtered sample paste is loaded with a spatula to fill the resulting well and is sealed on the top with Kapton film and tape. The plastic material of the slide should be chemically inert, and the sample material should not touch the glue on the Kapton tape, to avoid potential redox interactions. A sample sealed inside an anoxic chamber will keep its redox integrity for several hours under ambient conditions.

For samples with highly absorbing background materials, a thinner sample is preferable. For instance, if one is studying contaminant adsorption to iron oxides, about 99.9% of 10 keV X-rays will be absorbed in the iron oxide within the first 100 μm. Thin samples can be created from suspension by the filter mounting method. In this case, the suspension is filtered using a filter assembly for 1.1 or 2.5 cm diameter membranes. The supporting membrane and the filter cake (sample) are sealed together between Kapton film and tape. A drilled plastic slide as described earlier can be used as a rigid support under the membrane. The filter mounting method allows the creation of well-compacted, relatively uniform sample layers that can be thin or thick depending on the amount of suspension used. Filter membranes are made up of light materials, so their absorption of X-rays is usually not significant. Several membrane layers can be stacked to increase the thickness of the sample and to improve the detection efficiency.

Solutions: X-ray spectroscopy can also determine the predominant speciation of metals in solution, which can be useful for direct verification of thermodynamic speciation models. The limitation to such application is that relatively high concentrations are needed to obtain an EXAFS spectrum (typically 0.5–1.0 mM, but as low as 50 μM solutions of uranium have been measured; Kelly et al., 2007). Measurements on solutions can also be used for obtaining standards spectra, e.g., from a complexed analyte atom. A basic sample mounting method is the slide mounting described previously, which is suitable for fluorescence measurements of low-concentration samples or transmission measurements of higher-concentration samples. Variable-thickness solution cells have been developed (e.g., Ertel and Bertagnolli, 1993), which allow movement of the front or back window to change the sample thickness so that the optimal combination of absorption step size and acceptable X-ray absorption in the background matrix can be achieved at the beamline.

Radioactive samples: Most synchrotrons treat radioactive samples differently from other hazardous materials, so it is best to check the specific requirements before applying for beamtime. The additional steps needed are typically double containment of the sample (in case there is a leak from the primary container), special transport to and from the beamline, special radiation monitoring during the measurement, and absolutely no modification of the samples at the beamline. The first requirement can be met by encasing the sample mounts described earlier in an additional bag created by taping Kapton film around the sample or

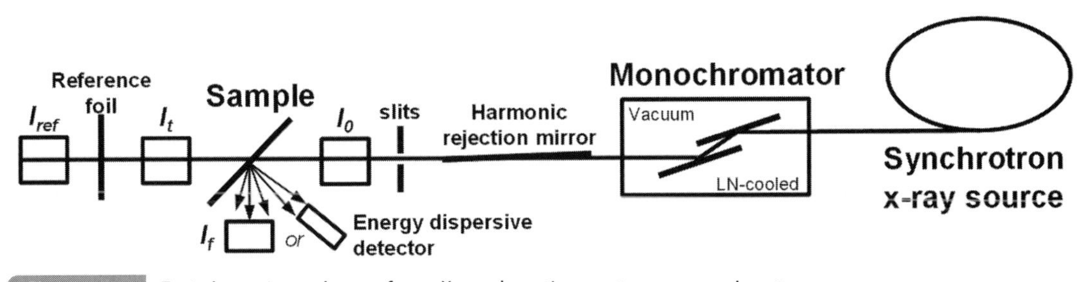

Figure 10.6 Typical experimental setup for an X-ray absorption spectroscopy experiment.

heat-sealing in a secondary bag. The second and third requirements do not entail changes in the setup but add a step of coordination with the Health Physics or Special Materials staff that could result in delays. Perhaps the most restrictive is the last requirement, because if a sample turns out to be unsuitable (e.g., the sample is too thick, or not enough analyte is present, or the holder cannot fit in the beamline equipment), nothing can be done at the beamline to cope with the situation. It is best to plan contingencies in advance (e.g., prepare several samples of different thicknesses or concentrations, or plan for access to a radioactive sample facility nearby where the samples can be manipulated).

10.3.3 Data Collection

The preparation of the beamline for an experiment will usually be done by the beamline staff, so we will only cover the basics here. The typical setup is shown in Figure 10.6. The polychromatic beam from the synchrotron is passed through a monochromator, which uses the Bragg reflection from two Si crystals to select a very narrow range of photon energies (typically within 0.5 eV). Multiples of the fundamental energy (a.k.a. harmonics) also satisfy the Bragg condition and will pass through the monochromator. Their removal is important and is often achieved by reflection from a metallic mirror (Pt or Rh) set at an angle to have high reflectivity for the fundamental energy and nearly zero reflectivity for the higher-energy harmonics. The incident beam size is then defined by slits. The intensity of the beam incident on the sample (I_0) is recorded by a gas ionization chamber. The gas mix in the I_0 chamber (a mixture of He, N_2, and/or Ar) is chosen so that it absorbs about 10–15% of the photons in the beam. The incident photon flux on the sample can be calculated by knowing the length of the ion chamber, the gas mix, and the current generated by the beam in the ion chamber (Ravel and Newville, 2005). The transmitted beam intensity, I_t, is also measured by a gas ionization chamber. The gas mix in I_t should absorb about 70–80% of the beam to ensure sufficient accuracy in the measurement of the transmitted X-rays and to allow the passage of some X-rays to the reference monitor, I_{ref}. The latter is used for in-line energy calibration of the setup by collecting a spectrum from a reference foil at the same time as the experimental spectrum.

The transmission spectrum of a sample is obtained as $\ln(I_0/I_t)$ measured at 0.5–4 eV intervals of the incident photon energy between approximately −100 eV and +900 eV of the edge energy. It is possible to measure the transmission spectrum when there is sufficient analyte in the beam to produce an edge step of about 0.1 or greater. As mentioned earlier, the absorption spectrum can be equivalently obtained by measuring the fluorescence intensity emitted from the sample and then calculated as I_f/I_0. This is typically done for samples with low concentration of the analyte (up to about 10 000 ppm), as higher concentrations may introduce self-absorption artifacts (more information to follow). In the setup described here, the fluorescence intensity I_f is monitored by an ionization chamber filled with a gas that absorbs nearly 100% of all photons (Ar is typically used; Kr can be used for higher efficiency). When detected by an ionization chamber, the fluorescence and scatter from all atoms in the sample are measured, so the fluorescence of interest will appear as an absorption edge on a relatively large background. It is also possible to use energy dispersive detectors to filter out the background fluorescence/scatter and to measure only the fluorescence coming from the element of interest. This improves the signal to noise ratio of the setup and can be beneficial in samples with very low concentrations of the analyte. However, the total count rate and the linear response range of energy dispersive detectors are typically lower than those of ionization chambers. From our experience with high-intensity undulator beamlines, an ionization chamber measurement of both the smooth background and the signal with greater statistical accuracy and greater linearity often outweighs the advantage of removing the elastic scatter and background fluorescence using a 4-element Si or a 16-element Ge energy dispersive detector. However, energy dispersive detectors enable the removal of artifacts in the data resulting from background fluorescence/scatter that is not smooth over the EXAFS energy range. In the end, the detection method that results in the lowest-noise artifact-free data over the shortest timeframe should be chosen, and this choice can vary between different beamlines and samples.

For samples that are in the intermediate concentration range (e.g., 5 000–20 000 ppm of analyte in the beam), it is often possible to obtain simultaneously transmission and fluorescence XAS data. Recording both channels at all times is recommended, as it is not known a priori which detection mode will yield better data for the experimental setup and sample. Having data from both channels can also be useful for identifying spectral artifacts, such as features from "glitches" in the incident intensity that did not get normalized out in one of the channels (when the ratio of the measured channel to incident intensity was taken). An important artifact to be aware of is the so-called "self-absorption" amplitude suppression in fluorescence detection. A facile way to detect self-absorption problems is to compare the data between the transmission and the fluorescence channels – the affected fluorescence XANES will show a smaller white-line amplitude and a shift to lower energy of the edge position relative to the transmission data; the fluorescence data will also have a smaller amplitude of the EXAFS oscillations. If the problem is detected at the beamline, a decision can be made to focus on collecting higher-quality transmission XAS data (by longer measurement or by creating thicker samples). If the artifact goes undetected, the interpretation of the fluorescence data will result in lower valence state and lower coordination number determinations. The interested reader is referred to several papers that describe the self-absorption phenomenon and propose post-measurement corrections (Booth and Bridges, 2005; Iida and Noma, 1993; Li et al., 2014; Troger et al., 1992). It should be emphasized that it is always better to record artifact-free data than to correct affected data.

The ejection of electrons during the X-ray absorption process can change the redox balance in a sample and cause unwanted transformations to occur. This effect is sometimes referred to as "radiation damage" or "radiation-induced change" and is observed more often when photo- or redox-sensitive materials are measured at higher-intensity beamlines. The typical manifestation is in relatively small changes in the measured spectra over time, but in more severe cases visual "burn marks" can appear on the samples. *The spectra should always be monitored for radiation-induced changes during data collection.* A convenient way to test for radiation damage is to measure a series of XANES scans on a fresh area of the sample. If only small changes are observed from scan to scan, it may still be possible to obtain artifact-free spectra if the changes in a single spectrum can be attributed to changes in less than 10% of the analyte (typical accuracy in valence state or coordination number determination is also about 10%). An obvious mitigation scheme for "mild" radiation damage is to measure spectra until the radiation-induced changes become significant and then move the beam to a fresh portion of the sample. Lowering the intensity of the incident beam may also help. For more severe cases of radiation damage (i.e., when spectral changes are significant during a single scan), freezing of the samples before measurement may slow down or prevent migration of free radicals and radiation-induced transformations. Specialized X-ray-accessible cryostats are available at most beamlines that allow sample temperature control down to −270 °C (liquid helium) or, more easily, down to −195 °C (liquid nitrogen). The temperature at which radiation-induced chemistry is mitigated is a matter of trial and error at the beamline, but a good starting point is −50 °C. Some samples may experience phase transitions as the temperature is lowered (e.g., the Verwey transition in magnetite at −153 °C) – such transitions should be avoided, as they could alter the speciation of the analyte. In addition, freezing produces ice in samples with high water content or may cause a colloidal sample to aggregate. While these transformations may or may not alter the speciation, they could introduce noise in the spectrum and create problems with data collection. EXAFS spectra are also temperature dependent, so spectra should be obtained at the same temperature from the standards to enable valid comparisons. Lowering the temperature decreases the thermal motion of the atoms and generally increases the amplitude of the EXAFS signal by decreasing the disorder in the coordination shells; i.e., the σ^2 parameter in Eq. (10.1). Despite having to account for these potential issues, freezing is a convenient way to mitigate radiation-induced chemistry and may be the only way to detect rapid radiation damage (i.e., transformations that are complete before the absorption edge energy is reached in a XANES scan). Another way to slow down radiation damage is to dry the samples, but this can cause changes in the speciation of the analyte.

10.4 Data Analysis

The analysis of spectra from a biogeochemical sample follows the same procedures as for any other sample, so we will only outline them here; the reader is referred to the monographs and the tutorials on the web for more details. Current computer codes that facilitate data analysis include Athena (Ravel and Newville, 2005), WinXAS (Ressler, 1998), MAX (Michalowicz et al., 2009), EXAFSPAK, SixPACK(Webb, 2005), and others.

The absorption spectrum of a sample is obtained from the measured intensities of the incident (I_0) and transmitted (I_t) or fluorescence (I_f) X-rays as $\ln(I_0/I_t)$ or I_f/I_0, respectively (Koningsberger and Prins, 1988). The resulting "raw" spectrum includes the smoothly varying absorption from the background matrix, the edge jump and XANES resulting from the excitation of the element of interest, and the fine EXAFS features overlaid on the edge jump (Figures 10.3 and 10.7). Data should be first aligned on the absolute energy axis using the reference channel. The edge position (E_0) should be chosen consistently between samples as the mid-point or as the inflection point in the rise of the edge. The background absorption is then removed by fitting a linear or quadratic polynomial to the pre-edge part of the data and subtracting this polynomial from the entire spectrum (Figure 10.7). The magnitude of the edge jump after pre-edge background subtraction is proportional to the number of analyte atoms sampled by the beam. For spectra to be compared between samples, the edge jump needs to be normalized to the same number of sampled atoms. This is done by fitting a quadratic polynomial to a region above the edge (about 50–300 eV) and dividing the data by the value of this polynomial at the edge position (defined as the step height). Normalized XANES spectra can now be compared with each other and with standards (Figures 10.2 and 10.7). The EXAFS part of the spectrum is extracted from the edge-normalized data by fitting a cubic spline polynomial to the post-edge part of the data (Figures 10.3 and 10.7). The flexibility of the spline curve should be adjusted in the analysis program to follow only the slowly varying background component in the data and not the higher-frequency oscillations that constitute the structural part of the data (different programs use different parameters to control flexibility). After the spline curve representing the background is subtracted, we obtain the EXAFS, $\chi(k)$. As discussed earlier, the Fourier-transform of $k^n\chi(k)$ (n = 1,2,3) data is used to provide a first-cut representation of the average local environment in the sample (Figures 10.4 and 10.7).

After these data reduction steps are carried out, there are several options for interpreting the spectrum. Fingerprinting, whereby the XANES or EXAFS spectra are compared with standards, is the most straightforward analysis of the data – if spectra are sufficiently similar, then the predominant speciation in these samples is the same. If the spectrum appears intermediate between two standards, then it is possible that the biogeochemical sample captured the partial transformation of one endmember phase into the other. An example of this may be the transformation of an Fe^{III} oxide into a secondary mineralization product during dissimilatory iron reduction or the gradual oxidation of a reduced species. The presence of isosbestic points (several intersection points between a measured spectrum and two endmembers) is a strong indication that only these two endmembers are involved in the transformation (Rui et al., 2013). The proportion of each endmember can be quantified by LC fits or by principal component analysis (PCA). LC analyses of XANES data to extract average valence state are usually easier to carry out, as the endmembers need to represent only the possible valence states in the sample and the speciation can only be approximate (e.g., O- vs. S- near-neighbor coordination). Several software packages and methods of varying sophistication exist for LC and PCA fitting

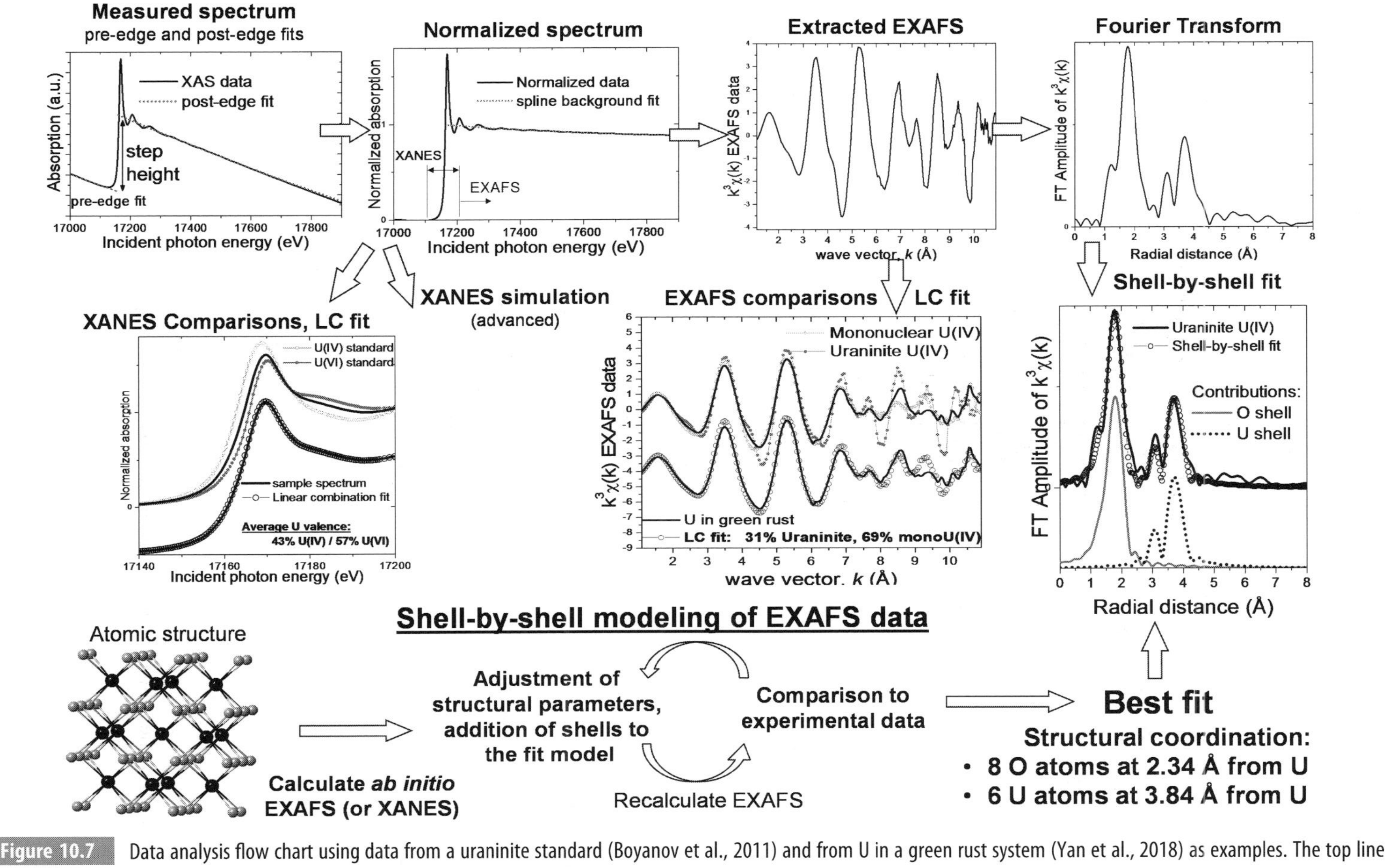

Figure 10.7 Data analysis flow chart using data from a uraninite standard (Boyanov et al., 2011) and from U in a green rust system (Yan et al., 2018) as examples. The top line of graphs illustrates the typical data reduction and extraction steps. The middle line illustrates analyses done on XANES and EXAFS data and the type of information that can be obtained. The bottom line outlines the steps involved in shell-by-shell modeling of EXAFS data.

(Frenkel et al., 2002; Ravel and Newville, 2005; Rossberg et al., 2003; Wasserman et al., 1999; Webb, 2005).

These analyses can be fruitful when standard spectra are available for one or more of the phases in the sample. In cases where the measured spectrum does not match known standards, numerical fits of the data with EXAFS calculations based on a molecular structure will be needed. Examples of such studies involve the adsorption mechanism of metals onto bacteria (Boyanov et al., 2003; Guine et al., 2006; Moon and Peacock, 2011) or the incorporation of contaminants into mineral hosts (Carvalho-E-Silva et al., 2003; Kelly et al., 2003; Mitsunobu et al., 2010; Nachtegaal and Sparks, 2003; Villalobos et al., 2005). Standards that directly represent the coordination geometry of an adsorption or incorporation complex are typically not available, so the interatomic distances and coordination numbers will need to be determined *de novo* from the EXAFS measurement and analysis. Shell-by-shell fitting refers to the incremental inclusion of coordination shells in an EXAFS fitting model and the refinement of their structural parameters against the measured experimental data (Figure 10.7). A substantial introduction to shell-by-shell fitting would not be possible here, so the reader is again referred to the more extensive monographs and web tutorials on the subject. Note that a deeper familiarity with EXAFS theory and significant experience with nonlinear parameter optimization are needed to avoid the numerous pitfalls that can result in incorrect interpretation of the data (which occurs more often than it should). The latter cautionary remark is not meant to discourage interested nonspecialist readers but rather, to make them aware of the "bumps" and "curves" on the road and to encourage them to seek advice from XAS specialists during their early attempts at shell-by-shell analysis. Most of the synchrotrons around the world offer introductory XAS courses, where such advice can be solicited.

10.5 Conclusions and Future

The application of XANES and EXAFS to geomicrobiology and environmental science has expanded greatly in the last decades and will continue to expand as the usefulness of molecular-scale information and the availability of beamtime increase. Current detection limits for EXAFS are on the higher end of what is directly relevant to environmental samples, so one area of growth will come with the increase in photon flux and the corresponding decrease in detection levels. Parallel with photon flux increase, new methods to mitigate radiation damage to the sample will need to be developed. A related area where progress is being made is the decrease in beam size and the development of spectroscopic mapping approaches. Getting to nanometer beam sizes will enable the probing of elemental concentrations and transformations at the scale of subcellular domains or small minerals, which are key to describing the high-reactivity components that drive the dynamics in an environmental system. Improvements and automation in mapping and data analysis are also likely to increase the interest and the user base of XAS, particularly as higher X-ray fluxes at synchrotrons generate larger data sets that will need analysis by nonspecialists.

10.6 References

Bargar, J.R., Reitmeyer, R., Lenhart, J.J., Davis, J.A., 2000a. Characterization of U(VI)-carbonato ternary complexes on hematite: EXAFS and electrophoretic mobility measurements. *Geochimica et Cosmochimica Acta*, 64(16): 2737–2749.

Bargar, J.R., Tebo, B.M., Villinski, J.E., 2000b. In situ characterization of Mn(II) oxidation by spores of the marine Bacillus sp strain SG-1. *Geochimica et Cosmochimica Acta*, 64(16): 2775–2778.

Bergengren, J., 1920. On spectra of absorption of phosphorus by X-ray. *Comptes Rendus Hebdomadaires Des Seances De L'Academie Des Sciences*, 171: 624–626.

Bernhard, G. et al., 2001. Uranyl(VI) carbonate complex formation: Validation of the Ca2UO2(CO3)(3)(aq.) species. *Radiochimica Acta*, 89(8): 511–518.

Beyenal, H. et al., 2004. Uranium immobilization by sulfate-reducing biofilms. *Environmental Science & Technology*, 38(7): 2067–2074.

Booth, C.H., Bridges, F., 2005. Improved self-absorption correction for fluorescence measurements of extended X-ray absorption fine-structure. *Physica Scripta*, T115: 202–204.

Boyanov, M.I. et al., 2011. Solution and microbial controls on the formation of reduced U(IV) species. *Environmental Science & Technology*, 45(19): 8336–8344.

Boyanov, M.I. et al., 2003. Adsorption of cadmium to *Bacillus subtilis* bacterial cell walls: A pH-dependent X-ray absorption fine structure spectroscopy study. *Geochimica et Cosmochimica Acta*, 67(18): 3299–3311.

Brinza, L. et al., 2014. Combining mu XANES and mu XRD mapping to analyse the heterogeneity in calcium carbonate granules excreted by the earthworm *Lumbricus terrestris*. *Journal of Synchrotron Radiation*, 21: 235–241.

Brooks, S.C. et al., 2003. Inhibition of bacterial U(VI) reduction by calcium. *Environmental Science & Technology*, 37(9): 1850–1858.

Brown, G.E., 1990. Spectroscopic Studies of Chemisorption Reaction Mechanisms at Oxide-Water Interfaces. In: Hochella, M.F., White, A.F. (Eds.), *Mineral-Water Interface Geochemistry*. Mineralogical Society of America, pp. 309–364.

Buchanan, B.B. et al., 1995. A XANES and EXAFS investigation of the speciation of selenite following bacterial metabolization. *Inorganic Chemistry*, 34(6): 1617–1619.

Bunker, G., 2010. *Introduction to XAFS: A Practical Guide to X-ray Absorption Fine Structure Spectroscopy*. Cambridge University Press, Cambridge.

Bunker, G., Stern, E.A., Blankenship, R.E., Parson, W.W., 1982. An x-ray absorption study of the iron site in bacterial photosynthetic reaction centers. *Biophysical Journal*, 37(2): 539–551.

Carvalho-E-Silva, M.L. et al., 2003. Incorporation of Ni into natural goethite: An investigation by X-ray absorption spectroscopy. *American Mineralogist*, 88(5–6): 876–882.

Catalano, J.G., Brown, G.E., 2005. Uranyl adsorption onto montmorillonite: Evaluation of binding sites and carbonate complexation. *Geochimica et Cosmochimica Acta*, 69(12): 2995–3005.

Charlet, L., Manceau, A., 1992. Insitu characterization of heavy-metal surface-reactions – the chromium case. *International Journal of Environmental Analytical Chemistry*, 46(1–3): 97–108.

Charnock, J.M. et al., 2000. Structural investigations of the Cu-A centre of nitrous oxide reductase from *Pseudomonas stutzeri* by site-directed mutagenesis and X-ray absorption spectroscopy. *European Journal of Biochemistry*, 267(5): 1368–1381.

Delaney, J.S., Dyar, M.D., Sutton, S.R., Bajt, S., 1998. Redox ratios with relevant resolution: Solving an old problem by using the synchrotron microXANES probe. *Geology*, 26(2): 139–142.

Denecke, M.A., Rothe, J., Dardenne, K., Lindqvist-Reis, P., 2003. Grazing incidence (GI) XAFS measurements of Hf(IV) and U(VI) sorption onto mineral surfaces. *Physical Chemistry Chemical Physics*, 5(5): 939–946.

Dent, A.J., Ramsay, J.D.F., Swanton, S.W., 1992. An EXAFS study of uranyl-ion in solution and sorbed onto silica and montmorillonite clay colloids. *Journal of Colloid and Interface Science*, 150(1): 45–60.

Eng, P.J., Rivers, M., Yang, B.X., Schildkamp, W., 1995. Micro-focusing 4 KeV to 65 KeV x-rays with bent Kirkpatrick-Baez mirrors. *X-Ray Microbeam Technology and Applications*, 2516: 41–51.

Ertel, T.S., Bertagnolli, H., 1993. EXAFS investigations of air and moisture sensitive liquid compounds – development of an appropriate sample holder with variable sample thickness and temperature control. *Nuclear Instruments & Methods in Physics Research Section B-Beam Interactions with Materials and Atoms*, 73(2): 199–202.

Etschmann, B.E. et al., 2014. Speciation mapping of environmental samples using XANES imaging. *Environmental Chemistry*, 11(3): 341–350.

Fendorf, S., Eick, M.J., Grossl, P., Sparks, D.L., 1997. Arsenate and chromate retention mechanisms on goethite .1. Surface structure. *Environmental Science & Technology*, 31(2): 315–320.

Fletcher, K.E. et al., 2010. U(VI) reduction to mononuclear U(IV) by *Desulfitobacterium* species. *Environmental Science & Technology*, 44(12): 4705–4709.

Foster, A.L., Brown, G.E., Tingle, T.N., Parks, G.A., 1998. Quantitative arsenic speciation in mine tailings using X-ray absorption spectroscopy. *American Mineralogist*, 83(5–6): 553–568.

Francis, A.J., Dodge, C.J., Lu, F.L., Halada, G.P., Clayton, C.R., 1994. XPS and XANES studies of uranium reduction by *Clostridium* Sp. *Environmental Science & Technology*, 28(4): 636–639.

Francis, A.J. et al., 2004. Uranium association with halophilic and non-halophilic bacteria and archaea. *Radiochimica Acta*, 92(8): 481–488.

Frenkel, A.I., Kleifeld, O., Wasserman, S.R., Sagi, I., 2002. Phase speciation by extended x-ray absorption fine structure spectroscopy. *Journal of Chemical Physics*, 116(21): 9449–9456.

Glasauer, S. et al., 2007. Mixed-valence cytoplasmic iron granules are linked to anaerobic respiration. *Applied and Environmental Microbiology*, 73(3): 993–996.

Guine, V. et al., 2006. Zinc sorption to three gram-negative bacteria: Combined titration, modeling, and EXAFS study. *Environmental Science & Technology*, 40(6): 1806–1813.

Hattori, T. et al., 2009. The structure of monomeric and dimeric uranyl adsorption complexes on gibbsite: A combined DFT and EXAFS study. *Geochimica et Cosmochimica Acta*, 73(20): 5975–5988.

Hettiarachchi, G.M., Scheckel, K.G., Ryan, J.A., Sutton, S.R., Newville, M., 2006. mu-XANES and mu-XRF investigations of metal binding mechanisms in biosolids. *Journal of Environmental Quality*, 35(1): 342–351.

Iida, A., Noma, T., 1993. Correction of the self-absorption effect in fluorescence x-ray-absorption fine-structure. *Japanese Journal of Applied Physics Part 1-Regular Papers Short Notes & Review Papers*, 32(6A): 2899–2902.

Ikeda, A. et al., 2007. Comparative study of uranyl(VI) and -(V) carbonato complexes in an aqueous solution. *Inorganic Chemistry*, 46(10): 4212–4219.

Ilton, E.S. et al., 2004. Heterogeneous reduction of uranyl by micas: Crystal chemical and solution controls. *Geochimica et Cosmochimica Acta*, 68(11): 2417–2435.

Jardine, P.M. et al., 1999. Fate and transport of hexavalent chromium in undisturbed heterogeneous soil. *Environmental Science & Technology*, 33(17): 2939–2944.

Kashiv, Y. et al., 2016. Imaging trace element distributions in single organelles and subcellular features. *Scientific Reports*, 6: 21437.

Kelly, S.D. et al., 2001. XAFS determination of the bacterial cell wall functional groups responsible for complexation of Cd and U as a function of pH. *Journal of Synchrotron Radiation*, 8: 946–948.

Kelly, S.D., Hasterberg, D., Ravel, B., 2008. Analysis of Soils and Minerals Using X-ray Absorption Spectroscopy. In: Ulery, A.L., Drees, L.R. (Eds.), *Methods of Soil Analysis, Part 5 -Mineralogical Methods*. Soil Science Society of America, Madison, WI.

Kelly, S.D., Kemner, K.M., Brooks, S.C., 2007. X-ray absorption spectroscopy identifies calcium-uranyl-carbonate complexes at environmental concentrations. *Geochimica et Cosmochimica Acta*, 71(4): 821–834.

Kelly, S.D. et al., 2003. Uranyl incorporation in natural calcite. *Environmental Science & Technology*, 37(7): 1284–1287.

Kemner, K.M., Kelly, S.D., 2007. Synchrotron-based Techniques for Monitoring Metal Transformations. In: Hurst, C.J. et al. (Eds.), *Manual of Environmental Microbiology*. ASM Press, Washington, DC, pp. 1183–1194.

Kemner, K.M. et al., 2004. Elemental and redox analysis of single bacterial cells by X-ray microbeam analysis. *Science*, 306(5696): 686–687.

Kim, C.S., Bloom, N.S., Rytuba, J.J., Brown, G.E., 2003. Mercury speciation by X-ray absorption fine structure spectroscopy and sequential chemical extractions: A comparison of speciation methods. *Environmental Science & Technology*, 37(22): 5102–5108.

Koningsberger, D.C., Prins, R., 1988. *X-Ray Absorption: Principles, Applications, Techniques of EXAFS, SEXAFS and XANES*. John Wiley and Sons, New York, NY.

Kwon, M.J. et al., 2014. Acid extraction overestimates the total Fe(II) in the presence of iron (hydr) oxide and sulfide minerals. *Environmental Science & Technology Letters*, 1(7): 310–314.

Lai, B. et al., 1992. Hard x-ray phase zone plate fabricated by lithographic techniques. *Applied Physics Letters*, 61(16): 1877–1879.

Latta, D.E., Kemner, K.M., Mishra, B., Boyanov, M.I., 2016. Effects of calcium and phosphate on uranium(IV) oxidation: Comparison between nanoparticulate uraninite and amorphous U-IV-phosphate. *Geochimica et Cosmochimica Acta*, 174: 122–142.

Latta, D.E., Mishra, B., Cook, R.E., Kemner, K.M., Boyanov, M.I., 2014. Stable U(IV) complexes form at high-affinity mineral surface sites. *Environmental Science & Technology*, 48(3): 1683–1691.

Li, W.-B. et al., 2014. Correction method for the self-absorption effects in fluorescence extended X-ray absorption fine structure on multilayer samples. *Journal of Synchrotron Radiation*, 21: 561–567.

Lindahl, P.A. et al., 1984. Nickel and iron EXAFS of f-420-reducing hydrogenase from *Methanobacterium thermoautotrophicum*. *Journal of the American Chemical Society*, 106(10): 3062–3064.

Lytle, F.W., Sayers, D.E., Stern, E.A., 1975. Extended x-ray-absorption fine-structure technique. 2. Experimental practice and selected results. *Physical Review B*, 11(12): 4825–4835.

Merroun, M.L. et al., 2005. Complexation of uranium by cells and S-layer sheets of *Bacillus sphaericus* JG-A12. *Applied and Environmental Microbiology*, 71(9): 5532–5543.

Michalowicz, A., Moscovici, J., Muller-Bouvet, D., Provost, K., 2009. MAX: Multiplatform Applications for XAFS. In: DiCicco, A., Filipponi, A. (Eds.), 14th International Conference on X-Ray Absorption Fine Structure. *Journal of Physics Conference Series*, 190: 012034.

Mitsunobu, S., Takahashi, Y., Terada, Y., Sakata, M., 2010. Antimony(V) incorporation into synthetic ferrihydrite, goethite, and natural iron oxyhydroxides. *Environmental Science & Technology*, 44 (10): 3712–3718.

Moon, E.M., Peacock, C.L., 2011. Adsorption of Cu(II) to *Bacillus subtilis*: A pH-dependent EXAFS and thermodynamic modelling study. *Geochimica et Cosmochimica Acta*, 75(21): 6705–6719.

Nachtegaal, M., Sparks, D.L., 2003. Nickel sequestration in a kaolinite-humic acid complex. *Environmental Science & Technology*, 37(3): 529–534.

O'Day, P.A., Brown, G.E., Parks, G.A., 1994. X-ray-absorption spectroscopy of cobalt(II) multi-nuclear surface complexes and surface precipitates on kaolinite. *Journal of Colloid and Interface Science*, 165(2): 269–289.

O'Loughlin, E.J., Kelly, S.D., Cook, R.E., Csencsits, R., Kemner, K.M., 2003. Reduction of uranium-(VI) by mixed iron(II/iron(III) hydroxide (green rust): Formation of UO2 nanoparticles. *Environmental Science & Technology*, 37(4): 721–727.

Paunesku, T., Vogt, S., Maser, J., Lai, B., Woloschak, G., 2006. X-ray fluorescence microprobe imaging in biology and medicine. *Journal of Cellular Biochemistry*, 99(6): 1489–1502.

Penner-Hahn, J.E., 2003. X-ray Absorption Spectroscopy, in: *Comprehensive Coordination Chemistry II*, pp. 159–186.

Polette, L.A. et al., 2000. XAS and microscopy studies of the uptake and bio-transformation of copper in *Larrea tridentata* (creosote bush). *Microchemical Journa*l, 65(3): 227–236.

Pushie, M.J., Pickering, I.J., Korbas, M., Hackett, M.J., George, G.N., 2014. Elemental and chemically specific X-ray fluorescence imaging of biological systems. *Chemical Reviews*, 114(17): 8499–8541.

Ravel, B., Newville, M., 2005. ATHENA, ARTEMIS, HEPHAESTUS: Data analysis for X-ray absorption spectroscopy using IFEFFIT. *Journal of Synchrotron Radiation*, 12: 537–541.

Ressler, T., 1998. WinXAS: A program for X-ray absorption spectroscopy data analysis under MS-Windows. *Journal of Synchrotron Radiation*, 5: 118–122.

Ressler, T., Wienold, J., Jentoft, R.E., Neisius, T., 2002. Bulk structural investigation of the reduction of MoO3 with propene and the oxidation of MoO2 with oxygen. *Journal of Catalysis*, 210(1): 67–83.

Rossberg, A., Reich, T., Bernhard, G., 2003. Complexation of uranium(VI) with protocatechuic acid – application of iterative transformation factor analysis to EXAFS spectroscopy. *Analytical and Bioanalytical Chemistry*, 376(5): 631–638.

Rui, X. et al., 2013. Bioreduction of hydrogen uranyl phosphate: Mechanisms and U(IV) products. *Environmental Science & Technology*, 47(11): 5668–5678.

Sayers, D.E., Stern, E.A., Lytle, F.W., 1971. New technique for investigating noncrystalline structures: Fourier analysis of extended x-ray–absorption fine structure. *Physical Review Letters*, 27 (18): 1204.

Scheinost, A.C. et al., 2008. X-ray absorption and photoelectron spectroscopy investigation of selenite reduction by Fe-II-bearing minerals. *Journal of Contaminant Hydrology*, 102(3–4): 228–245.

Schulze, D.G., Bertsch, P.M., 1995. Synchrotron X-ray techniques in soil, plant, and environmental research. *Advances in Agronomy*, 55: 1–66.

Scott, R.A., 1985. Measurement of metal-ligand distances by EXAFS. *Methods in Enzymology*, 117: 414–459.

Singer, D.M., Farges, F., Brown, G.E., 2009. Biogenic nanoparticulate UO2: Synthesis, characterization, and factors affecting surface reactivity. *Geochimica et Cosmochimica Acta*, 73(12): 3593–3611.

Stelling, O., 1925. Article with information on the connection between chemical constitution and K-x rays absorption spectra. IL.) Research on some phosphoric compounds. *Zeitschrift Fur Physikalische Chemie –Stochiometrie Und Verwandtschaftslehre*, 117(3/4): 161–174.

Stern, E.A., 1974. Theory of extended x-ray-absorption fine-structure. *Physical Review* B, 10(8): 3027–3037.

Stern, E.A., Bunker, B.A., Heald, S.M., 1980. Many-body effects on extended x-ray absorption fine-structure amplitudes. *Physical Review B*, 21(12): 5521–5539.

Stern, E.A., Sayers, D.E., Lytle, F.W., 1975. Extended x-ray-absorption fine-structure technique. 3. Determination of physical parameters. *Physical Review* B, 11(12): 4836–4846.

Sylwester, E.R., Hudson, E.A., Allen, P.G., 2000. The structure of uranium (VI) sorption complexes on silica, alumina, and montmorillonite. *Geochimica et Cosmochimica Acta*, 64(14): 2431–2438.

Templeton, A., Knowles, E., 2009. Microbial transformations of minerals and metals: Recent advances in geomicrobiology derived from synchrotron-based x-ray spectroscopy and x-ray microscopy, *Annual Review of Earth and Planetary Sciences*, 37(1): 367–391.

Templeton, A.S., Spormann, A.M., Brown, G.E., 2003. Speciation of Pb(II) sorbed by *Burkholderia cepacia*/goethite composites. *Environmental Science & Technology*, 37(10): 2166–2172.

Teo, B.K., 1986. *EXAFS: Basic Principles and Data Analysis*. Springer-Verlag, New York, NY.

Terzano, R. et al., 2010. Solving mercury (Hg) speciation in soil samples by synchrotron X-ray microspectroscopic techniques. *Environmental Pollution*, 158(8): 2702–2709.

Troger, L. et al., 1992. Full correction of the self-absorption in soft-fluorescence extended x-ray-absorption fine-structure. *Physical Review B*, 46(6): 3283–3289.

Vairavamurthy, A., Manowitz, B., Luther, G.W., Jeon, Y., 1993. Oxidation-state of sulfur in thiosulfate and implications for anaerobic energy-metabolism. *Geochimica et Cosmochimica Acta*, 57(7): 1619–1623.

Veeramani, H. et al., 2011. Products of abiotic U(VI) reduction by biogenic magnetite and vivianite. *Geochimica et Cosmochimica Acta*, 75(9): 2512–2528.
Villalobos, M., Bargar, J., Sposito, G., 2005. Mechanisms of Pb(II) sorption on a biogenic manganese oxide. *Environmental Science & Technology*, 39(2): 569–576.
Vodyanitskii, Y.N., 2013. Determination of the oxidation states of metals and metalloids: An analytical review. *Eurasian Soil Science*, 46(12): 1139–1149.
Vogt, S., Maser, J., Jacobsen, C., 2003. Data analysis for X-ray fluorescence imaging. *Journal de Physique IV*, 104: 617–622.
Waite, T.D., Davis, J.A., Payne, T.E., Waychunas, G.A., Xu, N., 1994. Uranium(VI) adsorption to ferrihydrite – application of a surface complexation model. *Geochimica et Cosmochimica Acta*, 58 (24): 5465–5478.
Waldo, G.S. et al., 1995. Formation of the ferritin iron mineral occurs in plastids – an x-ray-absorption spectroscopy study. *Plant Physiology*, 109(3): 797–802.
Wasserman, S.R., Allen, P.G., Shuh, D.K., Bucher, J.J., Edelstein, N.M., 1999. EXAFS and principal component analysis: A new shell game. *Journal of Synchrotron Radiation*, 6: 284–286.
Watson, J.H.P. et al., 2000. Structural and magnetic studies on heavy-metal-adsorbing iron sulphide nanoparticles produced by sulphate-reducing bacteria. *Journal of Magnetism and Magnetic Materials*, 214(1–2): 13–30.
Watson, J.H.P., Ellwood, D.C., 2003. The removal of the pertechnetate ion and actinides from radioactive waste streams at Hanford, Washington, USA and Sellafield, Cumbria, UK: The role of iron-sulfide-containing adsorbent materials. *Nuclear Engineering and Design*, 226(3): 375–385.
Webb, S.M., 2005. SIXpack: A graphical user interface for XAS analysis using IFEFFIT. *Physica Scripta*, T115: 1011–1014.
Williams, P.A. et al., 1999. The Cu-A domain of *Thermus thermophilus* ba(3)-type cytochrome c oxidase at 1.6 angstrom resolution. *Nature Structural Biology*, 6(6): 509–516.
Yan, S., Boyanov, M.I., Mishra, B., Kemner, K.M., O'Loughlin, E.J., 2018. U(VI) reduction by biogenic and abiotic hydroxycarbonate green rusts: Impacts on U(IV) speciation and stability over time. *Environmental Science & Technology*, 52(8): 4601–4609.
Zachara, J.M. et al., 2004. Chromium speciation and mobility in a high level nuclear waste vadose zone plume. *Geochimica et Cosmochimica Acta*, 68(1): 13–30.
Zachara, J.M. et al., 2007. Reduction of pertechnetate Tc(VII) by aqueous Fe(II) and the nature of solid phase redox products. *Geochimica et Cosmochimica Acta*, 71(9): 2137–2157.

11 Bacterial Surfaces in Geochemistry – How Can X-ray Photoelectron Spectroscopy Help?

MADELEINE RAMSTEDT, LAURA LEONE, AND ANDREY SHCHUKAREV

Abstract
Processes occurring at surfaces and interfaces are very important in environmental systems, necessitating surface-specific characterization tools that can help us understand processes at and specific properties of surfaces and interfaces, and their role in biogeochemical systems. This chapter describes the use and application of X-ray photoelectron spectroscopy (XPS) to study interfacial processes of relevance for geomicrobiology. Examples are given from studies determining cell wall composition, acid–base properties, cell surface charge, metal adsorption onto bacterial cells, and bacterial surface–induced precipitation of secondary minerals. As XPS is an ultrahigh-vacuum technique, several sample preparation methods have been applied to enable analysis of bacterial samples, including analysis of freeze-dried samples as well as frozen bacterial suspensions. These are described and discussed alongside advantages and disadvantages of different approaches, with a special focus on fast-freezing and the cryogenic technique.

11.1 X-ray Photoelectron Spectroscopy Basics

X-ray photoelectron spectroscopy (XPS), historically known as electron spectroscopy for chemical analysis (ESCA), is one of the most popular and heavily used surface analysis techniques. The popularity of XPS originates from its relatively simple theoretical background, data processing, and interpretation, giving data that are highly chemically informative. Moreover, the method has practical relevance for many applied problems, is applicable to virtually any type of sample, requires minimal or no sample preparation, and is relatively easy to access through a variety of commercially available laboratory electron spectrometers.

XPS is based on the photoelectric effect. When X-ray photons with known energy ($h\nu$) interact with the sample, electrons from constituting atoms are emitted. By measuring the kinetic energy (KE) of the emitted photoelectrons, it is possible to calculate their binding energy (BE) in accordance with the law of energy conservation:

$$h\nu = BE + KE + \phi \quad \text{(Eq.11.1)}$$

$$BE = h\nu - KE - \phi \quad \text{(Eq.11.2)}$$

where ϕ is the spectrometer work function (experimentally determined and constant for a given instrument). The binding energies for atomic core-level electrons (1s, 2s, 2p, 3s, 3p,

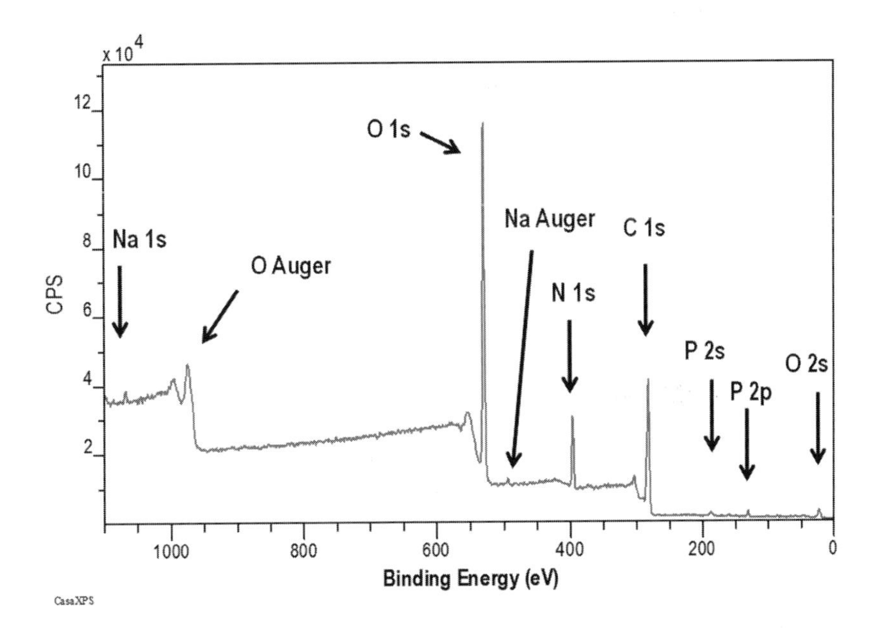

Figure 11.1 Survey XPS spectrum of a fast-frozen strain of *Pseudomonas aeruginosa*.

3d, etc.) are specific for each element in the Periodic Table. Individual lines from different elements may overlap, but if all core levels are taken into account, elemental identification can be performed unambiguously. Thus, the elemental chemical composition (except H and He) of the sample surface is easily determined from the *BE* values obtained by XPS. A standard survey XPS spectrum (Figure 11.1) represents a plot of the number of detected photoelectrons (counts per second [CPS]) against their binding energy (electron volts [eV]) and provides a characteristic "fingerprint" of the elemental composition at the surface of a sample.

The analysis depth in XPS (2–10 nm) is limited by the escape depth of photoelectrons from solids, ensuring a very high surface sensitivity of the technique. The depth of analysis is mostly dependent on the sample density as well as the *KE* of the measured photoelectron. Typically, it is 2–3 nm for metals, 3–6 nm for inorganic oxides (and most minerals), and 6–10 nm for organic compounds (e.g., bacteria) (Briggs and Grant, 2003). The intensity of the photoelectron line is proportional to the number of corresponding emitting atoms, and atomic concentrations of elements at the sample surface are routinely determined by dividing the peak area of specific photoelectron lines with tabulated relative atomic sensitive factors. The intensities measured from similar samples are reproducible with good precision, making the technique quantitatively very powerful. To exclude uncertainties in absolute intensity measurements (peak areas), atomic ratios between elements calculated from obtained atomic concentrations are routinely used. These ratios, which are directly related to the chemical formulas of individual compounds, provide the basis for quantification of XPS data. The selected analysis area can be in the range of several μm^2 up to cm^2, which gives the possibility for both small spot analysis and surface chemical mapping. The analytical sensitivity of XPS is exceptionally high, and 10% of a monolayer (or 0.05–0.1 atomic % surface concentration) can easily be detected and quantitatively measured.

The key feature of XPS, making the technique especially powerful for surface chemistry, is its direct and straightforward determination of the valence state of elements at the surface. This specific chemical information is obtained by measurements of the shifts in binding energies (so-called chemical shift) of atomic core levels caused by the change in valence electron density around the atom due to the formation of chemical bonds. As a result, a change in oxidation state, ligand electronegativity, coordination, protonation or hydrogen bonding, etc. can generally be observed. Experimentally determined *BE* values for different chemical compounds have been tabulated and are available in handbooks (Moulder et al., 1992) and databases (Beamson and Briggs, 1992; Naumkin et al., 2012). As an example of chemical shift due to changes in oxidation state, a high-resolution S 2p spectrum taken with fast-frozen bioleached residue of chalcopyrite, $CuFeS_2$, is shown in Figure 11.2. It illustrates different chemical species of sulfur at the surface of the sample (Khoshkhoo et al., 2014).

A detailed description of XPS basics and instrumentation, as well as practical and developing aspects, can be found in the book by Briggs and Grant (Briggs and Grant, 2003). For a shorter but comprehensive description, Ratner and Castner (2009) is recommended. Furthermore, the authors have recently published a perspective article as well as a methods paper on the use of cryo-XPS for studying interfaces between solid and liquid systems (Ramstedt and Shchukarev, 2016; Shchukarev and Ramstedt, 2017).

11.1.1 Sample Preparation Techniques for Cryogenic XPS

With respect to geomicrobiological samples (especially those investigating the mineral–bacteria interface), a major disadvantage in XPS is that the technique generally cannot operate in ambient conditions. Detection of photoelectrons requires ultrahigh vacuum (UHV). Therefore, XPS, similarly to most surface science techniques, is traditionally considered to be *ex situ*. The application of conventional XPS to wet samples is strongly limited by the high vapor pressure of water, which is not compatible with the UHV environment of an electron spectrometer. Possible experimental approaches to evade this restriction have been described in the literature (Shchukarev, 2006a, 2006b), including analysis in humid conditions (Salmeron and Schlögl, 2008) and analysis of aqueous solutions (Seidel et al., 2011; Winter and Faubel, 2006). Some synchrotron XPS beamlines, as well as dedicated ambient-pressure XPS instruments, can also analyze samples at near-normal pressure. The key problem for any investigation of the intact interfaces of solids in aqueous solutions by XPS is to develop a reliable and reproducible sample preparation protocol. The protocol should ideally produce samples appropriate for a UHV environment but at the same time provide an unaltered solid–solution interface with a solution layer that is thinner than the XPS analysis depth. Fast-freezing of the solid, in equilibrium with aqueous solution, to liquid nitrogen temperature can resolve the problem, since the solution side of the interface, predominantly water, becomes stabilized and also serves as a protective layer that keeps the interface close to intact.

For conventional laboratory spectrometers (i.e., not synchrotron based), two sample preparation techniques have been introduced that do not significantly alter the solution–solid interface and thus, can be considered as appropriate sample handling with respect to

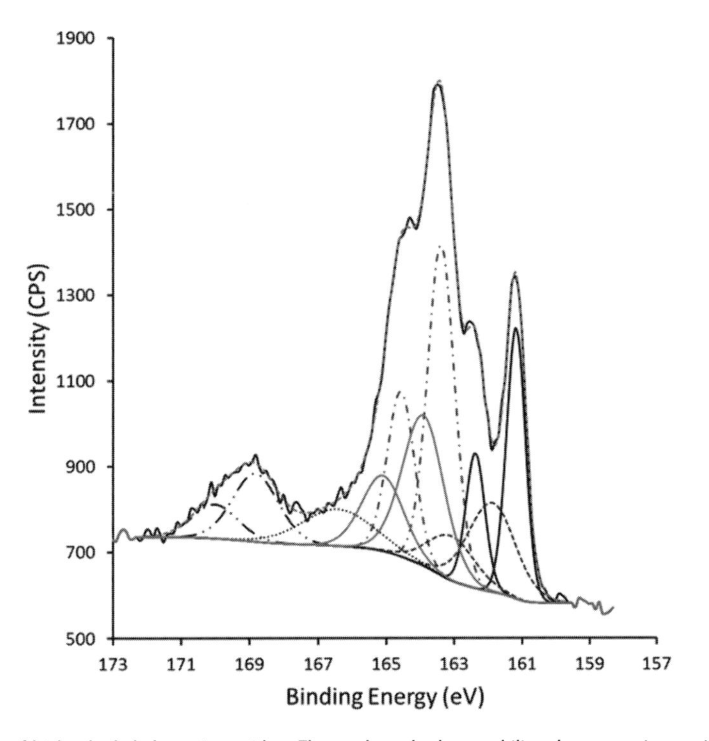

Figure 11.2 S 2p spectrum of bioleached chalcopyrite residue. The moderately thermophilic culture contains strains related to *Acidithiobacillus ferrooxidans, Acidithiobacillus caldus* C-SH12, *Sulfobacillus thermosulfidooxidans* AT-1, *"Sulfobacillus montserratensis"* L15, and an uncultured thermal soil bacterium. Each chemical state of S (ranging from S^{2-} to SO_4^{2-}) gives rise to a doublet due to spin-orbital splitting of 2p electrons (figure based on data from Khoshkhoo et al., 2014). Five doublets and one loss peak (dotted in grey) were used to fit the spectrum (broken grey line represents total fit and top solid black line the raw data).

investigating this interface. Based on well-known "freeze-drying" technology, deep freezing of hydrated samples followed by controlled ice sublimation inside the air lock of the electron spectrometer has been proposed (Ratner et al., 1978) and developed (Ratner, 1995) for XPS studies of surfaces with biomedical interest. A detailed sample preparation protocol for XPS analysis of frozen–hydrated samples was published by Castner and Ratner (Castner and Ratner, 2002). In that study, the ice sublimation step is followed by cryogenic XPS measurements. The reliability and reproducibility of the technique was recently improved by an experimentally established ice-sublimation temperature–time procedure (Delcroix et al., 2012). This sample preparation protocol was used for XPS investigation of the bacteria–water interface (Rouxhet and Genet, 2011). However, it is likely that surfaces of these freeze-dried bacteria differ from those of intact bacteria, where the solution at the interface remains.

In the early stages of XPS, Burger et al. developed a fast-freezing technique for XPS analysis of a drop of aqueous solutions on a "cold finger" (sample holder) cooled to liquid nitrogen temperature (Burger, 1978; Burger and Fluck, 1974; Burger et al., 1975, 1977). Of particular importance, this fast-freezing procedure was proved to preserve the chemical speciation of solutes (Burger and Fluck, 1974; Burger et al., 1975, 1977). However, the technique was not further explored until we, in our lab, modified and applied it to colloidal mineral suspensions in electrolytes (Ramstedt et al., 2002; Shchukarev and Sjöberg, 2005; Shchukarev et al., 2004). This modified fast-freezing sample preparation protocol has since been described and discussed in detail for XPS studies of, e.g., hematite suspensions (Shchukarev et al., 2007) and its use for bacterial cultures is described in this chapter. Cryogenic XPS with fast-frozen samples has been shown to be exceptionally useful in determination of the interface loading of electrolyte ions and its dependence on solution pH and ionic strength. Moreover, particle surface charge can be determined using the atomic ratio of electrolyte counter-ions (Shchukarev, 2006a), and interfacial protonation constants for adsorbed species are found to be remarkably different from their solution counterparts (Ramstedt et al., 2004; Shimizu et al., 2011). If fast-frozen samples are left inside the spectrometer to slowly dehydrate (hereafter called dehydrated fast-frozen), XPS measurements performed the next day at room temperature (in the same analysis location) can provide experimental evidence for important interfacial phenomena, such as specific adsorption, ion pair formation, and protonation of amine groups (Ramstedt et al., 2002, 2004; Shchukarev, 2006a; Shimizu et al., 2011), which are often impossible to directly observe using other analytical techniques. In addition, a drop in surface potential can sometimes be observed, caused by a collapse of the electrical double layer (Shchukarev and Sjöberg, 2005). The thickness of the solution layer at the interface has experimentally been estimated using cryo-XPS from changes in peak intensities and was found to be about 5 Å, which approximately corresponds to three water/solution layers (Ramstedt, 2004; Shchukarev and Sjöberg, 2005). Consequently, cryogenic XPS fills the gap between analysis of supernatant solutions and dry solids (bulk and surface) and provides important missing pieces of information about the solid–solution interface, thus ensuring a more complete description of most aqueous geomicrobial systems.

11.1.2 Freeze-drying Sample Preparation Techniques

Traditionally, lyophilization (freeze-drying) has been the only way to prepare microbial samples for XPS analysis. The procedure (in this chapter called external freeze drying) is well established, and the dried powders are easy to handle, store, and thereafter analyze under ultrahigh-vacuum conditions. Additionally, freeze-drying by controlled ice sublimation (hereafter called internal freeze drying) can be performed inside the spectrometer, as was described in Section 11.1.1. This latter method is preferable, since the use of an external laboratory freeze-dryer can lead to heavy contamination of the sample surface from residues left in the equipment or gaseous substances in the laboratory air. This is avoided if the sample is dehydrated inside the spectrometer. However, ice sublimation inside the spectrometer is experimentally complicated, time-consuming (Lukas et al., 1995), and not easily reproducible.

The application of freeze-drying is limited to samples suspended in pure water, or possibly from aqueous solutions of very low concentrations of dissolved constituents, i.e., much lower than concentrations generally observed in environments such as the sea or in body fluids.

Solutions with such high concentrations of dissolved salts will lead to relatively large amounts of precipitation. Thus, for most environmental and biological samples, external freezing followed by ice sublimation would result in a "thick" layer of nonvolatile solution components at the surface, e.g., inorganic salts and biomolecules. Considering the sampling depth of XPS, this artificial surface layer would make the interface inaccessible for measurements, and mainly solution components would be analyzed. Another important consequence of water/ice removal is mentioned earlier: collapse of the electrical double layer. This layer exists at almost all solid–aqueous solution interfaces, determines the distribution of different chemical species near the surface, and influences most interface processes.

In contrast to external freeze-drying, the fast-freezing technique can be applied to virtually any suspension, gel or colloid. During this procedure, solution components can also be observed in spectra, but as the amount of solution is significantly reduced, the interference will be smaller than with freeze-drying. Thus, for mineral suspensions this contribution is generally negligible, but for bacterial samples such interference can still complicate the data analysis to a great extent. This interference will be further discussed in Section 11.2.3.

11.2 How Can XPS Be of Help in Biogeochemistry?

11.2.1 Examples of How and Where the Technique Can Be Used

Many geochemical processes in the global cycling of elements are directly or indirectly influenced by bacteria and other microorganisms. For example, some bacteria photosynthesize, taking up CO_2 from the atmosphere and producing organic material and O_2. There are also bacteria that degrade organic matter by a range of different processes. In similar ways, bacteria affect the geological nitrogen, phosphorus, sulfur, and silica cycles, the cycling of redox-active elements such as Fe and Mn, and the cycling of trace elements. Bacteria also take part in weathering processes and processes producing secondary minerals, excrete substances that can erode minerals, and scavenge metals and other substances (Krumbein, 1983; Mills, 1998). Some of these processes have also been exploited by society, for example through bioleaching of sulfide minerals to recover valuable metals such as Cu, Zn, and Co (Brierley and Brierley, 2013; Olson et al., 2003).

XPS can be used to investigate several different aspects of geomicrobiological processes, such as how bacterial cells are influenced by their surroundings and alter their cell wall composition (Ramstedt et al., 2014), acid–base behavior of functional groups at the surface of the bacterial cell wall (Leone et al., 2007; Ojeda et al., 2008), cell surface charging behavior (Boonaert and Rouxhet, 2000), metal accumulation in, or at, the cell wall (Kang et al., 2015; Li et al., 2017; Ramstedt et al., 2014), secondary mineral precipitation at the cell wall, and bacterially induced mineral dissolution (Kalinowski et al., 2000). For most of these processes, XPS can give valuable insights, though complementary techniques are also often needed to better understand what is observed and to strengthen or reject the investigated hypotheses. Examples of how these processes can be investigated using XPS in combination with a few other selected analysis techniques are described in this section with a focus on geochemical applications. For further reading we refer the reader to three excellent reviews by van der Mei, Rouxhet, and Genet, which cover XPS analyses of bacteria in general (Genet et al., 2008; Rouxhet and Genet, 2011; van der Mei et al., 2000).

XPS analyses of bacterial cell walls will give information about elemental composition as well as information about oxidation state and to some extent functional groups, as described earlier. This gives the opportunity to detect the presence of, for example, metal ions at the surface of bacteria, investigate their oxidation state, and study accumulation under different conditions (Li et al., 2008; Ramstedt et al., 2014). It also enables comparisons of surface composition between bacterial strains and species (van der Mei et al., 2000) and/or how it is affected by environmental parameters (Ramstedt et al., 2014). In its most straightforward form, this can be done by comparing samples with respect to total content of, for example, C, N, O, and P (van der Mei et al., 2000).

11.2.1.1 Cell Wall Composition

In order to obtain a more detailed description of the chemical composition of the cell wall, i.e., beyond elemental composition, it is important to understand some aspects of the general composition of the bacterial cell surface. Traditionally, bacteria are described as either Gram-negative or Gram-positive. In general, the plasma membrane in Gram-positive bacteria is covered by a thick peptidoglycan layer (Figure 11.3) that also contains teichoic acids and lipoteichoic acids, and the Gram-negative cell wall has a periplasmic space between the plasma membrane and the lipopolysaccharide (LPS) membrane in which a thin peptidoglycan layer exists. Throughout the whole structure of the cell wall for both types there are proteins with different functions, such as porins, transporters, enzymes, etc. Outside the cell wall there can also be "coating structures" such as capsules (often polysaccharide or peptide rich) or S-layers (protein rich with some polysaccharides), as well as surface appendages such as flagella (protein rich) and pili (protein rich), the latter aiding the bacteria in different forms of motility (Beveridge, 2010; Young, 2010).

Since the depth of analysis in XPS is small, only the cell wall will be probed during analysis. (This is a great advantage compared with many other analysis techniques.) The depth of analysis has been estimated to be 2–5 nm for bacterial samples (Dufrêne et al., 1997) or, if assuming that the analysis depth is similar between a general organic material and

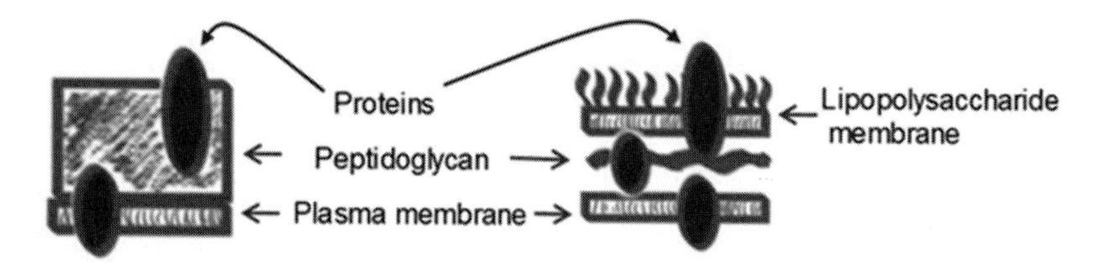

Figure 11.3 Simplified schematic showing the general overall organization of the cell wall in Gram-positive (left) and Gram-negative bacteria (right).

Table 11.1 Binding energies for C 1s electrons in biochemical compounds

Position (eV)[a]	Functional group	Example of reference compound
285.0	**C**-(CH)	Hydrocarbon
285.9	**C**-N, (C=O)-N-**C**	Amine, amide, peptide bond
286.5	**C**-O	Alcohol
286.6	(C=O)-O-**C**	Ester
287.9	**C**=O, O-**C**-O	Aldehyde, (hemi)acetal
288.1	(**C**=O)-NH-C	Amide, peptide bond
289.0	(**C**=O)-O-C, O=**C**-O$^-$	Ester or deprotonated carboxylate group
289.3	(**C**=O)-OH	Carboxylic acid (protonated)

(Beamson and Briggs, 1992; Rouxhet and Genet, 2011)
[a] Binding energy scale referenced to hydrocarbon at 285.0 eV

bacteria, 6–10 nm (Briggs and Grant, 2003). The depth of analysis mainly depends on the density of the surface material and its composition, as well as the kinetic energy of the emitted photoelectrons, as mentioned in Section 11.1. For Gram-negative bacteria, the analysis depth has been estimated to probe the lipopolysaccharide membrane and a part of the peptidoglycan layer underneath (Ramstedt et al., 2011). For Gram-positive bacteria, the analysis is only expected to probe the top of the peptidoglycan layer of the cell wall, since its thickness has been reported to be in the range of 15–30 nm (Vollmer et al., 2008). The presence of capsule, S-layer, pili, and/or flagella will be visible in XPS spectra. Since they will produce a kind of "coating" on the surface, this will alter the depth to which the cell wall is probed.

The building blocks (e.g., proteins, peptidoglycan, lipids, and polysaccharides) that form the bacterial cell wall are all different types of combinations of substances composed mainly of C, N, and O, with smaller amounts of elements such as S and P. The information from all these carbon-based materials will be summed up, e.g., in the C 1s spectrum, to give rise to a peak that is composed of components representing carbon atoms in different chemical environments (Table 11.1): for example, aliphatic C (with C or H as closest neighbor) at 285.0 eV, carbon with one single bond to O or N around 286.5 eV, and carbon double bonded to O with an N or O neighbor at 288.1 eV, e.g., in peptide bonds, or 289.0 eV in ester or deprotonated carboxylate groups.

In order to estimate the amount of the building blocks, two different methods have been presented in the literature. The first method involves using an equation system based on atomic ratios between C, N, and O obtained from analysis of a set of model constituents (Dufrêne et al., 1997; Rouxhet and Genet, 2011).

Equation system I:

$$[N/C]_{obs} = 0.279(C_{Pr}/C) \quad \text{(Eq.11.3)}$$

$$[O/C]_{obs} = 0.325(C_{Pr}/C) + 0.833(C_{PS}/C) \quad \text{(Eq.11.4)}$$

$$[C/C]_{obs} = 1 = (C_{Pr}/C) + (C_{PS}/C) + (C_{CH}/C) \quad \text{(Eq.11.5)}$$

where obs = observed, C_{Pr} corresponds to peptide C, C_{PS} = polysaccharide, and C_{CH} = aliphatic carbon (Dufrêne et al., 1997). A variety using only the C 1s region was also constructed using a similar approach and gave rise to the following equation system.

Equation system II:

$$[C_{288.1}/C]_{obs} = 0.279(C_{Pr}/C) + 0.167(C_{PS}/C) \quad \text{(Eq.11.6)}$$

$$[C_{286.5}/C]_{obs} = 0.293(C_{Pr}/C) + 0.833(C_{PS}/C) \quad \text{(Eq.11.7)}$$

$$[C_{285.0}/C]_{obs} = 0.428(C_{Pr}/C) + 1(C_{CH}/C) \quad \text{(Eq.11.8)}$$

However, the latter equation system is highly sensitive to the curve-fitting procedure of the individual experimentalist and does not take into account contributions from higher-energy components such as, e.g., protonated carboxylic acids around 289.3 eV. This could give rise to discrepancies between samples at acidic pH and alkaline pH, as the protonated forms of the carboxylic group are not accounted for in the second equation system. (The equations given are here indexed for a binding energy scale calibration that positions the aliphatic carbon contribution at 285.0 eV.)

To recalculate these ratios to an estimated weight fraction, the content of each substance can be divided as follows: $\frac{C_{Pr}/C}{43}$; $\frac{C_{PS}/C}{37}$; and $\frac{C_{HC}/C}{71.4}$ mmol carbon/g. Thereafter, the weights are summarized, and each substance weight is divided by the total sum to obtain a weight fraction in the analyzed volume (Rouxhet and Genet, 2011).

A second method, developed in our laboratory, is based on a model constructed through analysis of C 1s spectra of a large number of Gram-negative bacterial samples using multivariate curve resolution (Ramstedt et al., 2011). This analysis showed that the whole data set could be explained using three mathematically derived spectra that resemble lipid, peptide (protein and peptidoglycan), and polysaccharide. These theoretical model spectra can subsequently be used to predict the chemical composition of unknown samples, and this has been applied to describe chemical changes in the cell wall of Gram-positive *Bacillus subtilis* resulting from changes in surrounding pH and Zn^{2+} content (Ramstedt et al., 2014). Due to differences in cell wall composition between Gram-positive and Gram-negative bacteria, these mathematically derived components will describe slightly different substances in the cell wall between the two bacterial groups. In Gram-negative bacteria, the lipid component is mainly expected to describe lipids in the outer membrane, the polysaccharide component (sugar structures in the LPS membrane as well as any glycosylation of proteins), and finally, the peptide component – proteins and peptidoglycan. In Gram-positive bacteria, however, lipoteichoic acids will give rise to intensity interpreted as both lipid and polysaccharide, whereas teichoic acids and glycosylation of proteins are expected to mainly give rise to intensities interpreted by the model as polysaccharides. Lipids from the lipid membrane are only expected to be visible if this membrane protrudes through the peptidoglycan layer. Proteins and peptidoglycan will both be described by the peptide component as for Gram-negative bacterial samples (Ramstedt et al., 2011; Ramstedt et al., 2014).

The second model and the first equation system were used side by side on XPS data from dehydrated fast-frozen bacterial samples of *B. subtilis* cells exposed to changes in pH and

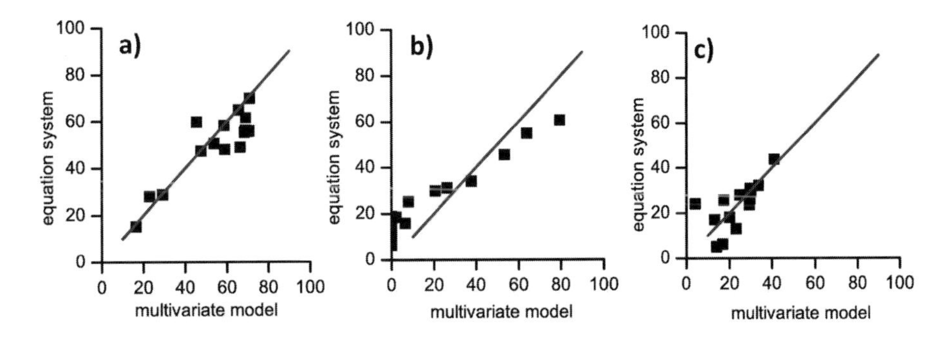

Figure 11.4 Correlation between fractions of a) peptide, b) lipid, and c) polysaccharide between data treatment using equation system 1 and the multivariate model. The solid line represents an ideal agreement between the two methods. (Reprinted with permission from Ramstedt et al., 2014, http://pubs.acs.org/doi/full/10.1021/la5002573.)

Zn^{2+} concentrations in saline (Ramstedt et al., 2014). A comparison between the two data-treatment approaches showed that although there were some small differences, they were comparable and could both be applied to describe the composition of all bacterial samples that have been dehydrated in different ways (Figure 11.4). However, for fast-frozen samples (hydrated), the model using the first equation system does not work, since water overshadows the oxygen content of the bacterial cell wall, thereby inhibiting the use of Eq. 11.4 and 11.5 (Ramstedt et al., 2014). Similar problems arise for other bacterial samples that have additional compounds contributing to the O1s or N1s spectra. If additional compounds are present that contain similar C functional groups as the principal constituents of the cell wall, these will be included in the overall analysis and thus cannot be distinguished. However, they might give rise to a detectable change in the apparent composition that can be used to, for example, follow the accumulation of this additional compound at the surface of the bacterial cell.

Comparisons of the results obtained using equation system 2 and the multivariate model show that there are also correlations between these two data treatment methods (Figure 11.5). Equation system 2 can be used on fast-frozen samples in cryo-XPS, and when we applied it to a range of *B. subtilis* samples, we found that it did not fully explain the C 1s peak with respect to a component at 289.3 ± 0.2 eV that is ignored in equation system 2. However, this component is also not completely described with the multivariate model (Figure 11.5d). Similarly to the correlations between the multivariate model and equation system 1, the best agreement was found for the polysaccharide component, whereas the equation systems seem to give lower percentages for samples with higher content of peptide and lipid in comparison with the multivariate method, i.e., giving rise to regression slopes of less than 1 in Figure 11.5a–c.

The ability to untangle the contributions from different substances in the C 1s spectra enables studies of how the cell wall of bacteria is affected by processes such as mutations and environmental conditions (Ramstedt et al., 2011, 2014). In this way, the cell wall composition of *B. subtilis* samples exposed to different pH values and Zn^{2+} concentrations was followed in saline (Ramstedt et al., 2014). For fast-frozen intact bacterial cells,

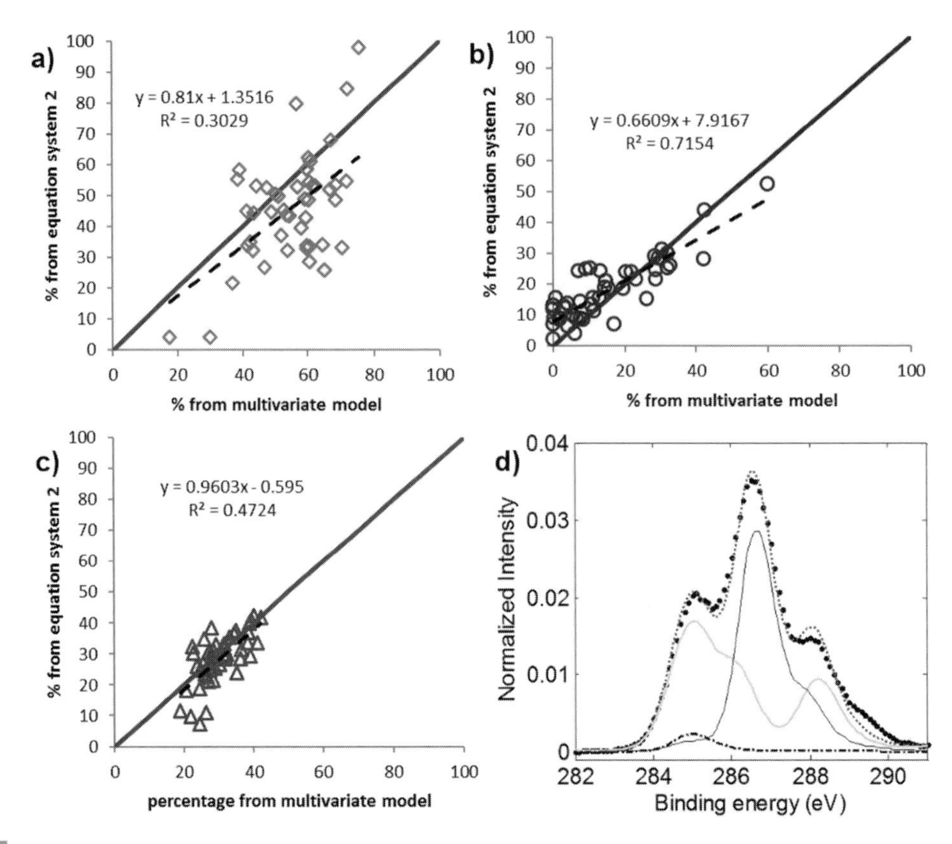

Figure 11.5 Correlation between fractions of a) peptide, b) lipid, and c) polysaccharide between data treatment using equation system 2 and the multivariate model. The solid line represents an ideal agreement between the two methods. The broken line represents the linear regression of all points. d) Example of a fit of the multivariate model to a fast-frozen sample at pH 4 with an "unexplained" shoulder around 289.3 eV (diamonds represent raw data, dotted line fit, grey solid line peptide component, black solid line polysaccharide component, and black dash-dot line lipid component).

the multivariate model showed that the content of peptide and polysaccharide decreased with increasing pH, whereas the content of lipid increased (Figure 11.6). This illustrates that for viable bacteria the surface composition is not static but a result of surface modifications induced probably by bacterial adaptations to the surrounding conditions. These changes can be followed using cryo-XPS, the multivariate model and equation system 2, and, to some extent, through atomic ratios between different elements at the surface. However, it is important to remember that ratios between elements can be misleading. For example, when elements are present in several substances that all change in relation to each other, the ratio might appear to be identical but the actual substance composition is very different.

Despite this, changes of, or differences in, the bacterial cell wall can often beneficially be visualized by plotting ratios between different elements (Busscher et al., 1994; Leone et al.,

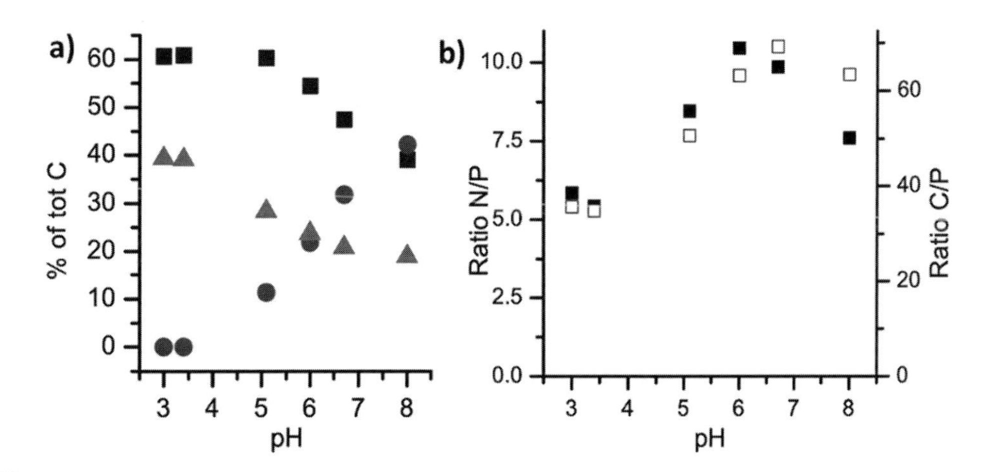

Figure 11.6 Changes in cell wall composition of *B. subtilis* following solution pH (fast-frozen samples). a) Cell wall composition from the multivariate model: squares represent peptide, triangles polysaccharide, and circles lipid. b) Variations in N/P (filled squares) and C/P atomic ratios (empty squares) with pH. (Reprinted with permission from Ramstedt et al 2014, http://pubs.acs.org/doi/full/10.1021/la5002573.)

2006); for example, total N /total P and total C /total P (Figure 11.6b). In Gram-positive bacteria, the nitrogen signal mainly arises from peptide components in the cell, whereas the phosphorus comes mainly from the teichoic acids and lipoteichoic acids. Carbon exists in most substances in the cell wall. Figure 11.6b shows that the amount of N and C relative to phosphorus increases from pH 3 to pH 7 and thereafter decreases, suggesting that there are differences in peptide composition with pH and also in the overall content or composition of (lipo)teichoic acids (Ellwood and Tempest, 1972). These differences can also be due to changes in metabolic secretion processes caused by changes in solution pH and/or compositional rearrangements that hide or expose constituents of the cell wall (Leone et al., 2006; Ramstedt et al., 2014). Changes in cell wall composition can also be illustrated by simply showing the difference in C 1s spectrum between samples (Figure 11.7).

11.2.1.2 Acid–Base Properties

Because XPS is sensitive to the changes in immediate molecular environment of the atom (the nearest neighbors), the nitrogen functional groups at the surface provide the possibility to follow the protonation by varying solution pH. The N 1s spectra for both freeze-dried and fast-frozen samples of *B. subtilis* bacteria exhibit two peaks (Figure 11.8a) (Leone et al., 2006) assigned to amide: neutral amine groups (=NH) at 400.0 eV and protonated amine groups ($=NH_2^+$) at 401.7 eV. The protonated amine found in freeze-dried bacterial samples indicated zwitterionic properties of the dry cell surface, assuming a proton transfer from carboxylate and/or phosphonate surface groups to neutral =NH sites (Leone et al., 2006). This phenomenon is common for amino acids and other organic molecules containing both amino and carboxylic groups in the structure when these substances are in solid state. Fast-frozen bacteria demonstrate a significant decrease in the atomic ratio of $=NH_2^+/=NH$ with

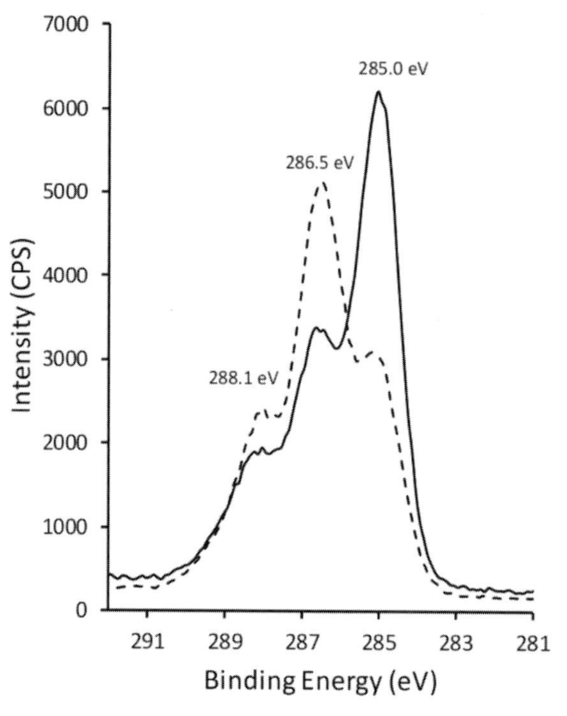

Figure 11.7 C 1s spectra for fast-frozen *B. subtilis* bacterial sample at pH 8 (solid line) and pH 3 (dashed line).

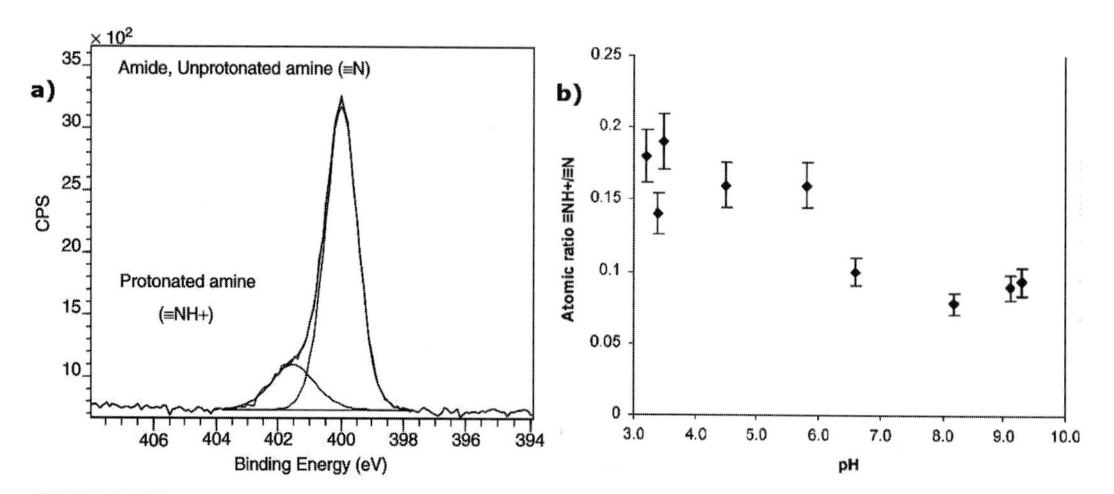

Figure 11.8 a) N 1s spectrum of externally freeze-dried *B. subtilis*, b) atomic ratio of protonated amine group to neutral amine/amide group as a function of pH (fast-frozen samples), in Figure 11.8a labelled as ≡NH+ and ≡N, respectively. (Reprinted with permission from Leone et al., 2006. Copyright © 2006 John Wiley & Sons, Ltd.) Similar behavior has been reported for N-(phosphonomethyl)glycine in solution as well as at goethite and manganite surfaces (Ramstedt et al., 2004), with complete loss of amine group protonation upon the dehydration of fast-frozen samples.

increasing pH (Figure 11.8b) (Leone et al., 2006). The shape of the curve resembles a typical titration curve defining the amine protonation constant to be close to pH 6.5.

11.2.1.3 Cell Surface Charge and Electrolyte Ions

Cryogenic XPS analysis of fast-frozen bacterial pastes makes it possible to follow the development of, and changes in, the bacterial cell surface charge in response to the pH of the surrounding medium and its ionic strength, metal ion loadings, etc. It has previously been shown for aqueous mineral suspensions that the atomic ratio of solution electrolyte ions at the interface (e.g., Na/Cl) is directly related to the charge of mineral particles (Ramstedt et al., 2002; Shchukarev, 2006a; Shchukarev et al., 2004; Shchukarev and Sjöberg, 2005). For a neutral surface, the atomic ratio of the counter-ions should be equal to 1. An excess of anions (atomic ratio <1) implies a positively charged surface, while a predominance of cations (atomic ratio >1) indicates a negative surface charge. At very acidic and very alkaline pH values, other interface reactions, such as ligand exchange and ion pair formation, are also observed and contribute to the accumulation of counter-ions at the mineral interface. However, these latter interfacial processes were not observed at the bacterial surfaces.

In Figure 11.9, the Na/Cl atomic ratios measured for fast-frozen samples prepared from both bacterial suspension and externally freeze-dried bacterial powders (equilibrated in 100 mM NaCl supporting electrolyte) are shown. At very acidic pH (pH 2), the Na/Cl atomic ratio at the surface of externally freeze-dried cells is less than 1, indicating a positive surface charge, presumably due to protonation of surface functional groups, e.g., carboxylic and phosphonate groups. With increased pH, these functional groups became increasingly deprotonated, causing the development of negative charge followed by accumulation of sodium ions at the surface to compensate the charge. Such pH dependence of the Na/Cl atomic ratio is also typically observed for mineral suspensions, where surface functional groups with well-defined protonation constants are responsible for the interaction with electrolyte counter-ions, forming, as a result, an electrical double layer (EDL). Moreover, apparent charge corresponds well to the measured ζ-potential and acid–base equilibrium modeled for externally freeze-dried *B. subtilis* (Leone et al., 2007). However, fast-frozen bacteria from cell culture behave differently (Figure 11.9). The Na/Cl atomic ratio is close to 1 in the pH range of 4–9, which should correspond to a near-neutral cell surface. At pH <4 (higher proton concentration), a significant increase in the Na^+ content is observed at the interface. Zeta potential measurements of vegetative *B. subtilis* cells measured as a function of pH at ionic strength 2 mM show that the cells have a small constant negative charge at pH 4.5–7.5 and become neutral at low pH (Ahimou et al., 2001). An increase in electrolyte concentration is known to decrease ζ-potential, which has also been reported for intact bacterial cells (Lin et al., 2006). Consequently, it can be expected that the surface charge of *B. subtilis* in 100 mM NaCl as supporting electrolyte would be close to zero. The Na/Cl atomic ratio >1 in the very acidic region formally indicates a negative charge of the surface and looks contradictory to the electrophoretic mobility of the cells. Our hypothesis explaining this observed dependence is a metabolic ion exchange reaction $H^+ \leftrightarrow Na^+$ proceeding at the interface, where the excess of sodium ions at the interface at low pH values simply compensates for proton uptake and therefore does not change the overall cell net charge (Leone et al., 2006). This reaction may

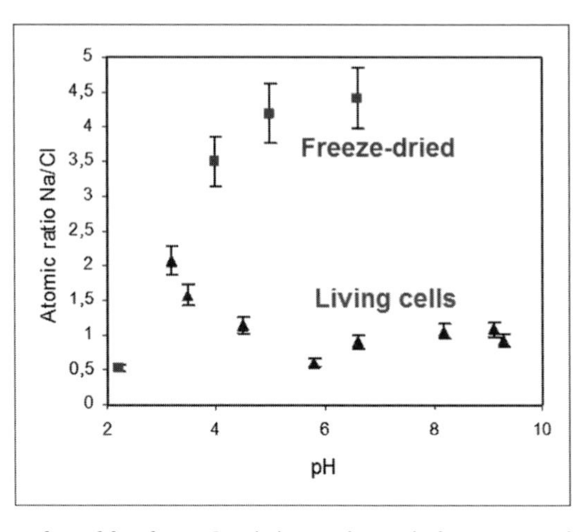

Figure 11.9 Na/Cl atomic ratio at the surface of fast-frozen *B. subtilis* samples made from resuspended externally freeze-dried cells (squares) and from resuspended freshly cultured cells (triangles). (Reprinted with permission from Shchukarev and Ramstedt 2017. Copyright © 2016 John Wiley & Sons, Ltd.)

perhaps also serve as a signaling pathway for cell reorganization to form spores (Ramstedt et al., 2014). The described experimental findings seem to reflect the ability of viable bacteria to regulate the immediate surroundings of the cell and its interfacial pH. Similar processes have previously been suggested in the literature at higher pH whereby bacteria can lower the near-surface pH by excreting protons, making the surface more positively charged during metabolism and increasing its ability to interact with anions (Kemper et al., 1993). The data at alkaline pH shown in Figure 11.9 support this hypothesis, as the Na/Cl ratio equals 1.

In the presence of Zn^{2+} ions, the Na/Cl atomic ratio increases with pH (Figure 11.10) but not as dramatically as observed for externally freeze-dried cells (without Zn^{2+}; Figure 11.9). This development of apparent negative cell surface charge follows zinc accumulation at the surface (Ramstedt et al., 2014). This suggests that Zn^{2+} uptake could lead to loss of the hypothesized proton exchange activity, perhaps due to an interruption of ion exchange channels and/or alterations of surface charges. Another cause could be precipitation of a zinc hydroxide, exposing a net negative surface charge that is consequently compensated for by sodium counter-ions.

11.2.1.4 Metal Adsorption

There are several methods for studying the scavenging of metal ions by bacterial cells, and it has been described that in order to fully account for bacterial metal ion uptake from solution, both biological and chemical processes need to be taken into consideration (Mirimanoff and Wilkinson, 2000). Therefore, a great advantage of the XPS technique is that it only detects metal ions present at the surface of bacteria, i.e., in the outer part of the cell wall. Metal ions that are accumulated inside the cell as a result of detoxification mechanisms will not be seen. Since XPS is sensitive to oxidation states, the fate of redox-active metal ions at the bacterial surface can be followed, e.g., as was done for Cr

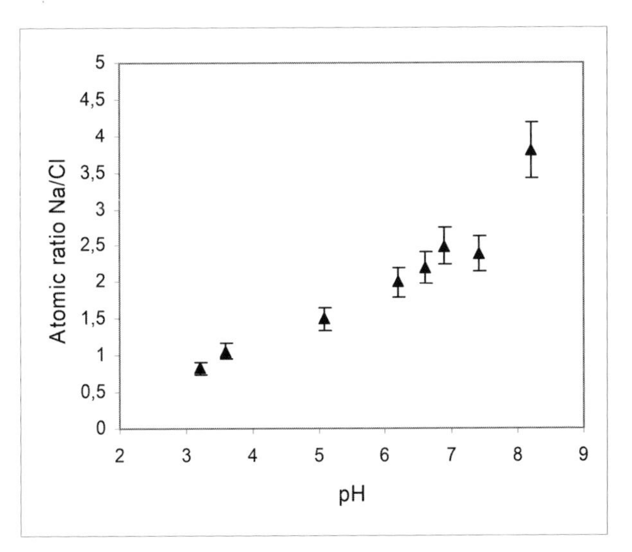

Figure 11.10 Na/Cl atomic ratio at the surface of fast-frozen *B. subtilis* (resuspended freshly cultured cells) exposed to Zn^{2+} ions (Zn^{2+}/dry weight biomass 0.12 mmol/g).

(III) at the surface of *Ochrobactrum anthropi* (Li et al., 2008), Cr(VI) at the surface of *Pseudomonas aeruginosa* (Kang et al., 2015), and U with *Bacillus* sp. (Li et al., 2017). However, despite the advantage of this technique to follow metal adsorption onto bacterial cells, there are only a few studies in the literature where XPS has been used for this purpose. The reason is probably that, traditionally, samples have needed to be dehydrated before analysis, and the technique has only fairly recently started to be used across the biological sciences and in geomicrobiology.

As an example of how metal accumulation can be studied, we will use a study performed in our laboratory that investigated the accumulation of Zn^{2+} onto the surface of Gram-positive *B. subtilis* over a pH range of 3–9 (Ramstedt et al., 2014) (Figure 11.11) using fast-frozen samples and cryo-XPS. It was found that Zn^{2+} was accumulated in a similar way as has been described (from solution data) for other metal ions and bacterial strains, e.g. by the Fein group (Yee and Fein, 2001, 2003). However, in this study, the low quantity of metal at the surface made it impossible to fully characterize the speciation of the adsorbed Zn^{2+} at the surface using XPS alone. Instead, to get a fuller picture, it would have been necessary to combine the XPS data with other complementary analytical techniques such as infrared (IR) spectroscopy and extended X-ray absorption fine structure (EXAFS) spectroscopy. These analysis techniques can provide information about which functional groups were involved in Zn^{2+} binding (possibly seen as a shift in peak positions in IR spectra) and describe the close coordination around the Zn^{2+} ions (from EXAFS). Despite the low concentration, we were able to follow metal accumulation at the surface using cryo-XPS on fast-frozen samples and simultaneously investigate how this accumulation affected the composition of the bacterial cell wall with respect to content of lipid, polysaccharide, and peptide (Ramstedt et al., 2014). It could also be expected that for systems where the metal

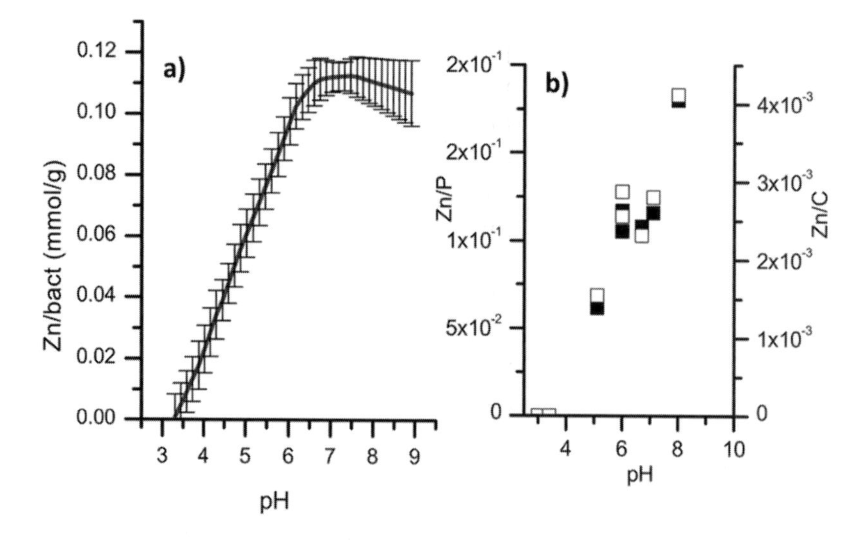

Figure 11.11 a) Adsorption of Zn^{2+} onto the surface of *B. subtilis* as a function of pH. Error bars represent standard deviation in four experiments. b) Zn^{2+} accumulation onto the bacterial surface measured using fast-frozen samples with cryo-XPS. The Zn/P atomic ratio is shown by filled squares and Zn/C with empty squares. A ratio of 0 represents no detected Zn at the surface. (Reprinted with permission from Ramstedt et al., 2014, http://pubs.acs.org/doi/full/10.1021/la5002573.)

uptake is greater, the speciation of the accumulated metal ions could be deduced to a greater extent using XPS analyses.

From EXAFS data of the above-described Zn accumulation at the surface of *B. subtilis*, it was possible to observe a heavy near neighbor to the Zn atom, and the only possible candidate was another Zn atom (unpublished data). Additionally, scattering from a lighter atom was present, possibly due to phosphorus in a higher coordination shell. This led us to the hypothesis that under these conditions a surface precipitate was formed, composed of a mixture of a $ZnHPO_4$ phase and a $Zn(OH)_2$ phase. Thermodynamic calculations showed that $Zn(OH)_2$ and $Zn_3(PO_4)_3$ precipitation is expected to occur at pH >7–8 and >4–6, respectively for the two solid phases (the latter calculated for equimolar content of Zn and phosphate), and this coincides with the observed adsorption edge (Figure 11.11a). Another possible explanation could be multiple interactions between Zn^{2+} and surface phosphonate groups in teichoic acids in the peptidoglycan layer, similar to what has been described for Mn^{2+} in *B. subtilis* (Kern et al., 2010).

11.2.1.5 Surface Precipitation of Secondary Minerals

Metal adsorption can also lead to surface-induced precipitation at the surface of bacterial cells, as was mentioned in the previous section. Several bacterial species have been reported to produce secondary minerals at the surface or in the vicinity of the bacterial cell; for example, manganese oxides formed at the surface of *Bacillus* sp., *Pseudomonas putida* and

Leptothrix sp. (Tebo et al., 2004). These processes should be possible to follow by XPS. Few examples exist in the literature, but one represents the accumulation of metal sulfides from Fe, Cr, Ni, and Mo by *Desulfovibrio* sp. bacteria in neutral and anoxic conditions (Clayton et al., 1994). However, care needs to be taken, since it is not easy to distinguish a bulk precipitate that is pelleted together with the bacteria during sample preparation from a coating at the surface of the bacterial cells. For surface coatings, one can expect a reduction of the signal originating from bacteria when the analysis depth is no longer able to probe into the bacterial cell. Unfortunately, a similar reduction in signal could be obtained in a mixture of bacteria and mineral where the amount of bacteria analyzed is simply diluted by the presence of mineral particles. Consequently, additional analysis techniques, for example optical or electron microscopy techniques (see Chapters 6 and 7), are needed to prove that the secondary minerals formed through the observed microbial processes were really coating the bacterial cells. Ideally, such analyses should be performed before the samples are sent to XPS.

11.2.2 Specific Methodology for Cryogenic XPS: Experimental Protocol and Tips

11.2.2.1 Spectrometer Preparation

To perform cryogenic XPS measurements using the fast-freezing technique with intact hydrated bacterial cell samples, two parts of the electron spectrometer have to be cooled to cryogenic temperature prior to sample loading: the sample manipulator in the analysis chamber and the claw of the sample transfer rod in the sample insertion chamber (entry lock).

11.2.2.2 Fast-freezing Procedure

A bacterial culture with carefully controlled surrounding solution is centrifuged at around 2880 *g* for 10–20 min to allow the bacterial pellet to be separated from the supernatant. The solution should have a minimum of interfering substances, which is why bacteria should be rinsed with, e.g., phosphate buffered saline (PBS) (as long as one is not specifically interested in the P content) or rinsed and equilibrated in saline solution at a controlled pH. Centrifugation is usually performed directly prior to the XPS experiment. After the supernatant is decanted, a small amount of the bacterial pellet is transferred using a pipette or a spatula onto the sample holder, which is immediately loaded into the spectrometer entry lock on the precooled (at least 20 min at 103 K) claw of the sample transfer rod. Decantation, wet pellet sampling, and mounting onto the sample holder are carried out outside the spectrometer while the entry lock is being vented using dry nitrogen ($P > 1$ bar). (Liquid nitrogen cooling of the claw can be stopped during the venting and resumed with the insertion chamber pumping.) After sample loading, the entry lock is closed, and the sample is kept under nitrogen atmosphere for 45 s on the precooled claw to ensure that the sample temperature is low enough for preventing water/ice losses during the chamber evacuation. Within a few seconds after the loading, instantaneous freezing of the bacterial culture can be visually observed. A typical fast-frozen bacterial sample is shown in Figure 11.12a. Sample preparation time (bacterial culture in contact with air outside the

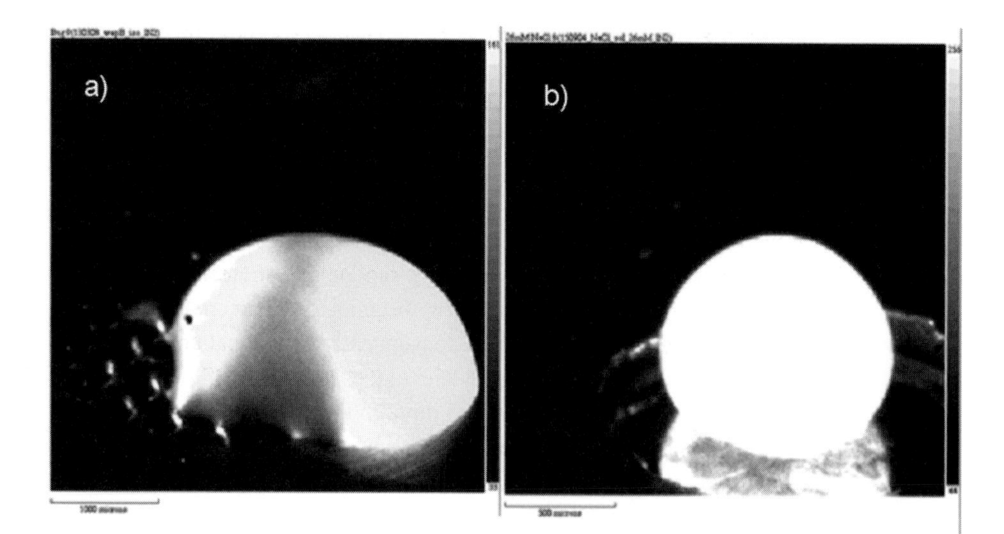

Figure 11.12 a) Image of fast-frozen cell pellet of bacteria on a sample holder with a grid. b) Fast-frozen drop of aqueous solution inside the spectrometer. Scale bar represents 1 mm in image a) and 0.5 mm in image b). Black spot on pellet in image a) is from dust on the lens of the camera and not part of the sample.

spectrometer) usually does not exceed 1 min. After freezing and cool conditioning of the sample and holder, the entry lock is pumped down to 4–5 × 10^{-5} Pa, and thereafter the frozen sample is transferred onto the precooled (115–116 K) sample manipulator in the analysis chamber. Measurements of fast-frozen liquid or noncentrifuged bacterial suspension can be made using the same procedure if the sample holder is precooled together with the claw in the first step. The liquid is then applied using a pipette directly onto the precooled sample holder inside the entry lock and freezes instantaneously (Figure 11.12b).

11.2.2.3 XPS Measurements

XPS spectra of the fast-frozen samples are recorded under liquid nitrogen cooling. Due to the X-ray irradiation and low-energy electrons from the charge-neutralizing system used, the manipulator temperature usually rises from 115 to 120 K during measurement. This temperature increase does not affect the vacuum in the analysis chamber (kept at 3–5 × 10^{-7} Pa), which indicates insignificant ice sublimation from the surface. The latter is evidenced by identical survey spectra routinely taken at the beginning and at the end of a cryo-XPS experiment. To ensure reproducibility of the measurements and uniform chemical composition of the sample, measurements can be replicated at several points of the sample. To observe differences between hydrated and dry surfaces, a fast-frozen sample can be left at the analysis position inside the analysis chamber overnight without cooling. Slow increase in the sample temperature results in gradual sublimation of water and other possible volatile species into the vacuum chamber, and if required, released gas molecules can be monitored by a built-in mass-spectrometer. The next day, compositional and chemical changes of the dehydrated fast-frozen surface can be studied using conventional room-temperature XPS.

11.2.3 Pros and Cons (Pitfalls) of Using the Cryo-XPS Technique in Various Circumstances with Respect to Analyzing Bacterial Cell Surfaces

11.2.3.1 Advantages

An advantage with analyzing bacteria using fast-freezing and cryo-XPS is that the sample preparation is relatively short, especially if compared with traditional microbiological assays involving different types of extraction steps. Bacteria are rinsed in a buffer (or saline) solution with controlled pH and centrifuged, and the pellet is applied to the sample holder directly. This also dramatically shortens the pretreatment compared with the freeze-drying procedures commonly used in conventional XPS. Furthermore, this procedure avoids additional carbon contamination, which has been reported for externally freeze-dried samples for XPS (van der Mei et al., 2000). It can also be assumed that the ubiquitous carbon contamination will be much lower when samples are analyzed as frozen pastes including small amounts of water, since water will create a hydrophilic environment that reduces the surface deposition of hydrophobic surface contaminants. Another advantage is that the sample still maintains water in its inner structures, and therefore it can be expected to be more similar to a hydrated system. Dehydration of bacterial cells can lead to shrinkage as the water is removed and also compositional changes at the surface (Ramstedt et al., 2014). Consequently, in a comparative study examining the composition of the surfaces when cells were externally freeze-dried using different procedures (e.g., poor temperature control), it was observed that the suboptimal procedures produced surfaces with an increased aliphatic carbon content (Genet et al., 2008). What these compositional changes are due to has not been elucidated, but it has been suggested that reorganization of macromolecules, migration of hydrophobic substances within the cell wall, and/or breakage of cells could be responsible for the observed changes (Ramstedt et al., 2014). X-ray degradation of organic substances during analysis can also influence the sample composition and may be an important contribution. However, during analyses under cryogenic temperatures, this degradation should be significantly reduced due to kinetic factors.

Another advantage of XPS is that, since it is a surface-sensitive analysis technique, spectra will not be influenced by metabolic changes inside the bacterial cell. Instead, only alterations occurring at the surface or on surface appendages will be visible in the spectra as long as the cells are intact. This means that, in studies investigating metal adsorption onto bacterial cells, one can obtain information about the amount of metal adsorbed or precipitated at the surface without interference from metal accumulation inside bacterial cells as a result of intracellular bacterial resistance mechanisms. This is a distinct advantage of the XPS technique in comparison with other analytical techniques that probe the entire bacterial cell, such as Fourier-transform infrared spectroscopy (FTIR) (Chapter 12) or EXAFS (Chapter 10).

11.2.3.2 Disadvantages

A disadvantage of using cryo-XPS, or conventional XPS, is that substances dissolved in the surrounding liquid of bacterial suspensions contribute to the XPS spectra, since they are present in the frozen solution layer or are deposited onto the sample during sample pretreatment (freeze-drying). Consequently, if the bacterial cells are rinsed with a buffer to ensure the same pH for all samples, buffer components can be expected to show up in XPS spectra. It is not possible to control the amount of liquid in the pastes, and consequently it is not possible to control the amount of soluble substances from the rinsing solution present in the frozen interface. However, since we analyze samples with only a few solution layers at the interface, the content of substances from, for example, a buffer will be very low. For rich media, on the other hand, the composition of this thin layer could make a substantial contribution to spectra. To avoid "contamination" of the spectra by buffer components, one can carefully adjust pH in saline solutions and remeasure the pH of the solution just before analysis instead of using buffers. Another problem arising from the rinsing procedure is that loosely attached substances of interest at the bacterial surface, for example, exopolymers, can be lost. These problems will also arise in conventional XPS on freeze-dried samples, with one additional challenge arising from the freeze-drying process. When water leaves the solution, the concentration of solutes is increased, which may, e.g., lead to large changes in pH. This means that in addition to solution compounds contributing to the XPS spectra, they can influence or react with the sample surface during dehydration. The larger quantity of solution during freeze-drying is also expected to create a much thicker deposition of solution compounds compared with what can be expected in fast-frozen samples.

A complication of analyzing bacterial cells using the fast-frozen technique is that the sample contains a fair amount of frozen water, which is inserted into the vacuum chamber of the XPS instrument. Consequently, it is of great importance to maintain a low temperature throughout the analysis period in order to obtain good data under ultrahigh vacuum, to avoid sample alteration, and not to expose the vacuum chamber of the XPS to water vapor. For this reason, it is important to supervise cryo-XPS measurements of fast-frozen samples carefully by controlling the vacuum and the cooling. If one wishes to leave the sample inside the spectrometer to reanalyze it at the same position the following day, it is very important to have maximum pumping capacity – for example, this can be obtained on some instruments by opening the gate valve to a sample preparation chamber with its own pumping system – and to switch off the high voltage during overnight dehydration. The key issue here is to minimize the sample amount and thus, the amount of water. Another practical issue for analyzing fast-frozen bacterial pellets relates to the adhesion of the sample to the sample holder. Care should be taken to maximize the contact between the sample and the sample holder, e.g., by placing the wet paste on a mesh on the sample holder (Figure 11.12a). As a positive side effect, the presence of this mesh will also increase the speed of initial fast-freezing. Another precaution can be to construct a "barrier" that holds the sample in place. This could be done using a special holder with a lower center area so that if the sample freezes into a sphere that can "roll," it will not roll off the holder (Figure 11.12b).

An important thing to keep in mind is that bacterial samples often are not uniform in surface morphology. The presence of large appendages such as flagella and pili will influence how "deep" into the bacterial cell the analysis is made. This can be both an advantage and a disadvantage. If we are interested in understanding how the surface of the bacterium interacts with its surroundings, it can be an advantage, since we "see" all the components in XPS spectra. However, if we are interested in studying the composition of just the outer cell wall, surface appendages will "contaminate" the data, and we might not analyze to the same depth of the cell wall between samples that differ in composition of surface appendages. Furthermore, it is important to note that XPS gives an average of the composition in the analysis area (with a Kratos Axis Ultra DLD spectrometer using hybrid mode with "slot" aperture, this corresponds to 0.7 mm × 0.3 mm). Consequently, if the sample is heterogeneous with areas of varying composition or consists of a mixture of bacterial cells with different surface composition, this will not be seen from XPS spectra. In general, XPS imaging is used to analyze heterogeneous samples and obtain information about sample heterogeneities. However, since the current lateral resolution of XPS imaging (for most conventional instruments) is larger than the size of the bacterial cell (~5 μm vs. ~1 μm), heterogeneities in bacterial cell walls or in mixed cultures of bacteria cannot be resolved.

XPS analyses, using both fast-freezing and freeze-drying techniques, give an average composition of the surface, and different data treatment methods have to be used to extract information about lipid, peptide, and polysaccharide content. However, more detailed information about each substance group, for example, what types of lipids are present, cannot be obtained using XPS. If such information is needed, the method has to be complemented with other techniques, for example, time-of-flight secondary ion mass spectrometry (ToF-SIMS) and/or matrix assisted laser desorption/ionization (MALDI). Both these techniques are widely used to determine the detailed surface composition of, for example, biomedical samples but have not been used to a large extent in combination with XPS to investigate bacterial surfaces.

11.3 Conclusions and Outlook

Bacterial surfaces play a significant but underestimated role in biogeochemistry. Consequently, it is of great interest to characterize and better understand processes occurring at and catalyzed by these surfaces. In this chapter we have described several biogeochemical processes that we believe could beneficially be studied using XPS, and especially using the fast-freezing technique and cryo-XPS. Today, commercial XPS instruments can be supplied with liquid nitrogen cooling in both the entry lock and the analysis chamber. Thus, it can be envisaged that the use of fast-freezing and cryo-XPS in the area of biogeochemistry will increase and we will see more studies investigating the role of bacterial surface chemistry in processes such as metal adsorption, surface-induced precipitation of minerals, biomineralization, redox cycling of metals, and mineral dissolution. Also, analyses of samples under reducing conditions are possible as long as the sampling, sample pretreatment, and loading into the spectrometer entry lock are conducted in an inert atmosphere, since the ultrahigh vacuum prevents sample oxidation.

The ability to maintain water at the surface of the bacterial sample enables important features of the solid–solution interface (e.g., the EDL, surface charge, and molecular orientations) to be maintained during analysis. This is of great importance, and consequently, it can be envisaged that the use of cryo-tools for XPS will increase, especially for analyses of samples that are naturally hydrated. With this increase, there will also most probably be further developments of methods for data treatment that can assist in determining complex chemical compositions of a variety of samples. It can also be assumed that the fast-freezing technique will, in the near future, be employed to an increasing extent also for other ultrahigh-vacuum surface analysis techniques such as ToF-SIMS and MALDI-SIMS. This will provide important complementary information that is lacking today. Consequently, a combination of these techniques in their cryogenic versions would enable a much more detailed description of bacterial surfaces in biogeochemical systems.

Developments can, furthermore, be expected both in sample preparation and in XPS instrumentation. Room-temperature analyses that could maintain the solid–solution interface in ultrahigh vacuum would be the ideal way to study this interface. The reported *in situ* environmental cell with thin graphene oxide windows transparent to low-energy photoelectrons (Kolmakov et al., 2011) could be an interesting route to this type of analysis in conventional XPS. This cell would not require cryogenic temperature to maintain the solid–solution interface and may be applicable to studies of biological and geomicrobiological samples. However, X-ray-induced degradation during analysis at room temperature could occur and should therefore be taken into consideration. The current progress in instrumentation (e.g., NanoESCA[1]) is producing instruments that reach a lateral resolution approaching one that allows chemical imaging of bacterial surface heterogeneity or mixed cultures. Using synchrotron excitation, this should already be possible today (ultimate imaging XPS resolution 100 nm) if a similar cryogenic XPS technique to that described in this chapter is applied. For larger types of cells, e.g., blood cells, the first work in this direction has been presented at room temperature (Skallberg et al., 2017). However, for bacterial samples, major issues relating to sample handling, mounting, cooling, exposure to the reduced pressure atmosphere, and sample degradation during the analysis need to be resolved before synchrotron analyses can be made on fast-frozen bacterial samples.

Research areas that probably will remain a challenge in geomicrobiology, even with improvements to today's XPS methodology, are, for example, complex biofilms or the interface between bacterial cells and mineral surfaces. Today, the only way we can investigate such interactions is to break apart sections and analyze their content; for example, removing the bacteria from a surface and investigating the "footprint" left on the mineral or performing some sort of sectioning of biofilms. However, the risk of sample contamination as well as large sample heterogeneities make these types of studies very challenging to interpret and difficult to reproduce, illustrating the importance of a reliable sample preparation protocol. The main issue with these complex samples is that the depth of analysis is small in XPS and we would like to investigate a large volume (as in a biofilm) or a hidden interface. Depth profiling using cluster ion guns could be a way to obtain this type of information while maintaining the chemical composition of the surface under sputtering. However, sampling and sample preparation are, and will remain, the "weakest link in the chain."

[1] www.scientaomicron.com/en/products/nanoesca-/instrument-concept

11.4 References

Ahimou, F., Paquot, M., Jacques, P., Thonart, P. and Rouxhet, P. G. 2001. Influence of electrical properties on the evaluation of the surface hydrophobicity of *Bacillus subtilis*. *J Microbiol Methods*, 45, 119–26.

Beamson, G. and Briggs, D. 1992. *High Resolution XPS of Organic Polymers: the Scienta ESCA300 Database*, Chichester, UK, Wiley.

Beveridge, T. J. 2010. *Bacterial Cells*, Chichester, UK, Wiley.

Boonaert, C. J. and Rouxhet, P. G. 2000. Surface of lactic acid bacteria: relationships between chemical composition and physicochemical properties. *Appl Environ Microbiol*, 66, 2548–54.

Brierley, C. L. and Brierley, J. A. 2013. Progress in bioleaching: part B: applications of microbial processes by the minerals industries. *Appl Microbiol Biotechnol*, 97, 7543–52.

Briggs, D. and Grant, J. T. 2003. *Surface Analysis by Auger and X-ray Photoelectron Spectroscopy*, Trowbridge, UK, The Cromwell Press, IM Publications and SurfaceSpectra Limited.

Burger, K. 1978. Charge correction in XPS-ESCA – bulk solvent as internal standard in study of quick-frozen solutions. *J Electron Spectros Relat Phenomena*, 14, 405–10.

Burger, K. and Fluck, E. 1974. X-ray-photoelectron spectroscopy (ESCA) investigations in coordination chemistry .1. Solvation of SBCL5 studied in quick-frozen solutions. *Inorg Nucl Chem Letters*, 10, 171–7.

Burger, K., Fluck, E., Binder, H. and Varhelyi, C. 1975. X-ray photoelectron-spectroscopy (ESCA) investigations in coordination chemistry .2. Study of outer sphere coordination and hydrogen bridge formation in cobalt(III) and nickel(II) complexes. *J Inorg Nucl Chem*, 37, 55–57.

Burger, K., Tschimarov, F. and Ebel, H. 1977. XPS-ESCA applied to quick-frozen solutions .1. Study of nitrogen-compounds in aqueous-solutions. *Electron Spectros Relat Phenomena*, 10, 461–5.

Busscher, H. J., Bialkowska-Hobrazanska, H., Reid, G., van der Kuijl-Booij, M. and van der Mei, H. C. 1994. Physicochemical characteristics of two pairs of coagulase-negative staphylococcal isolates with different plasmid profiles. *Colloids Surf, B*, 2, 73–82.

Castner, D. G. and Ratner, B. D. 2002. Biomedical surface science: foundations to frontiers. *Surf Sci*, 500, 28–60.

Clayton, C. R., Halada, G. P., Kearns, J. R., Gillow, J. B. and Francis, A. J. 1994. Spectroscopic study of sulfate reducing bacteria-metal ion interactions related to microbiologically influenced corrosion (MIC). *ASTM Spec Tech Publ*, 1232, 141–52.

Delcroix, M. F., Zuyderhoff, E. M., Genet, M. J. and Dupont-Gillain, C. C. 2012. Optimization of cryo-XPS analyses for the study of thin films of a block copolymer (PS-PEO). *Surf Interface Anal*, 44, 175–84.

Dufrêne, Y., Van der Wal, A., Norde, W. and Rouxhet, P. 1997. X-ray photoelectron spectroscopy analysis of whole cells and isolated cell walls of Gram-positive bacteria: comparison with biochemical analysis. *J Bacteriol*, 179, 1023–8.

Ellwood, D. C. and Tempest, D. W. 1972. Influence of culture pH on the content and composition of teichoic acids in the walls of *Bacillus subtilis*. *J Gen Microbiol*, 73, 395–402.

Genet, M. J., Dupont-Gillain, C. C. and Rouxhet, P. G. 2008. *XPS Analysis of Biosystems and Biomaterials*, New York, Springer Science+Business Media.

Kalinowski, B. E., Liermann, L. J., Brantley, S. L., Barnes, A. and Pantano, C. G. 2000. X-ray photoelectron evidence for bacteria-enhanced dissolution of hornblende. *Geochim Cosmochim Acta*, 64, 1331–43.

Kang, C., Wu, P., Li, Y., et al. 2015. Understanding the role of clay minerals in the chromium(VI) bioremoval by *Pseudomonas aeruginosa* CCTCC AB93066 under growth condition: microscopic, spectroscopic and kinetic analysis. *World J Microbiol Biotechnol*, 31, 1765–79.

Kemper, M. A., Urrutia, M. M., Beveridge, T. J., Koch, A. L. and Doyle, R. J. 1993. Proton motive force may regulate cell wall-associated enzymes of *Bacillus subtilis*. *J Bacteriol*, 175, 5690–6.

Kern, T., Giffard, M., Hediger, S., et al. 2010. Dynamics characterization of fully hydrated bacterial cell walls by solid-state NMR: evidence for cooperative binding of metal ions. *J Am Chem Soc*, 132, 10911–19.

Khoshkhoo, M., Dopson, M., Shchukarev, A. and Sandström, Å. 2014. Electrochemical simulation of redox potential development in bioleaching of a pyritic chalcopyrite concentrate. *Hydrometallurgy*, 144–145, 7–14.

Kolmakov, A., Dikin, D. A., Cote, L. J., et al. 2011. Graphene oxide windows for in situ environmental cell photoelectron spectroscopy. *Nat Nanotechnol*, 6, 651–7.

Krumbein, W. E. 1983. Microbial Geochemistry, Oxford, Blackwell Scientific Publications.

Leone, L., Ferri, D., Manfredi, C., et al. 2007. Modeling the acid-base properties of bacterial surfaces: a combined spectroscopic and potentiometric study of the Gram-positive bacterium *Bacillus subtilis*. *Environ Sci Technol*, 41, 6465–71.

Leone, L., Loring, J., Sjöberg, S., Persson, P. and Shchukarev, A. 2006. Surface characterization of the Gram-positive bacteria *Bacillus subtilis* – an XPS study. *Surf Interface Anal*, 38, 202–5.

Li, B., Pan, D., Zheng, J., et al. 2008. Microscopic investigations of the Cr(VI) uptake mechanism of living *Ochrobactrum anthropi*. *Langmuir*, 24, 9630–5.

Li, X., Ding, C., Liao, J., et al. 2017. Microbial reduction of uranium (VI) by Bacillus sp dwc-2: a macroscopic and spectroscopic study. *J Environ Sci (China)*, 53, 9–15.

Lin, D. Q., Zhong, L. N. and Yao, S. J. 2006. Zeta potential as a diagnostic tool to evaluate the biomass electrostatic adhesion during ion-exchange expanded bed application. *Biotechnol Bioeng*, 95, 185–91.

Lukas, J., Sodhi, R. N. S. and Sefton, M. V. 1995. An XPS study of the surface reorientation of statistical methacrylate copolymers. *J Colloid Interface Sci*, 174, 421–7.

Mills, A. L. 1998. The role of bacteria in environmental geochemistry, in Mills, A.L. (ed.) *The Environmental Geochemistry of Mineral Deposits, Part A: processes, Techniques, and Health Issues*, Littleton, CO, USA: Society of Economic Geologists Inc., 125–32.

Mirimanoff, N. and Wilkinson, K. 2000. Regulation of Zn accumulation by a freshwater Gram-positive bacterium (*Rhodococcus opacus*). *Environ Sci Technol*, 34, 616–22.

Moulder, J. F., Stickle, W. F., Sobol, P. E. and Bomben, K. D. 1992. *Handbook of X-ray Photoelectron Spectroscopy*, Eden Prairie, Minnesota, USA, Perkin-Elmer Corporation Physical Electronics Division.

Naumkin, A. V., Kraut-Vass, A., Gaarenstroom, S. W. and Powell, C. J. 2012. *NIST X-ray Photoelectron Spectroscopy Database*, http://srdata.nist.gov/xps/, U.S. Secretary of Commerce on behalf of the United States of America.

Ojeda, J. J., Romero-Gonzalez, M. E., Bachmann, R. T., Edyvean, R. G. J. and Banwart, S. A. 2008. Characterization of the cell surface and cell wall chemistry of drinking water bacteria by combining XPS, FTIR spectroscopy, modeling, and potentiometric titrations. *Langmuir*, 24, 4032–40.

Olson, G. J., Brierley, J. A. and Brierley, C. L. 2003. Bioleaching review part B: progress in bioleaching: applications of microbial processes by the minerals industries. *Appl Microbiol Biotechnol*, 63, 249–57.

Ramstedt, M. 2004. Chemical processes at the water-manganite (gamma-MnOOH) interface, Umeå, Sweden, Umeå University, PhD Thesis.

Ramstedt, M., Leone, L., Persson, P. and Shchukarev, A. 2014. Cell wall composition of *Bacillus subtilis* changes as a function of pH and Zn^{2+} exposure: insights from cryo-XPS measurements. *Langmuir*, 30, 4367–74.

Ramstedt, M., Nakao, R., Wai, S., Uhlin, B. and Boily, J. 2011. Monitoring surface chemical changes in the bacterial cell wall – multivariate analysis of cryo-X-ray photoelectron spectroscopy data. *J Biol Chem*, 286, 12389–96.

Ramstedt, M., Norgren, C., Sheals, J., Shchukarev, A. and Sjoberg, S. 2004. Chemical speciation of N-(phosphonomethyl)glycine in solution and at mineral interfaces. *Surf Interface Anal*, 36, 1074–7.

Ramstedt, M. and Shchukarev, A. 2016. Analysis of bacterial cell surface chemical composition using cryogenic X-ray photoelectron spectroscopy. In: Hong, H.-J. (ed.) *Bacterial Cell Wall Homeostasis: Methods and Protocols*. New York, NY: Springer New York.

Ramstedt, M., Shchukarev, A. V. and Sjoberg, S. 2002. Characterization of hydrous manganite (gamma-MnOOH) surfaces – an XPS study. *Surf Interface Anal*, 34, 632–6.

Ratner, B. D. 1995. Advances in the analysis of surfaces of biomedical interest. *Surf Interface Anal*, 23, 521–8.

Ratner, B. D. and Castner, D. G. 2009. Chapter 3, in *Electron Spectroscopy for Chemical Analysis*, Wiley.

Ratner, B. D., Weathersby, P. K., Hoffman, A. S., Kelly, M. A. and Scharpen, L. H. 1978. Radiation-grafted hydrogels for biomaterial applications as studied by ESCA technique. *J Appl Polym Sci*, 22, 643–64.

Rouxhet, P. and Genet, M. 2011. XPS analysis of bio-organic systems. *Surf Interface Anal*, 43, 1453–70.

Salmeron, M. and Schlögl, R. 2008. Ambient pressure photoelectron spectroscopy: a new tool for surface science and nanotechnology. *Surf Sci Rep*, 63, 169–99.

Seidel, R., Thürmer, S. and Winter, B. 2011. Photoelectron spectroscopy meets aqueous solution: studies from a vacuum liquid microjet. *J Phys Chem Lett*, 2, 633–41.

Shchukarev, A. 2006a. XPS at solid-aqueous solution interface. *Adv Colloid Interface Sci*, 122, 149–57.

Shchukarev, A. 2006b. XPS at solid-solution interface: experimental approaches. *Surf Interface Anal*, 38, 682–5.

Shchukarev, A., Boily, J. F. and Felmy, A. R. 2007. XPS of fast-frozen hematite colloids in NaCl aqueous solutions: I. Evidence for the formation of multiple layers of hydrated sodium and chloride ions induced by the {001} basal plane. *J Phys Chem C*, 111, 18307–16.

Shchukarev, A. and Ramstedt, M. 2017. Cryo-XPS: probing intact interfaces in nature and life. *Surf Interface Anal*, 49, 349–56.

Shchukarev, A., Rosenquist, J. and Sjöberg, S. 2004. XPS study of the silica–water interface. *J Electron Spectros Relat Phenomena*, 137–140, 171–6.

Shchukarev, A. and Sjöberg, S. 2005. XPS with fast-frozen samples: a renewed approach to study the real mineral/solution interface. *Surf Sci*, 584, 106–12.

Shimizu, K., Shchukarev, A. and Boily, J. F. 2011. X-ray photoelectron spectroscopy of fast-frozen hematite colloids in aqueous solutions. 3. Stabilization of ammonium species by surface (hydr)oxo groups. *J Phys Chem C*, 115, 6796–801.

Skallberg, A., Brommesson, C. and Uvdal, K. 2017. Imaging XPS and photoemission electron microscopy; surface chemical mapping and blood cell visualization. *Biointerphases*, 12, 02C408.

Tebo, B. M., Bargar, J. R., Clement, B. G., et al. 2004. Biogenic manganese oxides: properties and mechanisms of formation. *Annu Rev Earth Planet Sci*, 32, 287–328.

van der Mei, H., De Vries, J. and Busscher, H. 2000. X-ray photoelectron spectroscopy for the study of microbial cell surfaces. *Surf Sci Rep*, 39, 3–24.

Vollmer, W., Blanot, D. and De Pedro, M. A. 2008. Peptidoglycan structure and architecture. *FEMS Microbiol Rev*, 32, 149–67.

Winter, B. and Faubel, M. 2006. Photoemission from liquid aqueous solutions. *Chem Rev*, 106, 1176–211.

Yee, N. and Fein, J. 2001. Cd adsorption onto bacterial surfaces: A universal adsorption edge? *Geochim et Cosmochim Acta*, 65, 2037–42.

Yee, N. and Fein, J. 2003. Quantifying metal adsorption onto bacteria mixtures: a test and application of the surface complexation model. *Geomicrobiol J*, 20, 43–60.

Young, K. D. 2010. *Bacterial Cell Wall*, Chichester, Wiley.

12 Applications of Fourier-transform Infrared Spectroscopy in Geomicrobiology

JANICE P. L. KENNEY AND ANDRÁS GORZSÁS

Abstract

Fourier-transform infrared (FTIR) spectroscopy is a technique that measures the molecular-level vibrations in a material, such as a bacterial biofilm, to get a better understanding of the chemistry of the system. This technique is best used to observe changes in a system, e.g., how bacteria protonate and deprotonate as a function of pH or how contaminants sorb to minerals/bacteria, or for tracking the precipitation of a mineral or the breakdown of a contaminant in a system. It can also be used to identify the presence of a specific contaminant in a system, e.g., the presence of bacteria on an antimicrobial surface or the presence of pesticides in water. The following chapter will outline the different ways in which FTIR spectroscopy may be used to analyze a variety of samples in geomicrobiology. The techniques and their applicability are detailed, from individual sample recording (via diffuse reflectance measurements) to continuous monitoring of systems (using attenuated total reflectance measurements) and spatially resolved microspectroscopic analysis (either as imaging or as determining the positions for point sampling in a heterogeneous sample), and a general strategy for data handling is given, including the basics of some multivariate techniques. We will explain how to get the best possible data using each FTIR spectroscopic method, as well as how to best treat your data before analysis. Additionally, this chapter deals with understanding how to identify the representative FTIR bands for bacteria, and how those bands can change as a function of pH.

12.1 Introduction

Fourier-transform infrared (FTIR) spectroscopy is a nondestructive, fast, inexpensive, and sensitive technique that is easy to tailor to needs and even automate, from measurement to data processing. Thus, it is ideally suited to screening a large number of diverse samples with minimum cost, time, and workload while providing important (often key) information regarding the chemical composition of the sample. With microscopic accessories, even the spatial distribution of compounds can be studied and visualized (microspectroscopy) at micrometer level (Amarie et al., 2012). It is therefore not surprising that FTIR (micro) spectroscopy has been successfully applied in the field of geochemistry and geomicrobiology for the compositional analysis of a wide range of samples: from pure mineral identification (Ji et al., 2009; Madejová, 2003) to the identification/characterization of bacteria and their cell walls (Dziuba et al., 2007; Kenney et al., 2018; Leone et al., 2007; Ojeda et al.,

2008), from classifying minerals in sediment layers (Vaculíková and Plevová, 2005) to differentiating soils at different sampling sites (Haberhauer et al., 1998), to name but a few examples from the past decades. Many model systems have been set up in order to study the adhesion of bacteria to minerals (Elzinga et al., 2012; Parikh and Chorover, 2006; Parikh et al., 2014; Rong et al., 2008) and the adsorption or precipitation of metals on bacteria (Amarie et al., 2012; Huang and Liu, 2013; Kenney et al., 2018; Sar et al., 1999; Wei et al., 2011). Furthermore, model systems can be monitored via batch experiments (Alam et al., 2016; Borer et al., 2009; Hagvall et al., 2014; Kenney et al., 2018; Manning and Goldberg, 1996) or even via *in situ* experiments coupled to a titration setup (Elzinga and Kretzschmar, 2013; Kang et al., 2006; Krumina et al., 2016; Ojeda et al., 2008; Peak et al., 1999; Suci et al., 1994).

12.2 Theory of FTIR Spectroscopy

The 400–4000 cm^{-1} region of the electromagnetic spectrum is labelled the mid-infrared (mid-IR) range. When a molecule absorbs mid-IR radiation, it will increase the amplitude of its molecular vibrations (i.e., the displacement of the atoms from their equilibrium position), a process that in quantum mechanical terms equates to the transition of the molecular vibrational to a higher state (e.g., from the ground vibrational state to the first excited state, a so-called fundamental transition). FTIR (micro)spectroscopy is based on observing these transitions, i.e., the changes in the molecular vibrations of the sample. In practice, a detector measures the intensity difference between the original radiation (I0) and the radiation after interaction with the sample (I), and the resulting spectrum is the plot of intensity changes as a function of the energy of the light. Spectra can be plotted as either transmittance (T) or absorbance (Abs), where

$$T = I/I_0 \tag{Eq.12.1}$$

$$Abs = \log_{10}(1/T) \tag{Eq.12.2}$$

Abs spectra contain positive peaks (called "bands" in FTIR spectroscopic terminology) and are more common today, as they offer linear concentration dependency according to the Beer–Lambert law (Griffiths, 2002). Early works (especially before computerized spectrometers and the Fourier-transform revolution) used T more frequently, with negative bands and exponential correlation with concentration (Griffiths, 2002). Although bands are plotted as a function of the energy of the absorbed light, x-axis units are not expressed in terms of energy, or frequency, but uniquely as wavenumbers, with units of cm^{-1} (similarly to Raman spectroscopy) or as wavelength (in older works), with micrometers as units.

FTIR spectroscopy provides information primarily about the molecular bonds, symmetry, and structure of compounds in the sample. Therefore, it is able to inform about functional groups that are available (or that are changing; e.g., Alessi et al., 2014; Kenney et al., 2018; Leone et al., 2007) in a sample as well as to help identify changes in a sample, such as the precipitation of biogenic minerals (Alessi et al., 2014; Kenney et al., 2018; Meyer-Jacob et al., 2014). Generally speaking, FTIR spectra are the average of the entire portion of the sample that interacts with the beam. Therefore, to elucidate the differences between the bulk and the surface, care must be taken in sample preparation and choice of technique (discussed later). While the obtained spectra may not always be diagnostic for a particular compound (especially in a complex mixture like soils), classes of compounds are easy to differentiate (e.g., aromatic or aliphatic compounds, metals, etc.), and the way these classes interact with biogenic material, such as bacteria, can be monitored. This monitoring can be achieved either by the appearance of bands associated with the class of compound interacting with the biogenic material or by changes in the bands of the biogenic material themselves, or both (Alam et al., 2016; Holman et al., 2009; Papageorgiou et al., 2010; Parikh and Chorover, 2006). Most importantly, the overall chemical fingerprint of the sample can be determined sensitively not only qualitatively but also (semi)quantitatively (depending on the controls implemented and the availability of standards/calibration), although some limitations may apply (Reig et al., 2002; Vogel et al., 2008).

Admittedly, the specificity of the technique can be limited, especially in cases with poorly defined sample matrices, consisting of chemically and structurally similar compounds, producing overlapping spectral bands (e.g., differentiating the sorption of a metal to carboxyl groups in polysaccharides vs. lipopolysaccharides). However, this drawback is partly compensated by the speed and cost-effectiveness of the technique, coupled with its nondestructive nature, which allows the use of other, complementary techniques in direct connection to the FTIR (micro)spectroscopic measurements on the exact same samples (e.g., Raman (micro)spectroscopy, nuclear magnetic resonance (NMR), UV-visible (UV-VIS) or near-infrared (NIR) spectroscopy, X-ray absorption spectroscopy, mass spectrometry, etc.). The results from these measurements, some of which may be slower, more laborious, and more expensive, can be used to refine and/or confirm the interpretations of the FTIR techniques whenever specificity is suboptimal. In a similar manner, FTIR (micro)spectroscopy can be used to screen a large number of samples effectively in order to find and select the key samples or processes for further investigations.

This chapter explains the basics of FTIR spectroscopic theory and the methods used to analyze samples in geomicrobiology. For further reading on the subject, please consult the following texts: *Handbook of Vibrational Spectroscopy* (Chalmers and Griffiths, 2002), *Vibrational Spectroscopy for Tissue Analysis* (ur Rehman et al., 2013), and *Infrared and Raman Spectroscopic Imaging* (Salzer and Siesler, 2009). In addition, dedicated book chapters exist in a wide range of scientific disciplines, including detailed protocols (Baker et al., 2014; Gorzsás and Sundberg, 2014). Here, we do not give experimental protocols; instead, we focus on the (bio)geochemical applications of FTIR (micro)spectroscopy. Similarly, details of advanced data mining tools, such as two-dimensional correlation spectroscopy (Noda and Ozaki, 2005) and different multivariate image analysis techniques (Grahn and Geladi, 2007), are beyond the scope of the present chapter.

12.3 Interpretation of FTIR Spectral Bands

The position of the bands in the FTIR spectrum provides qualitative information based on the energy of the vibration. Since many vibrational modes involve the displacement of only a few atoms with the rest of the molecule remaining stationary, the position of a band is characteristic of a set of atoms and bonds (i.e., chemical functional groups). These positions are called "characteristic group frequencies" and refer to a range within which a particular vibration appears. The exact position depends on several factors, including bond strength and the reduced mass (i.e., overall mass relations) of the atoms involved in the molecule. In simple terms, it means that the rest of the molecule in which a particular functional group exists will influence the exact position of the band this functional group produces. Consequently, each infrared-active molecule has a unique FTIR spectroscopic fingerprint, which can be used for identification. However, in case of biological samples, such as bacteria, many of those bands are overlapping, rendering the interpretation of each absorbance peak complex. A compiled table of band assignments for bacterial cells can be found in Table 12.1 (Benning et al., 2004; Jiang et al., 2004; Kong and Yu, 2007; Movasaghi et al., 2008; Ojeda et al., 2008; Omoike and Chorover, 2004; Quilès et al., 2010; Schmitt and Flemming, 1998; Wang et al., 2010). The corresponding spectra are shown in Figure 12.1, with the letter at the peak of each band referring to the corresponding letter in Table 12.1. The spectra in Figure 12.1 were collected using an attenuated total reflectance (ATR) FTIR spectrometer (further described in Section 12.4.2.3) of a wet paste of bacterial cells (*Pseudomonas putida*) grown and washed according to Kenney and Fein (2011), and were measured as a function of the pH of the bacterial suspension. For more information on how the bacteria are prepared for analysis, see the supplementary online material.

Furthermore, since changing the mass of an atom in a molecular vibration has a profound effect on the band position, isotope exchange alters the spectra. One common use of this phenomenon is to employ D_2O instead of H_2O to study proteins and carboxyl groups, allowing clearer observations as the overlapping water bands are shifted from the protein amide I and II bands and the asymmetric $-COO^-$ (Omoike and Chorover, 2004). Similarly, using D_2O can also aid in resolving band overlap problems in the case of oxyanions such as arsenic (As-O) and chromium (Cr-O) adsorption to iron oxides, since their vibrational bands often overlap with the Fe-OH vibrations but not the Fe-OD vibrations (Carabante et al., 2009; Johnston and Chrysochoou, 2012). In general terms, however, positional shifts of vibrational bands do not only originate from isotopic changes but are more often the results of chemical or structural changes. For example, protonation and deprotonation of the hydroxyl functional groups on the bacterial surface can be tracked via lateral band shifts in the νO-H band with increasing pH (Figure 12.1). Lateral band shifts can also occur during protein alpha-helix/beta-sheet structural changes, the formation or breakage of H-bonds, or the adsorption or desorption of a heavier element. For example, when a bacterial cell adsorbs U onto a phosphate functional group on the bacterial cell wall, the phosphate group shifts to a lower wavenumber (also referred to as a redshift) (Alessi et al., 2014;

Table 12.1 The assignments of principal vibrational bands from biological samples, such as bacteria

Band	Wavenumber (cm^{-1})	Band Assignment	Groups
a	3515–3465	νO-H	Hydroxyl groups and water
b	3290–3260	νN-H, νO-H	Proteins (Amide A) and hydroxyl groups
c	3074	νN-H	Proteins (Amide B)
d	2964	νCH_3	Lipids and phospholipids
e	2931	νCH_2	Lipids and phospholipids
f	2886	νCH_3	Lipids and phospholipids
g	2865	νCH_2	Lipids and phospholipids
h	1750–1720	νC=O	Fatty acids and carboxylic groups
i	1643–1630	νC=O, δO-H	Proteins (Amide I), hydroxyl groups, and water
j	1548–1525	δN-H, νC-N	Proteins (Amide II)
k	1457	δCH_3, δCH_2	Lipids and phospholipids
l	1404	$\nu_s COO^-$	Carboxylic groups
m	1346–1276	δCH_3, δCH_2, νC-N, δN-H	Proteins (Amide III), fatty acids
n	1243–1226	ν_{as}P=O, δCOOH	Phosphate/phosphodiesters and carboxylic groups
o	1164	ν_sC-OH, νC-O, ν_{as}CO-O-C, νC-N	Carbohydrates
p	1126	ν_sP-O-C, ν_{as}C-O-C, νP-O-P	Phosphate/phosphodiesters and carbohydrates
q	1084	$\nu_s PO_2^-$	Phosphate/phosphodiesters
r	1064–950	ν_sC-OH, νC-O-C, $\nu_s PO_2^-$, ν_sC-O-P, νC-C, νC-N	Mixed vibrations: phosphates and carbohydrates (e.g., polysaccharides and DNA)

ν: stretching; δ: bending; s: symmetric; as: asymmetric.
Compiled after Schmitt and Flemming, 1998; Benning et al., 2004; Omoike and Chorover, 2004; Kong and Yu, 2007; Movasaghi et al., 2008.

Kenney et al., 2018). Thus, band positions and their changes in FTIR spectra can provide a wealth of qualitative information about the chemical composition of the sample.

The intensity of bands, on the other hand, can provide quantitative information according to the Beer–Lambert law:

$$T = 10^{-\varepsilon lc} \quad \text{(Eq.12.3)}$$

where ε is the molar absorptivity coefficient, l is the pathlength of the light in the sample, and c is the concentration of the absorbing compound. It is expressed in these terms:

$$Abs = \log_{10} T = -\varepsilon lc \quad \text{(Eq.12.4)}$$

meaning that the absorbance (and not the transmittance) is linearly correlated to the concentration. However, due to the sensitivity differences for different molecular vibrations (i.e., differences in molar absorption coefficients ε), direct comparison between band intensities is difficult. For instance, equal intensities of the –O-H and the –S-H bands do not mean that equal concentrations of –O-H and –S-H functional groups are present in the sample (or equal

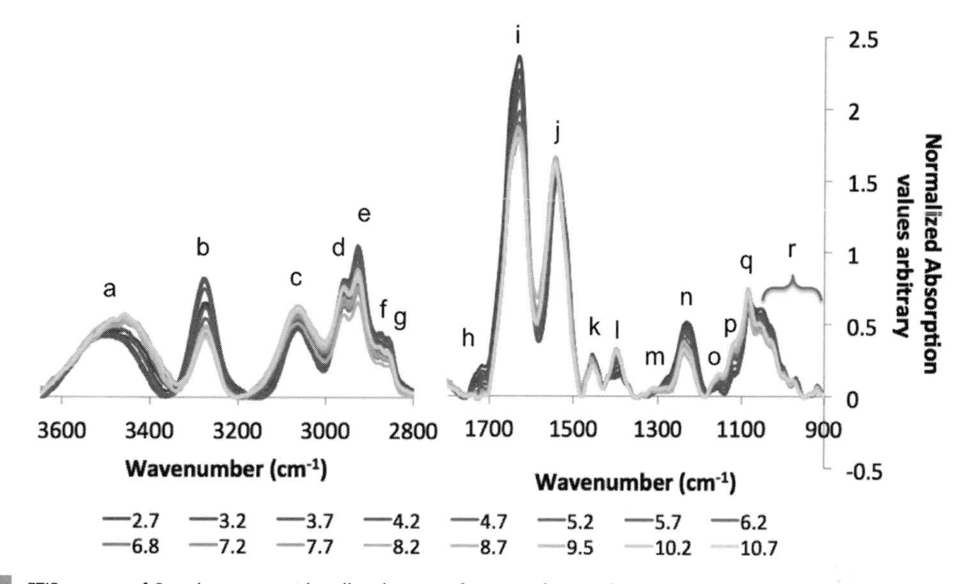

Figure 12.1 FTIR spectra of *Pseudomonas putida* cells taken as a function of pH in the range pH 2.7–10.7. The letters above spectral bands relate to the band letters in Table 12.1.

amounts of alcohols and thiols), because the –O-H vibrations are generally much stronger (having a higher ε) than the corresponding –S-H vibrations. When the same band is monitored, however, the intensity change can be used for determining concentration changes. For example, changes in the concentration of protonated versus deprotonated carboxyl groups can be seen in Figure 12.1 and elsewhere (Alessi et al., 2014; Kenney et al., 2018; Leone et al., 2007), where the νC=O band (or protonated carboxyl) at 1730 cm^{-1} decreases synchronously with an increase in the $\nu_s COO^-$ band (or deprotonated carboxyl) at 1404 cm^{-1}.

In ideal cases with no interfering factors, these intensities (or intensity changes) could even be translated to absolute concentrations via calibrations with known concentrations. In most practical cases, however, only relative concentrations (proportions) can be obtained, as natural and experimental variations often necessitate some kind of data normalization (see Section 12.5.1). This is often labelled as a semi-quantitative analysis. In simple systems, the normalized spectra can either be directly compared between samples, or an average can be taken of several spectra to represent sample groups (technical or natural replicates) and these averages can then be compared. In Figure 12.1, we normalized the spectra based on the δCH_3, δCH_2 band at 1457 cm^{-1}, because that peak changed the least as a function of pH in our samples. Other methods for comparison of bacterial samples have been to normalize based on the phosphate band at ~1084 cm^{-1} (Leone et al., 2007; Alessi et al., 2014) or the amide II band at ~1550 cm^{-1} (Benning et al., 2004; Marcotte et al., 2007). Normalization procedures always need to be justified to make the most scientific sense for a given dataset. It is important to keep in mind that normalizing a data set based on a certain band influences the interpretation of data (i.e., changes are observed with respect to the normalized band).

(Semi-)quantitative comparisons between spectra can be performed by evaluating individual band heights or band areas, or by creating a differential spectrum (i.e., a subtraction of one spectrum from another to identify the major differences between those samples). This approach, however, can oversimplify the problem and lead to erroneous results in complex, heterogeneous systems. One problem is that comparing average spectra ignores the variation between replicates, which could be due to variation in sample pH, preparation time, temperature, collection location, etc. Thus, not only is important information lost concerning the experimental setup, but it can also become difficult to assess whether the difference observed between sample groups is statistically significant or not (i.e., whether the variation within the averaged sample group is larger than the variation between multiple averaged sample groups). Another problem with evaluating single band intensities, however accurately performed, is that this may unnecessarily limit the available information. Due to the frequent overlap of FTIR spectral bands, especially when examining bacterial cells or heterogeneous matrices, individual bands may not be diagnostic. In those cases, spectral comparison should ideally be based on the entire spectrum to avoid missing information or misinterpretation of the results. Different multivariate tools are ideally suited for this task, as they can handle experimental variation without resorting to spectral averaging, while using all the spectral information (i.e., the full spectral range) simultaneously (discussed in Section 12.5.2).

A common mistake when using FTIR spectroscopy to study bacteria is the assumption that a peak observed in the spectrum equates to a functional group available for binding. However, this is only true if the functional group can be proved to be accessible in the system (e.g., by collecting a series of spectra as a function of pH). Bacteria consist of a collection of molecules that produce similar bands, and the infrared light often penetrates more deeply than a single bacterial cell surface, meaning that the entirety of a bacterium is analyzed and averaged to produce the recorded spectra. If, for example, the band in question changes as a function of pH, then it is affected by the changes to the system and therefore accessible as a functional group for binding.

12.4 Measurement Techniques

12.4.1 Transmission

Transmission mode is the most straightforward of the experimental setups: the infrared light is passed through the sample, and the amount of light that is absorbed by the sample is recorded. Naturally, it requires that a reliably measurable fraction of the incoming radiation still reaches the detector; i.e., while the sample should absorb enough infrared light to produce a signal, it should not absorb so much that no light can pass through. In practical terms for operators, no band should ideally have higher Abs values than 0.85–0.95 (i.e., a maximum 85–95% of the light should be absorbed at any given wavenumber in the

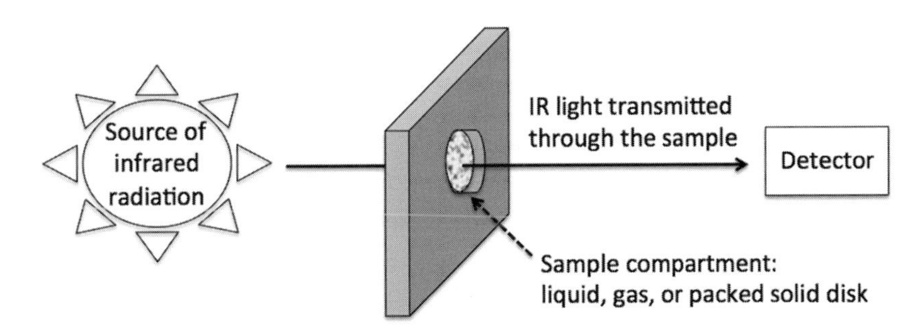

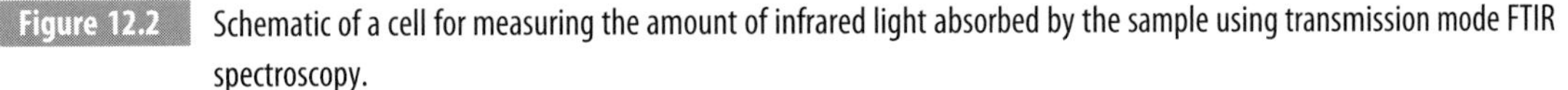
Figure 12.2 Schematic of a cell for measuring the amount of infrared light absorbed by the sample using transmission mode FTIR spectroscopy.

analyzed spectral region) if quantitative analyses are to be performed. The exact value also depends on the linear response range of the detector and the type of analysis/normalization; thus, these values should be considered only as a safe guideline. Samples can be diluted to lower the absorbance, e.g., by lowering the pressure for gases or by homogeneously mixing the sample with infrared-transparent compounds (even in the case of solids, often using KBr for dilution). Alternatively, the pathlength through which the infrared light needs to pass can be lowered (making sure, however, that enough light is still passing through to the detector to produce acceptable signal to noise ratios). For gases and liquids, a smaller pathlength can be achieved by using small-volume/thin sample containers (cuvettes, flow-through cells, etc.) (see Figure 12.2). For solid samples, however, this means either sectioning or the formation of a thin film/layer of the sample on an infrared-transparent carrier (typically CaF_2, BaF_2, NaCl, or KBr).

Accordingly, the main drawback of transmission FTIR spectroscopic measurements is the high absorbance of light by the measured samples, which may require lengthy initial measurement optimization and extensive sample preparation. Aqueous samples will be exceptionally difficult to measure, since water is very infrared active and will therefore absorb a significant amount of the light. In certain cases, the samples simply cannot be optimally prepared for transmission measurements in a cost- and time-effective manner with high reproducibility. However, if solid samples can be optimized for transmission measurements by means of, e.g., thin sections, they are directly usable for microspectroscopy as well, which provides the added benefit of spatially resolved information in cases of heterogeneous samples (see Section 12.4.3).

The key advantages of transmission measurements include the easy setup, resulting in high-throughput measurements (once optimal concentration/pathlength has been achieved during sample preparation) and good control over the sample quantities analyzed, and thus, straightforward quantification via the Beer–Lambert law with high reproducibility.

Transmission FTIR spectroscopy has been used to look at some important questions in geomicrobiology, including biomineralization (Madejová, 2003), bacterial identification (Dziuba et al., 2007), understanding how nutrient source changes microbial surfaces (Liu et al., 2016), and organic sorption or biodegradation (Alam et al., 2016; Bonhomme et al., 2003).

12.4.2 Reflection

An alternative to transmitting the light through the sample is to utilize different forms of reflection and thereby tailor the measurement modes to the samples at hand, rather than trying to adjust the sample for transmission measurements. The three main techniques in this category are diffuse reflectance, ATR, and grazing angle reflectance measurements. While the experimental setup is different for each, the appearance of spectra may be surprisingly similar to those obtained by simple absorption techniques (such as transmission mode measurements). This is particularly true for diffuse reflectance measurements. ATR measurements, however, produce spectra in which the bands are often shifted somewhat compared with pure absorption methods, and, more importantly, the relative intensities of spectral bands are markedly different (Bertie, 2002). Thus, ATR spectra cannot be directly compared with transmission (or even diffuse reflectance) spectra without correction.

12.4.2.1 Diffuse Reflectance

The common acronym for this technique is DRIFTS, which stands for diffuse reflectance infrared Fourier-transform spectroscopy. DRIFTS is particularly well suited to the analysis of powders and has gained popularity in recent years, replacing KBr pelleting techniques. This is mostly due to easier and more reproducible sample preparation compared with KBr pellets. However, DRIFTS works on entirely different optophysical bases than KBr pelleting. While KBr pelleting is a transmission technique, DRIFTS uses the diffusely reflected (scattered) light from the powder particles while eliminating the contribution of specular reflection from the particle surfaces (as the latter contains no chemical information) (see Figure 12.3). The resulting spectra, however, resemble transmission spectra more than reflectance spectra (lacking the positional shifts and relative intensity differences observed for ATR) (Bertie, 2002).

The major advantages of DRIFTS include its high sensitivity, speed, cost, and easy automation, making it ideal for high-throughput screening. While sample preparation involves homogenization and dilution with an infrared-transparent solid (often KBr for practical reasons), it is relatively straightforward and can be standardized. The main steps are manual grinding (with a mortar and pestle) or optional premilling (for samples too hard

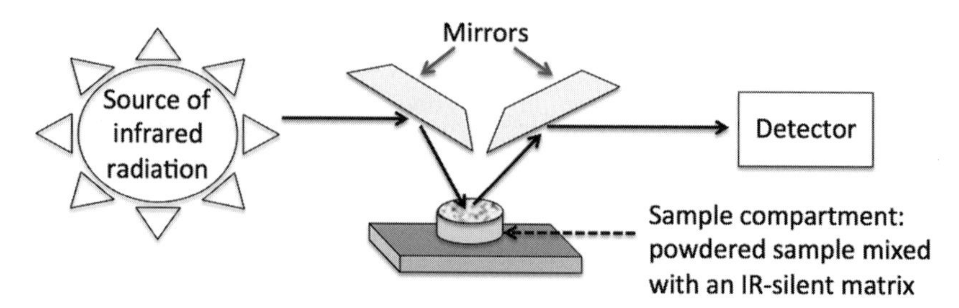

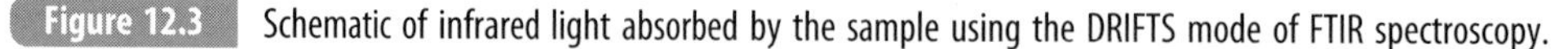
Figure 12.3 Schematic of infrared light absorbed by the sample using the DRIFTS mode of FTIR spectroscopy.

to manually grind) together with the infrared-transparent diluters to create a mixture. However, therein lies the first drawback of the technique: as a consequence of the addition of the diluter, the sample does not remain completely intact for further analysis. In addition, homogenization must be performed in a standardized way, since it can have profound effects on the final spectra. First, particle size affects the results (potentially increasing the contribution of specular reflection instead of diffuse reflection, or generally lowering signal to noise ratios). Perhaps more importantly, if intense milling/grinding is required, it may change the composition/structure of the chemical compounds in the sample – their degree of polymerization and crystallinity – and the generated heat can induce reactions (in extreme cases, even burning during milling). It should be mentioned that while the sensitivity of DRIFTS is certainly an advantage, it could also be considered a drawback: normal variation (due to sampling or biodiversity) may be substantial in the spectra. Such variation may necessitate rigorous spectral processing and powerful multivariate analysis before results can be interpreted.

DRIFTS has been used to study a range of geomicrobiological samples (which have first been dried and mixed with the infrared-silent material), such as bacterial or organic sorption to minerals, soils, or sediment (Gallé et al., 2004; Kang and Xing, 2007; Rosen and Persson, 2006; Rosen et al., 2010; Vogel et al., 2008), metal sorption to bacteria (Kamnev et al., 2006), and biomineralization (Shopska et al., 2013).

12.4.2.2 Grazing Angle Reflectance

In grazing angle reflectance measurements, a thin film of the sample is either applied or grown onto a highly reflective surface (e.g., gold, aluminum, silica, or stainless steel). The infrared light is projected onto this thin film of sample at a very low angle, ensuring a long path in the sample. The reflected infrared light that reaches the detector is analyzed to determine how much light is absorbed by the sample (Figure 12.4). The benefit of this technique is that very small quantities of samples can be analyzed. In geomicrobiology, this could be one of the drawbacks: since the light must be able to reflect off the reflective surface, minerals or biofilms must be thin enough to allow light to pass through to the reflective surface. This may

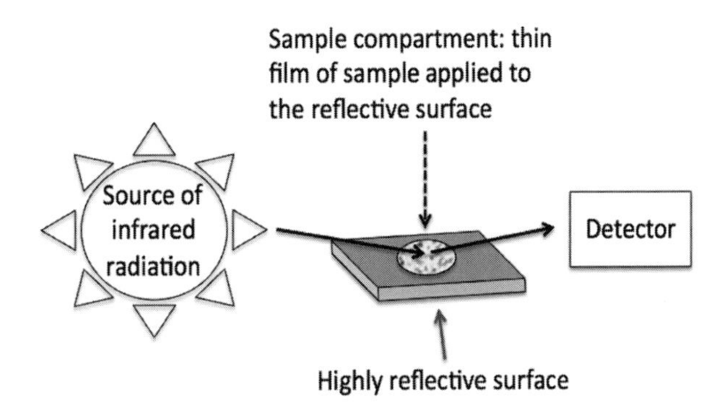

Figure 12.4 Schematic showing infrared light absorbed by the sample using the grazing angle reflectance mode of FTIR spectroscopy.

cause problems with analyzing environmental samples, as they are frequently too thick for this purpose in their natural state. Instead, samples are often applied to the surface as a thin coating. The easiest way to achieve this coating is to have the samples suspended in water (e.g., an aqueous suspension of bacteria or mineral powders) at low concentration and then drop an aliquot of the suspension onto the reflective sample holder, and allow it to dry completely. Samples can also be applied to the surface using a brush or a spin coater, or by dipping a coupon sample holder into a solution containing the sample, where the sample will coat the holder, or bacteria can grow onto the surface for analysis. Grazing angle FTIR spectroscopy is often used in medical biology when testing antimicrobial agents (Mohorčič et al., 2010; Yao et al., 2012), validating cleaning methods (Hamilton et al., 2005), and examining biofilm formation in geomicrobiology (Tapper, 1998).

12.4.2.3 Attenuated Total Reflectance

In ATR measurements, the sample is placed on an infrared-transparent crystal with high refractive index, called the internal reflection element (IRE). IREs are available in different sizes, shapes, types (offering single or multiple reflections), and materials (e.g., ZnSe, Ge, and diamond) to suit the measurement in terms of pH, pressure, and temperature sensitivity, available spectral regions (cutoff), and price. Most importantly, IREs are shaped in such a way that when the infrared radiation is totally internally reflected, an evanescent wave is created at the interface between the IRE and the sample, which then penetrates the sample and interacts with it, resulting in attenuation of the light (hence the name) (Figure 12.5). Since the evanescent wave rapidly (exponentially) decreases in intensity with the distance from the IRE, ATR is a surface-sensitive technique (i.e., the infrared light only probes a short distance above the IRE). The exact penetration depth depends on a number of factors, including the angle of incidence, the refractive index differences between the sample and

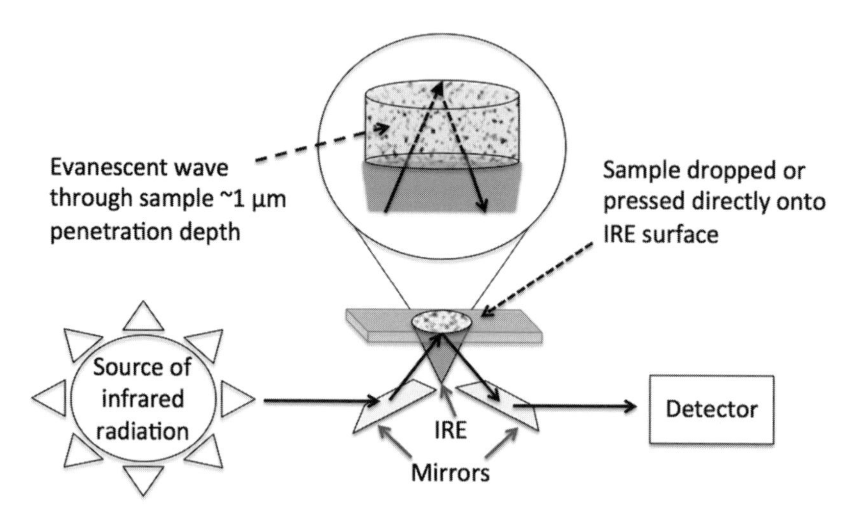

Figure 12.5 Schematic showing infrared light absorbed by the sample using ATR FTIR spectroscopy.

the IRE, and the wavelength of the light, but typically falls between 0.2 and 5 μm, with the potential for infrared light passing further still into the sample (Chan and Kazarian, 2007; Mirabella, 1983). Consequently, the sample is required to make good contact with the IRE. While forming a good contact is not a problem for gases and most liquids that wet the surface of the IRE, solid samples need to be deposited directly, by, e.g., drying an aqueous suspension of the sample, onto the IRE, or pressure must be applied to force the sample against the IRE. In the latter case, care must be taken that the IRE is not damaged by this pressure. In the case of minerals, good contact is generally hard to achieve unless the mineral is in powder form. In the case of bacteria, there are several ways of deposition (some may apply to mineral powders as well). First, a dry powder (e.g., freeze-dried for bacterial cells, dried and crushed for minerals) can be placed directly onto the IRE and pressure applied. Alternatively, an aqueous suspension of bacterial cells and/or powdered mineral can be prepared and allowed to dry on the surface of the crystal, either in the air (slower) or under N_2 flow (much faster). A drawback of the deposition technique, however, is that the drying of the environmental system in question may effectively change the system. If the system needs to remain hydrated, the best way to apply the sample is as a wet paste pressed against the IRE (Figure 12.5), which is one of the better ways to measure bacteria in environmental and laboratory-based systems. In this case, samples (which could include bacteria or powdered minerals) are centrifuged to remove as much of the free liquid supernatant as possible. This supernatant is then collected and run as a sample alongside the sample paste that was wetted with that fluid. Care must be taken between sample and supernatant measurements to avoid carryover contamination. In the case of suspected contamination, a new background should be taken of the empty IRE. This way of measuring both sample and supernatant allows the subtraction of the (often overwhelmingly large) signal from water (~1650 cm^{-1}) that generally plagues the FTIR spectra of wet samples. This subtraction process is discussed further in Section 12.5.1.

ATR measurements are also extremely well suited to continuous, *in situ* measurements in the form of either a flow-through arrangement (Loring et al., 2009; Ojeda et al., 2008) or a one-pot setup (Gillgren and Gorzsás, 2016; Hagvall et al., 2015; Krumina et al., 2016). In these cases, reactions can be followed in real time (with the temporal resolution often limited only by the recording speed of the spectrometer, which can exceed 100 spectra/s), as a function of a perturbation (pH change, addition of chemicals, addition of bacteria, etc.) and can be completely automated.

The advantages of working with wet pastes or continuous ATR FTIR spectroscopy is that the technique is noninvasive and nondestructive and allows rapid monitoring capacity. There is no sample preparation (no drying, milling, or mixing, as opposed to ATR measurements of solid powders or DRIFTS). Since the technique only probes the few micrometers directly in contact with the IRE, it renders irrelevant the contribution of anything beyond that zone. Thus, it allows direct measurements in aqueous solutions or suspensions by eliminating the problem of high water absorbance without the need for sectioning or thin-path cuvettes (as in transmission mode).

However, this surface sensitivity of the ATR technique may also be considered as a limitation in certain applications, where the surface properties are known to differ from the bulk, or in heterogeneous systems where signals from multiple processes (sorption to minerals or to bacteria) would be averaged into one sample spectrum. Moreover, ATR signals are usually weaker than the corresponding DRIFTS signals; thus, ATR detection limits are generally lower. To make things even more complicated, the penetration depth is wavelength dependent and thus nonuniform across the spectral range. One manifestation of this is the fact that band intensities are proportionally higher in the low-wavenumber region of the spectra and decrease gradually towards higher wavenumbers. Together with band positional shifts, these features make direct comparison of ATR FTIR spectra with other kinds of FTIR spectra difficult.

ATR FTIR spectroscopy has been used to analyze a variety of topics in geomicrobiology, including bacterial adhesion to minerals (Elzinga et al., 2012; Ojeda et al., 2008; Omoike et al., 2004; Parikh and Chorover, 2006; Parikh et al., 2014), adsorption of contaminants to bacteria and organic matter (Hagvall et al., 2014; Kenney et al., 2018; Poggenburg et al., 2018; Schmitt et al., 1995; Ueshima et al., 2008), effects of environmental stress on bacteria (Kamnev, 2008), and organic sorption to minerals, soils, or biochar (Kang and Xing, 2007; Kubicki et al., 1997; Madejová, 2003; Poggenburg et al., 2018; Yoon et al., 2004).

12.4.3 FTIR Microspectroscopy

Attaching a microscopy accessory to a FTIR spectrometer enables spatially resolved sampling and chemical profiling in heterogeneous samples (FTIR microspectroscopy). In the simplest case, the microscopy accessory is only used for precise sampling, i.e., making sure that spectra are only collected from specific areas of interest. For example, in a heterogeneous soil sample containing, e.g., minerals, fungus, a microbial matt, and detritus, where the bulk techniques described above would only provide an average of all components in the system, FTIR microspectroscopy enables the recording of individual spectra for each component, free from the contributions of the others, provided that a representative (i.e., "clean") spot can be localized for each. Alternatively, hyperspectral data can be collected over a defined sample area with a certain lateral resolution, from which chemical maps can be derived (imaging). Similarly to bulk techniques, several measurement modes are available for FTIR microspectroscopy (e.g., transmission, reflection, grazing angle, and ATR). In ATR and grazing angle modes, the sample is both seen and measured via an ATR or grazing angle objective, respectively, instead of the standard objective of the infrared microscope. While the ATR technique provides the highest spatial resolution (as a result of the high numerical aperture of the objective, partly due to the high refractive index of the IRE), it is often impractical in biogeochemical applications due to the requirement for perfect contact between the microscope ATR objective and the sample. While this is easily achieved for liquids, there is usually no need for spatially resolved measurements in such cases. In the case of solids, the force required for the contact between the IRE and the sample often distorts morphological features and may even damage certain samples. In the case of imaging, this practically negates the potential gain in lateral resolution via the ATR objective. In the case of point sampling, as opposed to imaging, frequent and rigorous cleaning is required between measurement positions even within the same sample to avoid carryover contamination, making it less practical.

Reflection mode FTIR microspectroscopy is similar to grazing angle measurements in that the sample is mounted on a carrier with highly reflective surface (e.g., gold or aluminum mirrors). A difference, however, is that the infrared light passes through the sample perpendicularly, not at a low angle, before being reflected back by the mirror and passing through the same path in the sample again in the reverse direction. Because the light passes through the sample twice before reaching the detector, the effective sample thickness (and thereby concentration) is doubled. However, low absorption is seldom a problem in biogeochemical samples. On the contrary, samples are generally too absorbing (especially when wet). Therefore, this increased thickness via the increased pathlengths of reflection and grazing angle measurements is often undesirable. Since reflection measurements are essentially identical to transmission measurements in every other sense, they will not be discussed separately here. Instead, we will focus on transmission FTIR microspectroscopy from here on.

In transmission mode, the infrared light passes through the sample much like the visible light does in standard microscopy setups, making this mode the most similar to visible microscopy. However, there are important differences, both theoretical and practical. First, while normal microscopes use lenses to focus the light, infrared microscopes use mirrors. This means that the magnification of FTIR microscopes cannot be as easily changed as in standard microscopes. In addition, standard glass microscopy slides cannot be used for FTIR microspectroscopy, as they absorb infrared light. Instead, samples should be mounted on infrared-transparent slides made of, e.g., BaF_2, CaF_2, ZnSe, or NaCl, which require more careful planning and handling, as they can be (more or less) water soluble, brittle, and expensive.

Another important difference lies in the available lateral resolution by table-top FTIR microscopes using standard mid-infrared light sources. FTIR microscopes using synchrotron radiation can focus the light on small areas and still achieve good signal strengths due to the high brilliance of the synchrotron infrared light. Synchrotron-assisted FTIR microspectroscopy has been used in geomicrobiology to image changes in biofilms after addition of pharmaceuticals (Holman et al., 2009), to examine the degradation of hydrocarbons in a microbial floc (Hazen et al., 2010), and to understand bacterial chromate reduction (Holman et al., 1999). However, synchrotron-based setups are understandably rare and thus will not be discussed further here. The maximum achievable lateral resolution using conventional mid-infrared sources depends on the technique, the sample, and even the detector (i.e., single-element (SE) vs. more advanced focal plane array (FPA) detectors). Both SE and FPA detectors are HgCdTe (MCT) devices, requiring liquid nitrogen cooling. The SE detector records a single spectrum of the sampled area. This spectrum therefore represents the average chemical composition of that area. Thus, the lateral resolution of SE detectors in practice means the size of the sampled area, which is determined by the applied knife-edge apertures. The exact size of the smallest applicable aperture is determined by the physical, chemical, and optical properties of the sample, as well as by the quality of the infrared source. However, as a general guideline, the smallest area that can safely and routinely be measured this way is ca. 50 μm × 50 μm, which is often impractical in geomicrobiology. Making the aperture size smaller results in increasing spectral distortions and decreased signal to noise ratio due to diffraction and limitations in light intensity.

In contrast to the SE detector, FPA detectors consist of arrays of miniature detector elements in the focal plane of the infrared radiation, each recording simultaneously and independently, similarly to pixels building up an image. This type of measurement is therefore called imaging, as opposed to mapping, which refers to techniques by which the spectra are recorded consecutively as an area is scanned. The simultaneous recording of all "pixels" improves sampling speed considerably, and since a unique spectrum is recorded for each "pixel" of the image, there is no need for apertures to improve lateral resolution or sampling precision, which results in better-quality spectra. FPAs are most often linear (e.g., 1×32) or square (e.g., 64×64), and their size determines the number of spectra recorded simultaneously (e.g., $64 \times 64 = 4\ 096$ spectra per image) as well as the area covered in a single image (since the physical dimensions of the individual detector elements and the magnification of the lens are fixed). The view field of a typical FPA detector with 64×64 detector elements is ca. 175 µm × 175 µm, meaning that the physical size of a detector element is about 2.7 µm × 2.7 µm. In reality, however, lateral resolution is considerably lower than the physical detector element size, due to diffraction limits based on the wavelength of the infrared light (Lasch and Naumann, 2006). Using the Rayleigh criterion, the minimum distance between features that can be resolved (Δx) can be described as $\Delta x \geq 0.61 l/NA$, where l is the wavelength and NA is the numerical aperture of the objective. Assuming a standard 20× Cassegrain-type objective with NA = 0.3, Δx falls in the range of 10–20 µm in the fingerprint region of the spectra (1500–500 cm^{-1}) and is only about 5 µm even at 4 000 cm^{-1}. It is worth emphasizing again that the practical lateral resolution is wavelength dependent and worse than what the physical detector sizes would allow. While ATR objectives can have considerably higher NA values and thus better lateral resolution, they are often impractical in biogeochemical applications, as explained earlier. The range of achievable lateral resolution with standard light sources is a considerable limitation in geomicrobiology, as it limits usability in samples with fine gradients (e.g., between minerals and organics/metals) and makes identification of differences around/within a single bacterial cell or small grouping of cells impossible at the time of writing.

Other drawbacks of FTIR microspectroscopy include laborious sample preparation (as results are greatly affected by sample quality) and measurement optimization, sensitivity to external disturbances (including vibrations in general, especially in the case of FPA detectors, thus requiring vibration-free environments), and more complicated and demanding data analysis processes (higher degree of experimental variation in the results, large data files to be processed, more complex spectral treatments, etc.). While recording times were considered a drawback for early FPA setups, improved electronics and computer systems now allow the recording of high-quality FPA images containing thousands of spectra in less than a minute.

The main advantage of FTIR microspectroscopy is that it allows specific probing in different areas of heterogeneous samples and can reveal the distribution of chemical compositional differences in both space and time (e.g., repeated recording of the same sample area during a process, such as fungal growth on mineral surfaces, etc.).

12.5 Data Handling

FTIR (micro)spectroscopic data have traditionally been evaluated by comparing the intensities (including band ratios) or positions (including band shifts) of certain bands among the collected spectra. This kind of analysis is easy to perform, and the results can

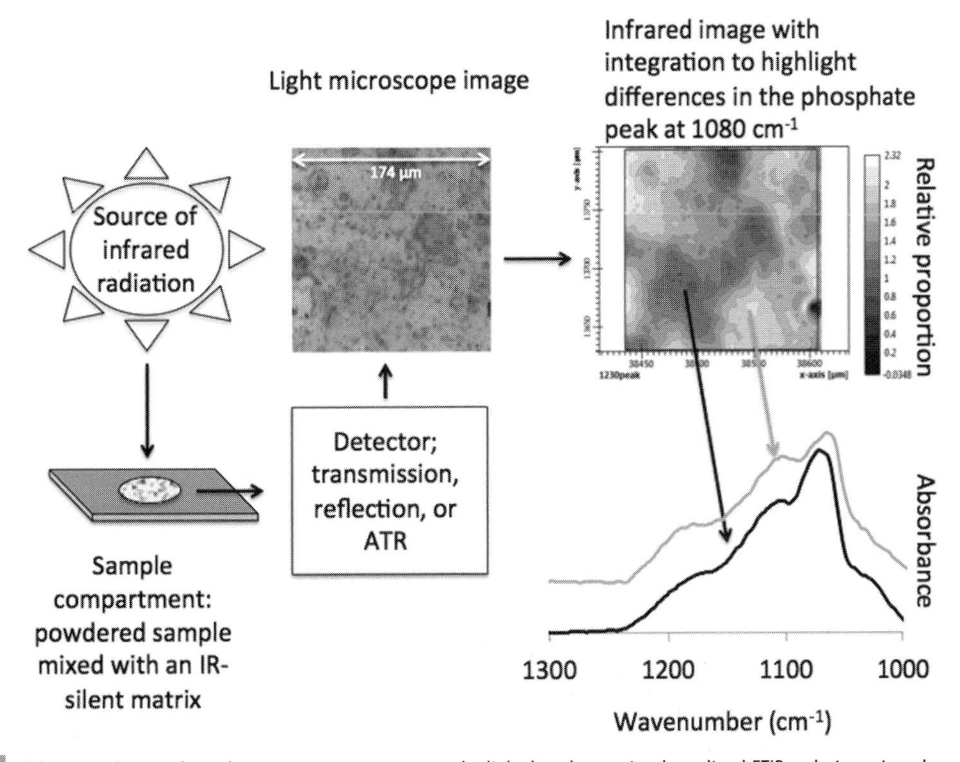

Figure 12.6 Schematic showing how the microscopy accessory can be linked to the previously outlined FTIR techniques in order to obtain visual and chemical data of the system.

be easy to interpret and visualize (in the form of "heat maps" in the case of image data; see the infrared image with phosphate integration in Figure 12.6). However, these results can often be suboptimal, especially with insufficient pretreatment data, since raw data can contain experimental artifacts that have the potential to plague or cause misleading results (e.g., due to light scattering from uneven samples, varying percentages of water in a wet paste, or variation in the amount of material in the path of the infrared light). While some of these problems could, in theory, be eliminated or minimized by careful spectral pretreatment and individual processing of each spectrum (which are potentially impractical for large data sets or images), others are inherent in the data. These inherent problems include the nonspecificity of single bands, natural biological variations, and complex spectral features (e.g., overlapping and shifting bands, etc.) due to complex chemical matrices. In this section, we will primarily deal with methods that enhance the information content of the collected spectra while minimizing noise (spectral pretreatment) and the effective mining of this information (spectral analysis). Guidelines to the interpretation of FTIR spectroscopic data and dealing with spectroscopic band assignments can be found in Section 12.3.

To illustrate the data analysis procedure, including pretreatment steps, we use an example system of monitoring bacterial cells (*Pseudomonas putida*) as a function of pH of the bacterial suspension. The spectra shown here were collected from a wet paste using an ATR FTIR spectrometer and trimmed to the 1800–900 cm^{-1} spectral region (the "fingerprint" region) for the final data analysis, as presented in Figure 12.1.

12.5.1 Preprocessing

In order to process wet pastes or aqueous samples using ATR FTIR spectroscopy, the contribution of water must be eliminated from the spectra. This can be achieved by subtracting the spectrum of the aqueous matrix from the spectrum of the sample, either post measurement (using, e.g., Excel), or directly during measurement by using the aqueous matrix for background measurement before measuring the sample spectrum. Using the aqueous matrix as background can provide better-resolved data if the experiment is conducted *in situ* at the spectrometer. However, for individual samples collected *ex situ*, it is best to subtract matrix- and pH-matched aqueous spectra (e.g., un-spiked groundwater, supernatant from aqueous suspension, matrix-matched background electrolyte solution) recorded separately. An example of subtraction of the supernatant from the wet bacterial paste at pH 2.7 from Figure 12.1 is shown in Figure 12.7. The subtraction was done by aligning the broad band at ~3350 cm^{-1}, since that band is host to major O-H vibrations from water. Alternatively, the band at ~1640 cm^{-1} can be used for aligning, as it allows better resolution in the "fingerprint" region of the spectrum. However, it is only possible if the samples do not have overlapping bands at that position, which is not the case for bacteria, since the amide I band of proteins severely overlaps with the water band in this region.

The principal steps for pretreatment are baseline correction and normalization, followed by optional smoothing. There are many techniques available for baseline correction, and practically all software that is used to collect FTIR spectra (e.g., OPUS, OMNIC) is equipped with at least one (often several) baseline correction functions. These predominantly operate with fixed baseline shapes based on either linear or polynomial functions to calculate the baseline between fixed (even user-defined) points in the spectra. Nevertheless,

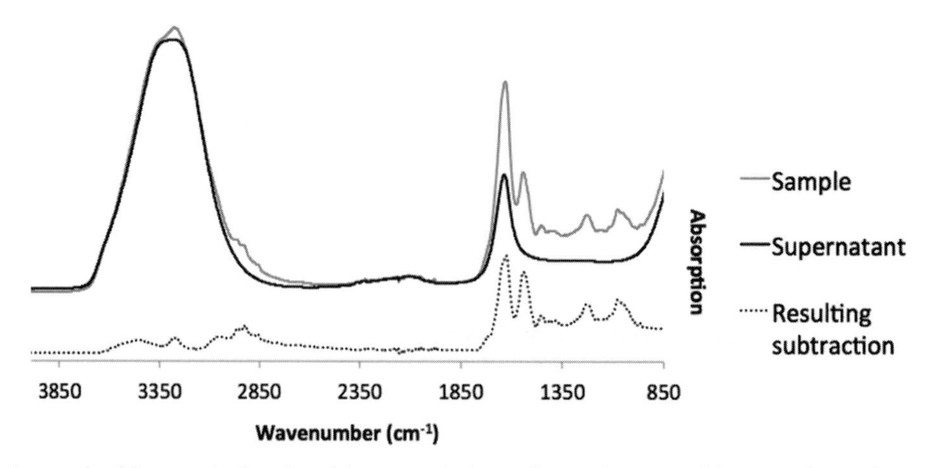

Figure 12.7 An example of the spectral subtraction of the aqueous background matrix (supernatant) from a wet bacterial paste.

such baselines are very often suboptimal and especially ill suited to series of spectra where changes among the spectra are significant (including spectra from different pixels of an image). They can also be hard to reproduce, since the baseline can depend greatly on the user-defined points in the spectrum. Thus, the best method for baseline correction of FTIR spectra is based on asymmetric least squares (AsLS) fitting as described by Eilers (2004), which eliminates all these problems. A detailed description of the use of AsLS baseline correction for vibrational spectra is available in the literature (Felten et al., 2015) and will not be elaborated further here. A free, open-source script with an easy-to-use graphical user interface, complete with example data sets and user manual, is also available online (www.kbc.umu.se/english/visp/download-visp/).

The next step in spectra treatment, traditionally, is normalization. Strictly speaking, this step is optional, but due to experimental setups, it is often needed (either to account for experimental variations or to bring intensities to a standardized level for comparison). The method of normalization depends on the data and the purpose of the analysis. In geochemistry and geomicrobiology, the most common normalization methods are offset normalization (e.g., Krumina et al., 2016) and area normalization to either a single band or the total spectral range (total area normalization) (e.g., Hagvall et al., 2014; Liu et al., 2016; Kenney et al., 2018).

Offset normalization, as the name suggests, only shifts the entire spectrum along the y-axis, so all spectra have a common point or range. This point or range must be in a region of the spectrum where no bands are located (i.e., pure baseline). Offset normalization is best suited for data sets in which few or no differences are expected during acquisition, such as when continuously collecting spectra as a function of time *in situ*. This normalization procedure changes the spectra the least, and it does not introduce any bias. However, if baseline correction is carried out correctly first, there is no real benefit from subsequent offset normalization.

Area normalization can be based on the total area of all the bands in the analyzed spectral region or on the area of a specific band. Total area normalization is the default option for samples that vary significantly (e.g., microorganisms grown with different carbon sources or from different geographic locations, or more heterogeneous field samples). Single band area normalization can be used when a certain band is known to be present in all spectra, provided that this band is also well resolved (does not overlap with other bands) and does not shift or broaden significantly between spectra. In practical terms, this necessitates significant control over the experimental conditions and limited changes between samples, or requires the addition of an internal standard that is inert and chemically very different from the system (to avoid band overlaps). Normalizing to such an additive can calibrate the concentrations as well, making the measurements fully quantitative. However, such additives are hard to find for real-life systems, and thus, often a band of an ever-present native compound is used instead. See Leone et al. (2007) for a description of why the phosphate peak was used to normalize their data set. It is important to keep in mind, however, that area normalization changes spectral intensities so that concentrations become relative (either to a band or to total proportions, based on the type of normalization). Thus, comparison between spectra can only reveal proportional (relative) changes, not absolute amounts, resulting in semi-quantitative measurements.

12.5.2 Data Analysis and Spectra Evaluation

Once all pretreatment procedures have been completed, the spectra may be evaluated qualitatively and (semi-)quantitatively. In the simplest of cases, the intensity of a band (or a ratio of bands) can be compared across spectra, including the extreme case of the complete absence of a band, which translates to undetectable quantities. In case of FTIR microspectroscopy, these intensity differences can be attributed to different areas of a heterogeneous sample (see Figure 12.6). However, as mentioned in Section 12.5, this kind of analysis is often suboptimal. Multivariate analysis techniques are better suited to handling spectroscopic data, especially in biological systems, where biological variation and environmental factors are superimposed on the experimental variation and changes are often subtle.

Since multivariate analysis techniques use the entire spectral information, they are better suited to revealing trends in the data set, finding correlations, and identifying different sources of variation (Stenlund et al., 2008). They are able to find outliers (samples that are the most different in the data set) and identify diagnostic spectroscopic profiles to classify samples by finding all spectral changes that significantly contribute to the differences between samples. The undisputed "workhorse" of multivariate analysis methods is principal component analysis (PCA; Mariey et al., 2001), which has the advantage of being fast, robust, and unsupervised. PCA essentially reduces the dimensionality of the spectral data set by introducing new variables (principal components) that are created by weighing the original variables against their correlation to the largest variation of the data set. PCA often provides an excellent overview of the system and its most prominent sources of variation and trends. PCA alone can be sufficient in cases where the data are of high quality and differences between samples are relatively clear and easy to separate from other sources of variation. One very elegant example of the application of PCA to FTIR spectra in the field of geochemistry comes from monitoring the pH dependence of arsenate coordination at the surface of goethite (Loring et al., 2009). However, the drawback of PCA is that it operates with abstract mathematical profiles (scores and loadings) instead of concentrations and spectra, which makes interpretation hard, and that it focuses solely on the largest variation in the data set irrespective of the scientific question at hand. Frequently in biological systems, the largest variation is not attributed to any specific scientific question but is a combination of experimental and environmental factors to various degrees. In these cases, more tailored analysis methods can be used, such as orthogonal projections to latent structures discriminant analysis (OPLS-DA; Driver et al., 2015; Trygg and Wold, 2002, 2014) or multivariate curve resolution – alternating least squares (MCR-ALS; de Juan et al., 2009; Jaumot et al., 2005; Pisapia et al., 2018).

Being a discriminant technique, OPLS-DA is always supervised, since the initial classification of the spectra needs to be performed before the created model can be applied to unclassified data (the prediction set). Like PCA, OPLS-DA operates with abstract mathematical representations in the form of scores and loadings. However, unlike PCA, OPLS-DA models filter variations between those that separate the defined classes (predictive components) and those that represent variation within the classes but do not contribute to separating them (orthogonal components). This improves interpretation and helps identify different sources of variation, leading to improved experiments and a better understanding of the system.

MCR-ALS, on the other hand, uses concentration and spectral profiles instead of scores and loadings, which is the natural way of working with (hyper)spectral data. It is an iterative method that can be used with minimal to no data manipulation or constraints (thereby leaving it largely unsupervised, although initial estimates and convergence criteria can still substantially influence results). Alternatively, the analysis can be refined by specific constraints, which may include a certain degree of supervision (such as forcing the model to set the concentration of certain compounds to zero in certain spectra, referred to as an equality constraint). However, most of these constraints are inherent in the data (e.g., non-negativity constraint for concentration: concentrations are naturally non-negative) or easily justifiable (such as setting the concentration to zero in pixels of an image where there is no sample, only the carrier). These constraints often result in better resolving capacity by limiting the ambiguity in the data. However, some ambiguity may still remain, and it most often manifests in the resolved components not being entirely "pure" (i.e., still containing contributions from more than one chemical compound to various degrees). Moreover, since it is an iterative method, outcomes may be strongly influenced by starting conditions (initial spectral and/or concentration profiles). And while results are considerably easier to interpret than in the case of PCA or OPLS-DA, it has to be remembered that MCR-ALS is designed to result in one single outcome (one set of spectral profiles and one set of concentration profiles). Alternative results that may describe the data (almost) equally well are never shown. Thus, users are encouraged to test different analysis parameters to explore the possible outcomes in order to make sure that the iteration stops at the absolute minimum and not in a potential local minimum.

The data in Figure 12.1 were analyzed using the MCR-ALS method, with an open-source script (www.kbc.umu.se/english/visp/download-visp/), and are presented in Figure 12.8.

Even though the resolved spectral profiles are similar, differences are noticeable, illustrating the resolving power of the technique (e.g., the small –C=O band above 1700 cm^{-1} in the COO-H profile). This excellent resolving power is further emphasized by the obtained concentration profiles, which show clear pH dependence of the components and distinct profiles for each, despite their spectral similarities. Note, however, that this particular data set is not sufficient to separate deprotonated COO^- and PO^-: they appear bundled in one common component.

Nevertheless, MCR-ALS is able to tease out more information from the data than could be achieved before, when only the deprotonation of the carboxyl group could be tracked by monitoring individual bands as a function of pH (e.g., Alessi et al., 2014).

12.6 Conclusions

This chapter focused on various FTIR (micro)spectroscopic techniques and their application in geomicrobiology research. FTIR spectroscopy is a useful tool for identifying changes in

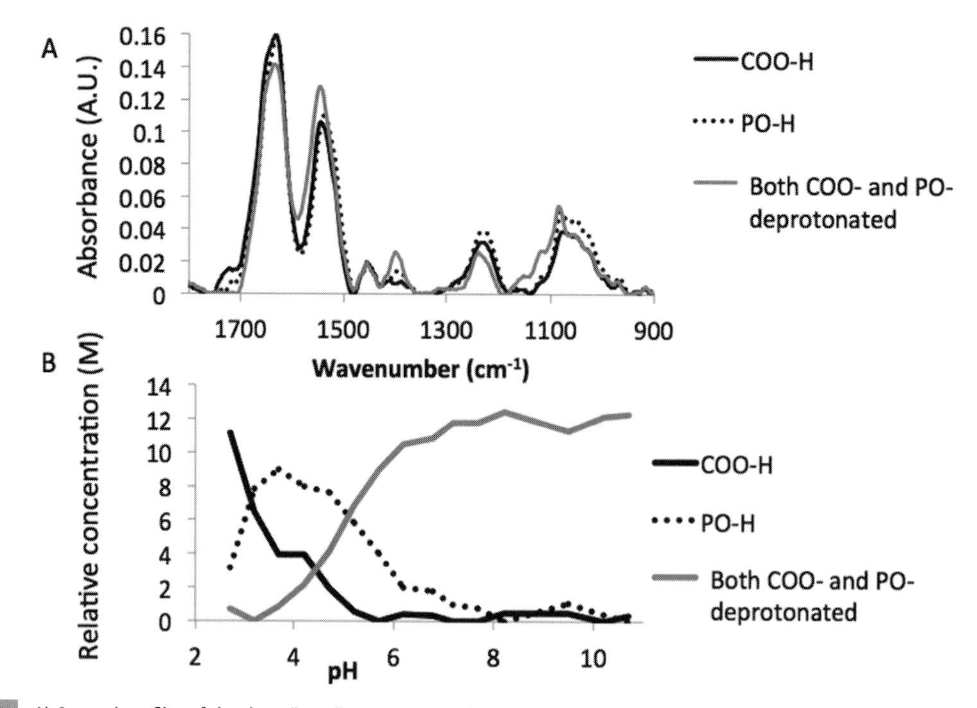

Figure 12.8 A) Spectral profiles of the three "pure" components –COO-H, PO-H, and the deprotonated component (where both COO^- and PO^- are deprotonated) as resolved by MCR-ALS with their normalized absorbance units (A.U.) on the x-axis. B) Relative abundance (concentration) of each component shown in A, as resolved by MCR-ALS.

a geomicrobiological system, such as dissolution/precipitation/degradation, etc., as a function of time, pH, ionic strength, and competing water chemistry effectively, sensitively, and nondestructively, without the need for labels, markers, or dyes. This chapter explained how to use common FTIR spectroscopic techniques (transmission, DRIFTS, grazing angle FTIR, ATR FTIR, and microspectroscopy) as well as their benefits and limitations. Additionally, this text explored data treatment options to maximize the information content and how to mine the data in the most effective way. A guide to identifying the bands present in the FTIR spectra of bacterial cells was also provided, assisting spectral interpretations.

12.7 Acknowledgments

The authors would like to thank the Vibrational Spectroscopy Core Facility (ViSp) at Umeå University for providing open access to the instrumentation for method, hardware, and software development. The support of the Department of Chemistry and the Chemical Biological Centre (KBC) at Umeå University towards ViSp is greatly acknowledged. We would also like to thank the Department of Earth Science and Engineering at Imperial College London for providing the funding to collect the data there and at ViSp at Umeå University via the Arthur Holmes Centenary Research Grant.

12.8 References

Alam, S.M., Cossio, M., Robinson, L., et al. (2016) Removal of organic acids from water using biochar and petroleum coke, *Environmental Technology & Innovation*, v. 6, pp. 141–151.

Alessi, D.S., Lezama-Pacheco, J.S., Stubbs, J.E., et al. (2014) The product of microbial uranium reduction includes multiple species with U(IV)–phosphate coordination, *Geochimica et Cosmochimica Acta*, v. 131, pp. 115–127.

Amarie, S., Keilmann, F., Zaslansky, P., Griesshaber, E. (2012) Nano-FTIR chemical mapping of minerals in biological materials, *Proceedings of Microscopy & Microanalysis*, v. 18, pp. 32–33.

Baker, M.J., Trevisan, J., Bassan, P., et al. (2014) Using Fourier transform IR spectroscopy to analyze biological materials, *Nature Protocols*, v. 9, pp. 1771–1791.

Benning, L.G., Phoenix, V.R., Yee, N., Tobin, M.J. (2004) Molecular characterization of cyanobacterial silicification using synchrotron infrared micro-spectroscopy, *Geochimica et Cosmochimica Acta*, v. 68, p. 729.

Bertie, J.E, (2002) Optical Constants. In *Handbook of Vibrational Spectroscopy* (Eds. J.M. Chalmers, P.R. Griffiths), John Wiley and Sons.

Bonhomme, S., Cuer, A., Delort, A-M., et al. (2003) Environmental biodegradation of polyethylene, *Polymer Degradation and Stability*, v 81, pp. 441–452.

Borer, P., Hug, S.J., Sulzberger, B., Kraemer, S.M., Kretzschmar, R. (2009) ATR-FTIR spectroscopic study of the adsorption of desferrioxamine B and aerobactin to the surface of lepidocrocite (γ-FeOOH), *Geochimica et Cosmochimica Acta*, v. 73, pp. 4661–4672.

Carabante, I., Grahn, M., Holmgren, A., Kumpiene, J., Hedlund, J. (2009) Adsorption of As (V) on iron oxide nanoparticle films studied by in situ ATR-FTIR spectroscopy, *Colloids and Surfaces A: Physicochemical and Engineering Aspects*, v. 346, pp. 106–113.

Chalmers, J.M., Griffiths, P.R. (2002) *Handbook of Vibrational Spectroscopy* (Eds. J.M. Chalmers, P.R. Griffiths), John Wiley and Sons.

Chan, K.L.A., Kazarian, S.G. (2007) Attenuated total reflection Fourier transform infrared imaging with variable angles of incidence: a three-dimensional profiling of heterogeneous materials, *Applied Spectroscopy*, v. 61, pp. 48–54.

de Juan, A., Maeder, M., Hancewicz, T., Duponchel, L., Tauler, R. (2009), Chemometric Tools for Image Analysis. In *Infrared and Raman Spectroscopic Imaging* (Eds. R. Salzer, HW. Siesler), Ch. 2, 65–106, Wiley-VCH Verlag GmbH & Co. KGaA.

Driver, T., Bajhaiya, A.K., Allwood, J.W., et al. (2015), Metabolic responses of eukaryotic microalgae to environmental stress limit the ability of FT-IR spectroscopy for species identification, *Algal Research*, v. 11, pp. 148–155.

Dziuba, B., Babuchowski, A., Nałęcz, D., Niklewicz, M. (2007) Identification of lactic acid bacteria using FTIR spectroscopy and cluster analysis, *International Dairy Journal*, v. 17, pp. 183–189.

Eilers, P.H. (2004) Parametric time warping, *Analytical Chemistry*, v. 76 (2), pp. 404–411.

Elzinga, E.J., Huang, J-H., Chorover, J., Kretzschmar, R. (2012) ATR-FTIR spectroscopy study of the influence of pH and contact time on the adhesion of *Shewanella putrefaciens* bacterial cells to the surface of hematite, *Environmental Science & Technology*, v. 46, pp. 12848–12855.

Elzinga, E.J., Kretzschmar, R. (2013) In situ ATR-FTIR spectroscopic analysis of the co-adsorption of orthophosphate and Cd(II) onto hematite, *Geochimica et Cosmochimica Acta*, v. 117, pp. 53–64.

Felten, J., Hall, H., Jaumot, J., et al. (2015) Vibrational spectroscopic image analysis of biological material using multivariate curve resolution-alternating least squares (MCR-ALS), *Nature Protocols*, v. 10, pp. 217–240.

Gallé, T., Van Lagen, B., Kurtenbach, A., Bierl, R. (2004) An FTIR-DRIFT study on river sediment particle structure: Implications for biofilm dynamics and pollutant binding, *Environmental Science & Technology*, v. 38, pp. 4496–4502.

Gillgren, T., Gorzsás, A. (2016) A one-pot set-up for real-time reaction monitoring by FTIR spectroscopy, *Wood Science and Technology*, v. 50, pp. 567–580.

Gorzsás, A., Sundberg, B. (2014) Chemical Fingerprinting of Arabidopsis Using Fourier Transform Infrared (FT-IR) Spectroscopic Approaches. In *Arabidopsis Protocols* (Eds. J.J. Sanchez-Serrano, J. Salinas), Humana Press.

Grahn, H.F., Geladi, P. (2007) *Techniques and Applications of Hyperspectral Image Analysis* (Eds. H.F. Grahn, P. Geladi). John Wiley and Sons Ltd.

Griffiths, P.R. (2002) Introduction to Vibrational Spectroscopy. In *Handbook of Vibrational Spectroscopy* (Eds.J.M. Chalmers, P.R. Griffiths), John Wiley and Sons.

Haberhauer, G., Rafferty, B., Strebl, F., Gerzabek, M.H. (1998) Comparison of the composition of forest soil litter derived from three different sites at various decompositional stages using FTIR spectroscopy, *Geoderma*, pp. 331–342.

Hagvall, K., Persson, P., Karlsson, T. (2014) Spectroscopic characterization of the coordination chemistry and hydrolysis of gallium(III) in the presence of aquatic organic matter, *Geochimica et Cosmochimica Acta*, v. 146, pp. 76–89.

Hagvall, K., Persson, P., Karlsson, T. (2015) Speciation of aluminum in soils and stream waters: the importance of organic matter, *Chemical Geology*, v. 417, pp. 32–43.

Hamilton, M.L., Perston, B.B., Harland, P.W., et al. (2005) Grazing-angle fiber-optic IRRAS for in situ cleaning validation, *Organic Process Research & Development*, v. 9, pp. 337–343.

Hazen, T.C., Dubinsky, E.A., DeSantis, T.Z., et al. (2010) Deep-sea oil plume enriches psychrophilic oil-degrading bacteria, *Science*, v. 330, p. 6001.

Holman, H-Y.N., Miles, R., Hao, Z., et al. (2009) Real-time chemical imaging of bacterial activity in biofilms using open-channel microfluidics and synchrotron FTIR spectromicroscopy, *Analytical Chemistry*, v. 81, pp. 8564–8570.

Holman, H-Y.N., Perry, D.L., Martin, M.C., et al. (1999) Real-time characterization of biogeochemical reduction of Cr(VI) on basalt surfaces by SR-FTIR imaging, *Geomicrobiology Journal*, v. 16, pp. 307–324.

Huang, W., Liu, Z. (2013) Biosorption of Cd(II)/Pb(II) from aqueous solution by biosurfactant-producing bacteria: Isotherm kinetic characteristic and mechanism studies, *Colloids and Surfaces B: Biointerfaces*, v. 105, pp. 113–119.

Jaumot, J., Gargallo, R., de Juan, A., Tauler, R. (2005) A graphical user-friendly interface for MCR-ALS: A new tool for multivariate curve resolution in MATLAB, *Chemometrics and Intelligent Laboratory Systems*, v. 76, pp. 101–110.

Ji, J., Ge, Y., Balsam, W., Damuth, J.E., Chen, J. (2009) Rapid identification of dolomite using a Fourier Transform Infrared Spectrophotometer (FTIR): A fast method for identifying Heinrich events in IODP Site U1308, *Marine Geology*, v. 258, pp. 60–68.

Jiang, W., Saxena, A., Song, B., et al. (2004) Elucidation of functional groups on Gram-positive and Gram-negative bacterial surfaces using infrared spectroscopy, *Langmuir*, v. 20, pp. 11433–11442.

Johnston, C.P., Chrysochoou, M. (2012) Investigation of chromate coordination on ferrihydrite by in situ ATR-FTIR spectroscopy and theoretical frequency calculations, *Environmental Science & Technology*, v. 46, pp. 5851–5858.

Kamnev, A.A. (2008) FTIR spectroscopic studies of bacterial cellular responses to environmental factors, plant-bacterial interactions and signalling, *Spectroscopy*, v. 22, pp. 83–95.

Kamnev, A.A., Tugarova, A.V., Antonyuk, L.P., et al. (2006) Instrumental analysis of bacterial cells using vibrational and emission Mossbauer spectroscopic techniques, *Analytica Chimica Acta*, v. 573–574, pp. 445–452.

Kang, S-Y., Bremer, P.J., Kim, K-W., McQuillan, A.J. (2006) Monitoring metal ion binding in single-layer *Pseudomonas aeruginosa* biofilms using ATR–IR spectroscopy, *Langmuir*, v. 22, pp. 286–291.

Kang, S., Xing, B. (2007) Adsorption of dicarboxylic acids by clay minerals as examined by in situ ATR-FTIR and ex situ DRIFT, *Langmuir*, v. 23, pp. 7024–7031.

Kenney, J.P.L., Ellis, T., Nicol, F.S., Porter, A., Weiss, D.J., (2018) The effect of bacterial growth phase and culture concentration on uranium removal from aqueous solution, *Chemical Geology*, v. 482, pp. 61–71.

Kenney, J.P.L., Fein, J.B. (2011) Importance of extracellular polysaccharides in proton and Cd binding to bacteria: a comparative study, *Chemical Geology*, v. 286 (3–4), pp. 109–117.

Kong, J., Yu, S. (2007) Fourier transform infrared spectroscopic analysis of protein secondary structures,*Acta Biochimica et Biophysica Sinica*, v. 39, pp. 549–559.

Krumina, L., Kenney, J.P.L., Loring, J., Persson, P. (2016) Desorption mechanisms of phosphate from iron oxide nanoparticle, *Chemical Geology*, v. 427, pp. 54–64.

Kubicki, J.D., Itoh, M.J., Schroeter, L.M., Apitz, S.E. (1997) Bonding mechanisms of salicylic acid adsorbed onto illite clay: An ATR–FTIR and molecular orbital study, *Environmental Science & Technology*, v. 31, pp. 1151–1156.

Lasch, P., Naumann, D. (2006) Spatial resolution in infrared microspectroscopic imaging of tissues, *Biochimica Biophysica Acta*, v. 1758, pp. 814–829.

Leone, L. et al. (2007) Modeling the acid-base properties of bacterial surfaces: A combined spectroscopic and potentiometric study of the Gram-positive bacterium *Bacillus subtilis*. *Environmental Science & Technology*, v.41 (18), pp. 6465–6471.

Liu, Y.X., Alessi, D.S., Owttrim, G.W., et al. (2016) Cell surface properties of cyanobacterium Synechococcus: Influences of nitrogen source, growth phase and N:P ratios, *Geochimica et Cosmochimica Acta*, v 187, pp. 179–194.

Loring, J.S., Sandström, M.H., Noren, K., Persson, P. (2009) Rethinking arsenate coordination at the surface of goethite, *Chemistry – A European Journal*, v. 15, pp. 5063–5072.

Madejová, J. (2003) FTIR techniques in clay mineral studies, *Vibrational Spectroscopy*, v. 31, pp. 1–10.

Manning, B.A., Goldberg, S. (1996) Modeling competitive adsorption of arsenate with phosphate and molybdate on oxide minerals, *Science Society of America Journal*, v. 60, pp. 121–131.

Marcotte, L., Kegelaer, G., Sandt, C., Barbeau, J., LaXeur, M. (2007) An alternative infrared spectroscopy assay for the quantification of polysaccharides in bacterial samples, *Analytical Biochemistry*, v. 361, pp. 7–14.

Mariey, L., Signolle, J.P., Amiel, C., Travert, J. (2001) Discrimination, classification, identification of microorganisms using FTIR spectroscopy and chemometrics, *Vibrational Spectroscopy*, v. 26, pp. 151–159.

Meyer-Jacob, C., Vogel, H., Boxberg, F., et al. (2014) Independent measurement of biogenic silica in sediments by FTIR spectroscopy and PLS regression, *Journal of Paleolimnology*, v.52, pp. 245–255.

Mirabella, F.M. (1983) Strength of interaction and penetration of infrared radiation for polymer films in internal reflection spectroscopy, *Journal of Polymer Science: Polymer Physics Edition*, v. 21, pp. 2403–2417.

Mohorčič, M., Jerman, I., Zorko, M., et al. (2010) Surface with antimicrobial activity obtained through silane coating with covalently bound polymyxin B, *Journal of Materials Science: Materials in Medicine*, v. 21, p. 2775.

Movasaghi, Z., Rehman, S., ur Rehman, I. (2008) Fourier transform infrared (FTIR) spectroscopy of biological tissues, *Applied Spectroscopy Reviews*, v. 43, pp. 134–179.

Noda, I., Ozaki, Y. (2005) Two-Dimensional Correlation Spectroscopy. In *Applications in Vibrational and Optical Spectroscopy*. John Wiley and Sons.

Ojeda, J.J., Romero-Gonzalez, M.E., Pouran, H.M., Banwart, S.A. (2008) In situ monitoring of the biofilm formation of *Pseudomonas putida* on hematite using flow-cell ATR-FTIR spectroscopy to investigate the formation of inner-sphere bonds between the bacteria and the mineral, *Mineralogical Magazine*, v. 72, pp. 101–106.

Omoike, A., Chorover, J. (2004) Spectroscopic study of extracellular polymeric substances from *Bacillus subtilis*: Aqueous chemistry and adsorption effects, *Biomacromolecules*, v. 5, pp. 1219–1230.

Omoike, A., Chorover, J., Kwon, K.D., Kubicki, J.D. (2004) Adhesion of bacterial exopolymers to r-FeOOH: Inner-sphere complexation of phosphodiester groups, *Langmuir*, v. 20, pp. 11108–11114.

Papageorgiou, S.K., Kouvelos, E.P., Favvas, E.P., et al. (2010) Metal–carboxylate interactions in metal–alginate complexes studied with FTIR spectroscopy, *Carbohydrate Research*, v. 345, pp. 469–473.

Parikh, S.J., Chorover, J. (2006) ATR-FTIR spectroscopy reveals bond formation during bacterial adhesion to iron oxide, *Langmuir*, v 22, pp. 8492–8500.

Parikh, S.J., Mukome, F.N.D., Zhang, X. (2014) ATR-FTIR spectroscopic evidence for biomolecular phosphorus and carboxyl groups facilitating bacterial adhesion to iron oxides, *Colloids and Surfaces B: Biointerfaces*, v. 119, pp. 38–46.

Peak, D., Ford, R.G., Sparks, D.L. (1999) An in situ ATR-FTIR investigation of sulfate bonding mechanisms on goethite, *Journal of Colloid and Interface Science*, v. 218, pp. 289–299.

Pisapia, C., Jamme, F., Duponchel, L., Ménez, B. (2018) Tracking hidden organic carbon in rocks using chemometrics and hyperspectral imaging,*Scientific Reports*, v. 8, p. 2396.

Poggenburg, C., Mikutta, R., Schippers, A., Dohrmann, R., Guggenberger, G. (2018) Impact of natural organic matter coatings on the microbial reduction of iron oxides, *Geochimica et Cosmochimica Acta*, v. 224, pp. 223–248.

Quilès, F., Humbert, F., Delille, A. (2010) Analysis of changes in attenuated total reflection FTIR fingerprints of *Pseudomonas fluorescens* from planktonic state to nascent biofilm state, *Spectrochimica Acta Part A: Molecular and Biomolecular Spectroscopy*, v. 75, pp. 610–616.

Reig, F.B., Adelantado, G.J.V., Moreno, M.C.M.M. (2002) FTIR quantitative analysis of calcium carbonate (calcite) and silica (quartz) mixtures using the constant ratio method. Application to geological samples, *Talanta*, v. 58, pp. 811–821.

Rong, X., Huang, Q., He, X., et al. (2008) Interaction of *Pseudomonas putida* with kaolinite and montmorillonite: A combination study by equilibrium adsorption, ITC, SEM and FTIR, *Colloids and Surfaces B: Biointerfaces*, v. 64, pp. 49–55.

Rosen, P., Persson, P. (2006) Fourier-transform infrared spectroscopy (FTIRS), a new method to infer past changes in tree-line position and TOC using lake sediment. *Journal of Paleolimnology*, v. 35, pp. 913–923.

Rosen, P., Vogel, H., Cunningham, L., et al. (2010) Fourier transform infrared spectroscopy, a new method for rapid determination of total organic and inorganic carbon and biogenic silica concentration in lake sediments. *Journal of Paleolimnology*, v. 43 (2), pp. 247–259.

Salzer, R., Siesler, H.W. (2009) Infrared and Raman Spectroscopic Imaging, Wiley-VCH Verlag GmbH & Co. KGaA.

Sar, P., Kazy, S.K, Asthana, R.K., Singh, S.P. (1999) Metal adsorption and desorption by lyophilized *Pseudomonas aeruginosa*, *International Biodeterioration & Biodegradation*, v. 44, pp. 101–110.

Schmitt, J., Flemming, H.C. (1998) FTIR-spectroscopy in microbial and material analysis, *International Biodeterioration & Biodegradation*, v. 41, pp. 1–11.

Schmitt, J., Nivens, D., White, D.C., Flemming, H.C. (1995) Changes of biofilm properties in response to sorbed substances – an FTIR-ATR study, *Water Science and Technology*, v. 32, pp. 149–155.

Shopska, M., Cherkezova-Zheleva, Z.P., Paneva, D.G., et al. (2013) Biogenic iron compounds: XRD, Mossbauer and FTIR study, *Central European Journal of Chemistry*, v. 11, pp. 215–227.

Stenlund, H., Gorzsás, A., Persson, P., Sundberg, B., Trygg, J. (2008) Orthogonal projections to latent structures discriminant analysis modeling on in situ FT-IR spectral imaging of liver tissue for identifying sources of variability, *Analytical Chemistry*, 80 (18), pp. 6898–6906.

Suci, P.A., Mittelman, M.W., Yu, F.P., Geesey, G.G. (1994) Investigation of ciprofloxacin penetration into *Pseudomonas aeruginosa* biofilms, *Antimicrobial Agents and Chemotherapy*, v. 38, pp. 2125–2133.

Tapper, R. (1998) The use of biocides for the control of marine biofilms, PhD thesis, University of Portsmouth, UK.

Trygg, J., Wold, S. (2002) Orthogonal projections to latent structures (O-PLS), *Journal of Chemometrics*, v. 16, pp. 119–128.

Ueshima, M., Ginn, R.R., Haack, E.A., Szymanowski, J.E.S., Fein, J.B., 2008. Cd adsorption onto *Pseudomonas putida* in the presence and absence of extracellular polymeric substances. *Geochimica et Cosmochimica Acta*, v. 24, pp. 5885–5895.

ur Rehman, I., Movasaghi, Z., Rehman, S. (2013) *Vibrational Spectroscopy for Tissue Analysis*. Series in Medical Physics and Biomedical Engineering, CRC Press.

Vaculíková, L., Plevová, E. (2005) Identification of clay minerals and micas in sedimentary rocks, *Acta Geodynamica et Geomaterialia*, v. 2, pp. 167–175.

Vogel, H., Rosén, P., Wagner, B., Melles, M., Persson, P. (2008) Fourier transform infrared spectroscopy, a new cost-effective tool for quantitative analysis of biogeochemical properties in long sediment records, *Journal of Paleolimnology*, v. 40, pp. 689–702.

Wang, H., Hollywood, K., Jarvis, R.M., Lloyd, J.R., Goodacre, R. (2010) Phenotypic characterisation of *Shewanella oneidensis* MR-1 under aerobic and anaerobic growth conditions by using Fourier transform infrared spectroscopy and high-performance liquid chromatography analyses, *Applied and Environmental Microbiology*, v. 76, pp. 6266–6276.

Wei, X., Fang, L., Cai, P., et al. (2011) Influence of extracellular polymeric substances (EPS) on Cd adsorption by bacteria, *Environmental Pollution*, v. 159, pp. 1369–1374.

Yao, J-W., Xiao, Y., Lin, F. (2012) The effect of various pH, ionic strength and temperature on papain hydrolysis of salivary film, *European Journal of Oral Sciences*, v. 120, pp. 140–146.

Yoon, T.H., Johnson, S.B., Musgrave, C.B., Brown, Jr, G.E. (2004) Adsorption of organic matter at mineral/water interfaces: I. ATR-FTIR spectroscopic and quantum chemical study of oxalate adsorbed at boehmite/water and corundum/water interfaces, *Geochimica et Cosmochimica Acta*, v. 68, pp. 4505–4518.

13 Mössbauer Spectroscopy

JAMES M. BYRNE AND ANDREAS KAPPLER

Abstract

Following the discovery almost half a century ago of the resonant emission and absorption of gamma rays in ^{191}Ir by Rudolf Mössbauer (Mossbauer, 1958), the effect that takes his name has been utilized by many studies to explore a wide range of elements and minerals. Recently, a renewed interest in the application of Mössbauer spectroscopy in geosciences, particularly with regard to nanomaterials and Mars (Klingelhöfer et al., 2003), has led to the technique becoming an increasingly important tool. This is especially true for ^{57}Fe mineralogy, where the technique is capable of selectively determining the oxidation state and mineralogy of Fe in samples of environmental soils and sediments, which might be dominated by a wide range of mineral phases that might not ordinarily be measurable with conventional techniques.

The purpose of this chapter is to give a general overview of Mössbauer spectroscopy to the non-specialist. While there are many elements that are able to undergo the Mössbauer effect, we focus on ^{57}Fe in particular in order to help encourage its use in the field of geomicrobiology. As such, many important factors will not be covered here, and the reader is encouraged to look at additional material to obtain a more comprehensive understanding of the theory of Mössbauer spectroscopy. Some example reviews include references (Dyar et al., 2006; Greenwood and Gibb, 1971; Gütlich et al., 2010, 2012; Murad and Cashion, 2004). Furthermore, when considering whether the technique is potentially useful in a study, it is always advisable to combine the results and observations obtained through Mössbauer spectroscopy with information obtained through diffraction, imaging, or synchrotron-based techniques to obtain a complete story of the nature of a sample. This chapter has been broken up into separate sections, including a brief introduction to the theory behind Mössbauer spectroscopy; the parameters that are obtained from each spectrum and provide the necessary information required for fitting; some experimental details about how measurements are carried out and should be prepared; and finally, the applications of the technique and how it can be used to determine the oxidation states and mineral identities of samples.

13.1 Introduction

The underlying theory behind the Mössbauer effect is often considered to be complex and intimidating to the nonexpert; however, a basic understanding of the principles is crucial in order to make sense of the information that can be obtained through its

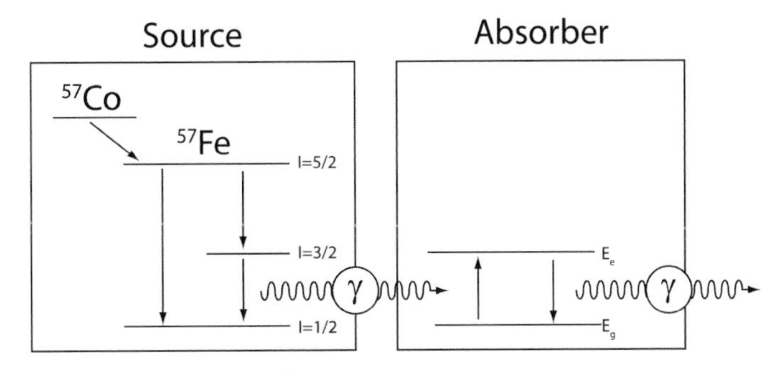

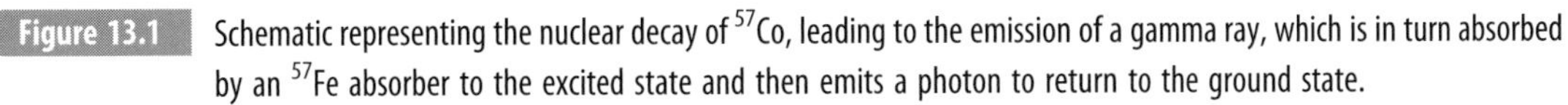
Figure 13.1 Schematic representing the nuclear decay of ^{57}Co, leading to the emission of a gamma ray, which is in turn absorbed by an ^{57}Fe absorber to the excited state and then emits a photon to return to the ground state.

application. The effect itself makes use of resonant absorption of γ-radiation by atomic nuclei. Essentially, a source with proton number (Z) and neutron number (N) emits a photon (γ-ray), which can then be absorbed by a nucleus with the same value of Z and N that is in its ground state (E_g). This absorber nucleus is now in the excited energy state (E_e) and transitions back to E_g by the re-emission of a photon of energy (14.4 keV). In the case of ^{57}Fe Mössbauer spectroscopy, a radioactive ^{57}Co source (often embedded in a rubidium matrix), which has a half-life of 270 days, undergoes nuclear decay to ^{57}Fe in the excited energy level (136 keV) with a nuclear spin state of I = 5/2 (Figure 13.1). The excited ^{57}Fe nucleus undergoes decay to the ground state (I = 1/2) either directly or via the 14.4 keV energy level (I=3/2), which having a half-life of ~100 ns, decays to the ground state through the emission of a γ-ray of 14.4 keV. It is this 14.4 keV γ-ray that is then absorbed by the absorber nucleus (i.e., the sample).

One of the most important requirements for the observation of the Mössbauer effect is the presence of recoilless absorption and emission of γ-radiation. As a simple analogy, consider the firing of a gun. When the trigger is fired, conservation of momentum causes the gun to recoil in the opposite direction to the projectile. The same effect takes place when a γ-ray of energy E_γ is emitted by a free nucleus with recoil energy $E_r = E_\gamma/(2Mc^2)$. In free gases or liquids, this energy E_r is several orders of magnitude higher than the natural linewidth of resonant absorption, and consequently, the Mössbauer effect cannot be observed. However, when the absorber is fixed in a solid, a fraction of the atoms will not undergo recoil, with the energy of momentum absorbed by the crystal lattice. This is known as the recoil-free fraction (*f*-fraction) and imposes restrictions on the type of sample that can be measured; i. e., liquid samples cannot be probed. However, it is possible to measure such samples provided they are frozen. In fact, decreasing the temperature of measurement also contributes to an increase in the *f*-fraction, thereby increasing the probability of the Mössbauer effect being observed. We will not go into any further depth in the description of the *f*-factor, except to say that it should be considered with care. Specifically, minerals have different *f*-factors, so they will undergo a different amount of absorption/emission at different temperatures. For instance, Fe(II) phases tend to have lower *f*-factors than Fe(III) phases, meaning that the ferric component could be overestimated with respect to the ferrous component (De Grave and Van Alboom, 1991). It can also cause problems when determining the relative abundances of different mineral phases, something that is regularly performed in geomicrobiology studies; however, measuring at low temperature (<~200 K) is generally sufficient to overcome the issue (De Grave and Van Alboom, 1991).

The emission and absorption profiles of the resonance lines can be described as having a Lorentzian lineshape with half width at half maximum (HWHM), which based on Heisenberg‘s uncertainty principle has a minimum value of 0.097 mm/s, although instrumental broadening will often contribute to a small increase in that value. The resonant shapes of the emission and absorption profiles are not single points, and a significant amount of overlap is required between the two for the energy profiles to produce a valid spectrum. For this we must exploit the Doppler shift, whereby the energy is related to the velocity (Figure 13.2). Therefore, Mössbauer spectra are recorded with the source (or sometimes absorber) moving back and forth in one direction. This causes a slight shift in the energy spectrum, which is detected and allows the emission and absorption profiles to

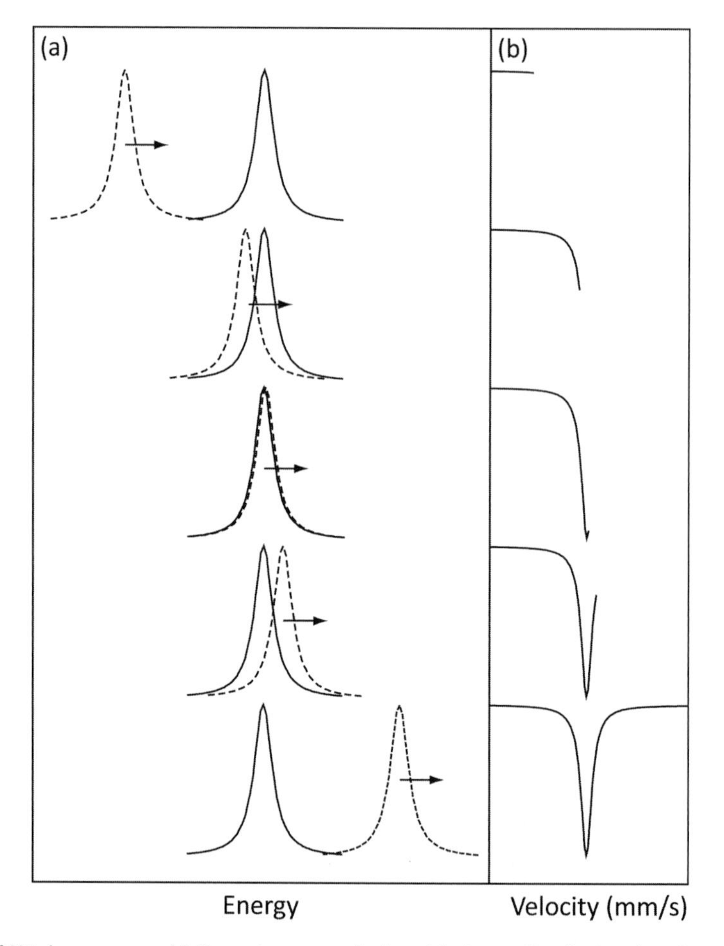

Figure 13.2 Formation of Mössbauer spectra. (a) The emitter moves back and forth to utilize the Doppler effect and shift the emission energy profile so that it overlaps with that of the absorber to create (b) a Mössbauer spectrum.

overlap. This is the reason why published Mössbauer spectra show velocity (mm/s) along the x-axis, as it corresponds to the velocity of the moving source.

The Mössbauer effect can be detected in a number of different elements; however, various factors, including the cost of the radioactive sources and the half-life, often result in the focus on the ^{57}Fe (14.4 keV) transition, with ^{57}Fe dominating Mössbauer-related publications (Murad, 2010). Other potential elements that can be measured include ^{119}Sn (23.8 keV), ^{121}Sb (37.1 keV), ^{125}Te (35.5 keV), ^{129}I (27.7 keV), ^{151}Eu (21.6 keV), ^{155}Dy (25.6 keV), ^{169}Tm (8.4 keV), and ^{181}Ta (6.2 keV).

13.2 Hyperfine Parameters

In the case of the emitter and absorber being in identical local coordination, the emission spectra will be observed as a single peak with Lorentzian lineshape. In reality, the absorber atom will have a different local environment from the emitter, resulting in the formation of hyperfine interactions, which are described as the Mössbauer parameters. Several different fitting programs exist; however, they all aim to apply the same thing, i.e., to determine the main parameters, which include isomer shift (IS), quadrupole splitting (ΔE_Q), and magnetic hyperfine splitting (B_{hf}) (Figure 13.3). From these three parameters, it is possible to determine the oxidation state of the target element, the mineral phase identity, or in some cases, the particle size of a sample.

13.2.1 Isomer Shift (IS, δ)

The isomer shift is determined from the relative distance from the center of the spectrum to the center of a reference material (calibration); e.g., an α-Fe(0) foil is usually measured for ^{57}Fe, measured in velocity (millimeters per second). While the parameter is frequently referred to as the isomer shift, that term only truly applies if the source (e.g., ^{57}Co) and absorber (e.g., ^{57}Fe) are both at the same temperature; otherwise, the term "center shift" is preferred. The IS arises due to the interaction of the electron cloud of an atom (specifically the s-electrons) with the nucleus, which is assumed to be a point charge. The s-electron density is dependent upon the bonding environment of the atom, resulting in different IS values for different compounds or minerals. The IS (δ) can be expressed using the mathematical formula

$$\delta = E_A - E_S = \frac{4\pi}{5} Ze^2(\rho_A - \rho_S)(R_e^2 - R_g^2) \quad \text{(Eq.13.1)}$$

where E_A and E_S correspond to the energy of the absorber and the source, respectively, ρ_A and ρ_S correspond to the electron density of the absorber and the source, and R_e and R_g correspond to the radii of the excited and ground-state nuclei, respectively. The ions Fe(III) and Fe(II) have five and six d-electrons, respectively, and as such, the additional d-electron in Fe(II) contributes towards a shielding of the s-electrons from the nucleus and leads to a decrease in contribution to ρ_A. The nuclear factor ($R_e^2 - R_g^2$) is negative for ^{57}Fe; hence, the isomer shift for Fe(II) is generally larger than that of Fe(III), as further discussed in Section

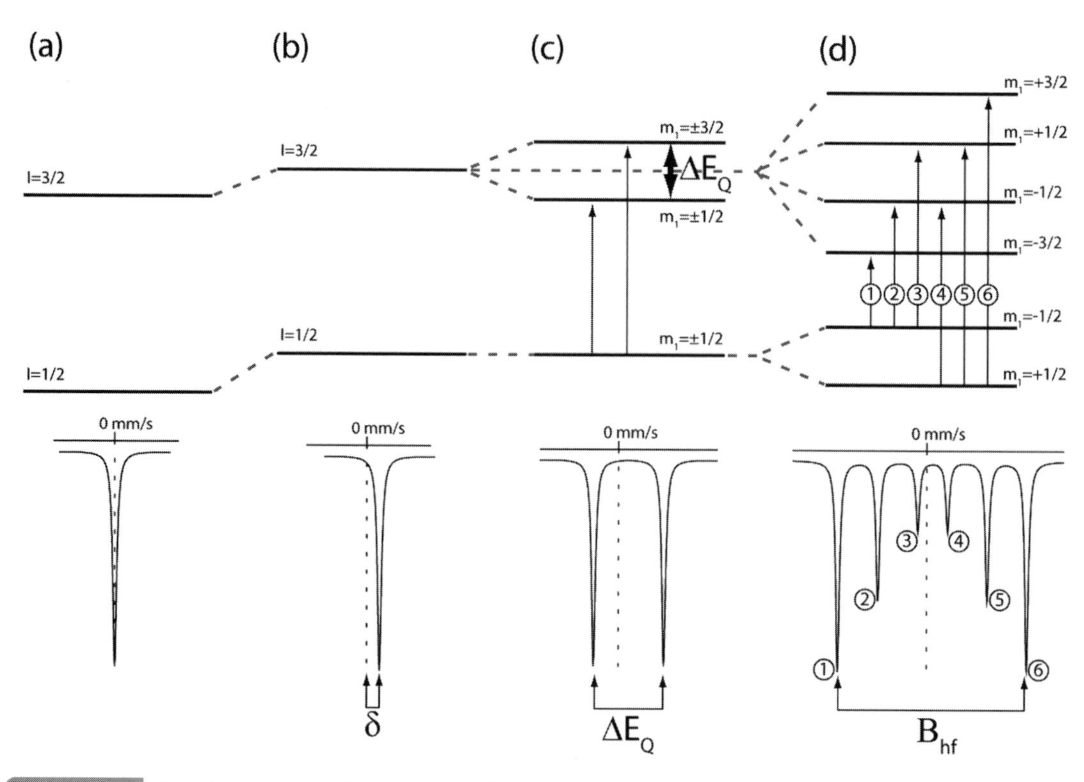

Figure 13.3 Hyperfine parameters observed in minerals undergoing the Mössbauer effect. (a) Free atom decaying from excited to ground state, (b) isomer shift results in shift of peak with respect to standard α-Fe(0), (c) quadrupole splitting due to asymmetry of absorber nucleus, (d) hyperfine field in a magnetically ordered material.

13.4.1. While α-Fe(0) is often the preferred calibration used to determine the IS, many older studies also used different materials that can be corrected against α-Fe(0) (Murad and Cashion, 2004). Furthermore, the IS is dependent upon the temperature of measurement, due to either magnetic or crystallographic ordering. The approximate shift in the IS at different temperatures compared with room temperature (295 K) is −0.22 mm/s at 600 K; +0.12 mm/s at 78 K; and +0.14 mm/s at 4 K (Murad and Cashion, 2004).

13.2.2 Quadrupole Splitting (ΔE_Q)

The quadrupole splitting parameter (ΔE_Q), also referred to as the electric quadrupole interaction, emerges as a result of an asymmetric (i.e., nonspherical) nucleus undergoing measurement. In short, an electric quadrupole interaction occurs if the nuclear spin state $I > 1/2$. For ^{57}Fe (as well as ^{119}Sn and ^{197}Au), the ground level (E_g) has spin state $I = 1/2$ and therefore cannot undergo splitting, whereas its excited energy level (E_e) has spin state $I = 3/2$ and splits into two degenerate states with magnetic spin quantum numbers ±3/2 and ±1/2 (Figure 13.3c). The ΔE_Q is observed due to the difference in energy between these degenerate levels and the ground state, with the magnitude and sign of ΔE_Q dependent upon the type of asymmetry present at the nucleus (either prolate or oblate), the oxidation state of the absorber nucleus, its spin state (i.e., high or low spin), and local atomic coordination. Generally speaking, high-spin Fe(III) atoms show lower quadrupole splittings than high-spin Fe(II) atoms, meaning that ΔE_Q is very important for the determination of oxidation state. Conversely, low-spin Fe(II) atoms tend to have very low values of ΔE_Q (see Section 13.4.1).

13.2.3 Hyperfine Field (H, Bhf)

The magnetic hyperfine field (B_{hf}) is a measure of the magnetic field at the atom, either as a result of intrinsic magnetism in the atom or through the application of an external field. It is a convenient parameter in the determination of mineral identity, with many minerals showing distinct hyperfine fields. When an atomic nucleus in the ground state ($I = 1/2$) exhibits magnetic ordering, each excited state splits into $2I + 1$ substates (Figure 13.3d), corresponding to two and four substates, respectively. This allows six energy transitions to take place, resulting in the characteristic sextet of a magnetically ordered ^{57}Fe Mössbauer spectrum, with the B_{hf} defined as the distance between the outermost peaks. The relative intensities of the peaks are related by $3{:}(4\sin 2\theta)/(1 + \cos 2\ \theta){:}1$, where θ corresponds to the angle between the nuclear spin and the incoming gamma ray. This means that the outermost and innermost lines (peaks 1, 3, 4, and 6) are always proportional to each other, whereas the remaining lines (peaks 2 and 5) can vary between 1 and 4. This value is most often averaged out to 2 for powdered samples, which have relative line intensities of 3:2:1:1:2:3 for peaks 1 to 6, respectively.

The hyperfine field may also undergo quadrupole–dipole interaction, known as the quadrupole shift (ε). In such cases, the six peaks of the hyperfine sextet are no longer equidistant, with peaks 1 and 6 moving in the opposite velocity direction to peaks 2, 3, 4, and 5 (Figure 13.3d). The quadrupole shift is related to the quadrupole splitting by $\varepsilon = \Delta E_Q/2$; however, in most cases, it is relatively small compared with B_{hf}. Nevertheless, it can often prove to be a useful parameter for mineral identification and can provide information on magnetic ordering, as with hematite or goethite (see Section 13.4.1).

13.2.4 Blocking Temperature

Though technically speaking not a hyperfine parameter, the blocking temperature (T_B) is a term often used to describe the transition of nanoparticles from being superparamagnetic to being magnetically ordered. Above T_B, the material is superparamagnetic, and thermal energy is sufficient to induce random orientation of the particles, leading to zero net magnetic moment. As temperature decreases, magnetic ordering takes place, leading to a positive net magnetic moment. During this transition, the Mössbauer spectrum of a sample will go from being a doublet (superparamagnetic) to a sextet having B_{hf} (magnetically ordered) with T_B related to particle size, or it can also be affected by metal substitution. Therefore, in some cases, T_B can be used to estimate particle size, although other more conventional methods are often more appropriate. In order to calculate T_B, several spectra must be obtained at several different temperatures, preferably with at least one showing only a doublet and one showing only a sextet. T_B is calculated as the temperature when each of the doublets and sextets corresponds to 50% of the relative spectral area. For example, Mikutta et al. were able to use the blocking temperature to distinguish ferrihydrite that had been coprecipitated in the presence and absence of the organic exopolysaccharide polygalacturonic acid (PGA) (Mikutta et al., 2008). The authors showed that the PGA led to a decrease in T_B, indicating a smaller particle size as compared with pure ferrihydrite.

13.3 Experimental

13.3.1 Sample Preparation

Sample preparation is an important issue in order to ensure that a good-quality spectrum is collected in a sufficiently rapid timeframe. The general criterion for a sample to undergo the Mössbauer effect is that it must be fixed so as to enable recoilless absorption and emission of γ-radiation. For transmission configuration, the sample must be thin and cover an area sufficiently large to prevent radiation "leakage," which might contribute to a poor signal to noise ratio. Ideally, there should be enough Fe present to obtain a usable spectrum within 1 day; however, this is often not possible with environmental samples, as simply adding more sample contributes to a lower count rate measured at the detector. It is therefore important to consider how much iron is present in a sample before carrying out a measurement in order to ascertain whether a reasonable spectrum can be obtained or not.

Powdered, dry, homogeneous samples are often the most practical to measure, as they can easily be loaded into small Fe-free sample holders (e.g., poly (methyl methacrylate)). In general, for transmission mode measurements, the thickness of the sample disc should be kept to a minimum and cover an area of 1 cm^2. The general rule of thumb is that a sample should contain a minimum of 5–10 mg/cm^2 of natural Fe, though attempts have been made to determine a more accurate sample amount (Rancourt et al., 1993). In some cases, there may be insufficient sample to cover an area of 1 cm^2, in which case it can be mixed with an inert Fe-free compound such as glucose and ground to make a homogeneous powder that covers the area.

Another sample preparation method, which avoids the requirement to dry the sample, is to use filtration. By passing an aqueous mineral precipitate through a filter paper (e.g., 0.45 μm pore size, Millipore), the solid fraction of a sample can be separated from the liquid fraction. This filter paper loaded with sample is then sealed between two layers of airtight Kapton tape ready for measurement. Filtration is useful for maintaining the oxidation state of the mineral and can help to avoid any mineralogical changes during drying. One drawback is the possibility of some water remaining in the sample, which can lead to a significant decrease in the recoilless fraction of the sample; however, this can easily be overcome by freezing the sample and measuring at <0 °C.

13.3.2 Measurement

The Mössbauer effect requires a moving source (to impart the Doppler shift) and a detector. In principle, the detectors can be placed behind the sample, as is the case with transmission geometry measurements, or behind the source, as for the miniaturized Mössbauer spectrometers (MIMOS II) on board Martian rovers, which do not have the advantage of someone to prepare their samples.

In the vast majority of laboratory-based studies, transmission mode is the preferred type of measurement. The sample is placed between the source and a detector, which is connected to a multi-channel analyzer that accumulates the recorded count rate. The source drive unit works much like a loudspeaker, with the source moved back and forth at a desired velocity, which for ^{57}Fe usually ranges from 4 to 12 mm/s. For environmental samples, it is sensible to use 12 mm/s to avoid missing important information, although this increases measurement time. The type of motion used to drive the source includes constant acceleration (i.e., saw tooth), fast flyback, and sinusoidal, with constant acceleration most commonly used in ^{57}Fe Mössbauer. Feedback loops ensure that the velocity signal of the drive unit is within acceptable accuracy (0.1%). For detection, most systems are equipped with gas-filled proportional detectors, usually Kr or Kr/CH_4.

Several alternatives to transmission mode exist, including the miniaturized MIMOS II, which was designed at the University of Mainz with the goal of sending a Mössbauer to Mars. This was achieved on both NASA rovers *Spirit* and *Opportunity*, which used MIMOS II to explore the Fe-mineralogy of many rocks on the Martian surface (Klingelhöfer et al., 2003). Back on Earth, efforts to apply the MIMOS II in the laboratory have yielded interesting results. For instance, Zegeye et al. were able to obtain time-resolved spectra showing the mineralogical transformation of lepidocrocite to magnetite by the Fe(III)-reducing bacteria *Shewanella putrefaciens* (Zegeye et al., 2011), while Markovski et al. used the MIMOS II to investigate *in situ* microbial reduction of ferrihydrite by *Geobacter sulfurreducens* (Markovski et al., 2017).

13.3.3 Temperature Control

Many environmental or microbially derived samples contain short range ordered mineral phases such as ferrihydrite or nanogoethite (van der Zee et al., 2003), which are not magnetically ordered at room temperature. While oxidation state can still be determined on these samples, it is difficult to identify the mineral, as many Fe minerals share similar parameters in the superparamagnetic state. This can be overcome by using low-temperature cryostats, which cool samples to between 4.2 and 295 K. These include simple bath cryostats, which have the drawback that the coolant (liquid N_2 for 77 K; liquid He for 4.2 K) can easily evaporate. Alternatively, closed-cycle systems can be used and have the advantage that the coolant is not lost, which is especially desirable for liquid helium temperature measurements. One drawback of such systems is the initial cost of the cryostat system and the amount of space required for the pump, transformer, and cryostat.

13.3.4 Data Processing

Many different approaches to fitting Mössbauer spectra exist, with several commercial and free programs available (e.g., CONFIT 2000 [Žák and Jirásková, 2006], WinNormos,

MössWinn, MossA [Prescher et al., 2012], and Recoil), each with advantages and disadvantages. The most fundamental aspects that are often sought after when fitting data are in the determination of IS in relation to a calibration foil, quadrupole splitting, and hyperfine field. Errors are associated with these parameters, and a goodness of fit χ^2 is used to indicate the quality of the fitting, although this can sometimes be paradoxically high if the signal to noise ratio is particularly good. Additional parameters that are also important include the linewidth (i.e., HWHM).

In the most general cases, Mössbauer spectra can be fit by a sum of Lorentzian peaks; for instance, a single peak requires just one Lorentzian, doublets require two Lorentzians, and so on. By varying the distances between these peaks, IS, ΔE_Q, and B_{hf} can be determined, provided that the sample has been calibrated against a reference material. The relative abundance of each phase present in a sample that contains a mixture of Fe minerals is determined from the total area under each spectral component (i.e., doublet, sextet, etc.). Instrumental broadening is an important characteristic of Mössbauer spectra and should always be considered during any fitting. The simplest approach to determine instrumental broadening is to perform a fit of the calibration foil using a series of three concentric doublets with Lorentzian lineshape and IS of 0.00 mm/s. The ΔE_Q of each doublet should approximate to 1.6, 6.2, and 10.6 mm/s, corresponding to peaks 1/6, 2/5, and 3/4, respectively. If the condition that the HWHM is 0.097 mm/s is relaxed, the doublets can then be fitted to the calibration curve. The HWHM including the instrumental broadening can be taken as the linewidth of the 3/4 doublet.

In many cases, such as in natural systems, particle size distributions or cation substitutions exist, which result in an incomplete fit when only Lorentzian lineshapes are considered. An alternative approach promoted by Rancourt and Ping (Rancourt and Ping, 1991) is to apply a Voigt-based lineshape, which is essentially a Gaussian distribution of Lorentzian lines. Using this fitting technique, the hyperfine field parameter can be varied while maintaining a constant IS and ΔE_Q and will be able to overcome the effect of widened sextets due to particle size distributions. A more developed fitting approach developed by Lagarec and Rancourt extends this procedure to include a distribution of the IS and ΔE_Q and is described as extended Voigt-based fitting (Lagarec and Rancourt, 1997).

13.4 Applications of Mössbauer Spectroscopy

13.4.1 Determination of Fe Oxidation State and Coordination

One of the most powerful aspects of ^{57}Fe Mössbauer spectroscopy is the ability to distinguish between oxidation states, most notably Fe(II) and Fe(III), and their local atomic coordination. This is particularly useful for soils and sediments, for which the determination of oxidation state by spectrophotometric techniques (e.g., Ferrozine; Stookey, 1970) requires complete dissolution of the solid phase in a strong acid, which is not always sufficiently accurate and may be further complicated in samples containing multiple redox-active elements (Klueglein and Kappler, 2012). Mössbauer spectroscopy, however, is a nondestructive technique, and provided that the necessary steps are taken to prevent abiotic oxidation, the sample can be further analyzed after the Mössbauer measurement with alternative techniques if required.

As indicated in Section 13.2.1, the IS provides very useful information in the determination of oxidation state and atomic coordination. Doublets with an IS less than ~0.25 mm/s can be considered to be most likely to be due to Fe(III) ions in tetrahedral coordination, whereas doublets with IS greater than ~0.29 mm/s but less than 0.60 mm/s can be considered to be most likely Fe(III) in octahedral coordination. Fe(II) ions have much higher IS, with values between 0.90 and 1.05 mm/s indicating tetrahedral Fe(II) and values between 1.05 and 1.20 mm/s indicating octahedral Fe(II) (Dyar et al., 2006). IS larger than 1.20 mm/s often suggests Fe(II) with eightfold coordination. Quadrupole splitting values range from a minimum up to values close to 4 mm/s. The quadrupole splitting values follow a similar pattern to IS for Fe(II) and Fe(III), with low values generally indicating Fe(III) oxidation state and higher values indicating Fe(II) oxidation state. A general rule of thumb is that a sample having IS < 0.8 mm/s combined with a $\Delta E_Q < 1$ mm/s is characteristic of Fe (III), whereas high-spin Fe(II) often has IS > 1.05 mm/s and $\Delta E_Q > 1.3$ mm/s. It is important to note, however, the case of low-spin octahedral Fe(II), which tends to show low IS and low ΔE_Q. For example pyrite (FeS_2) is known to contain an octahedrally coordinated low-spin ferrous atom covalently bonded to sulfur atoms and has ΔE_Q of 0.61 mm/s (Morice et al., 1969). Furthermore, potassium ferrocyanide ($K_4[Fe(CN)_6]$), another low-spin Fe(II) compound, has $\Delta E_Q = 0.00$ mm/s because of cubic symmetry, which is present due to the position of the six CN ligands surrounding the absorber nucleus; however, when one of the CN ligands is replaced, ΔE_Q dramatically increases (Gütlich et al., 2012). Thus, it is vitally important to consider the elemental composition of a sample when considering oxidation state. It is also worth noting that the symmetry of the quadrupole split doublet is most accurate in powdered samples. Asymmetry in the observed doublets can occur in single crystals, in imperfect powdered samples, or due to overlapping doublets in a multi-phased sample. In the case of samples with magnetic ordering, the quadrupole splitting tends to be smeared out due to the hyperfine field parameter; however, the IS remains a valid indicator of oxidation state. Figure 13.4 shows an example room temperature spectrum of a sample extracted from an As-contaminated water treatment sand filter in Hanoi, Vietnam (Nitzsche et al., 2015). It contains two ferrous and one ferric components, which are clearly visible. Analysis of their relative site populations can be shown to account to an overall Fe(II)/Fe (III) ratio of 0.50.

13.4.2 Isotopic Tracer Studies

The isotopic selectivity of Mössbauer spectroscopy lends itself to being used to look at mineral surfaces and isotopic exchange processes in ways that other techniques struggle to do. For example, Williams and Scherer synthesized goethite, hematite, and ferrihydrite from pure ^{56}Fe(III) without any ^{57}Fe, thus making the minerals invisible to Mössbauer (Figure 13.5) (Williams and Scherer, 2004). They reacted the minerals with a solution

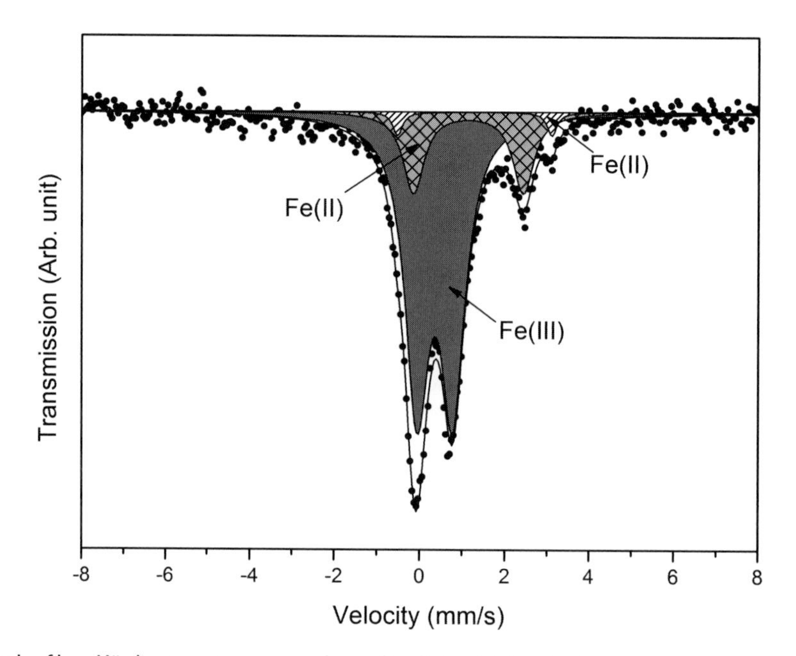

Figure 13.4 An example of how Mössbauer spectroscopy can be used to distinguish Fe oxidation states in a household sand filter measured at room temperature. The two widened doublets correspond to Fe(II), while the narrow doublet corresponds to Fe(III). (Adapted from Nitzsche et al., 2015 with permission.)

consisting of pure ^{57}Fe(II) and observed changes to the Mössbauer spectra before and after reaction. In their study, they were able to show that the Fe(II) in solution was reacting with the surface of the Fe(III) (oxyhydr)oxides, leading to electron transfer between the surface-adsorbed Fe(II) and the bulk minerals. These results were subsequently expanded by further studies, which probed the impact of changing concentrations on electron transfer (Larese-Casanova and Scherer, 2007) and also probed isotopic exchange between minerals and solutions (Gorski et al., 2012; Handler et al., 2009), indicating the potential usefulness of using ^{57}Fe in Mössbauer studies.

Isotopic tracer studies were also applied by Dippon et al. to explore the interactions of the nitrate-reducing Fe(II)-oxidizing bacterium *Acidovorax* sp. BoFeN1 with iron minerals. The authors used a liquid growth medium enriched with ^{57}Fe(II) to investigate whether magnetite could be produced via microbial Fe(II) oxidation in the presence of mineral nucleation sites (Dippon et al., 2012).

13.4.3 Mineral Identification

Another frequently used application of ^{57}Fe Mössbauer spectroscopy is mineral identification. Mössbauer can determine iron species in samples in which they might otherwise be dominated by other mineral phases and hence indistinguishable against the background. Another advantage over techniques such as X-ray diffraction (XRD) is the ability to measure short range ordered minerals (poorly crystalline phases) or samples that are X-ray amorphous.

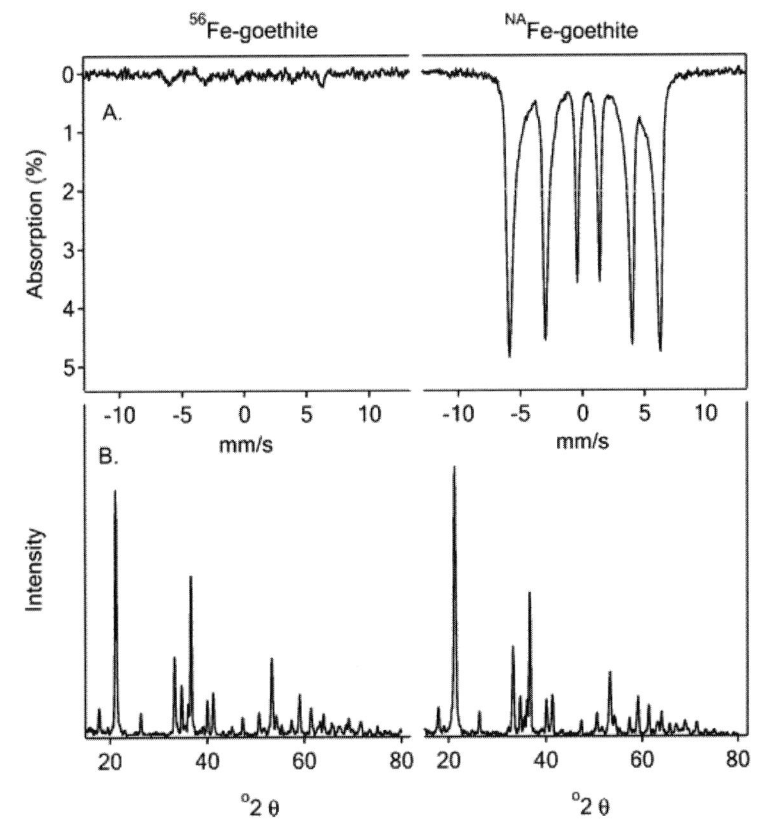

Figure 13.5 The ^{57}Fe selectivity of Mössbauer can be used to probe isotopic exchange reactions between minerals and aqueous solutions, or microbial-mineral mechanisms. (A) Mössbauer spectra at 295 K: ^{56}Fe goethite is invisible to the Mössbauer effect, whereas ^{NA}Fe (NA: natural abundance) goethite is not; (B) corresponding X-ray diffraction (XRD) patterns for the samples. (From Williams and Scherer, 2004, with permission.)

Furthermore, XRD cannot often distinguish between samples with similar lattices, e.g., spinels, while changes to the hyperfine parameters, coupled with shifts to the T_B or Verwey transition, can often be determined through the application of Mössbauer.

In many ways, the use of Mössbauer for mineral identification can be considered to be similar to XRD, in that each mineral possesses a basic fingerprint through which the hyperfine parameters (IS, ΔE_Q, and Bhf) can be determined and matched against reference values in order to identify the mineral under investigation. These references values are often found in the literature or via online databases. The most comprehensive database available is provided by the Mössbauer effect data center (www.medc.dicp.ac.cn/index.php). The Mineral Spectroscopy Database (www.mtholyoke.edu/courses/mdyar/database/) is home to an extensive library of thousands of Mössbauer spectra collected over many years by the research group of Prof. Darby Dyar, Mount Holyoke College. Alternatively, Mösstool (http://mosstool.com) has been designed to enable users to visualize how hyperfine parameters shape a spectrum, while the site also hosts a small Mössbauer database focused

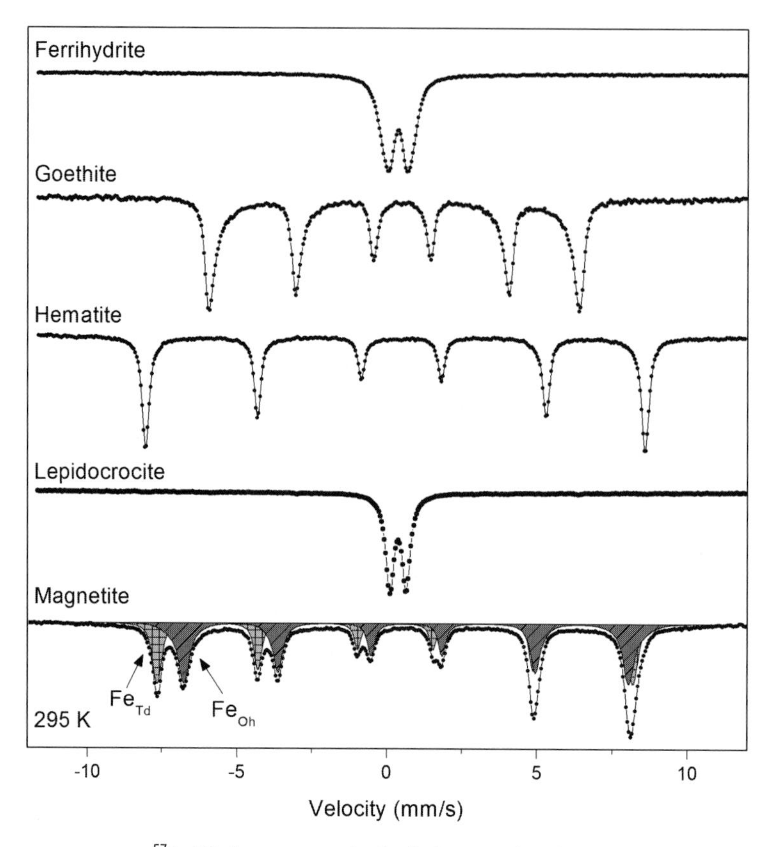

Figure 13.6 Example room-temperature ^{57}Fe Mössbauer spectra for ferrihydrite, goethite, hematite, lepidocrocite, and magnetite with the two different tetrahedral and octahedral lattice sites of magnetite, Fe_{Td} and Fe_{Oh}, highlighted.

towards environmental sciences. Nevertheless, despite the temptation to use Mössbauer as a fingerprinting method, it is preferable to apply multiple techniques for mineral determination where possible.

The following section contains details of a few selected iron minerals that are often observed in the field of geomicrobiology, including ferrihydrite, goethite, hematite, and magnetite. Some example spectra can be seen in Figure 13.6. We will briefly outline the type of behavior that can be observed for some of these mineral phases.

13.4.3.1 Ferrihydrite

Ferrihydrite is commonly observed to form due to microbial Fe(II) oxidation by nitrate-reducing, microaerophilic, or phototrophic Fe(II)-oxidizing bacteria (Swanner et al., 2015). It is also used as an electron acceptor for Fe(III)-reducing bacteria such as *G. sulfurreducens* or *S. oneidensis* (Dippon et al., 2015). In general, ferrihydrite occurs in two different forms, denoted 2-line and 6-line ferrihydrite, which correspond to the number of reflections observed in XRD diffractograms. Due to its very small crystal domain size, the exact structure of ferrihydrite has remained elusive for many decades, although the most recently accepted model by Michel et al. suggests that the mineral is dominated by octahedrally coordinated Fe(III) with some tetrahedral Fe(III) that makes up to 20% of the structure (Michel et al., 2007). This contrasts with the Drits model, which suggested that ferrihydrite had a mixture of hematite-like coordination and defects (Drits et al., 1993). Nevertheless, debate still surrounds this mineral, and Mössbauer spectroscopy has so far been unable to conclusively support the Michel model despite some indications of tetrahedral doublets (Murad, 1988; Pankhurst and Pollard, 1992). As a poorly crystalline (or short range ordered) mineral phase, ferrihydrite always displays a doublet when measured at room temperature, indicating that it is not magnetically ordered or at least has T_B well below 295 K. Room temperature IS and ΔE_Q vary between 0.3–0.5 and 0.65–0.80 mm/s, respectively. These values are influenced by substituted cations, such as aluminum or nickel, which cause distortions to the mineral structure (Chadwick et al., 1986), or even appear to show differences depending on mineral formation conditions, such as abiotic or biogenic processes (Eickhoff et al., 2014). As the measurement temperature decreases, ferrihydrite undergoes magnetic ordering, with the T_B commonly observed below 77 K, though this is affected by the presence of organics or impurities (Chen et al., 2015; Mikutta et al., 2008; Ziganshin et al., 2015). At 4.2 K, ferrihydrite is usually fully magnetically ordered, with a hyperfine field of ~48 T, a quadrupole shift close to zero, and an IS of ~0.48 mm/s (Eickhoff et al., 2014).

13.4.3.2 Goethite

Goethite (α-FeOOH) is extremely common in the environment and is composed of Fe(III) in octahedral coordination. It is frequently observed to form as a product of microbial Fe(II) oxidation (Hohmann et al., 2010; Larese-Casanova et al., 2010; Pantke et al., 2012) and has been seen to form when microbial reduction of ferrihydrite takes place in the presence of a bicarbonate buffer system over a protracted length of time (Byrne et al., 2011). The fully magnetically ordered goethite sextet is characterized by hyperfine field parameters that are close to those of ferrihydrite, which often makes distinction between the two minerals difficult; however, room-temperature B_{hf} of 38 T is often reported, with IS of 0.37 mm/s and quadrupole splitting of −0.26 mm/s (Murad, 2010). The low ΔE_Q is a useful indicator to distinguish it from ferrihydrite, which has a value close to zero when fully magnetically ordered. It is also important to consider superparamagnetic or nanogoethite, which is often reported in samples and is characterized by a doublet at room temperature. In fact, van der Zee et al. consider that nanogoethite plays a more dominant role in the environment than previously considered, particularly in lake and marine sediments (van der Zee et al., 2003). At liquid He temperatures, B_{hf} increases to a value of ~50.6 T with ΔE_Q −0.25 mm/s, while IS rises to ~0.49 mm/s (Murad, 2010).

13.4.3.3 Magnetite

Magnetite (Fe_3O_4) is a ferrimagnetic mineral which exists naturally in the environment and is often seen to be formed as a result of rock weathering. In recent decades, the formation of magnetite by various microorganisms has been described. This can take place within the cell, as in the case of magnetotactic bacteria (e.g., *Magnetospirillum magnetotacticum* strain MS-1), which form discrete islands of the magnetite within organic membranes (magnetosomes). These grains of magnetite are thought to be utilized for navigation (Frankel et al., 1983), although relatively few studies have taken advantage of the use of Mössbauer to probe their characteristics. Alternatively, dissimilatory iron(III)-reducing bacteria (DIRBs) can induce magnetite formation outside the cell wall through bioreduction of Fe(III) in ferrihydrite or lepidocrocite (Zegeye et al., 2011). It has also been shown that magnetite precipitation can be induced by microbial Fe(II) oxidation (Chaudhuri et al., 2001), and via the microbial oxidation of green rust (Miot and Etique, 2016; Miot et al., 2014).

Magnetite is a mixed-valent iron mineral with eight Fe(II) and eight Fe(III) ions in octahedral (Fe_{Oh}) coordination combined with eight Fe(III) in tetrahedral (Fe_{Td}) coordination and has the general formula $Fe(II)_{Oh}Fe(III)_{Td}Fe(III)_{Oh}O_4^{2-}$. Both Fe (II) and Fe(III) are magnetic ions with magnetic moments of 4 and $5\mu_B$, respectively. The magnetic moments of the Fe_{Oh} and Fe_{Td} lattice sites orientate themselves in antiparallel orientation, resulting in the magnetic moments of the Fe(III) effectively canceling each other out, leaving Fe(II) as the dominant factor in the magnetization of magnetite. The presence of the different lattice sites gives rise to the highly characteristic room-temperature Mössbauer spectrum of magnetite (Figure 13.6) comprising two overlapping sextets, which correspond to the $Fe^{3+}{}_{Td}$ site and the Fe (II)Fe(III)$^{2.5+}{}_{Oh}$ site. The combination of both octahedral ions is commonly attributed to electron hopping between the $Fe(II)_{Oh}$ and $Fe(III)_{Oh}$ sites at room temperature (Kündig and Steven Hargrove, 1969). At 140 K, stoichiometric magnetite in general has an Fe_{Oh} sextet with IS ~ 0.72 mm/s, $\varepsilon \sim -0.02$ mm/s, and B_{hf} 47.4 T, whereas the wider Fe_{Td} sextet, which is best observed by looking at the lowest-velocity peak on the magnetite spectrum, has a lower IS ~ 0.38 mm/s and $\varepsilon \sim 0.00$ mm/s but a higher B_{hf} of 50.2 T (Gorski and Scherer, 2010). As measurement temperature decreases below the Verwey transition (~121 K), the octahedral site splits into additional sextets, resulting in the low-temperature spectrum of magnetite having at least three different sextets. In fact, several studies have indicated that the magnetite Mössbauer spectrum could comprise as many as five different sextets (Srivastava et al., 1981). In principle, it is possible to determine the stoichiometry (i.e., Fe(II)/Fe(III) ratio) from magnetite spectra based on the relative areas of Fe_{Oh} and Fe_{Td} sextets (Gorski and Scherer, 2010):

$$Fe(II)/Fe(III) = (0.5 * Fe_{Oh})/(0.5 * Fe_{Oh} + Fe_{Td}) \qquad (Eq.13.2)$$

While Eq. (13.2) can be applied to spectra at room temperature, it is important to consider that many types of magnetite, such as biogenic magnetite, are often nanoparticulate and thus are not fully magnetically ordered at 295 K, which makes the calculation inaccurate. Particles with a diameter of less than ~30 nm are generally described as superparamagnetic and are not fully magnetically ordered at room temperature. As a result, spectra obtained for magnetite particles below this size threshold might consist of a combination of doublets and sextets all corresponding to the mineral. Measuring below the T_B can alleviate this issue and should result in the more commonly observed spectra. For this reason, many studies measure magnetite at 140 K, which is above the Verwey transition but below the blocking temperature of the mineral. This negates the requirement for complex multicomponent fitting and also avoids complications arising from the different *f*-factors (Section 13.1) of Fe_{Oh} and Fe_{Td} sites at high temperature (Gorski and Scherer, 2010). Mössbauer spectroscopy was recently applied to magnetite nanoparticles (~12 nm) that had undergone microbial Fe(II) oxidation by *Rhodopseudomonas palustris* TIE-1 followed by microbial Fe(III) reduction by *G. sulfurreducens* (Byrne et al., 2015). Using Eq. (13.2) on spectra obtained at 140 K, it was possible to identify the fact that the stoichiometry of the magnetite decreased from 0.46 to 0.42 after microbial oxidation before increasing again to 0.46 after microbial reduction.

Mössbauer spectroscopy can also be used to investigate elemental substitution in magnetite. This occurs frequently in the environment and has been shown for Zn- and Co-incorporated biogenic magnetite, whereby the incorporation of the different elements into either Fe_{Oh} and Fe_{Td} sites can cause a shift in IS and B_{hf} (Byrne et al., 2013, 2014).

13.4.3.4 Hematite

Hematite (α-Fe_2O_3) is often found as a major component of many different iron ores and banded iron formations throughout the world (Klein, 2005) and has also been detected as a major component of the iron minerals present on Mars (Klingelhöfer et al., 2004, 2006). Its role in supporting the growth and respiration of bacteria is still not fully understood. However, several studies have indicated that hematite can undergo a limited amount of microbial Fe(III) reduction, suggesting that it is potentially bioavailable (Cutting et al., 2009; Roden and Zachara, 1996). The mineral consists of a hexagonal arrangement of oxygen atoms in which two-thirds of the octahedral sites are occupied by Fe(III). Hematite possesses several different magnetic ordering types, including paramagnetism above the Curie temperature (T_C = 955 K), weak ferromagnetism above the Morin temperature (T_M ~ 264 K), and antiferromagnetism below T_M.

In general, the room-temperature Mössbauer spectrum of hematite displays a very wide sextet that is typically around 51.8 T with IS of 0.37 mm/s and quadrupole shift of −0.19 mm/s (Fysh and Clark, 1982; Murad, 2010; Murad and Schwertmann, 1986). In fully magnetically ordered hematite, the magnetic hyperfine field increases to a value of 54.2 T for bulk antiferromagnetic hematite when measured at 4.2 K. These values represent some of the highest B_{hf} in any of the iron (oxyhydr)oxides, thus making the identification of hematite relatively simple. Furthermore, there is often little difference between the hyperfine field at 77 K and at 4.2 K. Changes to the hyperfine field value can be an indicator that the sample is either poorly crystalline, has a small particle size, or contains impurities such as aluminum or titanium (Ericsson et al., 1986; Fysh and Clark, 1982; Murad and Schwertmann, 1986). Exposure to radiation has also been shown to alter the crystalline structure of the mineral (Brown et al., 2014). Interestingly, the type of magnetic ordering in hematite has a major impact on ε, which is approximately −0.20 mm/s when weakly ferromagnetic. In this state, the magnetic spin orientations are canted; however, as the temperature of measurement is decreased and passes through T_M, the angle of the spin canting decreases, leading to antiferromagnetism with layers of oppositely orientated magnetic moments. Below T_M, the ε increases to a value of 0.41 mm/s (Murad, 2010). The actual value of T_M is not well defined, as changes to the particle size, or the incorporation of additional elements, can cause the Morin transition to decrease in temperature (Ericsson et al., 1986) or even be suppressed in the case of very small particle size (>18 nm) (Kündig et al., 1966).

13.4.3.5 Lepidocrocite

The Fe(III) mineral lepidocrocite (γ-FeOOH) is often formed as a result of the oxidative weathering of iron-bearing minerals, although it has also been reported to form as a result of microbial Fe(II) oxidation (Fortin et al., 1993; Larese-Casanova et al., 2010; O'Loughlin et al., 2007) and is also commonly observed to co-occur with ferrihydrite in the environment (ThomasArrigo et al., 2014). Vollrath et al. used Mössbauer spectroscopy to demonstrate that the formation of iron minerals by the microaerophilic Fe(II)-oxidizer *Leptothrix cholodnii* Appels was dependent upon temperature of incubation, with lepidocrocite dominant when the cultures were incubated at 11 °C, whereas the mineral was dominated by ferrihydrite when incubated at 37 °C (Vollrath et al., 2013). While lepidocrocite is observed relatively easily by XRD, it is paramagnetic at room temperature, displaying a doublet with IS ~ 0.37 mm/s and ΔE_Q ~ 0.53 mm/s (Cornell and Schwertmann, 2003). This contrasts with ferrihydrite, which is poorly observed by XRD and yet is also characterized by a doublet at room temperature. As a result, both minerals are generally hard to distinguish using Mössbauer spectroscopy when measured at room temperature. However, in principle, the narrower ΔE_Q combined with a narrower linewidth of lepidocrocite can be used as an indicator of the mineral's identity. Due to its low Néel temperature, the mineral remains paramagnetic until below at least 77 K; however, at 4.2 K it shows a characteristic sextet with IS ~ 0.47 mm/s, ΔE_Q ~ 0.02 mm/s, and B_{hf} ~45.8 T, which is the lowest of all magnetically ordered Fe (oxyhydr)oxides (Cornell and Schwertmann, 2003; Hirt et al., 2002).

13.4.3.6 Green Rust

The term green rust (GR) is often used to describe a number of unstable hydroxyl salt minerals that comprise positively charged Fe(II)–Fe(III) sheets separated by layers of water molecules and anions, including SO_4^{2-}, CO_3^{2-}, and Cl^- (Bernal et al., 1959). The general formula for GR is $([Fe^{2+}{}_{(1-x)}Fe^{3+}{}_x(OH)_2]^{x+}\cdot[(x/n)A^{n-}\cdot(m/n)H_2O]^{x-})$ where A^{n-} denotes intercalated anions and x indicates the molar fraction of trivalent cation, which usually ranges from 0.25 to 0.33. GR has previously been shown to form as an intermediate phase during the microbial oxidation of aqueous Fe^{2+} to ferrihydrite (Pantke et al., 2012) and also as a precursor to magnetite during nitrate-dependent Fe(II) oxidation (Chaudhuri et al., 2001; Miot et al., 2014). The presence of both Fe(II) and Fe(III) adds to its suitability for measurement using Mössbauer spectroscopy, as the two sites are observable as distinct doublets; however, the rapid oxidation of GR in air can cause issues during the determination of its true Fe(II)/Fe(III) ratio and can also contribute to the emergence of multiple doublets. In order to limit this, many studies measure the samples at low temperature (77 K and 4.2 K), where abiotic oxidation takes place much more slowly (Cuttler et al., 1990). Mössbauer spectroscopy shows good suitability for the measurement of GRs, with the ability to distinguish between the different anion groups depending upon the hyperfine parameters and relative ratios of Fe(II)/Fe(III). For instance, Génin et al. summarized several studies indicating that stoichiometric GR(SO_4^{2-}) comprises two doublets corresponding to Fe(II) and Fe(III), with Fe(II)/Fe(III) of 1.9. GR(CO_3^{2-}), however, is best fitted with a third Fe(II) doublet, also with Fe(II)/Fe(III) of 1.9 (Génin et al., 1998). Finally, GR (Cl^-) is also observed to comprise three doublets, though with an Fe(II)/Fe(III) of 2.7. These values can vary depending upon the level of oxidation, and it is also important to apply caution to the interpretation of GR doublets, as their parameters are often indistinguishable from those belonging to Fe-bearing clays (Latta et al., 2012).

13.4.3.7 Siderite

Siderite ($FeCO_3$) is an Fe(II) mineral with a whitish color that has been observed to form due to the rapid reduction of ferrihydrite by *G. sulfurreducens* (Byrne et al., 2011) or in microbial biofilms (Sawicki and Brown, 1998) formed on an exposed granitic rock. Siderite does not show any magnetic ordering at room temperature and is characterized by a doublet with high IS and high ΔE_Q (Housley et al., 1968). The blocking temperature of siderite is reportedly very low, and it does not show magnetic ordering until below 38 K (Forester and Koon, 1969).

13.4.3.8 Vivianite

The extensive use of phosphate in liquid growth media to support microbial growth often results in the formation of the phosphate Fe(II) mineral vivianite, $Fe_3(PO_4)_2{\cdot}8H_2O$, during the addition of $FeCl_2$ into the medium. Furthermore, the mineral has been seen to form as a result of microbial reduction of magnetite (Dong et al., 2000). Vivianite is known to have a monoclinic structure and possesses Fe(II) at two distinct positions, Fe(1) and Fe(2), which are in relative proportions of 1:2. These two sites result in the emergence of two distinct well-split doublets (Forsyth et al., 1970; Greenwood and Gibb, 1971) at room temperature. Vivianite does not show any magnetic ordering until close to liquid He temperature (Gonser and Grant, 1967).

13.4.3.9 Mineral Identity Summary

There are a huge number of different iron minerals present in the environment. However, we have chosen to cover only some of the more common ones relevant for the field of geomicrobiology. Table 13.1 summarizes the parameters of a selected number of relevant minerals, although these values should only be taken as a general guide, as they do not reflect the potential impacts of impurities or particle size.

Table 13.1 Mössbauer parameters for some common Fe minerals found in the environment. Many minerals show different parameters depending upon the temperature of measurement, and so in some cases more than one set of parameters may be present. The reference corresponding to each set of mineral parameters is also included.

Mineral	T (K)	Phase	IS (mm/s)	ΔE_Q (mm/s)	B_{hf} (T)	R. A. (%)	Reference
Biogenic ferrihydrite	295	Fe(III)	0.37	0.76			(Eickhoff et al., 2014)
	5	Fe(III)	0.48	−0.04	47.4		(Eickhoff et al., 2014)
Goethite	295	Fe(III)	0.37	−0.26	38.0		(Murad, 2010)
	4.2	Fe(III)	0.48[a]	−0.25	50.6		(Murad, 2010)
Green rust (SO_4^{2-})	78	Fe(II)	1.27	2.88		66	(Génin et al., 1998)
		Fe(II)	0.47	0.44		34	
Green rust (CO_3^{2-})	78	Fe(II)	1.27	2.93		51	(Génin et al., 1998)
		Fe(II)	1.28	2.64		15	
		Fe(III)	0.47	0.42		34	
Green rust (Cl)	78	Fe(II)	1.26	2.80		36	(Génin et al., 1998)
		Fe(II)	1.27	2.55		37	
		Fe(III)	0.47	0.44		27	
Greigite	300	$Fe(III)_{Td}$	0.28	0.00	31.3		(Vandenberghe et al., 1992)
		$Fe(II)Fe(III)_{oh}$	0.43	−0.08	31.3		
	5	$Fe(III)_{Td}$	0.37	0.00	31.8		(Vandenberghe et al., 1992)
		$Fe(II)Fe(III)_{oh}$	0.71	−0.03	33.0		
Hematite	295	Fe(III)	0.37	−0.19	51.75		(Cornell and Schwertmann, 2003)
	4.2	Fe(III)	0.49	0.41	54.17		(Cornell and Schwertmann, 2003)
Lepidocrocite	295	Fe(III)	0.37	0.64			(Cornell and Schwertmann, 2003)
	4.2	Fe(III)	0.47	0.02	45.8		(Cornell and Schwertmann, 2003)
Magnetite	140	$Fe(III)_{Td}$	0.38	0.00	50.2	35.9	(Gorski and Scherer, 2010)
		$Fe(II)Fe(III)_{oh}$	0.72	−0.04	47.4	64.1	
Pyrite	295	Fe(II)	0.33[b]	0.61			(Morice et al., 1969)
	4.2	Fe(II)	0.42	0.60			(Wan et al., 2017)
Siderite	295	Fe(II)	1.24	1.80			(Housley et al., 1968)
	4.2	Fe(II)	1.36	2.06	18.4		(Forester and Koon, 1969)

Table 13.1 (cont.)

Mineral	T (K)	Phase	IS (mm/s)	ΔE_Q (mm/s)	B_{hf} (T)	R. A. (%)	Reference
Vivianite	80	Fe(II)	1.30[b]	2.59			(Gonser and Grant, 1967)
		Fe(II)	1.34[b]	3.16			
	5	Fe(II)	1.31[b]	2.59	26.8		(Gonser and Grant, 1967)
		Fe(II)	1.34[b]	3.18	13.5		

ΔE_Q: quadrupole splitting ($\Delta E_Q = 2\varepsilon$), where ε is quadrupole shift; B_{hf}: hyperfine field; IS: isomer shift; R.A.:– relative abundance.

[a]From Cornell and Schwertmann (2003).

[b]value recalculated for α-Fe(0).

13.5 Conclusion

The aim of this book chapter has been to provide an overview of some of the most crucial aspects to consider when starting to learn about Mössbauer spectroscopy. By the end of this chapter, you, the reader, should have a firm grasp of how the Mössbauer effect works, how to analyze the spectra that are produced as a result, and most importantly, how to best interpret the results of the data analysis. You should also now appreciate how to use Mössbauer spectroscopy for determining the oxidation state of an Fe-containing sample and to some degree, be able to identify the major mineral phases contained within your sample. However, caution must be applied when using Mössbauer for mineral identification, as its use as a fingerprinting technique is often beset with problems that arise due to the effect of many potential variables, including particle size, elemental substitution, the presence of organics, etc.

The future outlook for the use of Mössbauer spectroscopy in environmental geosciences, particularly geomicrobiology, is very positive, with an increasing number of international research groups using the technique. Nevertheless, it is important that the number of people who regularly analyze and interpret the data continues to grow. Hopefully, the material that has been described here will help to promote this growth.

13.6 References

Bernal, J. D., Dasgupta, D. R. and Mackay, A. L. 1959. The oxides and hydroxides of iron and their structural inter-relationships. *Clay Minerals Bulletin*, 4, 15.

Brown, A. R., Wincott, P. L., Laverne, J. A., et al. 2014. The impact of γ radiation on the bioavailability of Fe(III) minerals for microbial respiration. *Environmental Science & Technology*, 48, 10672–10680.

Byrne, J. M., Coker, V. S., Cespedes, E., et al. 2014. Biosynthesis of zinc substituted magnetite nanoparticles with enhanced magnetic properties. *Advanced Functional Materials*, 24, 2518–2529.

Byrne, J. M., Coker, V. S., Moise, S., et al. 2013. Controlled cobalt doping in biogenic magnetite nanoparticles. *Journal of the Royal Society Interface*, 10, 20130134.

Byrne, J. M., Klueglein, N., Pearce, C., et al. 2015. Redox cycling of Fe(II) and Fe(III) in magnetite by Fe-metabolizing bacteria. *Science*, 347, 1473–1476.

Byrne, J. M., Telling, N. D., Coker, V. S., et al. 2011. Control of nanoparticle size, reactivity and magnetic properties during the bioproduction of magnetite by Geobacter sulfurreducens. *Nanotechnology*, 22, 455709.

Chadwick, J., Jones, D. H., Thomas, M. F., Tatlock, G. J. and Devenish, R. W. 1986. A Mössbauer study of ferrihydrite and aluminium substituted ferrihydrites. *Journal of Magnetism and Magnetic Materials*, 61, 88–100.

Chaudhuri, S. K., Lack, J. G. and Coates, J. D. 2001. Biogenic magnetite formation through anaerobic biooxidation of Fe(II). *Applied and Environmental Microbiology*, 67, 2844–2848.

Chen, C., Kukkadapu, R. and Sparks, D. L. 2015. Influence of coprecipitated organic matter on Fe2+(aq)-catalyzed transformation of ferrihydrite: Implications for carbon dynamics. *Environmental Science & Technology*, 49, 10927–10936.

Cornell, R. M. and Schwertmann, U. 2003. *The Iron Oxides: Structure, Properties, Reactions, Occurrences and Uses*, Weinheim, Germany, Wiley-VCH.

Cutting, R. S., Coker, V. S., Fellowes, J. W., Lloyd, J. R. and Vaughan, D. J. 2009. Mineralogical and morphological constraints on the reduction of Fe(III) minerals by Geobacter sulfurreducens. *Geochimica et Cosmochimica Acta*, 73, 4004–4022.

Cuttler, A., Man, V., Cranshaw, T. and Longworth, G. 1990. A Mössbauer study of green rust precipitates: I. Preparations from sulphate solutions. *Clay Minerals*, 25, 289–301.

De Grave, E. and Van Alboom, A. 1991. Evaluation of ferrous and ferric Mössbauer fractions. *Physics and Chemistry of Minerals*, 18, 337–342.

Dippon, U., Pantke, C., Porsch, K., Larese-Casanova, P. and Kappler, A. 2012. Potential function of added minerals as nucleation sites and effect of humic substances on mineral formation by the nitrate-reducing Fe(II)-oxidizing strain *Acidovorax* sp. BoFeN1. *Environmental Science & Technology*, 46, 6556–6565.

Dippon, U., Schmidt, C., Behrens, S. and Kappler, A. 2015. Secondary mineral formation during ferrihydrite reduction by *Shewanella oneidensis* MR-1 depends on incubation vessel orientation and resulting gradients of cells, Fe^{2+} and Fe^{3+} minerals. *Geomicrobiology Journal*, 32, 868–877.

Dong, H., Fredrickson, J. K., Kennedy, D. W., et al. 2000. Mineral transformations associated with the microbial reduction of magnetite. *Chemical Geology*, 169, 299–318.

Drits, V., Sakharov, B., Salyn, A. and Manceau, A. 1993. Structural model for ferrihydrite. *Clay Minerals*, 28, 185–207.

Dyar, M. D., Agresti, D. G., Schaefer, M. W., Grant, C. A. and Sklute, E. C. 2006. Mössbauer spectroscopy of earth and planetary materials. *Annual Review of Earth and Planetary Sciences*, 34, 83–125.

Eickhoff, M., Obst, M., Schröder, C., et al. 2014. Nickel partitioning in biogenic and abiogenic ferrihydrite: The influence of silica and implications for ancient environments. *Geochimica et Cosmochimica Acta*, 140, 65–79.

Ericsson, T., Krishnamurthy, A. and Srivastava, B. K. 1986. Morin-transition in Ti-substituted hematite: A Mössbauer study. *Physica Scripta*, 33, 88.

Forester, D. W. and Koon, N. C. 1969. Mössbauer investigation of metamagnetic $FeCO_3$. *Journal of Applied Physics*, 40, 1316–1317.

Forsyth, J. B., Johnson, C. E. and Wilkinson, C. 1970. The magnetic structure of vivianite, $Fe_3(PO_4)_2.8H_2O$. *Journal of Physics C: Solid State Physics*, 3, 1127.

Fortin, D., Leppard, G. G. and Tessier, A. 1993. Characteristics of lacustrine diagenetic iron oxyhydroxides. *Geochimica et Cosmochimica Acta*, 57, 4391–4404.

Frankel, R. B., Papaefthymiou, G. C., Blakemore, R. P. and O'Brien, W. 1983. Fe3O4 precipitation in magnetotactic bacteria. *Biochimica et Biophysica Acta (BBA) – Molecular Cell Research*, 763, 147–159.

Fysh, S. and Clark, P. 1982. Aluminous hematite: A Mössbauer study. *Physics and Chemistry of Minerals*, 8, 257–267.

Génin, J.-M. R., Bourrié, G., Trolard, F., et al. 1998. Thermodynamic equilibria in aqueous suspensions of synthetic and natural Fe(II)–Fe(III) green Rusts: Occurrences of the mineral in hydromorphic soils. *Environmental Science & Technology*, 32, 1058–1068.

Gonser, U. and Grant, R. W. 1967. Determination of spin directions and electric field gradient axes in vivianite by polarized recoil-free γ-rays. *Physica Status Solidi (B)*, 21, 331–342.

Gorski, C. A., Handler, R. M., Beard, B. L., et al. 2012. Fe atom exchange between aqueous Fe^{2+} and magnetite. *Environmental Science & Technology*, 46, 12399–12407.

Gorski, C. A. and Scherer, M. M. 2010. Determination of nanoparticulate magnetite stoichiometry by Mössbauer spectroscopy, acidic dissolution, and powder X-ray diffraction: A critical review. *American Mineralogist*, 95, 1017–1026.

Greenwood, N. N. and Gibb, T. C. 1971. Mössbauer Spectroscopy, London, Chapman and Hall Ltd.

Gütlich, P., Bill, E. and Trautwein, A. X. 2010. *Mössbauer Spectroscopy and Transition Metal Chemistry: Fundamentals and Applications*, Berlin Heidelberg, Springer Science & Business Media.

Gütlich, P., Schröder, C. and Schünemann, V. 2012. Mössbauer spectroscopy—an indispensable tool in solid state research. *Spectroscopy Europe*, 24, 21.

Handler, R. M., Beard, B. L., Johnson, C. M. and Scherer, M. M. 2009. Atom exchange between aqueous Fe(II) and goethite: An Fe isotope tracer study. *Environmental Science & Technology*, 43, 1102–1107.

Hirt, A. M., Lanci, L., Dobson, J., Weidler, P. and Gehring, A. U. 2002. Low-temperature magnetic properties of lepidocrocite. *Journal of Geophysical Research: Solid Earth*, 107, EPM 5-1-EPM 5–9.

Hohmann, C., Winkler, E., Morin, G. and Kappler, A. 2010. Anaerobic Fe(II)-oxidizing bacteria show As resistance and immobilize As during Fe(III) mineral precipitation. *Environmental Science & Technology*, 44, 94–101.

Housley, R. M., Gonser, U. and Grant, R. W. 1968. Mössbauer determination of the Debye-Waller factor in single-crystal absorbers. *Physical Review Letters*, 20, 1279–1282.

Klein, C. 2005. Some Precambrian banded iron-formations (BIFs) from around the world: Their age, geologic setting, mineralogy, metamorphism, geochemistry, and origins. *American Mineralogist*, 90, 1473–1499.

Klingelhöfer, G., Morris, R. V., Bernhardt, B., et al. 2003. Athena MIMOS II Mössbauer spectrometer investigation. *Journal of Geophysical Research: Planets*, 108, 8067.

Klingelhöfer, G., Morris, R. V., Bernhardt, B., et al. 2004. Jarosite and hematite at Meridiani Planum from Opportunity's Mössbauer spectrometer. *Science*, 306, 1740–1745.

Klingelhöfer, G., Morris, R. V., De Souza, P. A., Jr., Rodionov, D. and Schröder, C. 2006. Two earth years of Mössbauer studies of the surface of Mars with MIMOS II. *Hyperfine Interactions*, 170, 169–177.

Klueglein, N. and Kappler, A. 2012. Abiotic oxidation of Fe(II) by reactive nitrogen species in cultures of the nitrate-reducing Fe(II) oxidizer *Acidovorax* sp. BoFeN1 – questioning the existence of enzymatic Fe(II) oxidation. *Geobiology*, 11, 396.

Kündig, W., Bömmel, H., Constabaris, G. and Lindquist, R. H. 1966. Some properties of supported small a-Fe_2O_3 particles determined with the Mössbauer effect. *Physical Review*, 142, 327–333.

Kündig, W. and Steven Hargrove, R. 1969. Electron hopping in magnetite. *Solid State Communications*, 7, 223–227.

Lagarec, K. and Rancourt, D. G. 1997. Extended Voigt-based analytic lineshape method for determining N-dimensional correlated hyperfine parameter distributions in Mössbauer spectroscopy. *Nuclear Instruments and Methods in Physics Research Section B: Beam Interactions with Materials and Atoms*, 129, 266–280.

Larese-Casanova, P., Haderlein, S. B. and Kappler, A. 2010. Biomineralization of lepidocrocite and goethite by nitrate-reducing Fe(II)-oxidizing bacteria: Effect of pH, bicarbonate, phosphate, and humic acids. *Geochimica et Cosmochimica Acta*, 74, 3721–3734.

Larese-Casanova, P. and Scherer, M. M. 2007. Fe(II) sorption on hematite: New insights based on spectroscopic measurements. *Environmental Science & Technology*, 41, 471–477.

Latta, D. E., Boyanov, M. I., Kemner, K. M., et al. 2012. Abiotic reduction of uranium by Fe(II) in soil. *Applied Geochemistry*, 27, 1512–1524.

Markovski, C., Byrne, J. M., Lalla, E., et al. 2017. Abiotic versus biotic iron mineral transformation studied by a miniaturized backscattering Mössbauer spectrometer (MIMOS II), X-ray diffraction and Raman spectroscopy. *Icarus*, 296, 49–58.

Michel, F. M., Ehm, L., Antao, S. M., et al. 2007. The structure of ferrihydrite, a nanocrystalline material. *Science*, 316, 1726–1729.

Mikutta, C., Mikutta, R., Bonneville, S., et al. 2008. Synthetic coprecipitates of exopolysaccharides and ferrihydrite. Part I: Characterization. *Geochimica et Cosmochimica Acta*, 72, 1111–1127.

Miot, J. and Etique, M. 2016. Formation and Transformation of Iron-Bearing Minerals by Iron(II)-Oxidizing and Iron(III)-Reducing Bacteria. In *Iron Oxides*, ed. Faivre, D., Weinheim, Germany, Wiley-VCH Verlag GmbH & Co. KGaA.

Miot, J., Li, J., Benzerara, K., et al. 2014. Formation of single domain magnetite by green rust oxidation promoted by microbial anaerobic nitrate-dependent iron oxidation. *Geochimica et Cosmochimica Acta*, 139, 327–343.

Morice, J. A., Rees, L. V. C. and Rickard, D. T. 1969. Mössbauer studies of iron sulphides. *Journal of Inorganic and Nuclear Chemistry*, 31, 3797–3802.

Mossbauer, R. L. 1958. Kernresonanzfluoreszenz von gammastrahlung in Ir-191. *Zeitschrift Fur Physik*, 151, 124–143.

Murad, E. 1988. The Mössbauer spectrum of "well"-crystallized ferrihydrite. *Journal of Magnetism and Magnetic Materials*, 74, 153–157.

Murad, E. 2010. Mössbauer spectroscopy of clays, soils and their mineral constituents. *Clay Minerals*, 45, 413–430.

Murad, E. and Cashion, J. 2004. Mössbauer Spectroscopy of Environmental Materials and their Industrial Utilization, USA, Kluwer Academic Publishers.

Murad, E. and Schwertmann, U. 1986. Influence of Al substitution and crystal size on the room-temperature Mössbauer spectrum of hematite. *Clays and Clay Minerals*, 34, 1–6.

Nitzsche, K. S., Lan, V. M., Trang, P. T. K., et al. 2015. Arsenic removal from drinking water by a household sand filter in Vietnam – Effect of filter usage practices on arsenic removal efficiency and microbiological water quality. *Science of the Total Environment*, 502, 526–536.

O'Loughlin, E. J.,Larese-Casanova,P.,Scherer,M. and Cook,R. 2007. Green rust formation from the bioreduction of γ–FeOOH (lepidocrocite): Comparison of several Shewanella species. *Geomicrobiology Journal*, 24, 211–230.

Pankhurst, Q. and Pollard, R. 1992. Structural and magnetic properties of ferrihydrite. *Clays and Clay Minerals*, 40, 268–272.

Pantke, C., Obst, M., Benzerara, K., et al. 2012. Green rust formation during Fe(II) oxidation by the nitrate-reducing *Acidovorax* sp. strain BoFeN1. *Environmental Science & Technology*, 46, 1439–1446.

Prescher, C., McCammon, C. and Dubrovinsky, L. 2012. MossA: A program for analyzing energy-domain Mossbauer spectra from conventional and synchrotron sources. *Journal of Applied Crystallography*, 45, 329–331.

Rancourt, D., McDonald, A., Lalonde, A. and Ping, J. 1993. Mössbauer absorber thicknesses for accurate site populations in Fe-bearing minerals. *American Mineralogist*, 78, 1–7.

Rancourt, D. G. and Ping, J. Y. 1991. Voigt-based methods for arbitrary-shape static hyperfine parameter distributions in Mössbauer spectroscopy. *Nuclear Instruments and Methods in Physics Research Section B: Beam Interactions with Materials and Atoms*, 58, 85–97.

Roden, E. E. and Zachara, J. M. 1996. Microbial reduction of crystalline iron(iii) oxides: Influence of oxide surface area and potential for cell growth. *Environmental Science & Technology*, 30, 1618–1628.

Sawicki, J. and Brown, D. 1998. Investigation of microbial–mineral interactions by Mössbauer spectroscopy. *Hyperfine Interactions*, 117, 371–382.

Srivastava, C. M., Shringi, S. N. and Babu, M. V. 1981. Mössbauer study of the low-temperature phase of magnetite. *Physica Status Solidi (A)*, 65, 731–735.

Stookey, L. L. 1970. Ferrozine – a new spectrophotometric reagent for iron. *Analytical Chemistry*, 42, 779–781.

Swanner, E. D., Wu, W., Schoenberg, R., et al. 2015. Fractionation of Fe isotopes during Fe (II) oxidation by a marine photoferrotroph is controlled by the formation of organic Fe-complexes and colloidal Fe fractions. *Geochimica et Cosmochimica Acta*, 165, 44–61.

ThomasArrigo, L. K., Mikutta, C., Byrne, J., et al. 2014. Iron and arsenic speciation and distribution in organic flocs from streambeds of an arsenic-enriched peatland. *Environmental Science & Technology*, 48, 13218–13228.

van der Zee, C., Roberts, D. R., Rancourt, D. G. and Slomp, C. P. 2003. Nanogoethite is the dominant reactive oxyhydroxide phase in lake and marine sediments. *Geology*, 31, 993–996.

Vandenberghe, R. E., De Grave, E., De Bakker, P. M. A., Krs,M. and Hus, J. J. 1992. Mössbauer effect study of natural greigite. *Hyperfine Interactions*, 68, 319–322.

Vollrath, S., Behrends, T., Koch, C. B. and Cappellen, P. V. 2013. Effects of temperature on rates and mineral products of microbial Fe(II) oxidation by *Leptothrix cholodnii* at microaerobic conditions. *Geochimica et Cosmochimica Acta*, 108, 107–124.

Wan, M., Schröder, C. and Peiffer, S. 2017. Fe(III):S(-II) concentration ratio controls the pathway and the kinetics of pyrite formation during sulfidation of ferric hydroxides. *Geochimica et Cosmochimica Acta*, 217, 334–348.

Williams, A. G. B. and Scherer, M. M. 2004. Spectroscopic evidence for Fe(II)–Fe(III) electron transfer at the iron oxide–water interface. *Environmental Science & Technology*, 38, 4782–4790.

Žák, T. and Jirásková, Y. 2006. CONFIT: Mössbauer spectra fitting program. *Surface and Interface Analysis*, 38, 710–714.

Zegeye, A., Abdelmoula, M., Usman, M., Hanna, K. and Ruby, C. 2011. In situ monitoring of lepidocrocite bioreduction and magnetite formation by reflection Mössbauer spectroscopy. *American Mineralogist*, 96, 1410–1413.

Ziganshin, A. M., Ziganshina, E. E., Byrne, J., et al. 2015. Fe(III) mineral reduction followed by partial dissolution and reactive oxygen species generation during 2,4,6-trinitrotoluene transformation by the aerobic yeast Yarrowia lipolytica. *AMB Express*, 5(8), 1–12.

PART V

MICROBIOLOGICAL TECHNIQUES

14 Lipid Biomarkers in Geomicrobiology: Analytical Techniques and Applications

JIASONG FANG, SHAMIK DASGUPTA, LI ZHANG, AND WEIQIANG ZHAO

Abstract

Lipid biomarker analysis is a useful tool for characterizing microbial communities in geomicrobiology. Phospholipid fatty acids (PLFA) are major components of microbial membranes, and analysis of these markers provides insight into microbial biomass, community structure, and metabolic processes. This article reviews the methods for extraction, fractionation, derivatization, and quantification of PLFA, as well as the interpretation of PLFA patterns for microbial community analysis in natural environmental systems. The discussion centers on the development, the subsequent modifications, and the advantages and limitations of the methods. Two case studies are given to illustrate the applications of intact phospholipid profiling (IPP) and PLFA in geomicrobiology. The recent developments and future directions of microbial signature lipid analysis are also discussed.

14.1 Introduction

The most challenging task facing geomicrobiologists today is the quantitative characterization of microbial communities. The classical culture-dependent microbial tests are inadequate for analysis of microbial communities in environmental samples, as <1% of all microscopically and/or molecular genetically detected cells can be isolated and cultured in a growth medium in the lab (e.g., Ciobanu et al., 2014; Schloss and Jo Handelsman, 2004). Analysis of phospholipid fatty acids (PLFA), together with DNA sequencing, is one of the currently accepted culture-independent methods for characterizing microbial communities.

Phospholipids are biomolecules present in cell membranes of all living organisms. They are amphiphilic molecules that contain a hydrophilic polar head group and a hydrophobic zone, i.e., the two fatty acyl chains (Figure 14.1). Microbial fatty acid chains are typically 12–22 carbons long and may be saturated, monounsaturated (containing one double bond), or polyunsaturated (containing up to six *cis*-double bonds). Fatty acids may have branched chains and contain other functional groups, epoxy, hydroxyl, or ring structures (cyclopropane, cyclohexyl, etc.) (Table 14.1). Bacteria typically synthesize saturated and monounsaturated fatty acids; the polyunsaturated fatty acids are typically found in cyanobacteria, psychrophilic bacteria, and piezophilic bacteria (Fang et al., 2010; O'Leary, 1962).

Developing techniques for the analysis of lipids dates back to the 1950s. Folch et al. (1957) first developed a method to extract lipids from animal tissues, whereby the

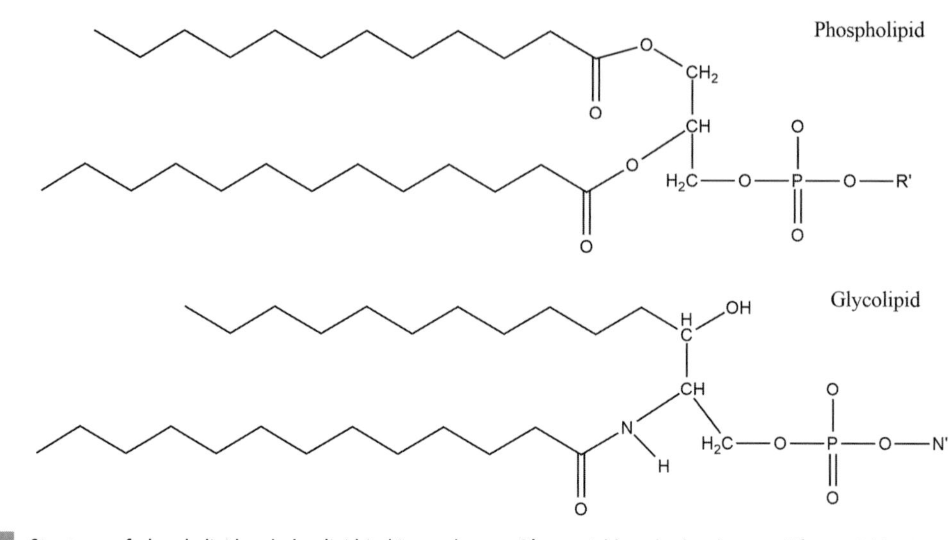

Figure 14.1 Structures of phospholipid and glycolipid in biomembranes. R′ = variable polar head group; N′ = variable sugar chain.

homogenized tissues are extracted for lipids in a $CHCl_3$/MeOH (2:1, v/v) solvent mixture, with washing with water or a salt buffer. The Bligh and Dyer method published in 1959 was similar to the Folch method in that a mixture of solvents, $CHCl_3$/MeOH/H_2O (1:2:0.8, v/v), was used for lipid extraction. These two methods set the foundation for lipid analysis. Since then, there have been modifications made to these classic analytical techniques. Noteworthy progress came about a decade later, when White and Frerman (1967) developed a procedure for the extraction and assay of lipids of the bacterium *Staphylococcus aureus*. The separation and fractionation of lipid components were carried out by means of silicic columns. The lipids collected included neutral lipids (chloroform fraction); glycolipids (acetone fractions); and phospholipids (methanol fraction). In addition, White and Frerman (1967) used thin-layer chromatography (TLC) for the separation of lipids prior to identification. Other modifications include the substitution of H_2O in the solvent mixture with phosphate buffer (White et al., 1979) or citrate buffer.

With rapidly developing analytical instrumentation, there were significant advances in the field of lipid analysis and applications in microbial ecology in the 1980s and early to mid-1990s (e.g., Dowling et al., 1986; Fang et al., 1996; Findlay et al., 1989; White, 1988). Vestal and White (1989) used lipids to study microbial biomass, community structure, metabolic activities, and nutritional status. White et al. (1993) introduced the signature lipid biomarker (SLB) technique (Figure 14.2). Since then, the SLB technique has been used widely to determine viable microbial biomass and community structure, as viable micro-organisms contain intact phospholipids and specific groups of microorganisms synthesize specific phospholipid fatty acids (PLFAs) and are thereby characterized by unique PLFA patterns. This work arguably laid the foundation for modern lipid biomarker analysis in

Table 14.1 Bacterial fatty acids

Common name	Systematic name	Formula
Saturated fatty acids		
Lauric	Dodecanoic	CH_3-$(CH_2)_{10}$-COOH
Myristic	Tetradecanoic	CH_3-$(CH_2)_{12}$-COOH
Palmitic	Hexadecanoic	CH_3-$(CH_2)_{14}$-COOH
Stearic	Octadecanoic	CH_3-$(CH_2)_{16}$-COOH
Arachidic	Eicosanoic	CH_3-$(CH_2)_{18}$-COOH
Behenic	Docosanoic	CH_3-$(CH_2)_{20}$-COOH
Unsaturated fatty acids		
	Tetradecenoic acid	CH_3-$(CH_2)_x$-CH=CH-$(CH_2)_y$-COOH (x + y = 10)
	Hexadecenoic acid	CH_3-$(CH_2)_x$-CH=CH-$(CH_2)_y$-COOH (x + y = 12)
Palmitoleic	*cis*-9-Hexadecenoic	CH_3-$(CH_2)_5$-CH=CH-$(CH_2)_7$-COOH
Palmitvaccenic	*cis*-11-Hexadecenoic	CH_3-$(CH_2)_3$-CH=CH-$(CH_2)_9$-COOH
	Octadecenoic	CH_3-$(CH_2)_x$-CH=CH-$(CH_2)_y$-COOH (x + y = 14)
Oleic	*cis*-9-Octadecenoic	CH_3-$(CH_2)_7$-CH=CH-$(CH_2)_7$-COOH
cis-Vaccenic	*cis*-11-Octadecenoic	CH_3-$(CH_2)_5$-CH=CH-$(CH_2)_9$-COOH
	Eicosenoic	CH_3-$(CH_2)_x$-CH=CH-$(CH_2)_y$-COOH (x + y = 16)
	Heneicosenoic	CH_3-$(CH_2)_x$-CH=CH-$(CH_2)_y$-COOH (x + y = 17)
	Docosenoic	CH_3-$(CH_2)_x$-CH=CH-$(CH_2)_y$-COOH (x + y = 18)
Cyclopropane fatty acids		
	Methylene-hexadecanoic	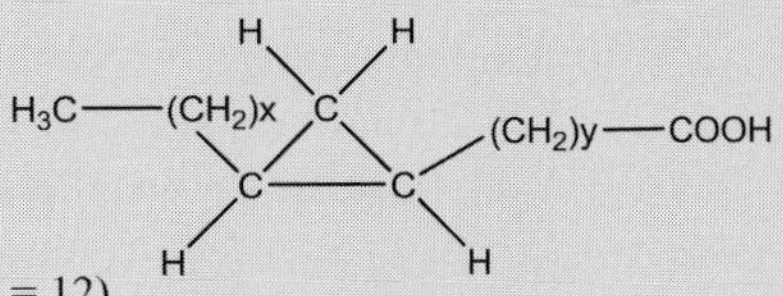 H H H_3C—$(CH_2)x$ C $(CH_2)y$—COOH C—C H H (x + y = 12)
Lactobacillic	cis-11,12-Methylene-octadecanoic	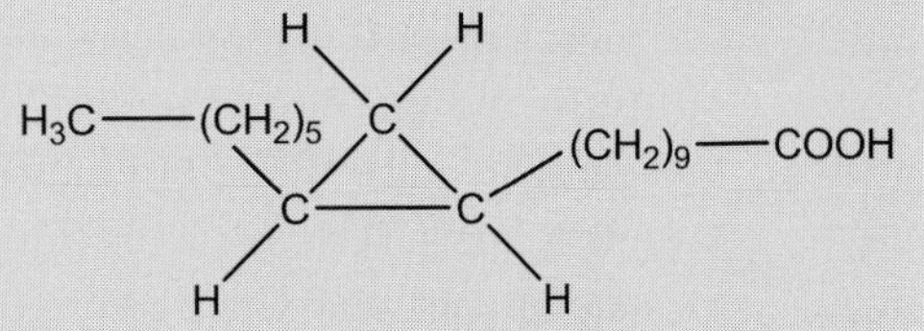 H H H_3C—$(CH_2)_5$ C $(CH_2)_9$—COOH C—C H H
Branched fatty acids		
	13-Methyltetradecanoic	CH_3-CH-$(CH_2)_{11}$-COOH | CH_3

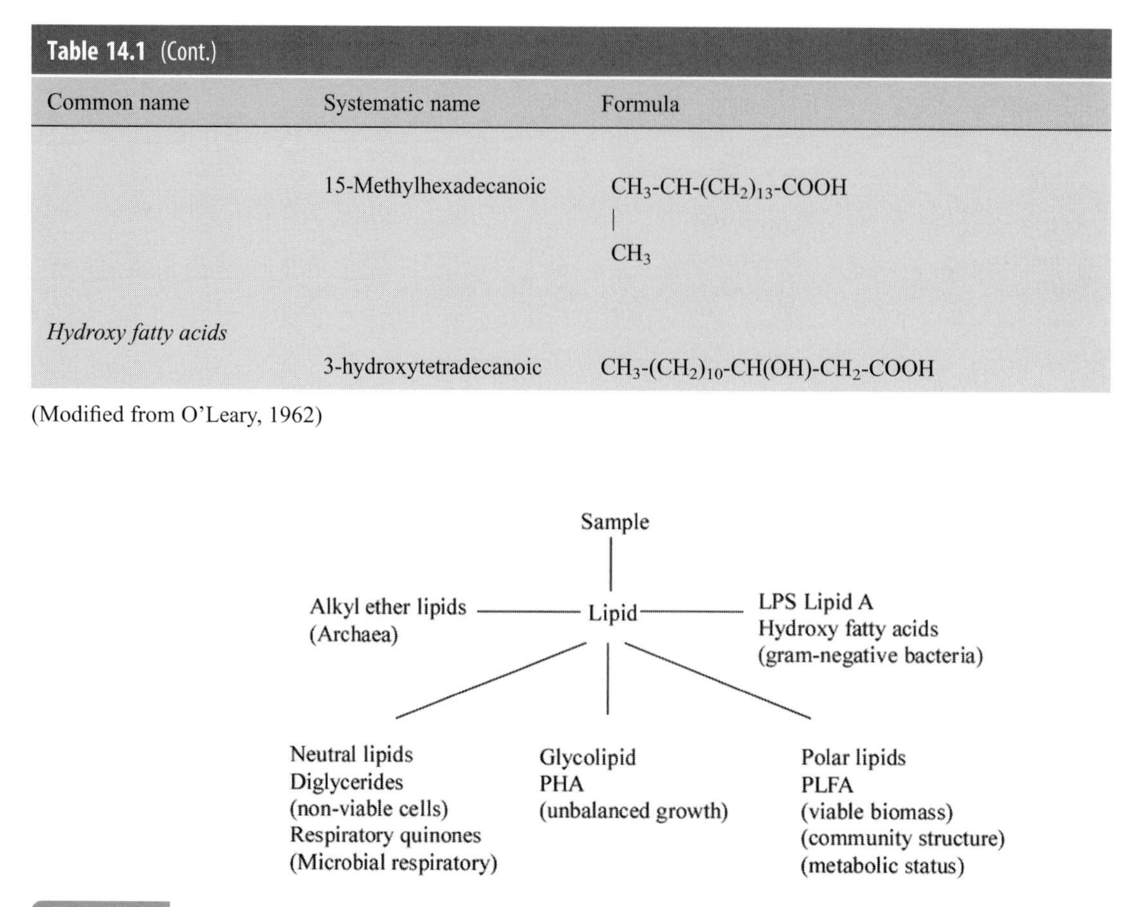

Table 14.1 (Cont.)

Common name	Systematic name	Formula
	15-Methylhexadecanoic	CH_3-CH-$(CH_2)_{13}$-COOH (with CH_3 branch on CH)
Hydroxy fatty acids		
	3-hydroxytetradecanoic	CH_3-$(CH_2)_{10}$-CH(OH)-CH_2-COOH

(Modified from O'Leary, 1962)

Figure 14.2 The signature lipid analysis as proposed by White et al. (1993). PHA is the bacterial storage lipid.

microbial ecology, biogeochemistry, and geomicrobiology. The use of PLFAs for microbial analysis was further optimized by using gas chromatography-mass spectrometry (GC-MS) (Tunlid and White, 1992; White, 1986, 1988). In addition, White et al. (1997) proposed that analysis of fatty acids of diglycerides provides an estimate of nonviable microbial biomass (Figure 14. 3). This is because diglycerides are formed from the hydrolysis of phospholipids, a process that produces diglycerides within minutes or hours after cell death or cell lysis, and the formed diglycerides contain the same signature fatty acids as the parent phospholipids. While PLFAs were mostly used to identify microbial groups, analysis of other lipids, such as sterols (for microeukaryotes, protozoa, and algae) or glycolipids (for phototrophs and Gram-positive bacteria), can give detailed information on eukaryotic communities (Fang et al., 2007).

While the SLB technique was mostly associated with GC-MS for identification and quantification of biomarkers, studies in the late 1990s and thereafter explored novel analytical methods for lipids. One of the first key studies was the use of liquid chromatography coupled with mass spectrometry for lipid analyses in geomicrobiology, pioneered by Fang and Barcelona (1998). The authors developed an analytical method utilizing

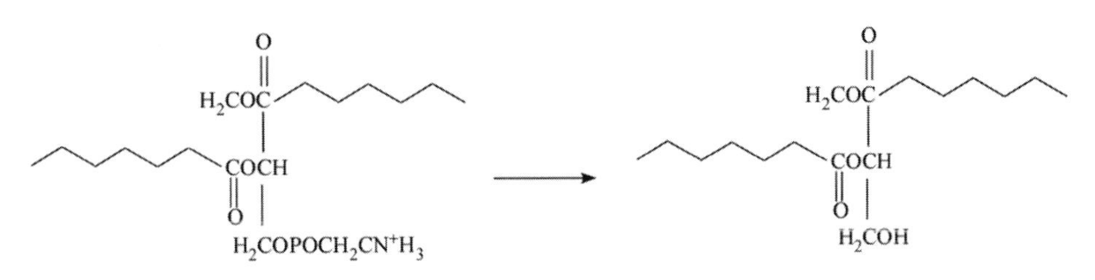

Figure 14.3 Microbial cell death and the conversion of phospholipids to diglycerides. (Modified from White and Ringelberg, 1997.)

liquid chromatography/electrospray ionization/mass spectrometry (LC/ESI/MS) for intact phospholipid profiling (IPP) of microbial lipids. This method permits coupled analysis of the polar head groups as well as the two fatty acid constituents of phospholipids. Unlike PLFA profiling, IPP reveals the variations in the head groups and the fatty acids, thus providing more accurate and detailed information about the microbial communities present. Overall, this technique allows unambiguous analysis of nonvolatile, thermally labile, and/or precharged molecules, such as phospholipids in their intact, endogenous form. ESI offers the advantages of compatibility with liquid chromatographs and low spectral background signals (Heller et al., 1987). Collision-induced dissociation (CID) of lipid molecules produces an array of fragment ions that helps in revealing the structural information of phospholipids (Murphy and Harrison, 1994). High sensitivity and good chromatographic separation are necessary when dealing with environmental samples containing much lower amounts of microbial biomass, for example, in aquifers (Smith et al., 1986). Fang and Barcelona (1998) applied the IPP technique for the structural determination and quantitative analysis of bacterial phospholipids of piezophilic bacteria and methanotrophs in two different studies. In one of the three published works of Fang et al. (2000), they compared the techniques of PLFA analysis by GC-MS and IPP by LC/ESI/MS and found IPP to be superior to the PLFA technique in the identification and differentiation of microbial groups. In their two other studies, they used the IPP technique in the biogeochemical study of extremely barophilic bacteria and the characterization of methanotrophic bacteria. The details of this technique are discussed in a later section.

14.2 Experimental Approach

In this section, the details of lipid extraction and other chemical procedures, including mass spectrometric analysis, will be presented in the form of easy-to-follow laboratory protocols. The techniques are largely from Fang and Findlay (1996) for bacterial fatty acids, with minor changes and adjustments made by the authors thereafter. Originally from Bligh and Dyer (1959), these techniques have been widely used for different kinds of samples, from sediments to microbial cells.

14.2.1 Materials Required

Equipment

- Gas chromatograph coupled with a mass spectrometer (GC-MS)
- Regular laboratory apparatus including vortex, centrifuge, ultrasonicator, nitrogen evaporator, solid phase extraction (SPE) vacuum manifold

Supplies

- Glass test tubes fitted with PTFE caps
- Silica columns for separation of lipids

Solutions and reagents

- Dichloromethane (DCM), methanol, chloroform, and acetone (high-performance liquid chromatography [HPLC] grade)
- 0.2 N KOH in methanol for transmethylation of phospholipid fatty acids
- N,O-bis(Trimethylsilyl trifluoroacetamide) (BSTFA) for trimethyl silane (TMS) derivatization of neutral lipids
- Internal standards 5α-cholestane and ethyl ester standard prepared in stocks and divided into working solutions (0.1 mg/ml) for analytical use

14.2.2 Laboratory Procedures

14.2.2.1 Lipid Extraction

1. Add 7.5 mL DCM, 15.0 mL methanol, and 5.0 mL 50 mM phosphate buffer to a 50 ml glass tube fitted with PTFE-lined screw cap.
   (a) Add solvents in the following order: buffer → DCM → methanol.
   (b) The ratio of microbial cells (dry weight) should be approximately 1:1 (i.e., for 5 mg of cells, add 5 mL of DCM in the initial solvent mixture).
2. Add sample, shake, and let stand in the dark overnight at 4 °C.
3. Split phases by adding 7.5 mL DCM and 7.5 mL deionized water. Shake and allow phases to separate. The DCM phase, which contains the lipids, will be the bottom phase. Allow sample to stand at 4 °C overnight.
4. Centrifuge sample at 250 × *g* for 2 minutes.
5. Remove the upper water–methanol phase with an aspirator.
6. Transfer the majority (e.g., ~14 of 15 mL) of the DCM to a Grade 2 V 8 μm medium flow filter paper and collect the solvent in a 15 mL glass tube. Rinse the filter with 3 × 0.75 mL of DCM, adding the rinse solvent to the collected extraction solvent. Evaporate the DCM with an N_2 evaporator. The dry material is considered the total extractable lipid. Redissolve the total lipid extract in 500 μL of a solvent mixture (hexane:DCM, 70:30, v/v) and vortex.

14.2.2.2 Separation of Lipids

The total lipid can be separated into different classes of lipids using an SPE vacuum manifold. This part of the experiment must be done in a fume hood.

1. Place miniature SPE columns (silica, 500 mg, 6 mL) into spigots on the glass tank of an SPE vacuum manifold.
2. Add 3 mL methanol and pull through the silica column at ~1 drop per second by gravity, assisted by applying a minimum pressure (5 psi). Stop flow when solvent reaches the head of the column.
3. Add 3 mL of chloroform, pressurize column, and force chloroform through column at 1 drop per second.
4. Add 3 mL of hexane and allow it to drip through by gravity. Stop flow when solvent reaches the head of the column.
5. Transfer the total lipid extract to the column with a clean, precombusted (5 hours at 450 °C) Pasteur pipette. Allow the solvent (and therefore the total lipid) to enter the column without solvent above the top of the column.
6. Wash the test tube twice with 150 μL hexane/DCM mixture (7:3 v/v), vortex, and transfer to the column.
7. Add 5 mL hexane and place a clean test tube under the column. Let hexane drip through the column at ~1 drop per second with appropriate vacuum. Stop flow when solvent reaches the head of the column. Remove the test tube – this is the hydrocarbon fraction.
8. Add 5 mL chloroform and place a clean test tube under the column. Let chloroform drip through the column at ~1 drop per second with appropriate vacuum. Stop flow when solvent reaches the head of the column. Remove the test tube – this is the neutral lipid fraction.
9. Add 5 mL acetone and pull through the column at ~1 drop per second. Stop flow when solvent reaches the head of the column. Remove the test tube – this is the glycolipid fraction.
10. Place a clean test tube under the column, add 5 mL of methanol, and pull through the column until the column dries. The methanol fraction contains phospholipids.

Notes and precautions

1. The column should never be allowed to go dry.
2. All fractions should be stored at −20 °C for future analysis.

14.2.2.3 Formation of Fatty Acid Methyl Esters (FAMEs)

Phospholipids are comprised of phosphoric esters of glycerol with two fatty acid residues esterified to the other hydroxyl groups. Derivatization of PLFAs involves conversion of fatty acids into their corresponding esters. This process increases the volatility of the fatty acids, thus providing a better separation and cutting down the time of spectrometric analysis (Brondz, 2002; Liu, 1994). The preparation of fatty acid methyl esters (FAMEs) is essentially the conversion of one ester into another (transesterification) by cleavage of an ester bond via an alcohol (Carvalho and Malcata, 2005); when such a reaction is carried out with methanol, the process of derivatization is termed transmethylation (Liu, 1994), as described in the following.

1. Dissolve the dry phospholipid in 0.5 mL of methanol:toluene (1:1, v/v).
2. Add 0.5 mL of 0.2 N KOH in methanol, seal the test tubes with a PTFE-lined screw cap, vortex, and heat at 37 °C for 15 minutes.
3. Cool to room temperature, add 0.5 mL of 0.2 N acetic acid, vortex (5 s), add 2 mL of DCM and 2 mL of deionized water, and vortex for 30 seconds.
4. Centrifuge at 250 × *g* for 5 min.
5. Transfer the DCM (bottom phase) to a clean test tube using a Pasteur pipette.
6. Add 1 mL DCM to the reaction test tube, vortex, and repeat steps 4 and 5.
7. Add 1 mL DCM; do not vortex or centrifuge. Transfer DCM.
8. Dry the DCM phase under a nitrogen evaporator. Redissolve in DCM and add ethyl ester internal standard.

Fatty acids are designated by the total number of carbon atoms:number of double bonds (i.e., a 16-carbon fatty acid is 16:0). The position of the double bond is indicated with a Δ number closest to the carboxyl end of the fatty acid molecule with the geometry of either *c* (*cis*) or *t* (*trans*). Fatty acid terminal methyl branching is indicated with *i* (*iso*) or *a* (*anteiso*), and mid-branching is indicated as the position of the methyl group from the carboxyl group of the fatty acid (e.g., 10Me16:0). Cyclopropane fatty acids are indicated with cy.

Notes and precautions

1. Make the methanol KOH solution fresh daily.
2. There is no delay between the addition of the acetic acid and the addition of the chloroform and water.

14.2.2.4 Purification of FAMEs

This procedure purifies the FAMEs in preparation for their separation and quantification using gas chromatography. The procedures utilize reverse-phase SPE technology using 3 mL glass columns.

1. Add 1 mL of dry JT Baker octadecyl resin to a 3 mL glass column fitted with a Teflon frit.
2. Add 2 mL of deionized water and pull through the column with vacuum. Continue flow until the column appears dry.
3. Add 2 mL of methanol and pull through the column at ~1 drop per second. Stop flow when solvent reaches the head of the column.
4. Add 1 mL of chloroform and allow it to drip through by gravity.
5. Add 2 mL of chloroform and stop flow. The column packing material will soon become translucent. First, dislodge any trapped bubbles by tapping the column and then

pressurize the column. Force the chloroform through the column at ~1 drop per second. Add 2 mL of chloroform and allow it to drip through by gravity.

6. Complete column conditioning by washing with 2 mL of acetonitrile:water. To prevent column drying, leave ~1 mL of acetonitrile:water on the top of the column and remove it just prior to addition of the FAMEs.
7. Add 250 μL of acetonitrile to the dry FAMEs. Vortex several times over the next 10 min.
8. Then add 250 μL deionized water, vortex, transfer to column, and allow the solvent to drip through the column.
9. Repeat steps 7 and 8 twice without the 10 min wait.
10. Wash the column with 1 mL of acetonitrile:water and allow the column to run dry. Wash the column with 200 μL hexane and allow the column to run dry. Dry the column by fitting the drying attachment and establishing a flow of N_2 for 5 minutes.
11. Rehydrate the packing by adding 2 mL of deionized water to the column and allow the column to run dry.
12. Place a clean test tube beneath the column, close the column, and add 750 μL of hexane:chloroform. Allow to stand for 2 minutes prior to initiating flow.
13. Wash column thrice with 0.5 mL hexane:chloroform.
14. Evaporate the hexane:chloroform in a nitrogen evaporator with a water bath set to 37 °C.

14.2.2.5 Determination of Double Bond Position in Monounsaturated Fatty Acids (MUFAs)

Monounsaturated fatty acids are biomarkers of Gram-negative bacteria, and polyunsaturated fatty acids are proxies indicative of microeukaryotes (Fang et al., 2010; White et al., 1993). This procedure of double bond position determination is adapted from Dunkleblum et al. (1985). The protocol is as follows:

1. Transfer 100 μL of FAME solution (in hexane) into a 2 mL glass vial.
2. Add 100 μL of dimethyl disulfide to the vial.
3. Add one drop of iodine solution (6% in diethyl ether). Shake the solution for 5 s.
4. Put the vial in a GC oven at 45 °C for 48 h.
5. Take out the vial from the GC oven and let it cool down for 3 min. Add 200 μL hexane.
6. Add 100 μL of 5% aqueous $Na_2S_2O_3$ to neutralize the excess iodine.
7. Take the hexane phase (top layer) and concentrate it to 30 μL for GC-MS analysis.

Notes and precautions

1. Iodine is a strong oxidant; handle with extreme care!
2. If water is present in the vial after step 7, use Na_2SO_4 to eliminate it by passing the solution through a glass column. The glass column is packed with glass wool and Na_2SO_4 on the top of the glass wool.

14.2.2.6 Hydrocarbon (HC) Fraction for GC-MS Analysis

1. Reduce the volume of the HC fraction slowly (no heat) using an N_2 evaporator to nearly dry (to about 10–20 μL left in the test tube).
2. Add 295 μL hexane and 5 μL 5α-cholestane internal standard.
3. Load samples in the autosampler tray for GC-MS analysis.

14.2.2.7 TMS Derivatization of Neutral Lipid (NL) Fraction

1. Dry neutral lipid fraction under N_2.
2. Within 24 hours before GC analysis, add 50 μL BSTFA to the dry NL fraction, and heat the lipid at 75 °C for 1 hour.
3. Remove the sample from the heating block, cool to room temperature, then dry under nitrogen, and add 295 μL of DCM and 5 μL internal standard solution (C_{18} fatty acid ethyl ester) for GC-MS analysis.

14.2.2.8 GC-MS Analysis and Identification of FAMEs

Samples are loaded in the GC-MS after extraction, separation, and necessary derivatization of the lipid fractions for the identification and quantification of biomarkers. The sample is injected in liquid form into the GC inlet, where it is vaporized and transported to the column through an inert carrier gas (helium). The oven temperature is programmed for optimized chromatography. Analytical separation of FAMEs is accomplished using a nonpolar fused-silica capillary column (e.g., DB-5 ms). The MS fragments the FAME to form various fragment ions, and a mass spectrum, displayed as relative abundance of each fragment ion with different mass/charge ratio (m/z), is generated. Individual fatty acids are identified from their mass spectra. Concentrations of individual compounds are obtained based on the GC-MS response of an analyte relative to that of an internal standard – C_{18} fatty acid ethyl ester for fatty acids and 5α-cholestane for HC and NL.

14.2.2.9 LC-MS Analysis of Phospholipids

The coupling of MS to a liquid chromatograph (LC) was an obvious extension from the use of GC-MS, considering that it is applicable for polar, nonvolatile, and thermally labile compounds. Fang et al. (2000b) performed the LC/ESI/MS analysis on a Hewlett Packard HP 1090 liquid chromatograph/HP 5989B single quadrupole mass spectrometer with an electrospray ionization interface. The LC was equipped with a 250 μL sample loop. A Zorbax (Hewlett Packard) C8 (150 mm × 4.6 mm, 5 μm) or C18 (150 mm × 3.2 mm, 5 μm) column was used for the chromatographic separation of phospholipids. A gradient solvent system, composed of 10 mM ammonium acetate (solvent A) and methanol (solvent B), was used with a flow rate of 0.3 mL/min. At the beginning of the gradient, the mobile phase was 50% of A and 50% of B for 2 min. Solvent B was increased to 80% at 20 min and 100% at 60 min. The mobile phase was then held isocratically for 5 min.

Mass spectrometers operate by converting the analyte molecules to an ionized state, with subsequent analysis of the ions and any fragment ions that are produced during the ionization process, on the basis of their *m/z*. The ESI source acts to fragment the ions and transfer them to the high-vacuum part of the mass spectrometer by a series of differing voltages. The electron spray needle can be maintained at a high positive potential for negative ionization, or vice versa. A hexapole ion guide in the MS source can enhance the efficiency of ion transfer and the sensitivity of the mass spectrometer. The concentrations of phospholipids are calculated based on the chromatographic area response of individual phospholipids with respect to that of an internal standard. The internal standard can be prepared in a way similar to that described in Section 14.2.1. The identification process is similar to that used for GC-MS identification.

14.2.3 Interpretation of Data from Phospholipid Fatty Acid Profiles

The presence of certain microbial species or groups can be determined from PLFA analysis. Unique fingerprints from environmental samples indicate particular microbial groups. In addition, the amount of recovered fatty acids is a measure of the total biomass in the system. Although this method gives us valuable information, caution must be exercised to avoid overinterpretation of the available data, as some of the PLFAs that are found in bacteria are also present in single-cell and multicellular eukaryotes, such as algae, fungi, and higher plants. Table 14.2 is a general overview of fatty acid biomarkers and their significance in geomicrobiological study.

14.3 Lipid Analysis in Geomicrobiology

14.3.1 Case Study 1: Lipid Analysis of Stromatolites and Biofilms Using GC-MS

One of the first major research studies in acid mine drainage (AMD) systems that used lipid profiling to decipher the resident microbial community structure and explore the biogeochemical interactions was conducted by Fang et al. (2007). They carried out lipid biomarker analysis of stromatolites in the Green Valley Site (GVS) AMD of western Indiana. This AMD site was unique, since both prokaryotes and eukaryotes were present. Biomarkers were identified indicating the presence of the microbial communities, which may have contributed to the formation of the stromatolites. Six morphologically distinct layers, based on microscopic and macroscopic examination (Brake and Hasiotis, 2008; Brake et al., 2002, 2005), were analyzed for their lipid profiles (Fang et al., 2007). Lipid analysis indicated the presence of layered microbial zones, comparable to communities that build marine stromatolites (Brake and Hasiotis, 2010). The abundance of such hydrocarbons as pristane, phytadienes, and *n*-alkenes (21:3 and 21:4) and PUFAs from biomarker analysis

Table 14.2 General interpretation of fatty acid biomarkers found in environmental samples

PLFA biomarker	Interpretation	Reference
Amount of PLFA	Abundance of living biomass in system	Kieft et al., 1994; White et al., 1993
iso and *anteiso* FAs (e.g., *i*15:0, *a*15:0, *i*17:0, *a*17:0)	Gram-positive bacteria	Zelles, 1997
OH FAs (e.g., 3OH 16:1, 3OH18:1); monounsaturated FAs (16:1ω7*c*, 18:1ω7*t*); cyclopropane FAs (e.g., *cy*7:0, *cy*19:0)	Gram-negative bacteria	Cavigelli et al., 1995; Zelles, 1997
Cyclopropane FAs	Starvation in Gram-negative bacteria	Ramos et al., 2002
Ratio of branched/mono-unsaturated fatty acids	Shift from Gram-positive to Gram-negative bacteria	Vestal and White, 1989; White et al., 1993
*i*17:1ω7*c, i*15:1ω7*c*, i19:1ω7*c*	*Desulfovibrio*	Edlund et al., 1985
10Me16:0, *cy*18:0 (ω7,8)	*Desulfobacter*	Dowling et al., 1986
trans-unsaturated fatty acids	Bacterial stress indicator	Heipieper et al., 2003
16:0 and 16:1 (equivalent proportions), 18:1ω7*c*/ω9*t*/ω12*t*	*Pseudomonas*	Haack et al., 1994
Mid-methyl-branched fatty acids (10Me16:0, 10Me17:0)	Actinomycetales	Frostegård et al., 1993
16:1ω8*c*,16:1ω5*c*	Methanotrophs	Nichols et al., 1987
16:1w5c	Cytophaga – Flavobacterium	Frostegård et al., 1993; Kelly et al., 1999
Polyunsaturated fatty acids	Presence of microeukaryotes; psychrophilic bacteria; piezophilic bacteria	Fang et al., 2000, 2003, 2007; White et al., 1997
PUFAs 16:2, 16:3, 22:5 and 22:6	Diatoms	Fang et al., 2006; Findlay and Watling, 1998; White et al., 1997
$18{:}2\Delta^{9,12}$ and $18{:}1\Delta^{9}c$	Fungi	Fang et al., 2007

clearly indicated that phototrophic microeukaryotes, including *E. mutabilis*, algae (diatoms), and fungi, were most dominant. The presence of terminal methyl-branched fatty acids and mid-methyl-branched fatty acids suggested the presence of Gram-positive and sulfate-reducing bacteria, respectively. Fungi appeared to also be an important part of the AMD microbial communities, as suggested by sterol profiles and the presence of PUFAs. Hydroxy fatty acids and C_{19} cyclopropane fatty acids were also detected, likely originating from acid-producing, acidophilic bacteria. The presence of archaea is indicated by abundant phospholipid ether-linked isoprenoid hydrocarbons (phytane and phytadienes). Figure 14.4 shows the variation of microbial biomass within the different layers of the stromatolite sample.

Dasgupta et al. (2012) studied lipid profiles of both benthic and floating biofilms in the GVS AMD site. Dominance of the photosynthetic microeukaryote *Euglena* was indicated

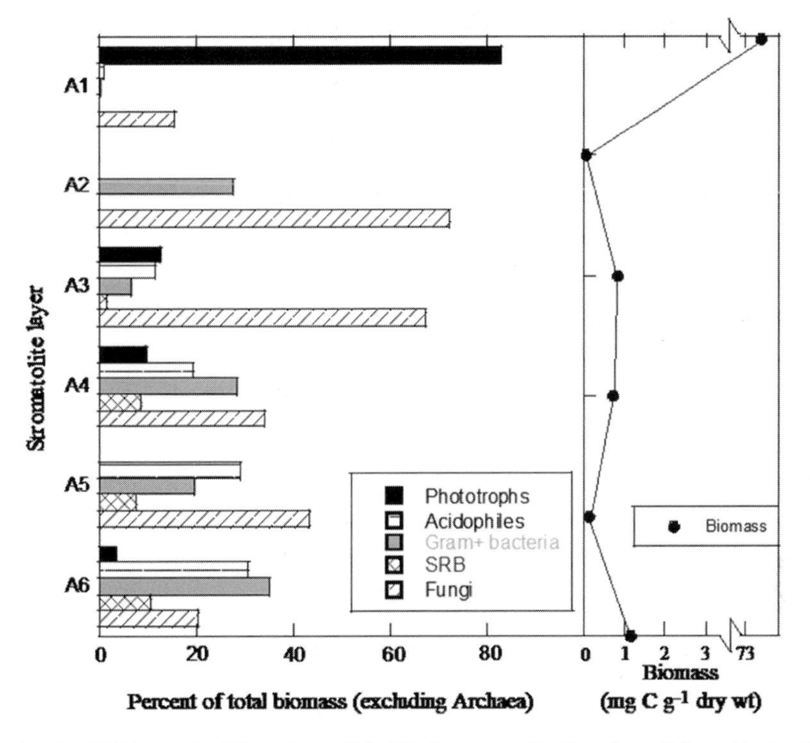

Figure 14.4 Variation in microbial biomass in different layers (A1–A6) of a stromatolite from Green Valley acid mine site in western Indiana. Microbial biomass was estimated based on the amount of signature fatty acids detected in each layer. (From Fang et al., 2007.)

by the detection of abundant phytadiene, phytol, phytanol, polyunsaturated *n*-alkenes, PUFAs, short-chain (C_{25-32}) wax esters (WEs), ergosterol, and tocopherols. Before their study, multiple polyunsaturated *n*-alkenes had never been detected in a single sample. In a later study by Dasgupta et al. (2013), the stable carbon isotope ratios of hydrocarbons and WEs indicated a carbon-limiting system in the AMD. Further, they proposed a reverse β-oxidation pathway mechanism by which *Euglena* possibly synthesized the ^{13}C-enriched wax esters in the AMD biofilms.

If we consider that eukaryotes evolved in the oceans, the biosynthesis of wax esters by eukaryotes possibly took place in an anoxic condition, considering that oxygenation of the deep ocean was a relatively late event, occurring around 580 million years ago (Canfield, 2005), much later than the fossil record of the first eukaryotes. The unique biochemical machinery to produce WEs may have allowed early eukaryotes to survive and diversify on early Earth, when the oceans were anoxic and sulfidic. On the other hand, sterol biosynthesis probably takes place in the cytosol of *Euglena* under oxic conditions (Dasgupta et al., 2012) from squalene (Ginger et al., 2010) via cycloartenol (Disch et al., 1998). Oxygen-dependent sterol biosynthesis may be a key step in eukaryotic evolution, since curvature mechanisms responsible for the dynamic character of the eukaryotic membrane are attributed to sterol biosynthesis (Summons et al.,

2006). *Euglena*, then, could behave both as a plant and as an animal in biosynthesizing sterols and WEs under oxic–anoxic conditions. The dual aerobic and anaerobic biosynthetic pathways of *Euglena* may be a response to survive the recurring anoxic and oxic conditions on primitive Earth. The microeukaryotes retained this mechanism of the conserved compartmentalization in their physiology to evolve and diversify in extreme conditions (Dasgupta et al., 2012).

In the studies described, lipid profiles were used to identify unique eukaryotic organisms in an extreme environment. Further, they helped to understand the biogeochemistry of an extremely acidic environment and to explore the primitive Earth when Fe-stromatolites were being formed and oxygenation of atmosphere taking place.

14.3.2 Case Study 2: Determining Intact Phospholipid Profiles of Methanotrophs Using LC-MS

Fang et al. (2000b) conducted a study combining lipid analysis with LC-MS. In their study, they determined the IPPs of seven methanotrophs from all three physiological groups I, II, and X, using LC/ESI/MS. The goal of this study was to demonstrate the potential of utilizing IPP to identify microbial groups. Following the protocol described for the LC-MS analysis, 28 IPPs were isolated from the seven methanotrophs. The bacteria analyzed in the study contained two major classes of phospholipids, PG (phosphatidylglycerol) and its derivatives phosphatidylmethylethanolamine (PME) and phosphatidyldimethylethanolamine (PDME). Fatty acids were either saturated or monounsaturated, with chain lengths from 14 to 18.

As a whole, this study successfully demonstrated the determination of phospholipid profiles using the LC-MS technique. Additionally, it presented a basis for the comparison and differentiation between functionally similar but taxonomically different bacteria based on IPP. Further, this study suggested that IPP can be more useful and effective in microbial chemotaxonomy (Fang et al., 2000a).

14.4 Advantages and Limitations of Lipid Analysis in Geomicrobiology

As mentioned before, the most significant advantage of lipid analysis lies in the fact that it avoids the disadvantages associated with culture-based methods, such as the inability to culture most of the cells, because optimal culture conditions have not yet been defined, or due to problems associated with the dependence on other microbes for growth or with differential growth rates. Unlike any other popular methods, lipid profiles give us excellent signature molecules called biomarkers, which can be linked to specific microbial species or groups of microorganisms. Not only do lipids give us information about microbial communities; they also indicate growth conditions, type of metabolic pathway, structure and physiology, stress-related changes, and paleoenvironmental conditions.

Nevertheless, it should be cautioned that this technique has certain limitations as well. Except for a handful of biomarkers, many fatty acids are common to multiple microorganisms. Additionally, care must be taken in lipid analysis, as lipids (e.g., PUFAs) are susceptible to oxidation. Furthermore, given that microbial biomarker composition can be altered by biotic and abiotic stresses, assigning a biomarker to specific microorganisms can also be difficult. Thus, caution must be exercised while working with environmental samples in interpreting specific biomarkers for certain microorganisms. For improved results, combining the technique of lipid analysis with genetic and isotopic methods is generally advised. Multivariate statistical methods (such as principal component analysis [PCA]) also act as an ally to lipid techniques for more accurate interpretation.

14.5 New Developments and Future Directions

This chapter described the analytical procedures of lipids as signature biomarkers in determining microbial biomass and community structure in environmental samples. These biomarkers are important biological components, and analysis of the compounds can provide information on the abundance and community structure of microbial communities in natural environments. In performing the analysis, care must be taken to avoid artifacts such as lipid oxidation, as many of the lipids, such as unsaturated fatty acids, are susceptible to oxidation. Lipid extracts or the fractionated lipids must be protected from air or lipid exposure, and samples should be stored in a refrigerator at every stage of the protocol. As aforementioned, caution must be exercised in interpreting lipid biomarker patterns for microbial community analysis in geomicrobiology.

Microbial signature lipid analysis has witnessed new and evolved techniques in extraction, fractionation, and quantification procedures during the last decade or so. For example, more recent techniques make use of the solvent methyl-*tert*-butyl ether (MTBE) (Matyash et al., 2008), accelerated solvent extraction (ASE), and supercritical fluid extraction (SFE) for enhanced extraction of lipids. Commercially available solid-phase extraction (SPE) cartridges for lipid fractionation have replaced thin-layer chromatography (TLC), a lengthy procedure with excessive solvent consumption. A newly developed field called "lipidomics," involving the analysis and interpretation of lipids, has emerged. Under the umbrella of metabolomics, lipidomics involves the identification and quantification of microbial metabolic pathways, networks of biosynthesis of microbial lipids and their functions, and interactions with other lipids and proteins. There have been several reviews in lipidomics and the associated analytical techniques in the last decade (e.g., Allwood et al., 2015; Li et al., 2013). Lipidomics was first put forward as a branch of metabolomics in 2003 (Han and Gross, 2003) and was defined as "the full characterization of lipid molecular species and of their biological roles with respect to expression of proteins involved in lipid metabolism and function, including gene regulation" (Spener et al., 2003). MS, nuclear magnetic resonance (NMR), and other spectroscopic techniques have been used for lipid characterization (Li at el., 2013). In terms of chromatographic methods, TLC, GC, ultra-performance liquid chromatography (UPLC), supercritical fluid chromatography (SFC), and capillary electrophoresis (CE), when combined with a mass spectrometer, can satisfy various analytical demands and provide a large amount of information on lipidomes (Li et al., 2013). In recent years, new advances in analytical technologies for lipidomics, such as new ionization techniques of MS (ESI), two-dimensional (2D) NMR, the improvement of MS imaging technology (such as high-resolution matrix-assisted laser desorption ionization [MALDI]), and the realization of two-dimensional liquid chromatography (2D LC) have significantly enhanced the investigation of lipidomics (Li et al., 2013). Atmospheric pressure (AP)-MALDI-MS/MS has been used to detect a set of *Bacillus* spore proteins, providing 60–75% species-specific differentiation based on their tryptic peptides (Nguyen and Russell, 2010). Lipid profiles from 16 different bacterial samples, deposited as single colonies, have been generated by desorption electrospray ionization (DESI) MS without prior sample preparation (Zhang et al., 2011a). Low temperature plasma (LTP) MS has been utilized for rapid characterization of microbial communities based on fatty acid ethyl ester (FAEE) profiles from intact cells without pretreatment (Zhang et al., 2011b). Nano-DESI is another API technique that has been applied successfully to MS characterization of microbial mixtures (Sandrin and Demirev, 2016), e.g., the analysis of *B. subtilis* and *Pseudomonas aeruginosa* strains directly from agar plates to perform comparative metabolomics (Rath et al., 2013). In terms of MS techniques, in addition to MALDI, secondary ion mass spectrometry (SIMS) has been applied at much higher resolution. A combined MALDI-SIMS imaging approach has been used to characterize microbial metabolites of biofilms (Lanni et al., 2014). It is predicted that these newly developed analytical techniques and instrumentation will cause great proliferation of the use of lipid biomarkers in geomicrobiology.

14.6 References

Allwood, J.W., AlRabiah, H., Correa, E., et al., 2015. A workflow for bacterial metabolic fingerprinting and lipid profiling: application to Ciprofloxacin challenged. *Escherichia coli*. *Metabolomics* 11 (2), 438–453.

Bligh, E.G. and Dyer, W.J., 1959. A rapid method of total lipid extraction and purification. *Canadian Journal of Biochemistry and Physiology* 37(8), 911–917.

Brake, J.M., Daschner, M.K. and Abbott, N.L., 2005. Formation and characterization of phospholipid monolayers spontaneously assembled at interfaces between aqueous phases and thermotropic liquid crystals. *Langmuir* 21(6), 2218–2228.

Brake, S.S. and Hasiotis, S.T., 2008. Eukaryote-dominated biofilms in extreme environments: overlooked sources of information in the geologic record. *Palaios* 23(3), 121–123.

Brake, S.S. and Hasiotis, S.T., 2010. Eukaryote-dominated biofilms and their significance in acidic environments. *Geomicrobiology Journal* 27(6–7), 534–558.

Brake, S.S., Hasiotis, S.T., Dannelly, H.K. and Connors, K.A., 2002. Eukaryotic stromatolite builders in acid mine drainage: implications for Precambrian iron formations and oxygenation of the atmosphere? *Geology* 30, 599–602.

Brondz, I., 2002. Development of fatty acid analysis by high-performance liquid chromatography, gas chromatography, and related techniques. *Analytica Chimica Acta* 465(1), 1–37.

Canfield, D.E., 2005. The early history of atmospheric oxygen: homage to Robert M. Garrels. *Annual Review of Earth and Planetary Science* 33, 1–36.

Carvalho, A.P. and Malcata, F.X., 2005. Preparation of fatty acid methyl esters for gas-chromatographic analysis of marine lipids: insight studies. *Journal of Agricultural and Food Chemistry* 53(13), 5049–5059.

Cavigelli, M.A., Robertson, G.P. and Klug, M.J., 1995. Fatty acid methyl ester (FAME) profiles as measures of soil microbial community structure. *Plant and Soil* 170, 99–113.

Ciobanu, D.A., Zawierucha, K., Moglan, I. and Kaczmarek, Ł., 2014. *Milnesium berladnicorum* sp. (Eutardigrada, Apochela, Milnesiidae), a new species of water bear from Romania. *ZooKeys* (429), 1.

Dasgupta, S., Fang, J., Brake, S.S., Hasiotis, S.T. and Zhang, L., 2012. Biosynthesis of sterols and wax esters by Euglena of acid mine drainage biofilms: implications for eukaryotic evolution and the early Earth. *Chemical Geology* 306, 139–145.

Dasgupta, S., Fang, J., Brake, S.S., Hasiotis, S.T. and Zhang, L., 2013. Stable carbon isotopic composition of lipids in Euglena-dominated biofilms from an acid mine drainage site: implications of carbon limitation, microbial physiology, and biosynthetic pathways. *Chemical Geology* 354, 15–21.

Disch, A., Schwender, J., Muller, C., Lichtenthaler, H.K. and Rohmer, M., 1998. Distribution of the mevalonate and glyceraldehyde phosphate/pyruvate pathways for isoprenoid biosynthesis in unicellular algae and the cyanobacterium Synechocystis PCC 6714. *Biochemical Journal* 333(2), 381–388.

Dowling, N.J.E., Widdel, F. and White, D.C., 1986. Phospholipid ester-linked fatty acid biomarkers of acetate-oxidizing sulfate-reducers and other sulfide-forming bacteria. Journal of General Microbiology 132, 1815–1825.

Dunkleblum, E.S., Tan, E. and Silo, P.J., 1985. Double-bond location in monounsaturated fatty acids by dimethyl disulfide derivatization and mass spectrometry: application to analysis of fatty acids in pheromone glands of four lepidoptera. *Journal of Chemical Ecology* 11, 265–277.

Edlund, A., Nichols, P.D., Roffey, R. and White, D.C., 1985. Extractable and lipopolysaccharide fatty acid and hydroxy fatty acid profiles from Desulfovibrio species. *Journal of Lipid Research* 26(8), 982–988.

Fang, J. and Barcelona, M.J., 1998. Structural determination and quantitative analysis of bacterial phospholipids using liquid chromatography/electrospray ionization/mass spectrometry. *Journal of Microbiological Methods* 33(1), 23–35.

Fang, J., Barcelona, M.J., Nogi, Y. and Kato, C., 2000a. Biochemical function and geochemical significance of novel phospholipids of the extremely barophilic bacteria from the Mariana Trench at 11,000 meters. *Deep-Sea Research Part I Oceanographic Research Papers* 147, 1173–1182.

Fang, J., Barcelona, M.J. and Semrau, J.D., 2000b. Characterization of methanotrophic bacteria on the basis of intact phospholipid profiles. *FEMS Microbiology Letters* 189(1), 67–72.

Fang, J., Chan, O., Kato, C., et al., 2003. Phospholipid fatty acid profiles of piezophilic bacteria from the deep sea. *Lipids* 38, 885–887.

Fang, J. and Findlay, R.H., 1996. The use of a classic lipid extraction method for simultaneous recovery of organic pollutants and microbial lipids from sediments. *Journal of Microbiological Methods* 27(1), 63–71.

Fang, J., Hasiotis, S.T., Das Gupta, S., Brake, S.S. and Bazylinski, D.A., 2007. Microbial biomass and community structure of a stromatolite from an acid mine drainage system as determined by lipid analysis. *Chemical Geology* 243, 191–204.

Fang, J., Uhle, M., Billmark, K., Bartlett, D.H. and Kato, C., 2006. Fractionation of carbon isotopes in biosynthesis of fatty acids by a piezophilic bacterium *Moritella japonica* strain DSK1. *Geochimica et Cosmochimica Acta* 70(7), 1753–1760.

Fang, J., Zhang, L. and Bazylinski, D.A., 2010. Deep-sea piezosphere and piezophiles: geomicrobiology and biogeochemistry. *Trends in Microbiology* 18(9), 413–422.

Findlay, R.H., King, G.M. and Watling, L., 1989. Efficacy of phospholipid analysis in determining microbial biomass in sediments. *Applied and Environmental Microbiology* 55(11), 2888–2893.

Findlay, R.H. and Watling, L., 1998. Seasonal variation in the structure of a marine benthic microbial community. *Microbial Ecology* 36(1), 23–30.

Folch, J., Lees, M. and Sloane-Stanley, G.H., 1957. A simple method for the isolation and purification of total lipids from animal tissues. *Journal of Biological Chemistry* 226(1), 497–509.

Frostegård, Å., Tunlid, A. and Bååth, E., 1993. Phospholipid fatty acid composition, biomass, and activity of microbial communities from two soil types experimentally exposed to different heavy metals. *Applied and Environmental Microbiology* 59(11), 3605–3617.

Ginger, M.L., McFadden, G.I. and Michels, P.A., 2010. Rewiring and regulation of cross-compartmentalized metabolism in protists. *Philosophical Transactions of the Royal Society of London B: Biological Sciences* 365(1541), 831–845.

Haack, S.K., Garchow, H., Odelson, D.A., Forney, L.J. and Klug, M.J., 1994. Accuracy, reproducibility, and interpretation of fatty acid methyl ester profiles of model bacterial communities. *Applied and Environmental Microbiology* 60(7), 2483–2493.

Han, X. and Gross, R.W., 2003. Global analyses of cellular lipidomes directly from crude extracts of biological samples by ESI mass spectrometry: a bridge to lipidomics. *Journal of Lipid Research* 44 (6), 1071–1079.

Heipieper, H.J., Meinhardt, F. and Segura, A., 2003. The *cis–trans* isomerase of unsaturated fatty acids in *Pseudomonas* and *Vibrio*: biochemistry, molecular biology and physiological function of a unique stress adaptive mechanism. *FEMS Microbiology Letters* 229(1), 1–7.

Heller, D.N., Cotter, R.J., Fenselau, C. and Uy, O.M., 1987. Profiling of bacteria by fast atom bombardment mass spectrometry. *Analytical Chemistry* 59(23), 2806–2809.

Kelly, J.J., Häggblom, M. and Tate, R.L., 1999. Changes in soil microbial communities over time resulting from one time application of zinc: a laboratory microcosm study. *Soil Biology and Biochemistry* 31(10), 1455–1465.

Kieft, T.L., Ringelberg, D.B. and White, D.C., 1994. Changes in ester-linked phospholipid fatty acid profiles of sub-surface bacteria during starvation and desiccation in a porous-medium. *Applied Environmental Microbiology* 60, 3292–3299.

Lanni, E.J., Masyuko, R.N., Driscoll, C.M., et al., 2014. MALDI-guided SIMS: multiscale imaging of metabolites in bacterial biofilms. *Analytical Chemistry* 86, 9139–9145.

Li, M., Yang, L., Bai, Y. and Liu, H., 2013. Analytical methods in lipidomics and their applications. *Analytical Chemistry* 86(1), 161–175.

Liu, K.S., 1994. Preparation of fatty acid methyl esters for gas-chromatographic analysis of lipids in biological materials. *Journal of the American Oil Chemists' Society* 71(11), 1179–1187.

Matyash, V., Liebisch, G., Kurzchalia, T.V., Shevchenko, A., and Schwudke, D., 2008. Lipid extraction by methyl-*tert*-butyl ether for high-throughput lipidomics. *Journal of Lipid Research* 49, 1137–1146.

Murphy, R.C. and Harrison, K.A., 1994. Fast atom bombardment mass spectrometry of phospholipids. *Mass Spectrometry Reviews* 13, 57–75.

Nguyen, J. and Russell, S.C., 2010. Targeted proteomics approach to species-level identification of *Bacillus thuringiensis* spores by AP-MALDI-MS. *Journal of American Society of Mass Spectrometry* 21, 993–1001.

Nichols, P.D., Mancuso, C.A. and White, D.C., 1987. Measurement of methanotroph and methanogen signature phospholipids for use in assessment of biomass and community structure in model systems. *Organic Geochemistry* 11(6), 451–461.

O'Leary, W.M. 1962. The fatty acids of bacteria. *Bacteriological Reviews* 26, 421–447.

Peters, K.E., Walters, C.C. and Moldowan, J.M., 2005. *The Biomarker Guide (Vol. 1).* Cambridge University Press, Cambridge, 249–250.

Ramos, J.L., Duque, E., Gallegos, M.T., et al., 2002. Mechanisms of solvent tolerance in gram-negative bacteria. *Annual Review of Microbiology* 56, 743–768.

Rath, C.M., Yang, J.Y., Alexandrov, T., and Dorrestein, P.C., 2013. Data-independent microbial metabolomics with ambient ionization mass spectrometry. *Journal of American Society of Mass Spectrometry* 24, 1167–1176.

Sandrin, T.R. and Demirev, P.A., 2017. Characterization of microbial mixtures by mass spectrometry. *Mass Spectrometry Reviews* 9999, 1–29.

Schloss, P.D. and Handelsman, J., 2004. Status of the microbial census. *Microbiology and Molecular Biology Reviews* 68(4), 686–691.

Smith, G.A., Nichols, P.D. and White, D.C., 1986. Fatty acid composition and microbial activity of benthic marine sediment from McMurdo Sound, Antarctica. *FEMS Microbiology Ecology* 2(4), 219–231.

Spener, F., Lagarde, M., Géloên, A. and Record, M., 2003. Editorial: What is lipidomics? *European Journal of Lipid Science and Technology* 105, 481–482.

Summons, R.E., Bradley, A.S., Jahnke, L.L. and Waldbauer, J.R., 2006. Steroids, triterpenoids and molecular oxygen. *Philosophical Transactions of the Royal Society B: Biological Sciences* 361 (1470), 951–968.

Tunlid, A. and White, C., 1992. Biochemical analysis of biomass, community structure, nutritional status, and metabolic activity of microbial communities in soil. In Bollag, J.M. and Stotzky, G. (Eds) Soil Biochemistry, Marcel Dekker, 229–262.

Vestal, J.R. and White, D.C., 1989. Lipid analysis in microbial ecology. *Bioscience* 39(8), 535–541.

White, D.C., 1986. Environmental effects testing with quantitative microbial analysis: Chemical signatures correlated with in situ biofilm analysis by FT/IR. *Environmental Toxicology* 1(3), 315–338.

White, D.C., 1988. Validation of quantitative analysis for microbial biomass, community structure, and metabolic activity. *Advances in Limnology* 31(1), 1–18.

White, D.C., Davis, W.M., Nickels, J.S., King, J.D. and Bobbie, R.J., 1979. Determination of the sedimentary microbial biomass by extractible lipid phosphate. *Oecologia* 40(1), 51–62.

White, D.C. and Frerman, F.E., 1967. Extraction, characterization, and cellular localization of the lipids of *Staphylococcus aureus*. *Journal of Bacteriology* 94(6), 1854–1867.

White, D.C., Meadows, P., Eglinton, G., et al., 1993. In situ measurement of microbial biomass. Philosophical Transactions of the Royal Society of London. *Series A: Physical and Engineering Sciences* 344, 59–67.

White, D.C., Pinkart, H.C. and Ringelberg, A.B., 1997. Biomass measurements: biochemical approaches. In Hurst, C.J., Knudson, G.R., McInerney, M.J., Stetzenbach,L.D., Walter, M.V. (Eds) Manual of Environmental Microbiology. DC: ASM Press, 91–101.

White, D.C. and Ringelberg, D.B., 1997. Utility of the signature lipid biomarker analysis in determining the in situ viable biomass, community structure, and nutritional/physiologic status of deep subsurface microbiota. In Amy PS and Haldeman DL (Eds) *The Microbiology of the Terrestrial Deep Subsurface*. New York: Elsevier Science Ltd; CRC, 119–136.

Zelles, L. (1997) Phospholipid fatty acid profiles in selected members of soil microbial communities. *Chemosphere* 35, 275–294.

Zhang, J.I., Costa, A.B., Tao, W.A. and Cooks, R.G., 2011b. Direct detection of fatty acid ethyl esters using low temperature plasma (LTP) ambient ionization mass spectrometry for rapid bacterial differentiation. *Analyst* 139, 3091–3097.

Zhang, J.I., Talaty, N., Costa, A.B., et al., 2011a. Rapid direct lipid profiling of bacteria using desorption electroscopy ionization mass spectrometry. *International Journal of Mass Spectrometry* 301, 37–44.

15 Phylogenetic Techniques in Geomicrobiology

DENISE M. AKOB, ADAM C. MUMFORD, DARREN S. DUNLAP, AND AMISHA T. PORET-PETERSON

Abstract

Molecular biological techniques have revolutionized the field of geomicrobiology by providing researchers with robust techniques for identifying microorganisms and characterizing microbial communities in a wide variety of environments. These techniques have freed researchers from the constraints of classical culture-based microbiology and allowed the discovery of previously unknown phylogenetic diversity of microorganisms. In this chapter, we discuss the theory, methods, and workflow for applying molecular techniques to identify and characterize microbial populations. Our chapter focuses on SSU rRNA gene-based approaches, guiding the reader from sample collection and gene amplification through bioinformatics and statistical analysis. The workflow presented has been successfully used to identify microbial populations and community dynamics in a wide variety of habitats to understand the interactions between microbes and their environment.

15.1 Introduction

Microorganisms live in nearly all terrestrial environments and are capable of diverse metabolisms that drive a wide variety of the biogeochemical reactions critical to global geochemical cycles. Reactions of importance for the study of geomicrobiology are catalyzed by Bacteria, Archaea, and microbial eukaryotes (fungi, algae, and protozoa) that are found in habitats ranging from shallow soils to deep below the seafloor (Akob and Küsel, 2011; Orcutt et al., 2011). Bacteria and Archaea represent the largest proportion of life's genetic diversity, and the total carbon represented by prokaryotes is estimated to be >60% of that represented by plants in terrestrial systems (Rosselló-Mora and Amann, 2001; Whitman et al., 1998). However, identifying the unseen diversity of microbial communities and understanding the roles of these unknown organisms in the environment has long posed a challenge to scientists.

Historically, the main approach to studying microbes was cultivation combined with identification based on morphology or physiology. However, microscopic counts of bacteria (lowercase bacteria refers to both Bacteria and Archaea) in environmental samples are always much higher than what is recovered by cultivation methods (<1% of total bacteria) (Amann et al., 1995; Staley and Konopka, 1985), suggesting that the majority of bacterial diversity remains undescribed and underestimated. As summarized in Akob and Küsel (2011), many bacteria are difficult or impossible to grow using available culture methods due to their variable growth strategies and unknown nutritional requirements. Even when cultivation is achieved, physiological characteristics provide only limited information for the identification of organisms obtained in pure culture (e.g., Rodriguez-R and Konstantinidis, 2014; Woese, 1987). The application of molecular techniques based on amplification and sequencing of nucleic acids revolutionized the field of microbiology by providing scientists with a suite of methods allowing the identification of microbial species and the estimation of microbial community diversity independently of cultivation.

Cultivation-independent methods utilize biomarkers contained in the nucleic acids, lipids, and proteins of organisms to identify microbes and estimate biodiversity. By using biomarkers extracted directly from environmental samples, cultivation-independent analyses of microbial diversity escape many of the biases inherent in cultivation-based methods (Amann et al., 1995; Olsen et al., 1986). In this chapter, we will give an overview of some of the most commonly used methods in cultivation-independent molecular biology.

Nucleic acid (DNA and RNA)-based approaches are the most commonly used methods for estimating microbial diversity in the environment. The development of techniques targeting small subunit ribosomal RNA molecules (SSU rRNA) revolutionized our understanding of the evolutionary relationships between Bacteria, Archaea, and Eukaryotes (Woese, 1987; Woese and Fox, 1977) (Figure 15.1A). Perhaps the most striking discovery was that the Bacteria and Archaea are distinct Domains of life evolved from a universal ancestor (Figure 15.1A). Since the seminal work of Woese (1987), SSU rRNA genes (DNA) and transcripts (RNA transcribed from a DNA template) have become the "gold standard" for classifying microorganisms and resolving the phylogenetic relationships between them.

The types of SSU rRNA primarily targeted vary in size based on Domain: 16S in Bacteria and Archaea and 18S in Eukaryotes (S = Svedberg units, a measure of sedimentation rates under centrifugation). These structural genes are useful targets for resolving phylogeny, as they are present in all organisms, are essential for protein synthesis, and therefore, are required for life. The function of the ribosome is dependent on the SSU rRNA transcript folding into the appropriate three-dimensional structure based upon base-pair matching; this constrains the molecular sequence, as incorrect assembly of the ribosome results in the inability to synthesize proteins (Lafontaine and Tollervey, 2001; Lecompte et al., 2002). These constraints force large portions of the molecule to be highly conserved while allowing some, less structurally important, sequences to be highly variable (Van de Peer et al., 1996). These hypervariable regions are short enough to fully sequence using next-generation sequencing (NGS) technology and have been shown to be as phylogenetically or taxonomically informative as full-length SSU rRNA sequences (Caporaso et al., 2011; Hugerth et al., 2014; Huse et al., 2008; Kim et al., 2011; Mahé et al. 2015).

Analysis of SSU rRNA sequences has provided microbiologists with an independent, standardized method for identifying bacterial species and studying the evolution of species (Woese, 1987). Although work has shown that other genes, e.g., rpoB, and particularly combinations of genes, also retain evolutionary data and allow resolution of microbial phylogenies to a finer degree (Dahllöf et al., 2000; Mollet et al., 1997; Santos and Ochman, 2004), the ease of SSU rRNA sequencing in the early days of molecular microbiology led to the rapid generation of vast quantities of SSU rRNA gene sequences. Since the publication

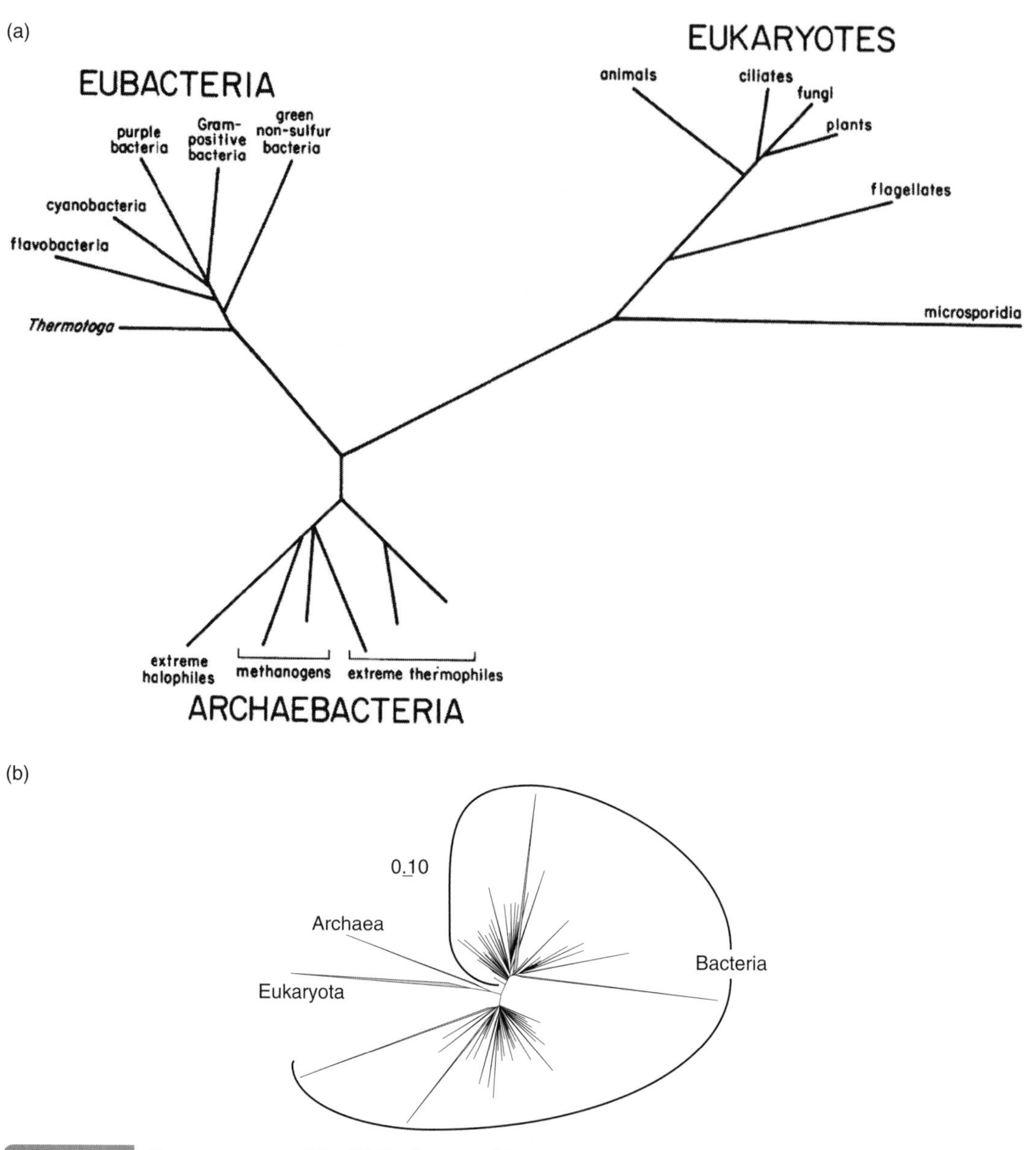

Figure 15.1 Phylogenetic trees of life. (A) The first tree of life based on 206 SSU rRNA sequences. (From Woese, 1987. Reproduced with permission from the American Society of Microbiology.) (B) The current tree of life derived from approximately 530 000 SSU rRNA sequences available from the SILVA release 119 (June 2014). The nomenclature of Eubacteria and Archaebacteria in the Woese tree has changed to Bacteria and Archaea, respectively.

of Woese's first universal phylogenetic tree, there were indications that the microbial world hosted a vast untapped microbial diversity (Woese, 1987). The widespread application of SSU rRNA gene-based analyses to microbiology revealed the great depth of unexplored phylogenetic diversity in the environment (as summarized in Hugenholtz et al., 1998), which continues to expand with each application of methods to new studies (Lynch and Neufeld, 2015). Despite the limited data set used to construct the first universal tree (Figure 15.1A), the overall topology of the tree of life has been maintained even with the addition of vast quantities of SSU rRNA gene sequences. In Figure 15.1B, we present the current tree of life based on the SILVA release 119, containing 534 968 nonredundant of SSU rRNA gene sequences (out of 4 346 367 total sequences) available as of July 2014 (Quast et al., 2013). The topology of this tree is not definitively resolved, and it is constantly being revised based on continual description of new microbial lineages in the International Journal of Systematic and Evolutionary Microbiology (http://ijs.sgmjournals.org/). The diversity of uncultivated microbes is well illustrated when one considers that only 11 269 sequences from all cultivated type strains comprise the Living Tree Project sequence database, compared with the over 500 000 nonredundant sequences contained in SILVA 119 (Yilmaz et al., 2014). Recent work using reconstructed genomes indicates the potential for the discoveries of new lineages and new definitions of microbial phylogeny (Hug et al., 2016).

15.2 Real-world Application of Phylogenetic Methods

Cultivation-independent techniques are among the most powerful analyses available to geomicrobiologists. These methods allow the characterization of microbial communities in a wide variety of samples independently of the limitations inherent in culture-based techniques. Community-level analysis falls under the umbrella of "metagenomics," which translates directly to "beyond the genome." Metagenomics is the study of genetic material directly from environmental samples and refers to both genome- and amplicon-based analyses (Handelsman, 2004). Genome-based metagenomics was first performed in 1998 by Handelsman et al.; DNA was extracted from soil microbial communities and then randomly fragmented, cloned, and sequenced to look for novel microbial products (Handelsman et al., 1998). Community metagenomes are used to reconstruct genomes from environmental microbial communities independently of cultivation (Albertsen et al., 2013; Gilbert and Dupont, 2011), which has led to advances in our understanding of community dynamics (Breitbart et al., 2002; Fierer et al., 2007; Whitaker and Banfield, 2006). In contrast to whole-genome metagenomics, amplicon-based analyses, such as those targeting SSU rRNA genes, focus solely on a specific gene target to assess community composition, and will be the primary focus of this chapter.

Microbial community characterization seeks to understand the structure and function of microbial communities by evaluating the (i) diversity (number of species and abundance) and (ii) identity of species. This can be carried out using a variety of methods that fall into two broad categories: sequencing and fingerprinting. Sequence-based techniques explore the differences between the genetic sequences of the individual organisms present in a sample, while fingerprinting techniques explore the genetic differences between the entire communities without necessarily seeking to identify the individual organisms present. All approaches are applied using a single generalized workflow (Figure 15.2), whereby nucleic

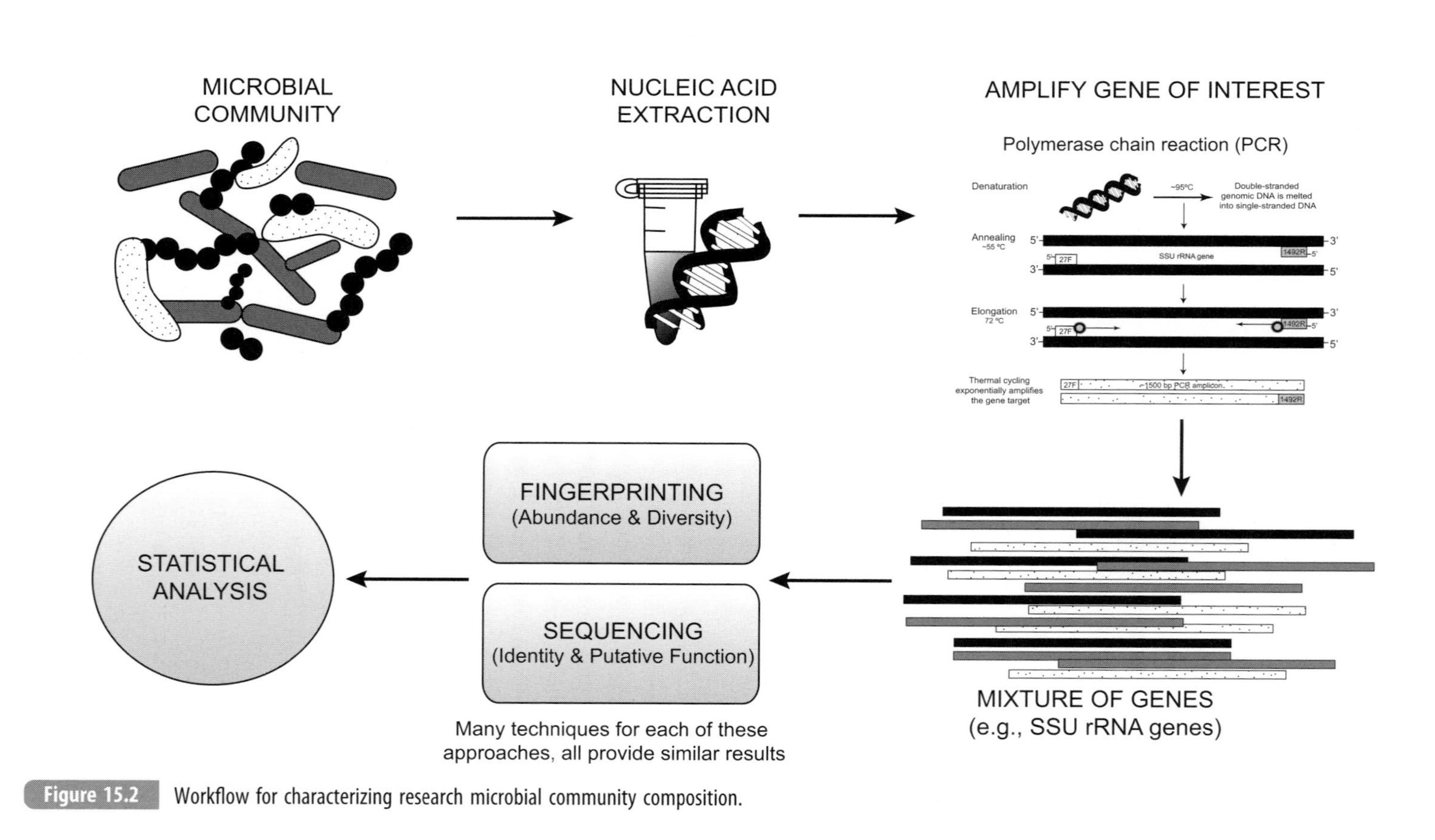

Figure 15.2 Workflow for characterizing research microbial community composition.

Table 15.1 Methods for profiling microbial community structure. Select references explaining the methods and applications relevant to geomicrobiology are presented.

Method	Fingerprinting (e.g., diversity)	Species identification	Pros	Cons	References
DGGE (denaturing gradient gel electrophoresis)	Yes	Indirect, limited	Low cost, easy analysis, applicable to any gene target	Challenging method, no "true" diversity and limited identity, low throughput, time intensive, lack of repeatability	(Green et al., 2009; Muyzer, 1999)
TRFLP (terminal restriction fragment length polymorphism)	Yes	Indirect, limited	Low cost, easy analysis, very repeatable, high throughput, applicable to any gene target	No "true" diversity and limited identity, time intensive	(Clement et al., 1998; Hartmann and Widmer, 2008; Kitts, 2001; Liu et al., 1997; Marsh, 1999; Osborn et al., 2000)
ARISA (automated ribosomal intergenic spacer analysis)	Yes	Indirect, limited	Low cost, easy analysis, very repeatable, high throughput	Limited bacterial internal transcribed spacer (ITS) databases	(Brown et al., 2005; Danovaro et al., 2006; Fischer and Pusch, 1999; Fisher and Triplett, 1999)
Cloning with Sanger sequencing	Limited	Yes	Larger reads, more taxonomic information, applicable to any gene target	Can miss low-abundance organisms, costly, time intensive	(Green and Sambrook, 2012; Leigh et al., 2010; Sanger and Coulson, 1975)
NGS (next-generation sequencing)	Yes – deep	Yes	High number of reads, both diversity and identity in one step, applicable to any gene target	Computationally intensive, challenging bioinformatics and quality assurance	(Caporaso et al., 2012; Degnan and Ochman, 2012; Liu et al., 2012; Quail et al., 2012; Sinclair et al., 2015)

acids are extracted from a microbial community, amplified through polymerase chain reaction (PCR), and then analyzed using either fingerprinting or sequencing techniques (Table 15.1).

In the following section, we will discuss the methods presented in the microbial community characterization workflow (Figure 15.2), beginning with extraction of nucleic acids (DNA or RNA), amplification of target genes, generation of data via fingerprinting and/or sequencing methods, and statistical analysis.

15.3 The Methods

15.3.1 Sample Collection and Preservation

Environmental and laboratory samples for geomicrobiological studies can come from many sources, such as soil, rock, water, or cultures, each requiring different collection and preservation methods. Soil, sediment, and rocks are collected in the field using sterile scoops and sterile containers such as Ziploc® bags, Whirl-Pak® bags, or centrifuge tubes. (Note: Any use of trade, product, or firm names is for descriptive purposes only and does not imply endorsement by the U.S. government.) For samples with a high water content, it is helpful to allow the solids to settle and decant the water prior to freezing. Water samples require more preparation in the field, as microbial biomass must be concentrated from the sample prior to freezing. In the field, biomass is routinely collected using filters, with the volume of water filtered dependent on the environment and the biomass. Filters for collection of microbial biomass can be comprised of a variety of materials and typically have a 0.2 or 0.45 μM pore size for optimal microbial cell recovery. They are available in a variety of formats (e.g., capsule, syringe, or membrane), allowing the researcher to select the best type for laboratory or field use. Filtration is often preferred over centrifugation, as large volumes of water with low biomass and/or high salt contents can be concentrated on a filter to collect more cell material while removing potential interferences from dissolved solids and humic materials. Filtration protocols must be paired with appropriate nucleic acid extraction protocols, particularly as some filter materials are not compatible with some extraction reagents and can dissolve, leaving behind inhibitors for downstream analysis. Centrifugation is more commonly used for microbial cultures in the laboratory, as the biomass is typically higher and cells pellet more easily; however, it is also possible to collect biomass from cultures on membrane filters when large volumes of culture media with low cell concentrations must be processed.

Stabilization conditions are vital for preserving nucleic acids; this is most easily accomplished by freezing samples as quickly as possible. For samples acquired in the field, it is generally advisable to transfer filters and/or solid samples to dry ice or use dry vapor shippers for transit to minimize degradation. DNA is relatively stable at −20 °C, but storage at −80 °C is preferred. Degradation of DNA will occur when samples are stored on wet ice or in a refrigerator, and repeated freeze-thaws can enhance degradation. RNA rapidly degrades if samples are not kept at −80 °C, and rapid, *in situ* freezing (e.g., on liquid nitrogen in the field) and long-term storage at −80 °C are necessary to properly preserve samples. Because dry ice or liquid nitrogen is often difficult or impossible to ship and bring into the field, a variety of reagents and products, such as RNALater® (Life Technologies, Grand Island, NY) and FTA® cards (Whatman, GE Healthcare, Little Chalfont, UK), have been developed for ambient-temperature preservation of nucleic acids. It is recommended to test a variety of these preservation methods for specific sample types beforehand, as previous studies have shown differences in effectiveness between these treatments depending on the specifics of individual samples and matrices.

15.3.2 Nucleic Acid Extraction

The extraction of nucleic acids from samples is the first step in the workflow of characterizing microbial communities. There are a wide variety of protocols available for this purpose, including both published protocols using readily available reagents and commercial kits containing all necessary supplies and reagents in a single box. The choice of method will depend on the desired on the type of nucleic acid (DNA or RNA), the sample matrix, the quantity of biomass in your sample, and the presence of PCR inhibitors in the matrix. In addition, the need for strict quality assurance and quality control (QA/QC), repeatability over time, cost, and the desired throughput play a part in choosing one method over another. While numerous studies have attempted to quantify the differences and biases between the various protocols (Feinstein et al., 2009; Herrera and Cockell, 2007; Inceoglu et al., 2010; Mahmoudi et al., 2011; Martin-Laurent et al., 2001; Miller et al., 1999; Vishnivetskaya et al., 2014; Wade and Garcia-Pichel, 2003), no consensus has yet been reached, and it is up to the researcher to determine the optimal method for a particular study and matrix. Regardless of the specific protocol chosen, all nucleic acid extractions share the following general steps:

Step 1: Cell lysis. In this step, the cells are lysed (broken open) via mechanical and/or mechanical means to expose the DNA. Mechanical processes include grinding, boiling, sonication, or bead beating the sample. Chemical lysis utilizes salts, detergents (such as sodium dodecyl sulfate [SDS] and cetyltrimethylammonium bromide [CTAB]), and/or enzymes (such as proteinase K and lysozyme) to disrupt the cell membrane. Most protocols for soil, sediment, and rock samples combine mechanical (bead beating or grinding) and chemical processes to homogenize the sample and enhance cell lysis (e.g., Hurt et al., 2001; Purdy, 2005; Zhou et al., 1996).

Step 2: Purification. The overall goal of the extraction is the purification of nucleic acids from organic and inorganic material contained in the environmental sample, e.g., soil, cellular debris, and humic substances. It is important to remove contaminating organic and inorganic matter, as it may inhibit downstream applications. Contaminants are removed with reagents that precipitate the contaminating organic and inorganic material, using dialysis to remove salts, or by utilizing nucleic acid-specific binding to silica or magnetic beads or membranes. The choice of purification method is highly dependent on the type of contaminants present in the sample, and in many cases, multiple purification steps and approaches are used. Classical methods of DNA extraction commonly utilize mixtures of phenol and chloroform as a reagent for separating water-soluble nucleic acids from proteins and many types of nonpolar compounds, but many newer protocols capture nucleic acids by selective binding onto beads or a membrane (Lever et al., 2015; Purdy, 2005). Many commercially available kits utilize the property of nucleic acids to bind to silica in the presence of chaotropic buffers (such as guanidine thiocyanate). Many commercially available DNA extraction kits are optimized around a variety of matrices (e.g., soil, water) and often include proprietary reagents designed to remove particular PCR-inhibiting compounds.

Step 3: Nucleic acid precipitation or recovery. After purification, nucleic acids are either in aqueous solution or bound to a solid phase such as beads or a membrane. The recovery of nucleic acids from an aqueous solution can be accomplished by addition of an alcohol (e.g., ethanol or isopropanol) to cause DNA or RNA to precipitate, allowing collection via centrifugation. This step also acts as an additional purification, as it further removes water-soluble salts. The resulting nucleic acid pellet can then be resuspended in high-purity sterile water or buffer (e.g., Tris-EDTA or Tris only buffers) to be used in downstream applications. Membrane- or bead-bound nucleic acids are eluted through the addition of high-purity sterile water or buffer followed by centrifugation. The selection of the appropriate resuspension solution or eluent is important, as compounds such as EDTA may interfere with downstream reactions (e.g., restriction digests, PCR).

15.3.3 Quantification and Quality Checking

After nucleic acid extraction, the isolated nucleic acids must be quantified to determine the net recovery from the samples. DNA and RNA concentrations are measured using UV absorbance, fluorometry, or gel electrophoresis. The concentration and purity of DNA are frequently assessed by measuring the UV absorbance of the sample with a spectrophotometer. Nucleic acids absorb UV light at 260 nm, while aromatic proteins absorb UV light at 280 nm. Canonically, high-quality samples of DNA and RNA have 260/280 ratios of approximately 1.8 to 2.0, and extracts that are contaminated with proteins or other compounds will have a ratio lower than this (Sambrook and Russell, 2001). Fluorometry utilizes DNA- or RNA-specific dyes and a fluorometer; the resulting signal will be proportional to the concentration of DNA in a sample but does not indicate whether contamination is present. Typically, commercially available kits are used for fluorometric measurements, such as Qubit™, Quant-iT™ PicoGreen®, or Quant-iT™ RiboGreen® assays (Invitrogen™, Carlsbad, CA).

Gel electrophoresis separates nucleic acids by size using an agarose gel placed in an electrical field. DNA on the gel is quantified by staining with a nucleic acid-binding dye and comparing the band intensity of the DNA with a sizing standard of known concentration. Nucleic acids can also be simultaneously quantified and sized on instruments such as an Agilent 2100 Bioanalyzer (Agilent Biotechnologies, Santa Clara, CA, USA) that combine fluorometry and gel electrophoresis. Quantification of extracts from low-biomass samples with gel electrophoresis and UV absorption will often result in negative or extremely low values, but often, enough DNA may be present for downstream amplification and analysis. Results obtained from such samples should be interpreted with care, as a recent study found that some commonly used molecular biology reagents and nucleic acid extraction kits may contain low levels of DNA (Salter et al., 2014). It is important to verify that DNA extracted from environmental samples is mostly intact, with a high molecular weight (e.g., migrating at or above the longest-size standard band of 10 or 20 kilobases [kb]) and minimal evidence of shearing whenever possible, with high-molecular-weight DNA being especially important for metagenomic and genomic sequencing. Furthermore, the use of highly degraded DNA in PCR reactions has been shown to interfere with polymerization and inhibit PCR (Golenberg et al., 1996).

15.3.4 Polymerase Chain Reaction

The next step in the amplicon-based microbial community analysis workflow is to amplify the nucleic acid sequence of interest (Figure 15.2) using PCR (Figure 15.3). PCR is a technique widely used in molecular biology for the exponential amplification of a few copies of a target nucleic acid sequence. PCR relies on the activity of the DNA polymerase enzyme to synthesize double-stranded DNA (dsDNA) from a single-stranded template. *Taq* polymerase, derived from *Thermus aquaticus* (Brock and Freeze, 1969), is the most commonly used polymerase, although several other DNA polymerases, e.g., *Pfu* polymerase, are also available (Terpe, 2013). Polymerases for laboratory uses are now produced commercially using recombinant technologies; they have high purity and predictable function, and undergo rigorous QA/QC to screen for microbial DNA contamination. DNA polymerases have different error rates or mutation frequencies (Terpe, 2013). The use of a "high-fidelity" or low-error-rate DNA polymerase is recommended to obtain products most accurately copied from their template sequences. The other main components of the PCR reaction are deoxynucleotidetriphosphates (dNTPs), buffer, and $MgCl_2$. The dNTPs, a mixture of dATP, dTTP, dCTP, and dGTP, are the building blocks of DNA, which the polymerase uses to synthesize the new DNA strand. Mg^{2+} is a necessary cofactor of DNA polymerase, and the buffer formulation provides the optimal pH and ionic strength for enzyme activity. Typically, *Taq* polymerase is supplied in a kit, including the enzyme, $MgCl_2$, and appropriate buffer.

DNA polymerase functions by adding dNTPs to single-stranded DNA (ssDNA) in a complementary manner (A to T, C to G) to form dsDNA. DNA polymerase must first bind to a short region of dsDNA on the template strand to begin synthesizing dsDNA from ssDNA. In PCR, primers (typically, 15–20-base pair [bp] oligonucleotides, Figure 15.3) are designed to be complementary to the sequence bracketing a gene of interest. By binding to the ssDNA upstream and downstream of the gene of interest, primers provide a starting point for DNA polymerase and allow selective amplification of the desired target while excluding other genes.

The choice of primer is dependent on the study and can target SSU rRNA, other highly conserved marker genes, or metabolic genes. Highly conserved marker genes include *rpoB*, encoding the beta subunit of RNA polymerase (Case et al., 2007; Dahllöf et al., 2000), and

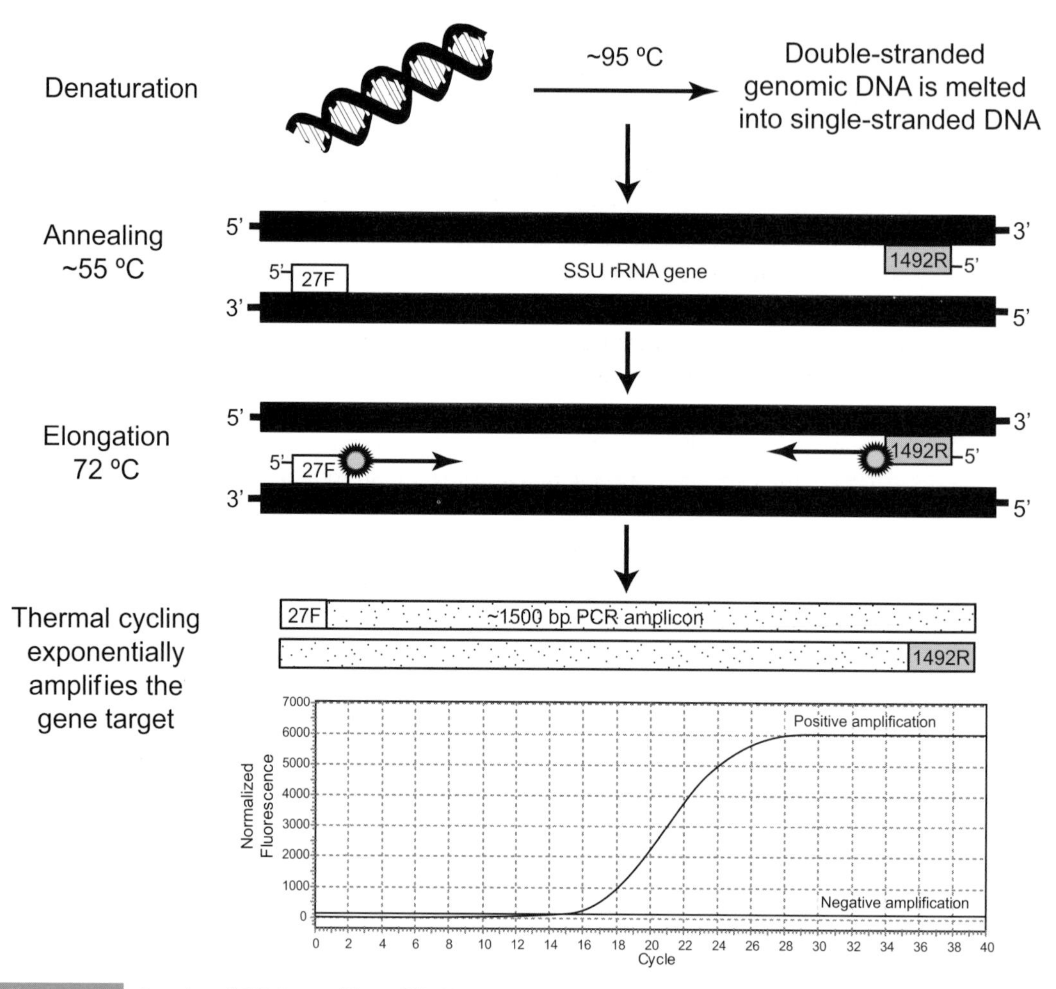

Figure 15.3 Overview of PCR for amplifying SSU rRNA genes with primers 27F and 1492 R.

gyrB, encoding the B subunit protein of DNA gyrase (Wang et al., 2007a). Often, alternative conserved marker genes are used in systems where SSU rRNA may not effectively distinguish particular strains of a species (Wang et al., 2007a). Metabolic genes, on the other hand, encode the proteins required for particular metabolic functions and can provide insight into the function and sometimes the identity of organisms, e.g., dissimilatory sulfite reductase (*dsr*) (Castro et al., 2000) or adenosine-5′-phosphatase (*aprA*) (Meyer and Kuever, 2007) genes for sulfate-reducing bacteria. Metabolic genes are often targeted because of their importance in biogeochemical cycles (e.g., methanogenesis, methane oxidation, nitrogen fixation, etc.), which allows the linking of microbial identity to function in environmental samples (Collavino et al., 2014; Gaby and Buckley, 2012; Luke and Frenzel, 2011; Narihiro and Sekiguchi, 2011).

Microbial community characterization studies generally use primer sets targeting the SSU rRNA gene to provide information on the whole microbial community. SSU rRNA primer sets include universal primers, which target both Bacteria and Archaea (e.g., Frank et al., 2008; Klindworth et al., 2012; Lane, 1991; Mao et al., 2012; Marchesi et al., 1998; Weisburg et al., 1991), domain-specific primers for Bacteria (e.g., Winsley et al., 2012), Archaea (e.g., Gantner et al., 2011; Kato et al., 2011), and Eukarya (Hugerth et al., 2014), and group-specific primer sets, which target a narrower phylogenetic group of organisms (e. g., Hall, 2008). In addition to SSU rRNA primer sets, there are primers that amplify large subunit (LSU) rRNA genes, useful for detection of eukaryotic algae and cyanobacteria and the internal transcribed spacer (ITS) region of fungal rRNA genes (Chaput et al., 2015 and references therein). The sheer number of available primer sets makes selection of rRNA gene primer sets seem daunting. probeBase (www.probebase.net) is an extremely useful curated online resource for selection of rRNA gene primers and probes (Greuter et al., 2016). This database contains information on primer specificity or coverage, annealing position, and other characteristics (e.g., sequence, GC content, etc.) and allows users to perform *in silico* matching against rRNA databases. Some analyses require amplification of nearly full-length SSU rRNA genes (~1 500 bp in length), whereas as others target short, partial sequence regions (~100–500 bp). Our group uses the SSU rRNA gene primers 27F (Johnson, 1994) and 1492 R (Lane, 1991) for applications requiring near-full-length bacterial sequences and the primers 515F and 806 R, targeting the hypervariable region 4 (V4), for high-throughput sequencing of bacterial and archaeal 16S rRNA genes (Caporaso et al., 2011).

An important consideration in optimizing amplification methods for different sample types is the inhibition of PCR. Because PCR requires many highly specific interactions during nucleic acid amplification, such as primer binding and elongation, a small interference in the reaction can cause an exponential decrease in PCR product or the outright failure of a PCR. Many factors can contribute to PCR inhibition, such as high GC content, secondary structures within the target nucleic acid, and inhibitory compounds not removed during nucleic acid extraction. Various chemical additives successfully reduce PCR inhibition from chemicals, high GC content, and secondary nucleic acid structures, including bovine serum albumin (BSA) and dimethyl sulfoxide (DMSO) (Nagai et al., 1998; Ralser et al., 2006). While a plethora of options exist for optimizing a PCR reaction, incorporating a well-established schedule of positive and negative controls can greatly reduce the amount of time spent changing every possible PCR variable. A single reaction is prepared for each environmental sample, and positive (a known amplifiable template) and negative (no template DNA) controls are always included to verify the success of the reaction and to ensure that no reagents were contaminated with DNA during preparation.

Regardless of target, every PCR cycle includes steps to denature the DNA, anneal the primers, and elongate the new DNA strand (Figure 15.3), and these steps are optimized for the primer set and enzyme used for amplification of the target gene. The PCR begins with an initialization step, with the heating of the reaction mix, typically to 95 °C, to denature (melt) the dsDNA to ssDNA. Thermal cycling then begins with denaturation, with the heating of the reaction mix to melt the DNA template and primers. The next step is annealing, with the cooling of reaction temperature to allow annealing of the primers to the ssDNA template. The annealing temperature is a static temperature in the range of 45–72 °C and is ~3–5 °C below the melting point (T_m) of the primers; this temperature can be thermodynamically predicted based on the primer sequence or empirically determined when validating a new primer set. After annealing, the polymerase binds to the primers, where it can begin DNA synthesis. The final cycling step is extension (or elongation), in which the temperature is raised, typically to 72 °C (depending on the polymerase), to allow the DNA polymerase to add complementary bases to the ssDNA template, resulting in the synthesis of a new dsDNA strand. The extension time depends both on the DNA polymerase used and on the length of the DNA fragment to be amplified. As a general guideline, DNA polymerase will add a thousand bases per minute. Under optimal conditions, at each extension step, the number of DNA target sequences doubles, leading to exponential amplification of the DNA target (Figure 15.3). After cycling (typically 30–40 cycles) is completed, the reaction is incubated at the extension temperature for a longer period of time (generally 5 to 10 minutes) in a final elongation step to ensure that any remaining ssDNA is fully complemented. A final hold at 4 °C is typically programmed into the thermal cycler to store the reaction until it can be used or frozen at −20 to −80 °C.

Agarose gel electrophoresis is performed after the PCR to verify the success of the reaction. Gel electrophoresis is used to assess whether PCR generated the anticipated DNA fragment, known as an amplicon, at the correct size, to check for nonspecific amplification, and to make sure that no contaminants were present in the reaction. Gel electrophoresis is performed on the PCR products and a size marker or DNA ladder containing DNA fragments on an agarose gel, which separates DNA by size. The size(s) of PCR products is/are determined by comparison with the size marker. A successful reaction generates a mixture of amplicons of the same gene that can be used in sequencing and fingerprinting approaches (Figure 15.2).

Under ideal conditions, PCR generates a mixture of amplicons that are (1) direct copies of the template DNA and (2) proportional to the abundance of the different sequences in the template DNA. After the growing use of PCR-based approaches to analyze complex microbial communities in the late 1990s, researchers recognized that these assumptions may not always hold true (Kanagawa, 2003; Polz and Cavanaugh, 1998; Suzuki and Giovannoni, 1996; Wintzingerode et al., 1997). The PCR may introduce biases in community composition through differential amplification of template sequences, generation of chimeric amplicons (i.e., a PCR product derived from two different template sequences), and introduction of base mutations (Kanagawa, 2003; Polz and Cavanaugh, 1998; Suzuki and Giovannoni, 1996; Wintzingerode et al., 1997). The use of higher input DNA concentrations (>5 ng; Kennedy et al., 2014; Polz and Cavanaugh, 1998), reductions in the number of cycles (Kanagawa, 2003; Suzuki and Giovannoni, 1996), and pooling of multiple PCR reactions (typically three reactions; Kanagawa, 2003; Polz and Cavanaugh, 1998) can minimize biases resulting from differential amplification. However, a recent study showed that using a higher template concentration is more effective at reducing bias than pooling PCR reactions (Kennedy et al., 2014). The use of "high-fidelity" DNA polymerases reduces the likelihood of base mutations (Terpe, 2013), and chimeric sequences can be screened out during bioinformatic processing. While it is not possible to completely remove these biases, users of PCR-based approaches should be aware of their potential effects on microbial community composition data (Wintzingerode et al., 1997).

In addition to DNA-based PCR, there are a variety of other PCR techniques, including reverse transcription PCR (RT-PCR) and quantitative PCR (qPCR). RT-PCR is the first step in amplifying RNA targets, which provide information about active communities or the activity of specific genes. In this reaction, the reverse transcriptase enzyme transcribes RNA to complementary DNA (cDNA), which is then subjected to PCR amplification prior to further analysis. Active communities and genes are best profiled using RNA targets, as RNA degrades more quickly than DNA, and its presence is correlated with active organisms and not remnant environmental DNA (e.g., dead cells). Studies comparing differences between active (SSU rRNA) and total (SSU rRNA genes) microbial communities in a uranium-contaminated subsurface environment showed significant differences between these populations (Akob et al., 2007). Other studies have monitored mRNA abundance and diversity to assess the metabolic activity of specific groups, e.g., sulfate- and iron-reducing bacteria (Akob et al., 2012).

15.3.5 Quantitative PCR

Quantification of nucleic acid targets, such as gene and transcript copy numbers, cannot be accurately determined from standard PCR, because it is not possible to detect differences in starting amounts of template. During qPCR reactions, the quantity of amplified PCR product is monitored in real time (hence, this technique is also referred to as real-time PCR), and the initial concentration of the target DNA in an unknown sample is estimated by comparing concentrations of the unknown sample during the amplification cycles with standards of known concentration. Extensive standardization is necessary to determine the limits of quantitation and the limits of detection for each qPCR assay. Fluorescent dyes that specifically bind dsDNA (e.g., SYBR® green) or primers with a TaqMan® probe are used to monitor the accumulation of PCR product over the amplification cycles. qPCR can be combined with reverse transcription (RT-qPCR) to analyze the expression of genes or to quantify live vs. dead cells. Refer to the reviews by Zhang and Fang (2006) and Smith and Osborn (2009) for a detailed overview of the different techniques, their requirements, and their limitations for environmental samples.

The previously mentioned factors for PCR inhibition are even more important to consider for qPCR, as a minor reduction in efficiency can cause exponential changes in the quantification of a specific gene. Additionally, many studies in geomicrobiology consist of extracting nucleic acids from sediment and soil samples, which are rich in inhibiting compounds such as humic acids, phenolic compounds, and heavy metals (Wilson, 1997). DNA extraction methods designed for sediments, as well as dilution of the template DNA, can help diminish the presence of inhibitory compounds (Hargreaves et al., 2013). Additionally, inhibition controls that include both the sample and a known standard quantity of template should be run for each sample to test for inhibition. Because small changes in method can have large impacts on qPCR results, Minimum Information for Publication of Quantitative Real-Time Experiments (MIQE) has been established as a guide to publishing data from qPCR methods to facilitate the standardization of methods between different laboratories (Bustin et al., 2009).

15.3.6 Profiling Community Structure: Fingerprinting and Sequencing

PCR produces a mixture of gene copies of a similar size that reflects the relative composition of the original microbial community. The next step in the workflow (Figure 15.2) is to separate out these amplicons so that they can be studied to provide information on microbial community structure (diversity and identity). Community profiling is performed using fingerprinting and/or sequencing methods (Table 15.1). The choice of method to use for a study depends on the research question. For example, if the aim of the study is to detect differences in overall diversity between two samples, then fingerprinting alone is a useful strategy. However, sequencing-based approaches are more appropriate if you are interested in knowing the differences in the taxonomies between microbial communities in multiple samples. For each of these aims, there are a variety of methods (Table 15.1), and each has its own tradeoff between cost, throughput, and resolution. However, the rapid advancement, decreasing cost, and increasingly wide availability of NGS technologies since the mid-2000s have changed the decision-making process, as NGS methods permit both fingerprinting and sequencing in one step. While the field of geomicrobiology is becoming increasingly reliant on NGS-based techniques, traditional sequencing (e.g., Sanger-based sequencing, described later) and fingerprinting methods are by no means obsolete (Hall, 2011; Stackebrandt et al., 2002; Tindall et al., 2010; van Dorst et al., 2014) and remain useful.

In the following sections, we will discuss the principles of each method listed in Table 15.1, and many studies will combine techniques for the comprehensive characterization of the microbial communities (e.g., Akob et al., 2008, 2014, 2015; Burkhardt et al., 2011; Priscu et al., 1999).

15.3.6.1 Fingerprinting Methods

Numerous techniques exist for fingerprinting microbial communities, all of which provide information on the diversity and relative abundance of microbial taxa between samples. Fingerprinting methods include denaturing gradient gel electrophoresis (DGGE), terminal restriction fragment length polymorphism analysis (TRFLP), and automated ribosomal intergenic spacer analysis (ARISA) (Table 15.1). For decades, these methods were the only tools available for assessing microbial diversity within uncultivable microbial communities. However, they have now been widely replaced by NGS methods. DGGE and TRFLP can be applied to any gene target; ARISA targets the ITS region between the SSU and LSU rRNA genes. ARISA-like approaches have been developed to assess the distribution of metabolic genes for ammonia oxidation (Norton et al., 2002) and methane oxidation (Tavormina et al., 2010) in environmental samples; these processes are catalyzed by multi-subunit monooxygenases encoded by genes arranged in operons with intergenic spacers.

DGGE combines PCR amplification with gel electrophoresis to separate amplicons of a similar size that have different DNA sequences, which affects the melting (denaturing) profile of each amplicon (Green et al., 2009; Muyzer, 1999). The banding pattern resulting from the DGGE can be interpreted to represent the relative number of microbial taxa (number of bands) and the relative abundance (band intensity) (Muyzer, 1999). Image processing software allows the analysis of differences in banding patterns between samples, which are then used to calculate diversity indices for the samples. Individual bands from a DGGE gel can also be picked and analyzed further to identify microbial species by either secondary DGGE separation or sequencing (e.g., Burkhardt et al., 2011; Castle and Kirchman, 2004). This approach is beneficial by adding an identification component to the fingerprinting approach, but it has also shown a downside to the method: microbial species with different DNA sequences can migrate to the same location within a DGGE gel (Crosby and Criddle, 2003). Therefore, the number of microbial taxa cannot be accurately resolved using this method, it does not provide a "true" measure of diversity, and the taxonomic classification of organisms is challenging under even the best of circumstances. DGGE was considered a rapid method for analyzing and comparing multiple samples, as up to 48 samples can be run at the same time depending on the apparatus used. Obtaining repeatable banding patterns is challenging; as a result, analyzing banding patterns is not trivial across multiple runs, and the resolution of the true number of taxa is limited at best.

TRFLP is another PCR-based technique, in which DNA is amplified with a fluorescently labeled primer, digested with restriction enzymes, and then size separated on an automated DNA sequencer (Clement et al., 1998; Kitts, 2001; Liu et al., 1997; Marsh, 1999; Osborn et al., 2000). Typically, the forward primer is fluorescently labeled at the 5′ end, and after restriction digest, this produces a mixture of labeled terminal restriction fragments (TRFs) that can be separated by size. Restriction digests can be performed with any type of restriction enzyme, although ones with a four-nucleotide recognition sequence are typically used (i.e., HaeIII, MspI, RsaI, etc.) (Kitts, 2001). Size separation occurs via gel electrophoresis, typically using capillary gel electrophoresis on an automated DNA sequencer. The output from the DNA sequencer is an electropherogram, which is analyzed manually or with automated fragment analysis programs such as T-REX (Culman et al., 2009) or TRAMPR (Fitzjohn and Dickie, 2007). These software programs calculate peak sizes (TRF length in bp) by comparing peaks with a DNA size standard and the area under each peak. Diversity indices and multivariate statistics can then be used to compare TRF patterns between samples, including presence/absence and relative abundance of peaks. As with DGGE, a single peak can correspond to multiple organisms; therefore, a peak does not necessarily represent a single microbial species, and the method often underestimates diversity in high-diversity samples (Crosby and Criddle, 2003; Orcutt et al., 2009). However, the benefit of TRFLP over DGGE is that TRFs can be directly referenced to sequence databases or sequence data from an individual study (e.g., Akob et al., 2008, 2015; Fennell et al., 2004; Zhang et al., 2008), it is reproducible between laboratories and samples (Osborn et al., 2000; Orcutt et al., 2009), and it is reliable (Hartmann and Widmer, 2008).

ARISA differs from TRFLP and DGGE, as it targets the ITS region between the SSU rRNA gene and the LSU rRNA gene, which has a highly variable sequence and length that are phylogenetically coherent. This means that organisms within the same clade have similar ITS sequences and bp lengths. PCR amplification is performed with an ITS-specific primer set that includes a fluorescently labeled forward primer (Fisher and Triplett, 1999), followed by separation using an automated electrophoresis system. The number of different bands indicates the number of species in the community, and the intensity of a band can be a proxy for determining relative abundance. ARISA may be advantageous over TRFLP, as it does not require a restriction digest and therefore has a faster throughput, and it can have a higher resolution (Danovaro et al., 2006). The length heterogeneity can be compared to databases; however, ITS databases are limited in comparison with the vast number of sequences in SSU rRNA gene repositories. As with DGGE and TRFLP, ARISA also suffers from the drawbacks that a single organism may have more than one peak in the community profile and unrelated organisms can also have similar spacer lengths (Crosby and Criddle, 2003), leading to inaccuracies in estimation of community diversity.

15.3.6.2 Sanger Sequencing

Until recently, sequence-based diversity studies relied on the DNA sequencing technique first described by Frederick Sanger in 1975 (Sanger and Coulson, 1975). This method produces long (up to ~800 bp) high-quality sequence reads but is limited in throughput. As a result, Sanger sequencing is now most commonly used when only a limited number of high-quality sequences are needed. For example, bidirectional Sanger sequencing is used to obtain the full-length SSU rRNA gene sequence of an isolate for detailed phylogenetic and taxonomic study. Sanger sequencing requires identical template DNA, which in the case of a mixed microbial community, must be separated by cloning.

Amplicon cloning is the process of inserting individual strands of dsDNA into a vector and transforming into a host bacterial cell, which clonally reproduces the inserted dsDNA as the host cell divides. This provides a method for separating out a mixture of amplicons into individual bacterial colonies containing clonal copies of the sequence of interest. More information and excellent, highly detailed descriptions of cloning can be found in *Molecular Cloning* (Green and Sambrook, 2012) as well as Leigh et al. (2010). Cloning vectors typically used for microbial community analysis are derived from bacterial plasmids – a relatively small circular strand of non-chromosomal dsDNA, containing an insertion site, an origin for replication, and a variety of genes useful in screening for successful insertion of the amplicon, including pathways for antibiotic resistance and metabolic markers such as beta-galactosidase. Cloning begins with the ligation of target amplicons in the presence of DNA ligase, which results in the formation of circular plasmids, each containing a single amplicon. Competent bacterial cells, frequently *Escherichia coli,* are then induced to take up an individual plasmid during a process called transformation. These cells are then grown as individual colonies by spread-plating on selective media containing the antibiotic as well as the substrate for the metabolic marker. Successfully transformed cells are selected by growth on the antibiotic-containing media (indicating successful uptake of the plasmid containing antibiotic resistance genes) as well as inactivity of the metabolic marker (indicating successful insertion of the amplicon into the plasmid), which generally results in a change to colony color. Many commercially available cloning vectors do not require screening via a metabolic marker and are termed "suicide vectors," as cells that take up plasmids without inserts do not grow (Bernard et al.,

1994). Successfully transformed cells can then be grown to high density and archived for later DNA extraction, or colonies can be directly PCR amplified. A variety of cloning kits, which include the vector, DNA ligase, competent cells, and all necessary reagents, are commercially available. For a history of the development of cloning vectors, including the incorporation of sequencing primers, antibiotic resistance genes, and metabolic markers, we refer the reader to Messing (2015).

While cloning has the potential to generate a theoretically limitless number of individual clones, the cost and logistics of sequencing more than several hundred clones can be prohibitive. This can lead to an underestimation of the total diversity in complex environmental samples and cause rare or low-abundance organisms to be overlooked (Table 15.1). However, in low-diversity samples, such as extreme environments or enrichment cultures, and pure cultures, the benefit of being able to obtain nearly full-length SSU rRNA gene sequences to recover a high level of taxonomic information is worth the cost and logistics.

15.3.6.3 Next-generation Sequencing

Advances in sequencing technologies over the past two decades, driven by both the technology derived from the sequencing of the first human genome and engineering breakthroughs in microfluidics, have significantly changed the field of genomics (Reis-Filho, 2009; Venter et al., 2001). While Sanger sequencing relies on an individual reaction for the sequencing of each nucleic acid strand, NGS, also known as massively parallel sequencing, overcomes this limitation by sequencing millions of nucleic acid strands simultaneously. In practice, this allows thorough, deep sequencing of a sample. This enables full genome sequencing and environmental microbial community analysis (amplicon or genome based) for a fraction of the cost and time of Sanger sequencing. However, NGS technologies have limitations. Depending on the sequencing method and platform used, the sequence read lengths are generally shorter than Sanger-generated sequences, potentially resulting in a lower phylogenetic resolution. Additionally, the large amount of sequence data (hundreds of thousands to millions of sequences per sample) produced requires significant computing power to analyze. Despite these limitations, NGS is now the standard approach for the generation of sequence data from environmental samples. Indeed, many universities and other institutions have fee-for-service facilities that offer a range of NGS options, described in the following.

A variety of different NGS platforms are currently in use, but all operate by attaching the nucleic acid strands to a solid surface and monitoring the synthesis of the complementary strands in real time using fluorescence or pH changes to detect nucleotide incorporation (Mardis, 2013). While each NGS platform uses proprietary chemistry and a unique workflow, the basic principles are as follows:

1. Amplify the target sequence using tagged primers that attach to a solid surface or bead.
2. Ligate the tagged nucleic acids to complementary surface-attached oligonucleotides.
3. Amplify the attached nucleic acid strands.
4. Detect and identify each added dNTP using a variety of methods, such as hydrogen atom release or fluorescent labels.

This reaction runs simultaneously on all nucleic acid molecules in the same reaction, resulting in up to 8 billion sequences output per run. Because many studies require much lower total sequence counts per sample to obtain the desired coverage, costs can be reduced by multiplexing many samples into a single run. Multiplexing combines multiple samples into the same NGS run by using barcode sequences added to the primers for each sample set (Bartram et al., 2011; Caporaso et al., 2011; Degnan and Ochman, 2012; Kozich et al., 2013; Mardis, 2008). Sequences can then be sorted post NGS into individual samples by identifying the barcode at the beginning of the sequence read.

The obtained sequence length, accuracy, run time, and number of reads vary between NGS platforms (Table 15.2). Sequence data from these platforms are prone to high error rates (Kozich et al., 2013). Errors most commonly arise from library preparation, problems sequencing through homopolymers (repeats of the same nucleotide in a DNA sequence), barcode switching, and low quality and/or miscalled bases (Bragg et al., 2013; Gilles et al., 2011; Kozich et al., 2013; Salipante et al., 2014; Schirmer et al., 2015). The first widely used platform for amplicon-based microbial community analysis was 454-pyrosequencing (454 Life Sciences, Roche, Branford, CT), which was based on the release of light during incorporation of a dNTP during synthesis (Ronaghi et al., 1998). However, the 454-pyrosequencing platform has been discontinued by its manufacturer as of 2016. Current platforms include Illumina MiSeq (San Diego, CA), Illumina HiSeq (San Diego, CA), Ion Torrent (Life Technologies, Frederick, MD), and PacBio (Pacific Biosciences, Menlo Park, CA). Illumina MiSeq (Degnan and Ochman, 2012; Sinclair et al., 2015) and Ion Torrent produce sequences with high total reads and short read lengths (Table 15.2), which is useful for microbial community analysis, where sequencing of a large number of sequences to obtain an adequate representation of the community is a high priority. Platforms with smaller total reads and longer read lengths, such as PacBio, allow *de novo* whole genome assembly. The NGS field is in a period of rapid growth with rapidly changing technology, and more detailed descriptions of the available platforms can be found in the scientific literature (Buermans and den Dunnen, 2014; Caporaso et al., 2012; Liu et al., 2012; Mardis, 2013; Quail et al., 2012). The latest NGS platforms are real-time single-molecule detection systems (Buermans and den Dunnen, 2014). Such a platform is available in the portable MinIon™ nanopore sequencer (Oxford Nanopore Technologies, Oxford, UK), but has yet to be thoroughly tested for read lengths and yields, error rates, and run times on a variety of sample types (Buermans and den Dunnen, 2014; Kilianski et al., 2015; Mikheyev and Tin, 2014).

Study Design and Sample Preparation for NGS. Before embarking on any NGS project, two aspects of study design – replication and metadata collection – require considerable attention. Technological advances in NGS make designing highly replicated studies with comprehensive coverage of microbial community diversity and dynamics possible (Ju and Zhang, 2015; Knight et al., 2012; Zhou et al., 2015). The collection of metadata (contextual information), such as GPS coordinates, temperature, pH, and other environmental parameters, is equally important. The type and extent of these data will be largely study dependent. Nonetheless, guidelines exist for reporting "minimum information" on environmental or experimental details for sequencing projects focused on marker genes (i.e.,

Table 15.2 Next-generation sequencing methods used for phylogenetic analysis of microbial communities.

Method	Average or Max Read Length	Max Reads	Pros	Cons	Typical Application
Roche 454	400–700 bp	0.7 million	Longer reads	Lower quality and number of reads	No longer supported
Illumina MiSeq	2 × 150 bp (paired ends)	25 million	High quality, more data	Second highest cost per bp	Amplicon-based community sequencing, metagenomics, *de novo* genomic sequencing
Illumina HiSeq	2 × 150 bp (paired ends)	400 million	Most data	Highest cost per bp	
Ion Torrent PGM	100–400 bp	60 million	Fastest turnaround time	Fewer and shorter reads than MiSeq	
PacBio	4 600–8 500 bp	47 000	Longest reads	Low number of reads	*De novo* genomic sequencing

SSU rRNA genes or metabolic genes), genomes, and metagenomes (Knight et al., 2012, and references therein). The most robust study designs couple replication with thorough compilation of metadata. This type of approach greatly aids in the analysis, interpretation, and/or explanation of observations derived from NGS data sets.

There are two issues of replication when considering study design: technical replication and biological replication. Technical replicates are derived from one sample and gauge the reproducibility of sample preparation and sequencing. Ideally, technical replicates should yield data sets that are indistinguishable from each other. Significant differences in technical replicates suggest sources of error arising from PCR, sequencing, or both (Kennedy et al., 2014; Poretsky et al., 2014). Technical replicates are often nested within biological replicates, which are independent samples from an experimental treatment or observational setting. Three biological replicates are considered sufficient for most experimental and observational studies as a "rule of thumb." However, this number should be dictated by the study aims or questions, resources for sample analyses, and other practical considerations. For example, many studies involving experimental manipulation use shifts in microbial community composition and/or structure as response variables. These studies would benefit from a high degree of replication (i.e., six biological replicates) to ascertain that changes in composition and/or structure are consistent responses. On the other hand, observational studies may require high-density sampling of a particular area or region with fewer biological replicates. One goal of an observational study may be to explain which environmental parameters best correlate with changes in microbial communities along a natural gradient. In this case, sequencing duplicate biological replicates obtained from numerous sites along the gradient may be a sufficient and cost-effective sampling strategy. The decision to sequence technical and/or biological replicates will be highly study dependent, yet researchers should incorporate both into study design, given the continually decreasing cost of sequencing multiple samples.

While significant differences in technical replicates will point to PCR and/or sequencing errors, these errors are most effectively tracked using a mock community of known composition (Edgar, 2013; Kozich et al., 2013; Poretsky et al., 2014; Salipante et al., 2014). A mock community can be purchased from suppliers or prepared in the laboratory and consists of genomic DNA preparations from a mixture of organisms with validated SSU rRNA sequences (i.e., obtained via Sanger sequencing or complete genome sequencing). Mock community genomic DNA is then used as an additional sample, prepared with the samples of interest, from PCR and sequencing through data analysis. Mock community analysis is used to estimate sequencing error rates by comparison of the obtained SSU rRNA sequences with the known sequences (Kozich et al., 2013; Salipante et al., 2014). A further goal of the mock community is to determine whether the number of observed taxa corresponds to the known number of taxa in the mock community. The presence of chimeric and/or spurious sequences will inflate the number of observed taxa, so an exact correspondence is not to be expected (Edgar, 2013; Kozich et al., 2013). Ideally, mock communities should be routinely included in sequencing runs, especially with the marked decrease in per sample sequencing costs.

Sample preparation for amplicon-based NGS (e.g., SSU rRNA genes or metabolic genes) starts with nucleic acid extraction (DNA or RNA converted to cDNA). For SSU rRNA gene analysis, amplification from extracted DNA, library preparation, and sequencing are frequently performed by a contract sequencing facility due to the prohibitive cost of the sequencing instruments. Sequencing facilities offer a variety of SSU rRNA gene primer set options targeting the hypervariable regions of the gene. The most popular option currently available involves sequencing of the V4 region on an Illumina MiSeq (Caporaso et al., 2012; Kozich et al., 2013). Alternatively, researchers may perform PCR amplification and library preparation steps in their own laboratories using commercially available kits or following protocols described in the literature (Bartram et al., 2011; Caporaso et al., 2012; Kozich et al., 2013). PCR amplification requires specialized primers with barcodes and NGS platform-specific adapters and other features (i.e., pad and linker sequences). PCR products must then be purified, pooled, and normalized before submission for sequencing. There are fewer fee-for-service facilities that offer NGS of LSU rRNA, fungal ITS, or metabolic genes. Hence, NGS projects with these marker genes as targets would require in-laboratory preparation (Collavino et al., 2014; Gaby and Buckley, 2012; Hugerth et al., 2014; Luke and Frenzel, 2011; Mahé et al., 2015).

15.3.7 Bioinformatics: Making Sense of It All

In the preceding steps of the workflow (Figure 15.2), large quantities of data are generated, representing the composition of the microbial populations in environmental samples. The next step is to assess the diversity and identity of these populations using bioinformatics and statistics. Bioinformatics approaches for Sanger- and NGS-derived data are outlined in the following, from QC of data through to taxonomic identification. While one of the goals for the analysis of both Sanger- and NGS-derived data is the taxonomic identification of the organisms in a sample, the steps needed to arrive at this point are quite different. In addition, the scale of NGS-derived data can allow robust fingerprinting and assessment of community diversity.

All current methods of microbial sequence analysis rely on the comparison of obtained data with curated databases of published sequences. Early in the development of sequencing technology, biomedical researchers recognized the need to organize sequencing data into publicly available databases to promote advances in genomic research, computer analysis, and communication of molecular biology information (Smith, 2013). In the United States, the GenBank database was established by the National Center for Biotechnology Information (NCBI) and contains annotated collections of all published (and many unpublished) nucleotide and protein sequences (Karsch-Mizrachi and Ouellette, 2001). GenBank is part of the International Nucleotide Sequence Database Collaboration, which also includes the DNA Data Bank of Japan (DDBJ) and the European Molecular Biology Laboratory (EMBL, UK) (Karsch-Mizrachi and Ouellette, 2001). Each of these repositories has separate methods of data submission and curation, but each center exchanges data daily, making the same identical database publicly available internationally (Karsch-Mizrachi and Ouellette, 2001). From these general databases, SSU rRNA-gene specific databases including the Ribosomal Database Project (RDP; Cole et al., 2003, 2007, 2009), Greengenes (DeSantis et al., 2006b), and SILVA (Pruesse et al., 2007; Quast et al., 2013) have evolved. These specialized databases are highly curated for sequence quality and accurate taxonomic placement of microorganisms and are regularly updated with data from GenBank, EMBL, and DDBJ. The availability of these databases and the vast quantities of SSU rRNA genes organized within them enable the rapid comparison and taxonomic assignment of new Sanger- and NGS-derived sequences. Indeed, one of the final steps in any microbial community analysis study is the deposition of sequences into a public database, and most journals require sequence data to be released to the scientific community before publication.

Approaches for Sanger-derived data. In the case of Sanger sequencing, data are returned from the automated DNA sequencer or the sequencing facility as an electropherogram consisting of a series of peaks corresponding to each base (A, T, C, or G) that eluted from the capillary over time. Taken together, these peaks represent the sequence of the targeted gene; however, these data must be quality checked and processed before further analysis. Along with the electropherogram, the sequencer also reports a "quality score" for each base, representing the confidence in the base assignment based primarily on the height of the peak above the baseline as well as the symmetry of the peak. Generally, peak heights and quality scores are highest at the beginning of a sequence but decrease after 700–800 bp until they become indistinguishable from background noise. The presence of DNA from more than one organism (i.e., from a contaminating organism in a culture thought to be of a single isolate) will result in a jumbled electropherogram and poor quality scores – such data should be discarded.

A wide variety of programs are available to view electropherograms, and most include methods for base-calling, basic sequence trimming, quality control, and sequence assembly (discussed later). Many of these programs are open-source or available free of charge, such as Sequence Scanner (Applied Biosystems, Foster City, CA) and FinchTV (www.geospiza.com/Products/finchtv.shtml). Others require a license fee, such as Geneious (Kearse et al., 2012), Sequencher® (Gene Codes Corporation, Ann Arbor, MI, USA, www.genecodes.com), and DNA Baser (www.dnabaser.com/). The programs perform the same functions but differ in their sequence viewer format, ability to perform batch commands, operating system requirements (some are universal, while others are Mac, UNIX, or PC only), and inclusion of sequence aligners and tree building algorithms.

The initial preparation of sequence data consists of two steps: quality trimming and removal of vector sequence. Quality trimming consists of removing bases that do not meet a threshold value set in the analysis software, and generally removes the low-quality data found at the very beginning and end of the electropherogram. Following quality trimming, the sequences of the cloning vector and/or cloning primers are removed. Many of the programs mentioned can perform these tasks automatically on batches of sequences. However, it is often advisable that the researcher directly examines the electropherogram following automated processing, as this allows them to define ambiguous bases and/or eliminate questionable data.

Following the initial processing of the sequence data, assembly of sequences may be required. As Sanger sequencing is only capable of obtaining ~800 bp per read, it becomes necessary to bidirectionally sequence long genes using both the forward and reverse primers. The reads are then assembled into contigs (overlapping reads of contiguous sequences) to reconstruct the full-length sequence using software programs such as those described earlier. Consensus sequences resulting from contig assembly can then be converted/exported as FASTA files (universal sequence files that can be used for downstream analysis in nearly all software packages), leaving the researcher with a single file containing the full-length sequence of the gene of interest. A final check of quality is to screen sequences for chimeras or amplicons derived from different template sequences. For full-length SSU rRNA gene sequences, commonly used programs include UCHIME (Edgar et al., 2011), Bellerophon (Huber et al., 2004), and DECIPHER (Wright et al., 2012).

A wide range of options are available to analyze a collection of quality checked, trimmed, and assembled sequences. These options range from the basic identification of a single isolate to detailed community analysis and correlation with environmental variables. The most basic option for the identification of an organism is BLAST, the Basic Local Alignment Search Tool (Altschul et al., 1990; Johnson et al., 2008) maintained by NCBI. Using a web-based interface, sequences are queried against the NCBI database, containing 181 336 445 sequences as of February 2015. The BLAST output is a report detailing which sequences in the database were most similar to the query as well as the statistical significance of these matches (% identity and e-values). BLAST is extremely fast and useful for querying small numbers of sequences, such as a small set of isolates from a study, but rapidly becomes cumbersome when large numbers of sequences are to be analyzed. As such, it is of limited utility when attempting to analyze complex microbial communities. In addition to using BLAST to search the GenBank database, newly obtained sequences can be compared with other databases, including RDP (Cole et al.,

2003, 2007, 2009), Greengenes (DeSantis et al., 2006b), and EzTaxon (Chun et al., 2007). The EzTaxon website uses the BLAST algorithm to search a database of type strains of prokaryotic species with validly published names, which is especially useful when classifying isolates.

Beyond basic sequence identification using BLAST, all phylogenetic analyses first require the alignment of sequences either to each other or to a reference database prior to any further analysis. Many algorithms exist for sequence alignment, and some are available in the standalone programs mentioned; a detailed discussion of all alignment algorithms is beyond the scope of this chapter, and therefore, we refer the reader to the references. Alignment assigns each base to a column without altering the sequence of bases, which in turn allows the comparison of the sequences and the scoring of their similarities on a column-by-column basis. In general, gap characters are introduced to bring sequences that contain insertions or deletions relative to one another into alignment at the same overall length. Following sequence alignment, it is then possible to make "apples to apples" comparisons of their similarity to each other. Alignments of sample sequences (nucleotides or amino acids) can be done using MAFFT (Katoh and Standley, 2013), CLUSTALX (Larkin et al., 2007), or MUSCLE (Edgar, 2004) as standalone programs. CLUSTALX and MUSCLE can be implemented within the phylogenetic analysis software MEGA (Tamura et al., 2013) or Geneious (Kearse, 2012). Ribosomal RNA-specific tools for the alignment of sequences against reference databases are also available. These include (1) the SINA aligner (Pruesse et al., 2012), which was designed and built to align sequences to the SILVA rRNA gene databases (Pruesse et al., 2007; Quast et al., 2013), (2) the NAST aligner (DeSantis et al., 2006a), which uses the Greengenes database (DeSantis et al., 2006b), and (3) INFERNAL, which accounts for RNA secondary structure (Nawrocki et al., 2009). SINA and NAST can also be used with customized databases (i.e., SSU rRNA genes from a particular group, such as methanogens).

More advanced alignment tasks involving tens to hundreds of sample sequences derived from clone libraries are frequently processed within the ARB software environment (Ludwig et al., 2004). Under continuous development since the mid-1990s, ARB provides a database and analysis environment tailored to SSU rRNA gene analysis. While any appropriately formatted database can be used within ARB, it has been designed in parallel with, and is best used with, the SILVA reference database. The alignment used in SILVA is unique in that it utilizes both the raw sequence as well as the inferred secondary structure of the 16S ribosome (Quast et al., 2013). Sequences can be aligned to this database using the SINA aligner as a standalone application, through an online interface, or by using the Fast Aligner tool contained in ARB (Pruesse et al., 2012). A detailed description of the SILVA database can be found in Quast et al. (2013), and a detailed description of the ARB software environment can be found in Ludwig et al. (2004). Along with its alignment and database tools, ARB provides an interface for various cutting-edge phylogenetic tree reconstruction tools, including RAxML (Stamatakis, 2014). Similar analyses can also be carried out using the tools available on the Ribosomal Database Project (RDP) website (Cole et al., 2003, 2009).

Phylogenetic tree reconstruction is one of the most useful tasks that can be carried out with a set of aligned sequences. In short, a phylogenetic tree is a visual representation of the mathematically calculated relatedness between sequences based on their alignment. The details behind the various algorithms used to generate these trees are beyond the scope of this article; however, much detail can be found in *Phylogenetic Trees Made Easy* (Hall, 2011), as well as *Principles of Microbial Diversity* (Brown, 2014). Once constructed, these trees provide useful insight into the evolutionary linkages of microorganisms obtained from widely varying environments. In addition, phylogenetic trees can be used to show the taxonomic and phylogenetic placement of new species of isolates; for details on the necessary data needed to construct trees for new species, we refer the reader to Tindall et al. (2010).

Approaches for NGS-derived data. Recently developed NGS methods allow the generation of thousands to millions of individual sequence reads per sample, and for the first time offer the potential for the complete sampling of a microbial community in any given environment. While the analytical methods associated with Sanger sequencing allowed the researcher to directly review and interact with the sequence data, the sheer scale of data generated by NGS methods renders this sort of detailed attention impossible. To respond to these challenges, several software packages have been developed with analytical pipelines capable of automated quality checking, aligning, and comparing the thousands to millions of sequences that are generated in a single study. Among these packages, mothur (Schloss et al., 2009), Quantitative Insights into Microbial Ecology (QIIME; Caporaso et al., 2010), and Ribosomal Database Project Tools (RDPTools; Cole et al., 2014) are widely used by geomicrobiologists. All three software packages are open-source and freely available for download and local installation on Windows, Linux, or UNIX operating systems. The RDP pipeline is also available via web interface for small data sets, and both mothur and QIIME can be run in high-performance computing environments to meet the computational demands of very large data sets. USEARCH is another analytical pipeline used by some geomicrobiologists (Edgar, 2013); however, current versions optimized for modern hardware require a license. Recently, VSEARCH, an open-source and free alternative to USEARCH, has been released (Rognes et al., 2015), allowing the algorithms behind USEARCH to take advantage of high-performance computing hardware. The mothur 454 (Schloss et al., 2011) or MiSeq (Kozich et al., 2013) and QIIME (Caporaso et al., 2010) analysis pipelines are particularly useful in that they provide a straightforward pathway from initial quality checking to clustering of sequences into operational taxonomic units (OTUs) and taxonomic classification. Additionally, both mothur and QIIME include a wide variety of statistical analysis tools for comparing microbial communities and visualizing those results. The respective websites for mothur and QIIME also include tutorials with example data sets, which can be downloaded and analyzed for training purposes.

The path from raw NGS data to useful inferences about microbial community composition and structure usually starts with examining the quality of the raw sequences. NGS libraries are often sequenced from both directions due to short read lengths (typically <250 bp), yielding forward and reverse reads or "paired-end" libraries. From most NGS platforms, raw sequence data are distributed in a format called FASTQ with the forward and reverse reads in separate files. FASTQ files contain quality or Phred scores for each base, which measure the probability of an incorrect base call (Cock et al., 2010). These scores typically range from 0 to 40, with a score of 30 indicating that a base call is 99.9% accurate. Nucleotides at the beginning of reads have the highest quality scores, while quality scores towards the end of reads tend to decline to values below 20, or less than 99% accuracy. Quality scores across reads can be visualized using standalone programs such as FastQC (Andrews, 2010; Schmieder and Edwards, 2011), FASTX-Toolkit (Hannon, 2010), and PRINSEQ (Schmieder and Edwards, 2011). FASTX-Toolkit and PRINSEQ also offer tools for filtering of sequences based on quality score information. This is a critical step in the analysis of NGS libraries, as the inclusion of low-quality sequences in final data sets inflates diversity estimates (i.e., number of OTUs) (Bokulich et al., 2013; Kunin et al., 2010). Most analysis pipelines incorporate quality score filtering as the first step of sequence processing and follow this general outline:

1. **Assembly of reads and removal of barcodes and primers.** For bidirectional NGS, the forward and reverse reads are first assembled or joined into a single read. Most assembly algorithms use quality score information to make base calls if there is a conflict between a base in the forward and reverse reads and will designate low-quality bases as ambiguous nucleotides (Ns). Outside the analysis pipelines (e.g., mothur, QIIME, USEARCH, etc.), there are several programs for merging paired-end reads. These include PANDAseq (Masella et al., 2012) and PEAR (Zhang et al., 2014), which were designed specifically for Illumina-generated sequences. Once reads are assembled, barcodes can be used to sort multiplexed libraries into individual samples and the forward and reverse primer sequences trimmed from the sequence ends. Barcode and/or primer trimming may be performed by the sequencing facility with data from demultiplexed samples provided as individual forward and reverse FASTQ files.
2. **Quality checking.** Despite the incorporation of quality information into the assembly step, there will be a subset of sequences that fail to merge into the correct size (i.e., little to no overlap between forward and reverse reads) or contain ambiguous bases and/or long stretches of the same nucleotide (e.g., homopolymers). These ambiguous and low-quality sequences should be removed from data sets before proceeding with clustering sequences into OTUs and taxonomic identification.
3. **Alignment to a database.** Sequences are first aligned to a reference database (e.g., SILVA, Greengenes, or RDP for SSU rRNA) before clustering into OTUs in some analysis pipelines. A number of different alignment algorithms are available as standalone programs and within analysis pipelines (e.g., BLAST, NAST, INFERNAL, etc.). This step also provides another check on data set quality, as the reference alignment can be trimmed to include only regions of interest (i.e., the V4 region of the 16S rRNA gene), and sequences that do not align well can be discarded.
4. **Chimera removal.** It is important to detect and remove chimeric sequences derived from the combination and amplification of DNA from different organisms into a single amplicon. If not removed, these sequences may mistakenly be interpreted as representative of novel diversity (Hugenholtz and Huber, 2003). Algorithms to detect chimeras include those that search sequences against a reference database (e.g., ChimeraSlayer or UCHIME) or perform reference-independent (*de novo*) detection, which identifies

chimeras based on comparison of sequences within a data set starting with the most abundant reads (e.g., UCHIME) (Edgar et al., 2011; Haas et al., 2011).

5. **OTU assignment.** A microbial species can be operationally defined as organisms sharing >97% SSU rRNA gene identity (Stackebrandt and Goebel, 1994). While binning of sequences at a 0.03 cutoff (>97% similarity) has become common practice, it should be noted that OTUs clustered at a distance of 0.03 may represent two or more different microbial species. Other cutoff distances can be used to cluster OTUs based on the needs and aims of the researcher. However, there are no "hard" rules for correspondence between taxonomic ranks and cutoff distances, and any observed correspondence may change between phylogenetic groups due to differential mutation rates of rRNA genes (Koeppel and Wu, 2013; Schloss and Westcott, 2011). Analysis pipelines offer different clustering methods, broadly categorized as *de novo* or reference based, for generating OTUs. For example, mothur includes an option for *de novo* OTU clustering using a distance matrix of the aligned sequences (Schloss et al., 2009). However, the processing of a distance matrix for very larger data sets (i.e., millions of sequences) may not be computationally practical using this approach. In response, mothur now includes an OTU clustering method that first groups sequences by taxonomy and then generates OTUs from the smaller distance matrices of the binned sequences (Schloss and Westcott, 2011). Other *de novo* OTU clustering approaches also use distances between aligned sequences to generate OTUs but do so without the initial alignment of sequences to a reference database (e.g., USEARCH; UCLUST, available in QIIME; Edgar, 2013). QIIME offers a reference-based clustering option called closed-reference OTU picking, which aligns sequences to reference sequences (e.g., Greengenes) and uses the distance between sequences of interest and references to generate OTUs (Caporaso et al., 2010). This approach is not recommended for environmental samples with microbial communities composed of organisms not highly represented in databases and potentially novel organisms. QIIME also includes a hybrid strategy (open-reference picking) that performs closed-reference and *de novo* clustering in tandem (Caporaso et al., 2010). These various OTU clustering methods may yield different results on the same data set. It is important to understand the caveats of each method and report which approach was used for OTU generation. Finally, representative sequences for each OTU can be retrieved from data sets for follow-up analyses, such as taxonomic assignment and phylogenetic reconstruction.
6. **Taxonomic assignment.** In this step, OTU sequences are compared with a reference database to assign taxonomy (e.g., SILVA, Greengenes, and RDP). OTUs can then be grouped by taxonomic affiliations to conduct "phylotype"-based analyses of sequence data sets. Similarly to other steps, there are multiple options available for the taxonomic classification of OTUs. The most popular option is the RDP naïve Bayesian classifier (Wang et al., 2007b), which is available online (http://rdp.cme.msu.edu/) or embedded in analysis pipelines (e.g., mothur, QIIME, and USEARCH). This tool assigns taxonomy based on the probability that an OTU sequence matches a database sequence and provides a confidence estimate for the match (Wang et al., 2007b). The Bayesian classifier can be used with any of the available databases, and this choice may influence OTU taxonomic classifications.

For example, RDP includes fewer candidate phyla than SILVA and Greengenes, so OTUs may be designated as unclassified at the phylum level using RDP but as one of the numerous candidate phyla when using the other databases. Additionally, RDP includes taxonomic ranks to genus, whereas Greengenes and SILVA include species names. Alternatives to the Bayesian classifier include the k-Nearest Neighbor search algorithm (Cole et al., 2005), BLAST, and Global Alignment for Sequence Taxonomy (Huse et al., 2008).

7. **Statistical analysis.** The processing of SSU rRNA gene or other amplicon-based NGS data sets yields several products that can be analyzed: (1) an OTU table consisting of the number of sequences each OTU represents for every sample, (2) taxonomic classifications of the OTUs, and (3) representative OTU sequences. These data can be used to calculate diversity indices, to determine phylogenetic similarity, and in the statistical integration of microbial community composition and geochemistry. Tools for the calculation of diversity indices and other statistical analyses are included in mothur and QIIME but are also available in other software, such as the R environment for statistical computing (R Core Team, 2015). In addition to basic statistical functions, users may install R packages that import and manipulate NGS data sets and perform specific statistical tests, including multivariate analyses. Phyloseq is an increasingly popular R package used to import OTU and taxonomy files generated from mothur, QIIME, and other analysis pipelines into R and offers tools for data manipulation and statistical analyses as well (McMurdie and Holmes, 2013). Beyond R, programs such as EstimateS (Colwell, 2013), PAST (Hammer et al., 2001), and vegan (Oksanen et al., 2015) are freely available to estimate diversity metrics. We discuss statistical analysis of NGS data sets in more detail later.

The methods described thus far focus on OTU-based approaches, but there is an alternative technique for analysis of NGS data sets that does not rely on clustering sequences into OTUs. This approach is called phylogenetic or evolutionary placement of sequences into a reference tree (Berger et al., 2011; Matsen et al. 2010). Prior to placement, NGS reads are quality checked, trimmed, and chimeras removed as described earlier. Phylogenetic placement can be performed using software packages such as pplacer (Matsen et al., 2010), RAxML (Berger et al., 2011; Stamatakis, 2006), and Phylosift (Darling et al., 2014). Briefly, a fixed phylogenetic tree of aligned reference sequences is reconstructed using likelihood methods for tree building available in programs such as RAxML (Stamatakis, 2006) and FastTree (Price et al., 2010). NGS reads are then placed in the fixed phylogenetic tree via alignment to the reference alignment. Sequences are taxonomically assigned based on their placement and abundance information displayed as thickened branches. In particular, pplacer as a standalone program (or embedded in Phylosift) offers downstream statistical tools for comparison of microbial communities analyzed via phylogenetic placement (Darling et al., 2014; Matsen et al., 2010). Phylogenetic placement is particularly useful for studies that require confident phylogenetic/taxonomic assignments, such as the analysis of cyanobacterial 16S rRNA genes, which are insufficiently curated in the existing rRNA databases (Sudek et al., 2015), or metabolic genes (Dupont et al., 2015).

15.3.8 Statistical Analysis

Statistical estimators of biodiversity, adopted from macroecological studies, are utilized by geomicrobiologists to determine the efficiency of cultivation-independent sampling techniques and the differences between microbial populations (Bent and Forney, 2008; Bohannan, 2003; Hughes and Bohannan, 2004; Hughes et al., 2001). Almost all diversity estimates are sensitive to sampling effort or, in the case of NGS data sets, the number of sequences obtained per sample (Bowman et al., 1971; Soetaert and Heip, 1990). This number often differs greatly between samples, with library sizes ranging from thousands to greater than ten thousand sequences. To minimize this problem, it is recommended to normalize (or rarefy) the number of sequences such that an equal number are used across samples to calculate diversity estimates (Gihring et al., 2012; Schloss et al., 2011). A good exercise to understand the influence of sampling effort on diversity estimates is to compare these metrics for original and normalized NGS libraries. For example, the number of observed OTUs per sample is often used as a diversity metric, and this value tends to increase with the number of sequences obtained. Another important caveat to consider in the estimation of microbial diversity metrics from 16S rRNA data sets is the influence of rRNA gene copy number. The rRNA gene copy number can range from 1 to 15 in Bacteria and Archaea, and microbes with higher rRNA gene copy number may be overrepresented (Crosby and Criddle, 2003; Stoddard et al., 2015). While this is not part of standard analysis pipelines, researchers are recognizing the need to account for rRNA gene copy number variation in diversity metrics in studies of Bacteria and Archaea (Kembel et al., 2012) and Eukaryotes, where SSU rRNA copy number variation spans a much wider range (Darby et al., 2013). We describe some standard measures of biodiversity as they relate to NGS data sets in the remainder of this section.

Although NGS platforms have allowed researchers to conduct deep sequencing of samples, it is rare to completely sequence all members in a microbial community. Percent coverage (Begon et al., 1990) and rarefaction analysis (Heck et al., 1975; Holland, 2003) are two measures used to estimate alpha diversity (species richness) in samples and gauge whether sequencing efforts were sufficient to capture the community diversity. Percent coverage is usually calculated as Good's coverage ($C = 1 - s/n$), where s is the number of OTUs observed once (singleton OTUs) and n is the total number of individuals in a sample. Good's coverage values range from 0 to 1, with low values suggesting incomplete sampling and high values indicating that most OTUs were observed more than once in a data set. In rarefaction analysis, OTUs are retrieved from random subsamples of sequences. The number of OTUs is plotted against the number of sequences in a stepwise fashion to obtain a visual representation of sampling effort (rarefaction curve). In a completely sampled community, a rarefaction curve will plateau as sampling effort increases, indicating that no additional OTUs are found with greater sequencing depth. This rarely occurs in diverse microbial communities, while most members of simple microbial communities may be sampled using NGS. Percent coverage calculations and rarefaction analyses can be done in mothur or QIIME or by using other statistical software.

Numerous indices exist for describing alpha diversity within a community. Nonparametric richness estimators, such as Chao1 and abundance-coverage estimator (ACE) (Chao, 1984; Chao and Lee, 1992), predict species richness in unequally sampled sites, treatments, or habitats. Chao1 and ACE are based on the probability of observing rare and abundant species, and researchers have noted that estimates are dependent on sampling effort (Gihring et al., 2012; Hughes et al., 2001; Schloss and Handelsman, 2004). Microbial community diversity can be described using the Shannon–Wiener (*H'*), Inverse Simpson's (1/D), and Fisher's alpha indices. While most indices consider both species richness and abundances, the Simpson's Evenness Index ($E_{1/D}$) is a more explicit metric for the relative distribution of species within a community. These diversity measures are also known to vary with sample size (Bowman et al., 1971; Soetaert and Heip, 1990), highlighting the need for normalization (Gihring et al., 2012). Because it is beyond this chapter to exhaustively review diversity metrics, we refer the reader to *Measuring Biological Diversity* (Magurran, 2004).

Sequence data can also be used to compare phylogenetic relatedness between samples as a measure of beta diversity. Generally, this approach calculates phylogenetic distance between samples based on their sequence composition and uses this distance to cluster samples and determine whether structural (dis)similarities exist between communities. Several options exist for conducting this type of analysis, with Unifrac as one of the most widely used methods (Hamady et al., 2010; Lozupone and Knight, 2005; Lozupone et al., 2011). Unifrac can be implemented in mothur, QIIME, Phyloseq, and the R package Picante (Kembel et al., 2010). Picante also offers other phylogeny-based diversity estimates, including the mean phylogenetic distance and mean nearest taxon distance for both within- and between-community comparisons (phylocom; Webb et al., 2008). Beta diversity can also be assessed by direct comparison of microbial communities (i.e., presence/absence of particular taxa, differences in alpha diversity measures, etc.), but incorporation of phylogenetic information may provide greater insight into the evolutionary and ecological factors that shape microbial communities.

Other types of analyses are possible with NGS data sets. Recently, it has been pointed out that microbial community composition data sets (OTU abundances and taxonomic affiliations) are analogous to gene expression data sets (transcript levels of genes). Both data types are essentially count tables of varying sample sizes or depth of coverage, which makes statistical techniques developed for the analysis of gene expression data suitable for microbial community composition data (McMurdie and Holmes, 2014; Paulson et al., 2013). These methods have been developed to detect differentially transcribed genes without the need to normalize library sizes via rarefaction (McMurdie and Holmes, 2014). There are several commonly used R packages available to perform differential abundance analysis on microbial community data sets: edgeR (Robinson et al., 2009), DESeq2 (Love et al., 2014), and metagenomeSeq (Paulson et al., 2013). The R package phyloseq includes functions for converting OTU abundances and taxonomic assignment into objects that can be analyzed by these packages (McMurdie and Holmes, 2014). This type of analysis is most suited to the detection of responsive OTUs in studies involving experimental manipulation of highly similar microbial communities (Corman et al., 2016). The statistical tests implemented use data from all libraries to calculate normalized counts of OTUs, so performing this type of analysis on very different microbial communities is not recommended (McMurdie and Holmes, 2014).

One of the great challenges in geomicrobiology arises from the drive to understand the interactions between the microbial communities and the environment in which they live. The integration of these very different types of data can be obtained by utilizing multivariate analyses first developed for macroecology. These statistical tools allow a variety of comparisons to be made between the microbial communities present at different sites (i.e., clustering-based approaches such as non-metric dimensional scaling [NMDS]) and in some cases utilize environmental parameters to attempt to explain the variability among the communities (i.e., canonical correspondence analysis and redundancy analysis). An excellent description of the statistical methods and their applications can be found in the review "Multivariate analyses in microbial ecology" (Ramette, 2007), the book *Numerical Ecology with R* (Borcard et al., 2011), and a recently constructed online resource, GUSTA ME (Buttigieg and Ramette, 2014; Ramette, 2007). Many of the tools needed to compare communities with each other are contained within mothur and QIIME (Caporaso et al., 2010; Kozich et al., 2013; Schloss et al., 2009), while those needed to integrate environmental data with community data can be found within the vegan package for R (Okansen et al., 2014; R Core Team, 2014) or PAST (Hammer et al., 2001). There is also licensed software, such as Canoco and PRIMER-E, designed specifically to conduct multivariate statistical analyses.

15.4 Recent Advances and Conclusions

Technologies for microbial community analyses are rapidly expanding, the application of whole-genome reconstruction has led to the discovery of new microbial lineages (Baker et al., 2006; Evans et al., 2015), and it is postulated that continued study could lead to the discovery of a fourth Domain of life (Woyke and Rubin, 2014). This new era of discovery is being propelled by the development of "omics" technologies, whereby studies are interrogating the metagenomes, metatranscriptomes, and metaproteomes of environmental samples. Omics approaches are aimed at characterizing the entire genetic information contained within an environmental community, independently of any gene primer sets, which are known to overlook various groups (Eisen, 2007; Venter et al., 2004). Metagenomes provide information on the presence of microbial functional groups in the environment, independently of actual activity or viability of specific organisms (Moran, 2009). The characterization of mRNA transcripts and proteins directly from environmental samples, called metatranscriptomics and metaproteomics, respectively, is able to overcome this limitation by directly analyzing the activity of a community at a specific place and time, eliminating the need to choose functional genes and markers directly. Metatranscriptomics targets the short-lived mRNA produced during the transcription of active genes and can provide a snapshot of the microbial activity (Madsen, 2005; Moran et al., 2013). Metaproteomics, on the other hand, focuses on analyzing the expressed proteins directly, which reduces the complexity further by eliminating transcribed mRNA intermediates

(Hanson et al., 2014). These methods have provided significant insights into microbial ecology and biogeochemistry, including contaminated soils and wastewater (Benndorf et al., 2007), acid mine drainage (Ram et al., 2005; Wrighton et al., 2012), and marine systems (Hewson et al., 2010; Morris et al., 2010; Poretsky et al., 2009).

The methods and technologies described earlier have revolutionized the fields of environmental microbiology and geomicrobiology by providing researchers with robust techniques for assessing the diversity and phylogeny of microbial populations. SSU rRNA gene-based approaches have successfully been used to identify microbial populations living in extreme habitats, e.g., mofettes (Beulig et al., 2014), metal- and radionuclide-contaminated environments (e.g., Akob et al., 2011, 2014, 2015; Burkhardt et al., 2010; Fabisch et al., 2013), and karstic environments, e.g., caves (Rusznyak et al., 2012) and aquifers (Herrmann et al., 2015). Polyphasic studies combining gene-based community and geochemical data with robust statistical analyses have allowed researchers to more closely resolve the impact of microbial populations on their environment and the impact of the environment on population dynamics. The application of phylogenetic methods has also exponentially expanded our ability to characterize and catalog the microbial world as we know it.

15.5 References

Akob, D. M., T. Bohu, A. Beyer, et al. (2014). Identification of Mn(II)-oxidizing bacteria from a low-pH contaminated former uranium mine. *Applied and Environmental Microbiology* **80**(16): 5086–5097, Doi:10.1128/AEM.01296-14.

Akob, D. M., I. M. Cozzarelli, D. S. Dunlap, E. L. Rowan and M. M. Lorah (2015). Organic and inorganic composition and microbiology of produced waters from Pennsylvania shale gas wells. *Applied Geochemistry* **60**: 116–125, Doi:http://dx.doi.org/10.1016/j.apgeochem.2015.04.011.

Akob, D. M., L. Kerkhof, K. Küsel, et al. (2011). Linking specific heterotrophic bacterial populations to bioreduction of uranium and nitrate in contaminated subsurface sediments by using stable isotope probing. *Applied and Environmental Microbiology* **77**(22): 8197–8200, Doi:10.1128/AEM.05247-11.

Akob, D. M. and K. Küsel (2011). Where microorganisms meet rocks in the Earth's Critical Zone. *Biogeosciences* **8**(12): 3531–3543, Doi:10.5194/Bg-8-3531-2011.

Akob, D. M., S. H. Lee, M. Sheth, et al. (2012). Gene expression correlates with process rates quantified for sulfate- and Fe(III)-reducing bacteria in U(VI)-contaminated sediments. *Frontiers in Microbiology* **3**: 280, Doi:10.3389/fmicb.2012.00280.

Akob, D. M., H. J. Mills, T. M. Gihring, et al. (2008). Functional diversity and electron donor dependence of microbial populations capable of U(VI) reduction in radionuclide-contaminated subsurface sediments. *Applied and Environmental Microbiology* **74**(10): 3159–3170, Doi:10.1128/AEM.02881-07.

Akob, D. M., H. J. Mills and J. E. Kostka (2007). Metabolically active microbial communities in uranium-contaminated subsurface sediments. *FEMS Microbiology Ecology* **59**(1): 95–107, Doi:10.1111/fem.2007.59.issue-1.

Albertsen, M., P. Hugenholtz, A. Skarshewski, et al. (2013). Genome sequences of rare, uncultured bacteria obtained by differential coverage binning of multiple metagenomes. *Nature Biotechnology* **31**(6): 533–538, Doi:10.1038/nbt.2579.

Altschul, S. F., W. Gish, W. Miller, E. W. Myers and D. J. Lipman (1990). Basic local alignment search tool. *Journal of Molecular Biology* **215**: 403–410.

Amann, R., W. Ludwig and K.-H. Schleifer (1995). Phylogenetic identification and in situ detection of individual microbial cells without cultivation. *Microbiological Reviews* **59**(1): 143–169.

Andrews, S. (2010). "FastQC: A quality control tool for high throughput sequence data," from www.bioinformatics.babraham.ac.uk/projects/fastqc/.

Baker, B. J., G. W. Tyson, R. I. Webb, et al. (2006). Lineages of acidophilic archaea revealed by community genomic analysis. *Science* **314**(5807): 1933–1935, Doi:10.1126/science.1132690.

Bartram, A. K., M. D. J. Lynch, J. C. Stearns, G. Moreno-Hagelsieb and J. D. Neufeld (2011). Generation of multimillion-sequence 16S rRNA gene libraries from complex microbial communities by assembling paired-end Illumina reads. *Applied and Environmental Microbiology* **77**(11): 3846–3852, Doi:10.1128/aem.02772-10.

Begon, M., J. L. Harper and C. R. Townsend (1990). *Ecology: Individuals, Populations and Communities*. Oxford, Blackwell Scientific Publications.

Benndorf, D., G. U. Balcke, H. Harms and M. von Bergen (2007). Functional metaproteome analysis of protein extracts from contaminated soil and groundwater. *The ISME Journal* **1**(3): 224–234, Doi:10.1038/ismej.2007.39.

Bent, S. J. and L. J. Forney (2008). The tragedy of the uncommon: understanding limitations in the analysis of microbial diversity. *The ISME Journal* **2**(7): 689–695, Doi:10.1038/ismej.2008.44.

Berger, S. A., D. Krompass and A. Stamatakis (2011). Performance, accuracy, and Web server for evolutionary placement of short sequence reads under maximum likelihood. *Systematic Biology* **60**(3): 291–302, Doi:10.1093/sysbio/syr010.

Bernard, P., P. Gabarit, E. M. Bahassi and M. Couturier (1994). Positive-selection vectors using the F plasmid ccdB killer gene. *Gene* **148**(1): 71–74, Doi:http://dx.doi.org/10.1016/0378-1119(94)90235-6.

Beulig, F., V. B. Heuer, D. M. Akob, et al. (2014). Carbon flow from volcanic CO_2 into soil microbial communities of a wetland mofette. *The ISME Journal* **9**(3): 746–759, Doi:10.1038/ismej.2014.148.

Bohannan, B. (2003). New approaches to analyzing microbial biodiversity data. *Current Opinion in Microbiology* **6**(3): 282–287, Doi:10.1016/S1369-5274(03)00055-9.

Bokulich, N. A., S. Subramanian, J. J. Faith, et al. (2013). Quality-filtering vastly improves diversity estimates from Illumina amplicon sequencing. *Nature Methods* **10**(1): 57–59, Doi:10.1038/nmeth.2276.

Borcard, D., F. Gillet and P. Legendre (2011). *Numerical Ecology with R*. New York, NY, Springer-Verlag.

Bowman, K. O., K. Hutcheson, E. P. Odum and L. R. Shenton (1971). Comment on the distribution of indices of diversity. In *Statistical Ecology* G. P. Patil, E. C. Peilou and W. E. Waters. Philadelphia, PA, USA, Pennsylvania State University Press.

Bragg, L. M., G. Stone, M. K. Butler, P. Hugenholtz and G. W. Tyson (2013). Shining a light on dark sequencing: characterising errors in Ion Torrent PGM data. *PLoS Computational Biology* **9**(4): e1003031, Doi:10.1371/journal.pcbi.1003031.

Breitbart, M., P. Salamon, B. Andresen, et al. (2002). Genomic analysis of uncultured marine viral communities. *Proceedings of the National Academy of Sciences* **99**(22): 14250–14255, Doi:10.1073/pnas.202488399.

Brock, T. D. and H. Freeze (1969). Thermus aquaticus gen. n. and sp. n., a nonsporulating extreme thermophile. *Journal of Bacteriology* **98**(1): 289–297.

Brown, J. W. (2014). *Principles of Microbial Diversity*. Washington, DC, ASM Press.

Brown, M. V., M. S. Schwalbach, I. Hewson and J. A. Fuhrman (2005). Coupling 16S-ITS rDNA clone libraries and automated ribosomal intergenic spacer analysis to show marine microbial diversity: development and application to a time series. *Environmental Microbiology* **7**(9): 1466–1479, Doi:10.1111/j.1462-2920.2005.00835.x.

Buermans, H. P. J. and J. T. den Dunnen (2014). Next generation sequencing technology: advances and applications. *Biochimica et Biophysica Acta (BBA) – Molecular Basis of Disease* **1842**(10): 1932–1941, Doi:http://dx.doi.org/10.1016/j.bbadis.2014.06.015.

Burkhardt, E.-M., D. M. Akob, S. Bischoff, et al. (2010). Impact of biostimulated redox processes on metal dynamics in an iron-rich creek soil of a former uranium mining area. *Environmental Science & Technology* **44**(1): 177–183, Doi:10.1021/es902038e.

Burkhardt, E.-M., S. Bischoff, D. M. Akob, G. Buechel and K. Küsel (2011). Heavy metal tolerance of Fe(III)-reducing microbial communities in contaminated creek bank soils. *Applied and Environmental Microbiology* **77**(9): 3132–3136, Doi:10.1128/AEM.02085-10.

Bustin, S. A., V. Benes, J. A. Garson, et al. (2009). The MIQE guidelines: minimum information for publication of quantitative real-time PCR experiments. *Clinical Chemistry* **55**(4): 611–622, Doi:10.1373/clinchem.2008.112797.

Buttigieg, P. L. and A. Ramette (2014). A guide to statistical analysis in microbial ecology: a community-focused, living review of multivariate data analyses. *FEMS Microbiology Ecology* **90**(3): 543–550, Doi:10.1111/1574-6941.12437.

Caporaso, J. G., J. Kuczynski, J. Stombaugh, et al. (2010). QIIME allows analysis of high-throughput community sequencing data. *Nature Methods* **7**(5): 335–336, Doi:10.1038/nmeth.f.303.

Caporaso, J. G., C. L. Lauber, W. A. Walters, et al. (2011). Global patterns of 16S rRNA diversity at a depth of millions of sequences per sample. *Proceedings of the National Academy of Sciences* **108** (Supplement 1): 4516–4522, Doi:10.1073/pnas.1000080107.

Caporaso, J. G., C. L. Lauber, W. A. Walters, et al. (2012). Ultra-high-throughput microbial community analysis on the Illumina HiSeq and MiSeq platforms. *The ISME Journal* **6**(8): 1621–1624, Doi: www.nature.com/ismej/journal/v6/n8/suppinfo/ismej20128s1.html.

Case, R. J., Y. Boucher, I. Dahllöf, et al. (2007). Use of 16S rRNA and rpoB genes as molecular markers for microbial ecology studies. *Applied and Environmental Microbiology* **73**(1): 278–288, Doi:10.1128/aem.01177-06.

Castle, D. and D. L. Kirchman (2004). Composition of estuarine bacterial communities assessed by denaturing gradient gel electrophoresis and fluorescence in situ hybridization. *Limnology and Oceanography: Methods* **2**(9): 303–314, Doi:10.4319/lom.2004.2.303.

Castro, H. F., N. H. Williams and A. Ogram (2000). Phylogeny of sulfate-reducing bacteria. *FEMS Microbiology Ecology* **31**(1): 1–9.

Chao, A. (1984). Nonparametric estimation of the number of classes in a population. *Scandinavian Journal of Statistics* **11**(4): 265–270, Doi:10.2307/4615964.

Chao, A. and S.-M. Lee (1992). Estimating the number of classes via sample coverage. *Journal of the American Statistical Association* **87**(417): 210–217, Doi:10.2307/2290471.

Chaput, D. L., C. M. Hansel, W. D. Burgos and C. M. Santelli (2015). Profiling microbial communities in manganese remediation systems treating coal mine drainage. *Applied and Environmental Microbiology* **81**(6):2189–2198, Doi:10.1128/AEM.03643-14.

Chun, J., J. H. Lee, Y. Jung, et al. (2007). EzTaxon: a web-based tool for the identification of prokaryotes based on 16S ribosomal RNA gene sequences. *International Journal of Systematic and Evolutionary Microbiology* **57**: 2259–2261, Doi:10.1099/ijs.0.64915-0.

Clement, B. G., L. E. Kehl, K. L. DeBord and C. L. Kitts (1998). Terminal restriction fragment patterns (TRFPs), a rapid, PCR-based method for the comparison of complex bacterial communities. *Journal of Microbiological Methods* **31**(3): 135–142, Doi:http://dx.doi.org/10.1016/S0167-7012(97)00105-X.

Cock, P. J. A., C. J. Fields, N. Goto, M. L. Heuer and P. M. Rice (2010). The Sanger FASTQ file format for sequences with quality scores, and the Solexa/Illumina FASTQ variants. *Nucleic Acids Research* **38**(6): 1767–1771, Doi:10.1093/nar/gkp1137.

Cole, J. R., B. Chai, R. J. Farris, et al. (2005). The Ribosomal Database Project (RDP-II): sequences and tools for high-throughput rRNA analysis. *Nucleic Acids Research* 33(Database issue): D294–296, Doi:10.1093/nar/gki038.

Cole, J. R., B. Chai, R. J. Farris, et al. (2007). The Ribosomal Database Project (RDP-II): introducing myRDP space and quality controlled public data. *Nucleic Acids Research* **35**(suppl 1): D169–D172, Doi:10.1093/nar/gkl889.

Cole, J. R., B. Chai, T. L. Marsh, et al. (2003). The Ribosomal Database Project (RDP-II): previewing a new autoaligner that allows regular updates and the new prokaryotic taxonomy. *Nucleic Acids Research* **31**(1): 442–443.

Cole, J. R., Q. Wang, E. Cardenas, et al. (2009). The Ribosomal Database Project: improved alignments and new tools for rRNA analysis. *Nucleic Acids Research* 37(Database issue): D141–145, Doi:10.1093/nar/gkn879.

Cole, J. R., Q. Wang, J. A. Fish, et al. (2014). Ribosomal Database Project: data and tools for high throughput rRNA analysis. *Nucleic Acids Research* **42**(D1): D633–D642, Doi:10.1093/nar/gkt1244.

Collavino, M. M., H. J. Tripp, I. E. Frank, et al. (2014). nifH pyrosequencing reveals the potential for location-specific soil chemistry to influence N-2-fixing community dynamics. *Environmental Microbiology* **16**(10): 3211–3223, Doi:10.1111/1462-2920.12423.

Colwell, R. K. (2013). EstimateS: statistical estimation of species richness and shared species from samples. User's Guide and application published at http://purl.oclc.org/estimates.

Corman, J. R., A. T. Poret-Peterson, A. Uchitel and J. J. Elser (2016). Interaction between lithification and resource availability in the microbialites of Río Mesquites, Cuatro Ciénegas, México. *Geobiology* **14**(2): 176–189, Doi:10.1111/gbi.12168.

Crosby, L. D. and C. S. Criddle (2003). Understanding bias in microbial community analysis techniques due to rrn operon copy number heterogeneity. *Biotechniques* **34**(4): 790–794, 796, 798 passim.

Culman, S. W., R. Bukowski, H. G. Gauch, H. Cadillo-Quiroz and D. H. Buckley (2009). T-REX: software for the processing and analysis of T-RFLP data. *BMC Bioinformatics* **10**: 171, Doi:10.1186/1471-2105-10-171.

Dahllöf, I., H. Baillie and S. Kjelleberg (2000). rpoB-based microbial community analysis avoids limitations inherent in 16S rRNA gene intraspecies heterogeneity. *Applied and Environmental Microbiology* **66**(8): 3376–3380, Doi:10.1128/aem.66.8.3376-3380.2000.

Danovaro, R., G. M. Luna, A. Dell'Anno and B. Pietrangeli (2006). Comparison of two fingerprinting techniques, terminal restriction fragment length polymorphism and automated ribosomal intergenic spacer analysis, for determination of bacterial diversity in aquatic environments. *Applied and Environmental Microbiology* **72**(9): 5982–5989, Doi:10.1128/aem.01361-06.

Darby, B. J., T. C. Todd and M. A. Herman (2013). High-throughput amplicon sequencing of rRNA genes requires a copy number correction to accurately reflect the effects of management practices on soil nematode community structure. *Molecular Ecology* **22**(21): 5456–5471, Doi:10.1111/mec.12480.

Darling, A. E., G. Jospin, E. Lowe, et al. (2014). PhyloSift: phylogenetic analysis of genomes and metagenomes. *PeerJ* **2**: e243, Doi:10.7717/peerj.243.

Degnan, P. H. and H. Ochman (2012). Illumina-based analysis of microbial community diversity. *The ISME Journal* **6**(1): 183–194, Doi:www.nature.com/ismej/journal/v6/n1/suppinfo/ismej201174s1.html.

DeSantis, T. Z., Jr., P. Hugenholtz, K. Keller, et al. (2006a). NAST: a multiple sequence alignment server for comparative analysis of 16S rRNA genes. *Nucleic Acids Research* **34**: W394–399, Doi:10.1093/nar/gkl244.

DeSantis, T. Z., P. Hugenholtz, N. Larsen, et al. (2006b). Greengenes, a chimera-checked 16S rRNA gene database and workbench compatible with ARB. *Applied and Environmental Microbiology* **72** (7): 5069–5072, Doi:10.1128/aem.03006-05.

Dupont, C. L., J. P. McCrow, R. Valas, et al. (2015). Genomes and gene expression across light and productivity gradients in eastern subtropical Pacific microbial communities. *The ISME Journal* **9** (5): 1076–1092, Doi:10.1038/ismej.2014.198.

Edgar, R. C. (2004). MUSCLE: multiple sequence alignment with high accuracy and high throughput. *Nucleic Acids Research* **32**(5): 1792–1797, Doi:10.1093/nar/gkh340.

Edgar, R. C. (2013). UPARSE: highly accurate OTU sequences from microbial amplicon reads. *Nature Methods* **10**(10): 996–998, Doi:10.1038/nmeth.2604.

Edgar, R. C., B. J. Haas, J. C. Clemente, C. Quince and R. Knight (2011). UCHIME improves sensitivity and speed of chimera detection. *Bioinformatics* **27**(16): 2194–2200, Doi:10.1093/bioinformatics/btr381.

Eisen, J. A. (2007). Environmental shotgun sequencing: its potential and challenges for studying the hidden world of microbes. *PLoS Biology* **5**(3): e82, Doi:10.1371/journal.pbio.0050082.

Evans, P. N., D. H. Parks, G. L. Chadwick, et al. (2015). Methane metabolism in the archaeal phylum Bathyarchaeota revealed by genome-centric metagenomics. *Science* **350**(6259): 434–438, Doi:10.1126/science.aac7745.

Fabisch, M., F. Beulig, D. M. Akob and K. Küsel (2013). Surprising abundance of Gallionella-related iron oxidizers in creek sediments at pH 4.4 or at high heavy metal concentrations. *Frontiers in Microbiology* **4**: 390, Doi:10.3389/fmicb.2013.00390.

Feinstein, L. M., W. J. Sul and C. B. Blackwood (2009). Assessment of bias associated with incomplete extraction of microbial DNA from soil. *Applied and Environmental Microbiology* **75** (16): 5428–5433, Doi:10.1128/AEM.00120-09.

Fennell, D. E., S. K. Rhee, Y. B. Ahn, M. M. Haggblom and L. J. Kerkhof (2004). Detection and characterization of a dehalogenating microorganism by terminal restriction fragment length polymorphism fingerprinting of 16S rRNA in a sulfidogenic, 2-bromophenol-utilizing enrichment. *Applied and Environmental Microbiology* **70**(2): 1169–1175.

Fierer, N., M. Breitbart, J. Nulton, et al. (2007). Metagenomic and small-subunit rRNA analyses reveal the genetic diversity of Bacteria, Archaea, fungi, and viruses in soil. *Applied and Environmental Microbiology* **73**(21): 7059–7066, Doi:10.1128/aem.00358-07.

Fischer, H. and M. Pusch (1999). Use of the [C14]leucine incorporation technique to measure bacterial production in river sediments and the epiphyton. *Applied and Environmental Microbiology* **65**(10): 4411–4418.

Fisher, M. M. and E. W. Triplett (1999). Automated approach for ribosomal intergenic spacer analysis of microbial diversity and its application to freshwater bacterial communities. *Applied and Environmental Microbiology* **65**(10): 4630–4636.

Fitzjohn, R. and I. Dickie (2007). TRAMPR: an R package for analysis and matching of terminal-restriction fragment length polymorphism (TRFLP) profiles. *Molecular Ecology Notes* **7**(4): 583–587, Doi:10.1111/men.2007.7.issue-4.

Frank, J. A., C. I. Reich, S. Sharma, et al. (2008). Critical evaluation of two primers commonly used for amplification of bacterial 16S rRNA genes. *Applied and Environmental Microbiology* **74**(8): 2461–2470, Doi:10.1128/aem.02272-07.

Gaby, J. C. and D. H. Buckley (2012). A comprehensive evaluation of PCR primers to amplify the nifH gene of nitrogenase. *PLoS ONE* **7**(7): e42149, Doi:10.1371/journal.pone.0042149.

Gantner, S., A. F. Andersson, L. Alonso-Saez and S. Bertilsson (2011). Novel primers for 16S rRNA-based archaeal community analyses in environmental samples. *Journal of Microbiological Methods* **84**(1): 12–18, Doi:10.1016/j.mimet.2010.10.001.

Gihring, T. M., S. J. Green and C. W. Schadt (2012). Massively parallel rRNA gene sequencing exacerbates the potential for biased community diversity comparisons due to variable library sizes. *Environmental Microbiology* **14**(2): 285–290, Doi:10.1111/j.1462-2920.2011.02550.x [doi].

Gilbert, J. A. and C. L. Dupont (2011). Microbial metagenomics: beyond the genome. *Annual Reviews of Marine Science* **3**: 347–371, Doi:10.1146/annurev-marine-120709-142811.

Gilles, A., E. Meglécz, N. Pech, et al. (2011). Accuracy and quality assessment of 454 GS-FLX Titanium pyrosequencing. *BMC Genomics* **12**(1): 1–11, Doi:10.1186/1471-2164-12-245.

Golenberg, E. M., A. Bickel and P. Weihs (1996). Effect of highly fragmented DNA on PCR. *Nucleic Acids Research* **24**(24): 5026–5033, Doi:10.1093/nar/24.24.5026.

Green, M. R. and J. Sambrook (2012). *Molecular* Cloning: A Laboratory Manual. Cold Spring Harbor, NY, Cold Spring Harbor Laboratory Press.

Green, S. J., M. B. Leigh and J. D. Neufeld (2009). Denaturing gradient gel electrophoresis (DGGE) for microbial community analysis. In *Microbiology of Hydrocarbons, Oils, Lipids, and Derived Compounds*. K. N. Timmis. Heidelberg, Germany, Springer: 4137–4158.

Greuter, D., A. Loy, M. Horne and T. Ratteil (2016). probeBase – an online resource for rRNA-targeted oligonucleotide probes and primers: new features 2016. *Nucleic Acids Research* **44**(D1): D586–D589, Doi:10.1093/nar/gkv1232.

Haas, B. J., D. Gevers, A. M. Earl, et al. (2011). Chimeric 16S rRNA sequence formation and detection in Sanger and 454-pyrosequenced PCR amplicons. *Genome Research* **21**(3): 494–504, Doi:10.1101/gr.112730.110.

Hall, B. G. (2008). *Phylogenetic* Trees Made Easy: A How-To Manual. Sunderland, MA, Sinauer Associates.

Hall, B. G. (2011). Phylogenetic Trees Made Easy: A How-To Manual. Sunderland, MA, Sinauer Associates.

Hamady, M., C. Lozupone and R. Knight (2010). Fast UniFrac: facilitating high-throughput phylogenetic analyses of microbial communities including analysis of pyrosequencing and PhyloChip data. *The ISME Journal* **4**(1): 17–27, Doi:10.1038/ismej.2009.97.

Hammer, O., D. Harper and P. Ryan (2001). PAST: paleontological statistics software package for education and data analysis. *Palaeontologia Electronica* **4**(1): 1–9.

Handelsman, J. (2004). Metagenomics: application of genomics to uncultured microorganisms. *Microbiology and Molecular Biology Reviews* **68**(4): 669–685, Doi:10.1128/mmbr.68.4.669-685.2004.

Handelsman, J., M. R. Rondon, S. F. Brady, J. Clardy and R. M. Goodman (1998). Molecular biological access to the chemistry of unknown soil microbes: a new frontier for natural products. *Chemistry & Biology* **5**(10): R245–R249, Doi:http://dx.doi.org/10.1016/S1074-5521(98)90108-9.

Hannon, G. J. (2010). "FASTX-Toolkit: FASTQ/A short-reads pre-processing tools," from http://hannonlab.cshl.edu/fastx_toolkit/.

Hanson, B. T., I. Hewson and E. L. Madsen (2014). Metaproteomic survey of six aquatic habitats: discovering the identities of microbial populations active in biogeochemical cycling. *Microbial Ecology* **67**(3): 520–539, Doi:10.1007/s00248-013-0346-5.

Hargreaves, S. K., A. A. Roberto and K. S. Hofmockel (2013). Reaction- and sample-specific inhibition affect standardization of qPCR assays of soil bacterial communities. *Soil Biology and Biochemistry* **59**: 89–97, Doi:http://dx.doi.org/10.1016/j.soilbio.2013.01.007.

Hartmann, M. and F. Widmer (2008). Reliability for detecting composition and changes of microbial communities by T-RFLP genetic profiling. *FEMS Microbiology Ecology* **63**(2): 249–260, Doi:10.1111/j.1574-6941.2007.00427.x.

Heck, K. L., G. Vanbelle and D. Simberloff (1975). Explicit calculation of rarefaction diversity measurement and determination of sufficient sample size. *Ecology* **56**(6): 1459–1461.

Herrera, A. and C. S. Cockell (2007). Exploring microbial diversity in volcanic environments: A review of methods in DNA extraction. *Journal of Microbiological Methods* **70**(1): 1–12.

Herrmann, M., A. Rusznyak, D. M. Akob, et al. (2015). Large fractions of CO_2-fixing microorganisms in pristine limestone aquifers appear to be involved in the oxidation of reduced sulfur and nitrogen compounds. *Applied and Environmental Microbiology* **81**(7): 2384–2394, Doi:10.1128/AEM.03269-14.

Hewson, I., R. S. Poretsky, H. J. Tripp, J. P. Montoya and J. P. Zehr (2010). Spatial patterns and light-driven variation of microbial population gene expression in surface waters of the oligotrophic open ocean. *Environmental Microbiology* **12**(7): 1940–1956, Doi:10.1111/j.1462-2920.2010.02198.x.

Holland, S. M. (2003). "Analytic Rarefaction 1.3 User's guide and application," from www.uga.edu/strata/software/anRareReadme.html.

Huber, T., G. Faulkner and P. Hugenholtz (2004). Bellerophon: a program to detect chimeric sequences in multiple sequence alignments. *Bioinformatics* **20**(14): 2317–2319, Doi:10.1093/bioinformatics/bth226.

Hug, L. A., B. J. Baker, K. Anantharaman, et al. (2016). A new view of the tree of life. *Nature Microbiology* **1**: 16048, Doi:10.1038/nmicrobiol.2016.48.

Hugenholtz, P., B. M. Goebel and N. R. Pace (1998). Impact of culture-independent studies on the emerging phylogenetic view of bacterial diversity. *Journal of Bacteriology* **180**(18): 4765–4774.

Hugenholtz, P. and T. Huber (2003). Chimeric 16S rDNA sequences of diverse origin are accumulating in the public databases. *International Journal of Systematic and Evolutionary Microbiology* **53** (Pt 1): 289–293, Doi:10.1099/ijs.0.02441-0.

Hugerth, L. W., E. E. L. Muller, Y. O. O. Hu, et al. (2014). Systematic design of 18S rRNA gene primers for determining eukaryotic diversity in microbial consortia. *PLoS ONE* **9**(4): e95567, Doi:10.1371/journal.pone.0095567.

Hughes, J. and B. J. M. Bohannan (2004). Application of ecological diversity statistics in microbial ecology. *Molecular Microbial Ecology Manual* 7.01: 1321–1344.

Hughes, J., J. Hellmann, T. Ricketts and B. Bohannan (2001). Counting the uncountable: statistical approaches to estimating microbial diversity. *Applied and Environmental Microbiology* **67**(10): 4399–4406.

Hurt, R., X. Qiu, L. Wu, et al. (2001). Simultaneous recovery of RNA and DNA from soils and sediments. *Applied and Environmental Microbiology* **67**(10): 4495–4503.

Huse, S. M., L. Dethlefsen, J. A. Huber, et al. (2008). Exploring microbial diversity and taxonomy using SSU rRNA hypervariable tag sequencing. *PLoS Genetics* **4**(11): e1000255, Doi:10.1371/journal.pgen.1000255.

Inceoglu, O., E. F. Hoogwout, P. Hill and J. D. van Elsas (2010). Effect of DNA extraction method on the apparent microbial diversity of soil. *Applied and Environmental Microbiology* **76**(10): 3378–3382, Doi:10.1128/AEM.02715-09.

Johnson, J. L. (1994). Similarity analysis of rRNAs. In *Methods for General and Molecular Bacteriology*. P. E. Gerhardt, W. W. A. Wood and N. R. Krieg. Washington, DC, American Society of Microbiology: 683–700.

Johnson, M., I. Zaretskaya, Y. Raytselis, et al. (2008). NCBI BLAST: a better web interface. *Nucleic Acids Research* **36**: W5–W9, Doi:10.1093/nar/gkn201.

Ju, F. and T. Zhang (2015). Experimental design and bioinformatics analysis for the application of metagenomics in environmental sciences and biotechnology. *Environmental Science & Technology* **49**(21): 12628–12640, Doi:10.1021/acs.est.5b03719.

Kanagawa, T. (2003). Bias and artifacts in multitemplate polymerase chain reactions (PCR). *Journal of Bioscience and Bioengineering* **96**(4): 317–323, Doi:http://dx.doi.org/10.1016/S1389-1723(03)90130-7.

Karsch-Mizrachi, I. and B. F. F. Ouellette (2001). The GenBank Sequence Database. In *Bioinformatics: A Practical Guide to the Analysis of Genes and Proteins*. A. D. Baxevanis and B. F. F. Ouellette, New York, John Wiley & Sons, Inc.: 45–63.

Kato, S., T. Itoh and A. Yamagishi (2011). Archaeal diversity in a terrestrial acidic spring field revealed by a novel PCR primer targeting archaeal 16S rRNA genes. *FEMS Microbiology Letters* **319**(1): 34–43, Doi:10.1111/j.1574-6968.2011.02267.x.

Katoh, K. and D. M. Standley (2013). MAFFT Multiple Sequence Alignment Software Version 7: Improvements in performance and usability. *Molecular Biology and Evolution* **30**(4): 772–780, Doi:10.1093/molbev/mst010.

Kearse, M., Moir, R., Wilson, A., et al. (2012). Geneious Basic: an integrated and extendable desktop software platform for the organization and analysis of sequence data. *Bioinformatics* **28**(12): 1647–1649.

Kembel, S. W., P. D. Cowan, M. R. Helmus, et al. (2010). Picante: R tools for integrating phylogenies and ecology. *Bioinformatics* **26**: 1463–1464.

Kembel, S. W., M. Wu, J. A. Eisen and J. L. Green (2012). Incorporating 16S gene copy number information improves estimates of microbial diversity and abundance. *PLOS Computational Biology* **8**(10): e1002743, Doi:10.1371/journal.pcbi.1002743.

Kennedy, K., M. W. Hall, M. D. J. Lynch, G. Moreno-Hagelsieb and J. D. Neufeld (2014). Evaluating bias of Illumina-based bacterial 16S rRNA gene profiles. *Applied and Environmental Microbiology*, Doi:10.1128/aem.01451-14.

Kilianski, A., J. L. Haas, E. J. Corriveau, et al. (2015). Bacterial and viral identification and differentiation by amplicon sequencing on the MinION nanopore sequencer. *GigaScience* **4**(1): 1–8, Doi:10.1186/s13742-015-0051-z.

Kim, M., M. Morrison and Z. Yu (2011). Evaluation of different partial 16S rRNA gene sequence regions for phylogenetic analysis of microbiomes. *Journal of Microbiological Methods* **84**(1): 81–87, Doi:http://dx.doi.org/10.1016/j.mimet.2010.10.020.

Kitts, C. L. (2001). Terminal restriction fragment patterns: a tool for comparing microbial communities and assessing community dynamics. *Current Issues in Intestinal Microbiology* **2**(1): 17–25.

Klindworth, A., E. Pruesse, T. Schweer, et al. (2012). Evaluation of general 16S ribosomal RNA gene PCR primers for classical and next-generation sequencing-based diversity studies. *Nucleic Acids Research* **41**(1): 1–11, Doi:10.1093/nar/gks808.

Knight, R., J. Jansson, D. Field, et al. (2012). Unlocking the potential of metagenomics through replicated experimental design. *Nature Biotechnology* **30**(6): 513–520, Doi:10.1038/nbt.2235.

Koeppel, A. F. and M. Wu (2013). Surprisingly extensive mixed phylogenetic and ecological signals among bacterial Operational Taxonomic Units. *Nucleic Acids Research* **41**(10): 5175–5188, Doi:10.1093/nar/gkt241.

Kozich, J. J., S. L. Westcott, N. T. Baxter, S. K. Highlander and P. D. Schloss (2013). Development of a dual-index sequencing strategy and curation pipeline for analyzing amplicon sequence data on the MiSeq Illumina sequencing platform. *Applied and Environmental Microbiology* **79**(17): 5112–5120, Doi:10.1128/AEM.01043-13.

Kunin, V., A. Engelbrektson, H. Ochman and P. Hugenholtz (2010). Wrinkles in the rare biosphere: pyrosequencing errors can lead to artificial inflation of diversity estimates. *Environmental Microbiology* **12**(1): 118–123, Doi:10.1111/j.1462-2920.2009.02051.x.

Lafontaine, D. L. J. and D. Tollervey (2001). The function and synthesis of ribosomes. *Nature Reviews Molecular Cell Biology* **2**: 514–520, Doi:10.1038/35080045.

Lane, D. J. (1991). 16S/23S rRNA sequencing in E. coli. In *Nucleic Acid Techniques in Bacterial Systematics*. E. Stackebrandt and M. Goodfellow. New York, NY, John Wiley & Sons: 115–175.

Larkin, M. A., G. Blackshields, N. P. Brown, et al. (2007). Clustal W and Clustal X version 2.0. *Bioinformatics* **23**(21): 2947–2948, Doi:10.1093/bioinformatics/btm404.

Lecompte, O., R. Ripp, J. C. Thierry, D. Moras and O. Poch (2002). Comparative analysis of ribosomal proteins in complete genomes: an example of reductive evolution at the domain scale. *Nucleic Acids Research* **30**(24): 5382–5390, Doi:10.1093/nar/gkf693.

Leigh, M. B., L. Taylor and J. D. Neufeld (2010). Clone libraries of ribosomal RNA gene sequences for characterization of bacterial and fungal communities. In *Handbook of Hydrocarbon and Lipid Microbiology*. K. Timmis, Berlin, Springer Berlin Heidelberg: 3969–3993.

Lever, M. A., A. Torti, P. Eickenbusch, et al. (2015). A modular method for the extraction of DNA and RNA, and the separation of DNA pools from diverse environmental sample types. *Frontiers in Microbiology* 6: 476, Doi:10.3389/fmicb.2015.00476.

Liu, L., Y. Li, S. Li, et al. (2012). Comparison of next-generation sequencing systems. *Journal of Biomedicine and Biotechnology* **2012**: 11, Doi:10.1155/2012/251364.

Liu, W. T., T. L. Marsh, H. Cheng and L. J. Forney (1997). Characterization of microbial diversity by determining terminal restriction fragment length polymorphisms of genes encoding 16S rRNA. *Applied and Environmental Microbiology* **63**(11): 4516–4522.

Love, M. I., W. Huber and S. Anders (2014). Moderated estimation of fold change and dispersion for RNA-seq data with DESeq2. *Genome Biology* **15**(12): 550, Doi:10.1186/s13059-014-0550-8.

Lozupone, C. and R. Knight (2005). UniFrac: a new phylogenetic method for comparing microbial communities. *Applied and Environmental Microbiology* **71**(12): 8228–8235, Doi:10.1128/aem.71.12.8228-8235.2005.

Lozupone, C., M. E. Lladser, D. Knights, J. Stombaugh and R. Knight (2011). UniFrac: an effective distance metric for microbial community comparison. *The ISME Journal* **5**(2): 169–172, Doi:10.1038/ismej.2010.133.

Ludwig, W., O. Strunk, R. Westram, et al. (2004). ARB: a software environment for sequence data. *Nucleic Acids Research* **32**(4): 1363–1371.

Luke, C. and P. Frenzel (2011). Potential of pmoA amplicon pyrosequencing for methanotroph diversity studies. *Applied and Environmental Microbiology* **77**(17): 6305–6309, Doi:10.1128/Aem.05355-11.

Lynch, M. D. J. and J. D. Neufeld (2015). Ecology and exploration of the rare biosphere. *Nature Reviews Microbiology* **13**(4): 217–229, Doi:10.1038/nrmicro3400.

Madsen, E. L. (2005). Identifying microorganisms responsible for ecologically significant biogeochemical processes. *Nature Reviews Microbiology* **3**(5): 439–446.

Magurran, A. E. (2004). *Measuring Biological Diversity*. Malden, MA, Blackwell Publishing.

Mahé, F., J. Mayor, J. Bunge, et al. (2015). Comparing high-throughput platforms for sequencing the V4 region of SSU-rDNA in environmental microbial eukaryotic diversity surveys. *Journal of Eukaryotic Microbiology* **62**(3): 338–345, Doi:10.1111/jeu.12187.

Mahmoudi, N., G. F. Slater and R. R. Fulthorpe (2011). Comparison of commercial DNA extraction kits for isolation and purification of bacterial and eukaryotic DNA from PAH-contaminated soils. *Canadian Journal of Microbiology* **57**(8): 623–628, Doi:10.1139/w11-049.

Mao, D. P., Q. Zhou, C. Y. Chen and Z. X. Quan (2012). Coverage evaluation of universal bacterial primers using the metagenomic datasets. *BMC Microbiology* **12**: 66, Doi:10.1186/1471-2180-12-66.

Marchesi, J. R., T. Sato, A. J. Weightman, et al. (1998). Design and evaluation of useful bacterium-specific PCR primers that amplify genes coding for bacterial 16S rRNA. *Applied and Environmental Microbiology* **64**(2): 795–799.

Mardis, E. R. (2008). The impact of next-generation sequencing technology on genetics. *Trends in Genetics* **24**(3): 133–141, Doi:10.1016/j.tig.2007.12.007.

Mardis, E. R. (2013). Next-generation sequencing platforms. *Annual Review of Analytical Chemistry* **6**(1): 287–303, Doi:doi:10.1146/annurev-anchem-062012-092628.

Marsh, T. (1999). Terminal restriction fragment length polymorphism (T-RFLP): an emerging method for characterizing diversity among homologous populations of amplification products. *Current Opinion in Microbiology* **2**(3): 323–327.

Martin-Laurent, F., L. Philippot, S. Hallet, et al. (2001). DNA extraction from soils: old bias for new microbial diversity analysis methods. *Applied and Environmental Microbiology* **67**(5): 2354–2359, Doi:10.1128/aem.67.5.2354-2359.2001.

Masella, A. P., A. K. Bartram, J. M. Truszkowski, D. G. Brown and J. D. Neufeld (2012). PANDAseq: paired-end assembler for illumina sequences. *BMC Bioinformatics* **13**(1): 31, Doi:10.1186/1471-2105-13-31.

Matsen, F. A., R. B. Kodner and E. V. Armbrust (2010). pplacer: linear time maximum-likelihood and Bayesian phylogenetic placement of sequences onto a fixed reference tree. *BMC Bioinformatics* **11** (1): 538, Doi:10.1186/1471-2105-11-538.

McMurdie, P. J. and S. Holmes (2013). phyloseq: An R package for reproducible interactive analysis and graphics of microbiome census data. *PLOS ONE* **8**(4): e61217, Doi:10.1371/journal.pone.0061217.

McMurdie, P. J. and S. Holmes (2014). Waste not, want not: Why rarefying microbiome data is inadmissible. *PLOS Computational Biology* **10**(4): e1003531, Doi:10.1371/journal.pcbi.1003531.

Messing, J. (2015). Microbiology spurred massively parallel genomic sequencing and biotechnology. *Microbe (Washington, D.C.)* **9**: 271–277.

Meyer, B. and J. Kuever (2007). Molecular analysis of the diversity of sulfate-reducing and sulfur-oxidizing prokaryotes in the environment, using aprA as functional marker gene. *Applied and Environmental Microbiology* **73**(23): 7664–7679, Doi:10.1128/AEM.01272-07.

Mikheyev, A. S. and M. M. Y. Tin (2014). A first look at the Oxford Nanopore MinION sequencer. *Molecular Ecology Resources* **14**(6): 1097–1102, Doi:10.1111/1755-0998.12324.

Miller, D. N., J. E. Bryant, E. L. Madsen and W. C. Ghiorse (1999). Evaluation and optimization of DNA extraction and purification procedures for soil and sediment samples. *Applied and Environmental Microbiology* **65**(11): 4715–4724.

Mollet, C., M. Drancourt and D. Raoult (1997). rpoB sequence analysis as a novel basis for bacterial identification. *Molecular Microbiology* **26**(5): 1005–1011, Doi:10.1046/j.1365-2958.1997.6382009.x.

Moran, M. A. (2009). Metatranscriptomics: eavesdropping on complex microbial communities. *Microbe (Washington, D.C.)* **4**: 329–335.

Moran, M. A., B. Satinsky, S. M. Gifford, et al. (2013). Sizing up metatranscriptomics. *The ISME Journal* **7**(2): 237–243, Doi:10.1038/ismej.2012.94.

Morris, R. M., B. L. Nunn, C. Frazar, et al. (2010). Comparative metaproteomics reveals ocean-scale shifts in microbial nutrient utilization and energy transduction. *The ISME Journal* **4**(5): 673–685, Doi:10.1038/ismej.2010.4.

Muyzer, G. (1999). DGGE/TGGE a method for identifying genes from natural ecosystems. *Current Opinion in Microbiology* **2**(3): 317–322.

Nagai, M., A. Yoshida and N. Sato (1998). Additive effects of bovine serum albumin, dithiothreitol, and glycerol on PCR. *Biochemistry and Molecular Biology International* **44**(1): 157–163.

Narihiro, T. and Y. Sekiguchi (2011). Oligonucleotide primers, probes and molecular methods for the environmental monitoring of methanogenic archaea. *Microbial Biotechnology* **4**(5): 585–602, Doi:10.1111/j.1751-7915.2010.00239.x.

Nawrocki, E. P., D. L. Kolbe and S. R. Eddy (2009). Infernal 1.0: inference of RNA alignments. *Bioinformatics* **25**(10): 1335–1337, Doi:10.1093/bioinformatics/btp157.

Norton, J. M., J. J. Alzerreca, Y. Suwa and M. G. Klotz (2002). Diversity of ammonia monooxygenase operon in autotrophic ammonia-oxidizing bacteria. *Archives of Microbiology* **177**(2): 139–149, Doi:10.1007/s00203-001-0369-z.

Okansen, J., Blanchet, F.G., Kindt, R., et al. (2014). vegan: Community Ecology Package. R package version 2.2–1.

Oksanen, J., F. G. Blanchet, R. Kindt, et al. (2015). vegan: Community Ecology Package.

Olsen, G. J., D. J. Lane, S. J. Giovannoni, N. R. Pace and D. A. Stahl (1986). Microbial ecology and evolution: A ribosomal RNA approach. *Annual Review of Microbiology* **40**(1): 337–365, Doi: doi:10.1146/annurev.mi.40.100186.002005.

Orcutt, B., B. Bailey, H. Staudigel, B. M. Tebo and K. J. Edwards (2009). An interlaboratory comparison of 16S rRNA gene-based terminal restriction fragment length polymorphism and sequencing methods for assessing microbial diversity of seafloor basalts. *Environmental Microbiology* **11**(7): 1728–1735, Doi:10.1111/j.1462-2920.2009.01899.x.

Orcutt, B. N., J. B. Sylvan, N. J. Knab and K. J. Edwards (2011). Microbial ecology of the dark ocean above, at, and below the seafloor. *Microbiology and Molecular Biology Reviews* **75**(2): 361–422, Doi:10.1128/mmbr.00039-10.

Osborn, A. M., E. R. B. Moore and K. N. Timmis (2000). An evaluation of terminal-restriction fragment length polymorphism (T-RFLP) analysis for the study of microbial community structure and dynamics. *Environmental Microbiology* **2**(1): 39–50, Doi:10.1046/j.1462-2920.2000.00081.x.

Paulson, J. N., O. C. Stine, H. C. Bravo and M. Pop (2013). Differential abundance analysis for microbial marker-gene surveys. *Nature Methods* **10**(12): 1200–1202, Doi:10.1038/nmeth.2658.

Polz, M. F. and C. M. Cavanaugh (1998). Bias in template-to-product ratios in multitemplate PCR. *Applied and Environmental Microbiology* **64**(10): 3724–3730.

Poretsky, R., L. M. Rodriguez-R, C. Luo, D. Tsementzi and K. T. Konstantinidis (2014). Strengths and limitations of 16S rRNA gene amplicon sequencing in revealing temporal microbial community dynamics. *PLoS ONE* **9**(4): e93827, Doi:10.1371/journal.pone.0093827.

Poretsky, R. S., I. Hewson, S. Sun, et al. (2009). Comparative day/night metatranscriptomic analysis of microbial communities in the North Pacific subtropical gyre. *Environmental Microbiology* **11** (6): 1358–1375, Doi:10.1111/j.1462-2920.2008.01863.x.

Price, M. N., P. S. Dehal and A. P. Arkin (2010). FastTree 2 – approximately maximum-likelihood trees for large alignments. *PLOS ONE* **5**(3): e9490, Doi:10.1371/journal.pone.0009490.

Priscu, J. C., E. E. Adams, W. B. Lyons, et al. (1999). Geomicrobiology of subglacial ice above Lake Vostok, Antarctica. *Science* **286**(5447): 2141–2144, Doi:10.1126/science.286.5447.2141.

Pruesse, E., J. Peplies and F. O. Glockner (2012). SINA: accurate high-throughput multiple sequence alignment of ribosomal RNA genes. *Bioinformatics* **28**(14): 1823–1829, Doi:10.1093/bioinformatics/bts252.

Pruesse, E., C. Quast, K. Knittel, et al. (2007). SILVA: a comprehensive online resource for quality checked and aligned ribosomal RNA sequence data compatible with ARB. *Nucleic Acids Research* **35**(21): 7188–7196, Doi:10.1093/nar/gkm864.

Purdy, K. J. (2005). Nucleic acid recovery from complex environmental samples. *Methods in Enzymology* **397**: 271–292, Doi:10.1016/S0076-6879(05)97016-X.

Quail, M., M. Smith, P. Coupland, et al. (2012). A tale of three next generation sequencing platforms: comparison of Ion Torrent, Pacific Biosciences and Illumina MiSeq sequencers. *BMC Genomics* **13**(1): 341.

Quast, C., E. Pruesse, P. Yilmaz, et al. (2013). The SILVA ribosomal RNA gene database project: improved data processing and web-based tools. *Nucleic Acids Research* **41**(D1): D590–D596, Doi:10.1093/nar/gks1219.

R Core Team (2014). *R: A Language and Environment for Statistical Computing*. Vienna, Austria, R Foundation for Statistical Computing.

R Core Team (2015). *R: A Language and Environment for Statistical Computing*. Vienna, Austria, R Foundation for Statistical Computing.

Ralser, M., R. Querfurth, H. J. Warnatz, et al. (2006). An efficient and economic enhancer mix for PCR. *Biochemical and Biophysical Research Communication* **347**(3): 747–751, Doi:10.1016/j.bbrc.2006.06.151.

Ram, R. J., N. C. VerBerkmoes, M. P. Thelen, et al. (2005). Community proteomics of a natural microbial biofilm. *Science* **308**(5730): 1915–1920, Doi:10.1126/science. 1109070.

Ramette, A. (2007). Multivariate analyses in microbial ecology. *FEMS Microbiology Ecology* **62**(2): 142–160, Doi:10.1111/j.1574-6941.2007.00375.x.

Reis-Filho, J. S. (2009). Next-generation sequencing. *Breast Cancer Research* 11 (Suppl 3): S12, Doi:10.1186/Bcr2431.

Robinson, M. D., D. J. McCarthy and G. K. Smyth (2009). edgeR: a Bioconductor package for differential expression analysis of digital gene expression data. *Bioinformatics* **26**(1): 139–140, Doi:10.1093/bioinformatics/btp616.

Rodriguez-R, L. M. and K. T. Konstantinidis (2014). Bypassing cultivation to identify bacterial species. *Microbe (Washington, D.C.)* **9**: 111–118.

Rognes, T., F. Mahé, T. Flouri, D. McDonald and P. Schloss (2015). vsearch: VSEARCH 1.4.0 [Data set]. *Zenodo*, Doi:10.5281/zenodo.31443.

Ronaghi, M., M. Uhlén and P. Nyrén (1998). A sequencing method based on real-time pyrophosphate. *Science* **281**: 363–365.

Rosselló-Mora, R. and R. Amann (2001). The species concept for prokaryotes. *FEMS Microbiology Reviews* **25**(1): 39–67.

Rusznyak, A., D. M. Akob, S. Nietzsche, et al. (2012). Calcite biomineralization by bacterial isolates from the recently discovered pristine karstic herrenberg cave. *Applied and Environmental Microbiology* **78**(4): 1157–1167, Doi:10.1128/Aem.06568–11.

Salipante, S. J., T. Kawashima, C. Rosenthal, et al. (2014). Performance comparison of Illumina and Ion Torrent next-generation sequencing platforms for 16S rRNA-based bacterial community profiling. *Applied and Environmental Microbiology* **80**(24): 7583–7591, Doi:10.1128/aem.02206-14.

Salter, S. J., M. J. Cox, E. M. Turek, et al. (2014). Reagent and laboratory contamination can critically impact sequence-based microbiome analyses. *BMC Biology* **12**(1): 1–12, Doi:10.1186/s12915-014-0087-z.

Sambrook, J. and D. W. Russell (2001). *Molecular Cloning – A Laboratory Manual*. New York, Cold Spring Harbor Laboratory.

Sanger, F. and A. R. Coulson (1975). A rapid method for determining sequences in DNA by primed synthesis with DNA polymerase. *Journal of Molecular Biology* **94**(3): 441–448, Doi:http://dx.doi.org/10.1016/0022-2836(75)90213-2.

Santos, S. R. and H. Ochman (2004). Identification and phylogenetic sorting of bacterial lineages with universally conserved genes and proteins. *Environmental Microbiology* **6**(7): 754–759, Doi:10.1111/j.1462-2920.2004.00617.x.

Schirmer, M., U. Z. Ijaz, R. D'Amore, et al. (2015). Insight into biases and sequencing errors for amplicon sequencing with the Illumina MiSeq platform. *Nucleic Acids Research* **43**(6): e37, Doi:10.1093/nar/gku1341.

Schloss, P. D., D. Gevers and S. L. Westcott (2011). Reducing the effects of PCR amplification and sequencing artifacts on 16S rRNA-based studies. *PLoS ONE* **6**(12): e27310.

Schloss, P. D. and J. Handelsman (2004). Status of the microbial census. *Microbiology and Molecular Biology Reviews* **68**(4): 686–691, Doi:10.1128/MMBR.68.4.686-691.2004.

Schloss, P. D., S. Westcott, T. Ryabin, et al. (2009). Introducing mothur: open-source, platform-independent, community-supported software for describing and comparing microbial communities. *Applied and Environmental Microbiology* **75**(23): 7537–7541.

Schloss, P. D. and S. L. Westcott (2011). Assessing and improving methods used in operational taxonomic unit-based approaches for 16S rRNA gene sequence analysis. *Applied and Environmental Microbiology* **77**(10): 3219–3226, Doi:10.1128/AEM.02810-10.

Schmieder, R. and R. Edwards (2011). Quality control and preprocessing of metagenomic datasets. *Bioinformatics* **27**(6): 863–864, Doi:10.1093/bioinformatics/btr026.

Sinclair, L., O. A. Osman, S. Bertilsson and A. Eiler (2015). Microbial community composition and diversity via 16S rRNA gene amplicons: Evaluating the Illumina platform. *PLoS ONE* **10**(2): e0116955, Doi:10.1371/journal.pone.0116955.

Smith, C. J. and A. M. Osborn (2009). Advantages and limitations of quantitative PCR (qPCR)-based approaches in microbial ecology. *FEMS Microbiology Ecology* **67**(1): 6–20, Doi:10.1111/j.1574-6941.2008.00629.x.

Smith, K. (2013). *A Brief History of NCBI's Formation and Growth*. The NCBI Handbook [Internet]. Bethesda, MD, National Center for Biotechnology Information.

Soetaert, K. and C. Heip (1990). Sample-size dependence of diversity indexes and the determination of sufficient sample-size in a high-diversity deep-sea environment. *Marine Ecology Progress Series* **59**(3): 305–307, Doi:10.3354/Meps059305.

Stackebrandt, E., W. Frederiksen, G. M. Garrity, et al. (2002). Report of the ad hoc committee for the re-evaluation of the species definition in bacteriology. *International Journal of Systematic and Evolutionary Microbiology* **52**(3): 1043–1047, Doi:10.1099/ijs.0.02360-0.

Stackebrandt, E. and B. M. Goebel (1994). Taxonomic note: A place for DNA-DNA reassociation and 16S rRNA sequence analysis in the present species definition in bacteriology. *International Journal of Systematic Bacteriology* **44**(4): 846–849, Doi:10.1099/00207713-44-4-846.

Staley, J. T. and A. Konopka (1985). Measurement of in situ activities of nonphotosynthetic microorganisms in aquatic and terrestrial habitats. *Annual Review of Microbiology* **39**(1): 321–346, Doi: doi:10.1146/annurev.mi.39.100185.001541.

Stamatakis, A. (2006). RAxML-VI-HPC: maximum likelihood-based phylogenetic analyses with thousands of taxa and mixed models. *Bioinformatics* **22**(21): 2688–2690, Doi:10.1093/bioinformatics/btl446.

Stamatakis, A. (2014). RAxML version 8: a tool for phylogenetic analysis and post-analysis of large phylogenies. *Bioinformatics* **30**(9): 1312–1313, Doi:10.1093/bioinformatics/btu033.

Stoddard, S. F., B. J. Smith, R. Hein, B. R. K. Roller and T. M. Schmidt (2015). rrnDB: improved tools for interpreting rRNA gene abundance in bacteria and archaea and a new foundation for future development. *Nucleic Acids Research* 43(Database issue): D593–D598, Doi:10.1093/nar/gku1201.

Sudek, S., R. C. Everroad, A. L. Gehman, et al. (2015). Cyanobacterial distributions along a physicochemical gradient in the Northeastern Pacific Ocean. *Environmental Microbiology* **17**(10): 3692–3707, Doi:10.1111/1462-2920.12742.

Suzuki, M. T. and S. J. Giovannoni (1996). Bias caused by template annealing in the amplification of mixtures of 16S rRNA genes by PCR. *Applied and Environmental Microbiology* **62**(2): 625–630.

Tamura, K., G. Stecher, D. Peterson, A. Filipski and S. Kumar (2013). MEGA6: Molecular Evolutionary Genetics Analysis Version 6.0. *Molecular Biology and Evolution* **30**(12): 2725–2729, Doi:10.1093/molbev/mst197.

Tavormina, P. L., W. Ussler, III, S. B. Joye, B. K. Harrison and V. J. Orphan (2010). Distributions of putative aerobic methanotrophs in diverse pelagic marine environments. *The ISME Journal* **4**(5): 717–717.

Terpe, K. (2013). Overview of thermostable DNA polymerases for classical PCR applications: from molecular and biochemical fundamentals to commercial systems. *Applied Microbiology and Biotechnology* **97**(24): 10243–10254, Doi:10.1007/s00253-013-5290-2.

Tindall, B. J., R. Rosselló-Móra, H.-J. Busse, W. Ludwig and P. Kämpfer (2010). Notes on the characterization of prokaryote strains for taxonomic purposes. *International Journal of Systematic and Evolutionary Microbiology* **60**(1): 249–266, Doi:10.1099/ijs.0.016949-0.

Van de Peer, Y., S. Chapelle and R. De Wachter (1996). A quantitative map of nucleotide substitution rates in bacterial rRNA. *Nucleic Acids Research* **24**(17): 3381–3391, Doi:10.1093/nar/24.17.3381.

van Dorst, J., A. Bissett, A. S. Palmer, et al. (2014). Community fingerprinting in a sequencing world. *FEMS Microbiology Ecology* **89**(2): 316–330, Doi:10.1111/1574-6941.12308.

Venter, J. C., M. D. Adams, E. W. Myers, et al. (2001). The sequence of the human genome. *Science* **291**(5507): 1304–1351, Doi:10.1126/Science.1058040.

Venter, J. C., K. Remington, J. F. Heidelberg, et al. (2004). Environmental genome shotgun sequencing of the Sargasso Sea. *Science* **304**(5667): 66–74, Doi:10.1126/science.1093857.

Vishnivetskaya, T. A., A. C. Layton, M. C. Y. Lau, et al. (2014). Commercial DNA extraction kits impact observed microbial community composition in permafrost samples. *FEMS Microbiology Ecology* **87**(1): 217–230, Doi:10.1111/1574-6941.12219.

Wade, B. D. and F. Garcia-Pichel (2003). Evaluation of DNA extraction methods for molecular analyses of microbial communities in modern calcareous microbialites. *Geomicrobiology Journal* **20**(6): 549–561, Doi:10.1080/713851168.

Wang, L.-T., F.-L. Lee, C.-J. Tai and H. Kasai (2007a). Comparison of gyrB gene sequences, 16S rRNA gene sequences and DNA–DNA hybridization in the *Bacillus subtilis* group. *International Journal of Systematic and Evolutionary Microbiology* **57**(8): 1846–1850, Doi:10.1099/ijs.0.64685-0.

Wang, Q., G. M. Garrity, J. M. Tiedje and J. R. Cole (2007b). Naïve Bayesian classifier for rapid assignment of rRNA sequences into the new bacterial taxonomy. *Applied and Environmental Microbiology* **73**(16): 5261–5267, Doi:10.1128/aem.00062-07.

Webb, C. O., D. D. Ackerly and S. W. Kembel (2008). Phylocom: software for the analysis of phylogenetic community structure and trait evolution. *Bioinformatics* **24**(18): 2098–2100, Doi:10.1093/bioinformatics/btn358.

Weisburg, W. G., S. M. Barns, D. A. Pelletier and D. J. Lane (1991). 16S ribosomal DNA amplification for phylogenetic study. *Journal of Bacteriology* **173**(2): 697–703.

Whitaker, R. J. and J. F. Banfield (2006). Population genomics in natural microbial communities. *Trends in Ecology & Evolution* **21**(9): 508–516, Doi:http://dx.doi.org/10.1016/j.tree.2006.07.001.

Whitman, W., D. Coleman and W. Wiebe (1998). Prokaryotes: The unseen majority. *Proceedings of the National Academy of Sciences of the United States of America* **95**(12): 6578–6583.

Wilson, I. G. (1997). Inhibition and facilitation of nucleic acid amplification. *Applied and Environmental Microbiology* **63**(10): 3741–3751.

Winsley, T., J. M. van Dorst, M. V. Brown and B. C. Ferrari (2012). Capturing greater 16S rRNA gene sequence diversity within the domain Bacteria. *Applied and Environmental Microbiology* **78**(16): 5938–5941, Doi:10.1128/aem.01299-12.

Wintzingerode, F. V., U. Gobel and E. Stackebrandt (1997). Determination of microbial diversity in environmental samples: pitfalls of PCR-based rRNA analysis. *FEMS Microbiology Reviews* **21**(3): 213–229.

Woese, C. R. (1987). Bacterial evolution. *Microbiological Reviews* **51**(2): 221–271.

Woese, C. R. and G. E. Fox (1977). Phylogenetic structure of the prokaryotic domain: the primary kingdoms. *Proceedings of the National Academy of Sciences* **74**(11): 5088–5090, Doi:10.1073/pnas.74.11.5088.

Woyke, T. and E. M. Rubin (2014). Searching for new branches on the tree of life. *Science* **346**(6210): 698–699, Doi:10.1126/science.1258871.

Wright, E. S., L. S. Yilmaz and D. R. Noguera (2012). DECIPHER, a search-based approach to chimera identification for 16S rRNA sequences. *Applied and Environmental Microbiology* **78**(3): 717–725, Doi:10.1128/aem.06516-11.

Wrighton, K. C., B. C. Thomas, I. Sharon, et al. (2012). Fermentation, hydrogen, and sulfur metabolism in multiple uncultivated bacterial phyla. *Science* **337**(6102): 1661–1665, Doi:10.1126/science.1224041.

Yilmaz, P., L. W. Parfrey, P. Yarza, et al. (2014). The SILVA and "All-species Living Tree Project (LTP)" taxonomic frameworks. *Nucleic Acids Research* **42**(D1): D643–D648, Doi:10.1093/nar/gkt1209.

Zhang, J., K. Kobert, T. Flouri and A. Stamatakis (2014). PEAR: a fast and accurate Illumina Paired-End reAd mergeR. *Bioinformatics* **30**(5): 614–620, Doi:10.1093/bioinformatics/btt593.

Zhang, R., V. Thiyagarajan and P. Qian (2008). Evaluation of terminal-restriction fragment length polymorphism analysis in contrasting marine environments. *FEMS Microbiology Ecology* **65**(1): 169–178, Doi:10.1111/fem.2008.65.issue-1.

Zhang, T. and H. H. Fang (2006). Applications of real-time polymerase chain reaction for quantification of microorganisms in environmental samples. *Applied Microbiology and Biotechnology* **70**(3): 281–289, Doi:10.1007/s00253-006-0333-6.

Zhou, J., Z. He, Y. Yang, et al. (2015). High-throughput metagenomic technologies for complex microbial community analysis: open and closed formats. *mBio* **6**(1): pii: e02288-14, Doi:10.1128/mBio.02288-14.

Zhou, J. Z., M. A. Bruns and J. M. Tiedje (1996). DNA recovery from soils of diverse composition. *Applied and Environmental Microbiology* **62**(2): 316–322.

Index

absorbance/absorption
  absorption edges, 241, 242, 243, 246, 252
  atomic spectroscopic techniques, 17–18
  colorimetric assay, 14, 15
  ferrozine method, 12
  FTIR spectroscopy, 289, 291, 292, 295, 299
  nucleic acids, 44, 368
  phosphatase assay, 40
  spectrophotometric techniques, 8
  X-ray absorption coefficient, 239, 241, 243–246, 249
  X-ray absorption spectra, 239
acidification
  ICP-AES, 19
  inorganic carbon analysis, 20, 21
  MC-ICP-MS, 107
  sample collection, preservation, 4, 5, 8
adsorption, 67, 89
  cadmium, zinc, 76
  contaminants, ATR, 300
  ion/ion exchange chromatography, 15
  isothermal titration calorimetry, 68–69, 71
  kinetics, 67
  metals, 63–64, 239
    FTIR spectroscopy, 289
    XPS, 276–278
  potentiometric titration, 80–81
  protonation models, 87–88
  sample collection, 6
  XPS, 266
aerobes, purifying, 27
agar dilution, 28
alcohol
  ethanol
    dehydration
      FISH, 195, 196, 199
      SEM, TEM, 150, 153, 159, 176–177
    drying, SEM, 153
    fixation, FISH, 191, 194
    Gram stain, 34
    precipitation, 43
    slurry mount, pXRD, 223
    sterilization, 5, 6, 27, 42
  isopropanol
    precipitation, 43
    sterilization, 5
  isopropyl
    sterilization, 6
  nucleic acid extraction, 43, 44
  nucleic acid precipitation, 368
  transesterification, 348
AM-AFM (amplitude modulation atomic force microscopy). *See* atomic force microscopy
amines, amino acids, amino groups, 73, 89, 273, 383
  Bacillus subtilis, 273
  chromatography, 24
  ion/ion exchange chromatography, 15
  protonation, 74, 266
  TEM, 168
amorphous materials
  carbon-based, 171
  extracellular polymeric substances, 154
  iron oxhydroxides, 110
  pXRD, 225, 227
  ultra-thin sections, 181
amplification, 361, *See also* PCR, qPCR
  ARISA, 375
  clonal, solid-phase, 51
  gel electrophoresis, 372
  NGS, 380
  TRFLP, 375
anaerobes, purifying, 27–28
anti-capillary tweezer, 168, 170, 171
argon, 4, 19, 111, 157, 251, 252
ARISA (automated ribosomal intergenic spacer analysis), 46, 374, 375–376
atomic force microscopy (AFM), 122, 124–143
  amplitude modulation (AM), 127
  frequency modulation (FM), 127
atomic spectroscopic techniques, 17–19
  atomic absorption spectrometry (AAS), 17–18
  inductively coupled plasma-atomic emission spectrometry (ICP-AES), 18–19
ATP (adenosine triphosphate), 40
ATR (attenuated total reflectance). *See* reflections

Bacillus licheniformis, 76
Bacillus subtilis, 27, 69, 71, 75, 270–278, 356
background subtraction, 135, 227, 254
bands, spectral, 289, 290, 291–294
  ATR vs. transmission, 296
baseline correction, 304

beamlines, 239, 240, 241, 248, 249, 250–253, 264
Beer-Lambert law, 8, 12, 13, 289, 292, 295
bias potential, 123
bias voltage, 122
binding energy, 137, 240, 262–264, 269, 270
bioassays, 39–41
bioavailability, 63, 243
biodiversity, 46, 297, 361, 388–390
biofilms, 89, 150, 151, 196
  FISH, 191
  fixation, SEM, 152
  formation, 137
  FTIR (micro)spectroscopy, 298, 301
  functional groups, 79
  grazing angle reflectance, 297
  MALDI, 356
  microfossils, 156
  optical density, 29
  siderite in, 331
  TEM, 170, 182
  ultra-thin sections, 173
  XPS, 284
  X-ray techniques, 239
bioinformatics, 52, 204, 365, 372, 380–387
biological communities, 188
biomass, 342
  environmental samples with low, 345
  isothermal titration calorimetry, 65–66, 69
  measure of, recovered fatty acids, 351
  microbial growth, 29, 30
  nucleic acid quantification, 369
  PFA fixation, 193
  potentiometric titration, 79
  pXRD, 227
  sample collection, 366
biomineral precipitation, 239
biomineralization, 79, 137, 173, 181, 215, 229, 283, 295, 297
bioreactor, 26, 27
blanks
  acid, 101
  elemental analysis, 23
  potentiometric titration, 85
  reagent, 12
block trimming, 178
Bragg reflection, 251
Bragg's law, 217, 218, 231
Brunauer-Emmett-Teller (BET) analysis, 233
buffering capacity, 79, 82, 85, 89
buffers
  chaotropic, 368
  colorimetric assay, 10
  ferrozine method, 11
  FISH, 191, 192, 193, 194, 195, 196
    oligonucleotide, 198, 199
  ion/ion exchange chromatography, 16
  lysis, 43, 45
  PCR, 369
  phosphate, lipid extraction, 346
  SEM, TEM, 151, 170
  XPS, 282

calibration
  atomic force microscopy, 135, 139
  binding energy scale, 270
  cantilever, 137–139
  curve, 22, 23, 24, 82
    atomic absorption spectrometry, 18
    colorimetric assay, 13, 15
    ferrozine method, 11
    ICP-AES, 19
    ion/ion exchange chromatography, 16
    Mössbauer spectroscopy, 322
    pXRD, 221
    spectrophotometric techniques, 9
    total organic carbon, 22
  inter-laboratory, 104
  isomer shift, 317
  pXRD, 228
  X-ray spectroscopy, 251
calorimetry, 64–76, 89
  isothermal titration, 65–76
cantilever, 124–133, 137–139, 140–141
carbon
  non-purgeable organic (NPOC), 20
  purgeable organic (POC), 20
  total organic (TOC), 19–22
carbonates
  CO_2 measurement, 38
  FISH, 190
  in inorganic carbon, 20
  inorganic, removing, 82
  ion/ion exchange chromatography, 15
  MC-ICP-MS, 100
  potentiometric titration, 89
  protonation, 68
  surface complexation models, 88
carboxyl groups, carboxylic acids, 73, 75, 89, 270, 273, 291, 307, 348
CARD-FISH (catalyzed reporter deposition fluorescence in situ hybridization). *See* FISH
catalysts, 239
cell damage, 79, 84, 181
cell density, 66, 69, 182
cell structure, 33, 157, 268
cell walls
  Bacillus subtilis without, 27
  fixation, FISH, 192
  FTIR spectroscopy, 288
  functional groups, 79, 84, 87
  Gram-positive vs. -negative, 34, 45
  potentiometric titration, 83, 86
  rigid, 194, 196
  XPS, 267–279, 283

charge-coupled device (CCD), 220
chemical composition
  FTIR (micro)spectroscopy, 288, 292, 301, 302
  pXRD, 228, 229, 232
  whole-mount samples, 151, 154
  XPS, 263, 268, 270, 280, 284
chemical fingerprint, 263, 290
chemical mapping, 263
chemolithotrophic bacteria, 156
chemostat, 26, 27
chlorides, 15, 101
chromatography
  gas, 24–25
    mass spectrometry (GC-MS), 344
  high performance liquid (HPLC), 24–25
  ion/ion exchange, 9, 15–17, 99, 100–101
  liquid, 111
    2-D, 356
    electrospray ionization/mass spectrometry (LC/ESI/MS), 46, 345
    mass spectrometry (LC-MS), 344, 350
  thin-layer (TLC), 355
chromium, 12–13, 99, 291
clay, 108, 156, 225, 331
CO_2, 4, 5, 19–22, 37–39, 67
collision-induced dissociation (CID), 345
colony, 26, 28, 29, 376
  colony-forming unit (CFU), 30
colorimetric assay, 9–15, 45, 240
  ferrozine method, 10–12
community structure, 53, 342, 351, 365, 374–380
conductive probe, atomically sharp, 123
conductivity, 15, 241
  detectors, 16, 22, 25
constant capacitance model, 87
constant current mode, 124
constant height mode, 124
contact mode, 123, 124, 126–127
contamination, 25
  atomic force microscopy, 130, 134
  ATR, 299
  bioaerosol, 50
  carbon, 21, 281
  CO_2, 67, 69
  contaminant solubility, toxicity, 243
  contaminant transformations, 239
  DNA, RNA, 42, 44, 367
  FTIR microspectroscopy, 300
  hygroscopic water, 22
  industrial waste, 12
  ion/ion exchange chromatography, 101
  MC-ICP-MS, 101
  preferred orientation, pXRD, 225
  sample collection, 5–6, 7
  SEM, TEM, 157, 175, 176
  XPS, 266, 282, 283, 284
coordination
  arsenate, 306
  atomic, 245, 317, 319
    iron, 322–323
  metal ion, 64
  metal-ligand, 239
  metals within proteins, 239
  molecular, 239
  numbers, 239, 252, 256, 264
  shells, 244, 245, 246, 253, 256, 278
cryostats, 253, 321
crystal habit, 216
crystalline samples, 215
  iron, 110
  poly-, 215, 216, 218, 221, 249
crystallography, 166, 216
cultivation, 26–33
culture
  batch, 26
  continuous, 26
  planktonic cells, 81
cyanobacteria, 106, 170, 230, 341, 371, 387, *See also* Synechococcus
cycling
  iron, 9
  metals, 93–111, 283
  nutrient, 36, 39
  redox active elements, 267
  trace elements, 267

data analysis
  FTIR spectroscopy, 302–307
  metagenomics, metatranscriptomics, 52
  Mössbauer spectroscopy, 321–322
  PLFA profiles, 351
  potentiometric titration, 85–87
  pXRD, 226–234
  X-ray spectroscopy, 253–256
deflection
  laser beam, 129
  lateral, 126
  vertical, 126
dehydration. *See also* alcohol, ethanol
  Gram stain, 34
  metal complexation, 75
  TEM, 176–177
  XPS, 266, 270, 277, 281
deposition
  ATR, 299
  fast-freezing, cryogenic XPS, 281
  freeze-drying, XPS, 282
  polished block, 159
  polished thin-section, 157
  whole-mount sample, 154
deprotonation, 83, 86, 88, *See also* protonation
  functional groups, 275
  hydroxyl groups, 291

deprotonation (cont.)
phosphates, 74
detectors
0-D, 219
2-D charge-coupled device (CCD), 220
amperometric, 16
atomic absorption spectrometry, 17
atomic force microscopy, 131
backscatter, SEM, 155
electrical conductivity, 16
electron capture, 25
energy dispersive, 247, 252
flame ionization (FID), 25
flame photometric, 25
focal plane array (FPA), 301
FTIR (micro)spectroscopy, 289, 294, 301–302
ICP-AES, 18
LINK X-ray EDS, 181
Mössbauer spectroscopy, 320
nitrogen phosphorus, 25
nitrogen, total, 21
non-dispersive infrared (NDIR), 20, 22
photoionization, 25
position-sensitive (PSD), 220, 221
pXRD, 218–221
refractive index, 25
scintillation, 219
SEM, 149, 154
single element, 301
solid-state, 220
thermal conductivity (TCD), 25
UV, UV-Vis, 16, 25
DFS (dynamic force spectroscopy). *See* force spectroscopy
DGGE (denaturing gradient gel electrophoresis). *See* gel electrophoresis
diffractograms, 218, 222, 223, 228
diffuse reflectance. *See* DRIFTS
disordered materials, 239
dissolution, 63, 79, 110, 142, 162, 267
DNA (deoxyribonucleic acid). *See* nucleic acids
doped materials, 226, 239
Doppler shift, 316, 320
double layer model, 87
doublets, 246, 319, 322
ferrihydrite, 327, 330
green rust, 331
lepidocrocite, 330
magnetite, 328
siderite, 331
sulfur, 265
vivianite, 331
DRIFTS (diffuse reflectance infrared Fourier transform spectroscopy). *See* reflections
drying
critical point, 153, 154
elemental analysis, 22
hybridized samples, FISH, 195
radiation damage, 253
sample preservation, 5

edge position, 254
electric quadrupole interaction. *See* quadrupole splitting
electrical double layer (EDL), 266, 267, 275
electrolytes
isothermal titration calorimetry, 66–67, 69
potentiometric titration, 79, 81, 85
XPS, 266, 275–277
electromagnetic lenses, 149
electromagnetic radiation, particles, 17, 130, 216
electromagnetic spectrum, 8, 216, 289
electron acceptors, 39, 95, 106
electron backscatter diffraction (EBSD) analysis, 162
electron beam
backscatter electron mode, 155
cell damage by, 181
irradiation by, 150
SAED, 173
secondary electron mode, 154
SEM, 149
TEM, 166
ultra-thin sections, 178
vs. FIB, 161
electron donors, 106
electron gun, 149, 166
electron microscopy, 148, 189, 246, 279
sample preparation, vs. X-ray spectroscopy, 241
scanning (SEM), 34, 148–164, 167
and pXRD, 229
cryogenic, 150
serial block-face, 150
transmission (TEM), 34, 149, 161, 166–185
and pXRD, 233
grids, 168, 170–171
electron shell, 240
electron spectrometry, 262, 264, 265, 279
electron spectroscopy for chemical analysis (ESCA). *See* X-ray photoelectron spectroscopy
electrophoretic mobility, 275
electrostatic force microscopy (EFM), 127
electrostatic sector, 98
elemental composition, analysis, 22–24, 149, 181, 246–247, 263, 268, 323
embedding, 177–178, 200
energy dispersive spectroscopy (EDS), 150, 161, 168, 181
enrichment, 26–27
cultures, amplicon cloning, 377
inoculation, 27
isolation, 27–29
enrobing, 173–176
enthalpy, 64, 71
complexation, 72, 74–76
protonation, 71, 73–74

entropy, 64, 73–74, 75
equilibrium, 79, 153
evolutionary placement. *See* phylogenetic placement
EXAFS (extended X-ray absorption fine structure) spectroscopy. *See* X-ray spectroscopy
exothermic reactions
  cadmium, zinc adsorption, 76
  calorimetric data, 69
  carbonate protonation, 68
  carboxylic acid protonation, 73
  complexation, 75
  hydroxide neutralization, 67
  microbial surface protonation, 73
  Nitrosopumilis maritimus protonation, 71
  phosphate protonation, 74
  thiol protonation, 74
extracellular polymeric substances (EPS), 150, 154, 171
exudates, 63, 68, 79, 80, 89

Faraday's constant, 86
fatty acid methyl esters (FAMEs), 347–349, 350
ferromagnetism/anti-ferromagnetism, 329
ferrozine method, 10–12, 102, 107, 109
FIB. *See* focused ion beam milling
field samples, 3–8
fingerprinting methods, 46, 246, 254, 363, 374–376, *See also* ARISA; DGGE; TRFLP
FISH (fluorescence in situ hybridization), 188–209
  biorthogonal noncanonical amino acid tagging (BON-CAT), 189
  catalyzed reporter deposition (CARD), 189, 197, 201–203
  combinatorial labelling and spectral imaging (CLASI), 189
  double labelling of oligonucleotide probes (DOPE), 189, 196, 207
  gene, 189, 204
  locked nucleic acid (LNA), 189, 196
  magneto-, 189
  micro autoradiography (MAR), 189
  microscopy, 195, 208
  multilabelled oligonucleotides (MiL), 189, 196, 207
  oligonucleotide, 188, 196–201
  peptide nucleic acid (PNA), 189, 191, 196
  phage, 189
  polynucleotide, 189, 197, 207
  quantum dot, 189, 191
FITEQL, 85–87
fixation, 33, 35
  FISH, 190, 191–194
  SEM, 150–151, 152
  TEM, 168, 170
flow cytometry, 189, 195, 208
fluorescence. *See also* FISH; PCR; X-ray fluorescence
  ARISA, 376
  microscopy, 188, 189, 206
    dyes. *See also* gene probing
  NGS, 51, 377
  pXRD, 219
  TRFLP, 375
fluorescent screen, 166, 170
fluorometry, 368
FM-AFM (frequency modulation atomic force microscopy). *See* atomic force microscopy
focused ion beam (FIB) milling, 161–162, 184
force spectroscopy, 140–141
  dynamic (DFS), 141
Formvar film, 168, 170, 171, 181
Fourier transform, 138, 245–246, 254
Fourier-transform infrared (FTIR) spectroscopy, 8, 73, 281, 288–308, *See also* reflections
  microspectroscopy, 288–308
    synchrotron-assisted, 301
fractionation, 96, 101
  iron isotopes, 104, 106–107, 108, 110
  lipid, 355
  lithium isotopes, 93
  magnesium isotopes, 95
  molybdenum isotopes, 95
freeze-drying/freeze-dried samples, 265, 266–267, 273, 281, 282
freezing
  fast-/fast-frozen samples, 264–266, 267, 271–275, 279, 282, 284
  Mössbauer spectroscopy, 315
  nucleic acid preservation, 366
  radiation damage, 253
  sample collection, 7
    FISH, 190
FTIR spectroscopy. *See* Fourier-transform infrared spectroscopy
functional groups, 64, 79, 87, 89, 269, 271, 273, 277
  acid-base behaviour, XPS, 267
  deprotonation, 83, 275
  fatty acids, 341
  FTIR spectroscopy, 289, 291, 294
  protonation, 275
    models, 86
  TEM, 168

gas chromatography-mass spectrometry (GC-MS), 344, 350
gel electrophoresis, 44, 48, 49, 368, 372
  denaturing gradient (DGGE), 46, 374–375
  sodium dodecyl sulfate polyacrylamide (SDS-PAGE), 45
  TRFLP, 375
gene primers, 369–371
gene probing, 46–47
  database, 371
  FISH, 190, 194, 195, 203–207

gene probing (cont.)
  fluorescently labelled, 188, 207
    oligonucleotide FISH, 199
  horse-radish peroxidase (HRP) labelled, 202, 207
  microarrays, 46–47
  qPCR and, 48
Gibbs energy, 63–64, 72, 86
glovebox, 10, 67
glow discharge system, 170
glutaraldehyde ($C_5H_8O_2$)
  FISH, 191
  SEM, 150–151, 152
  TEM, 168, 170, 181
goethite, 111, 274, 306, 319, 323, 326, 327
Gram stain, 34–35
  Gram-negative bacteria, 191, 349
  Gram-negative prokaryotes, 192, 194
  Gram-positive bacteria, 45, 196, 273, 344
  Gram-positive prokaryotes, 192, 194
  Gram-positive vs. -negative bacteria cell walls, 268–269, 270
groundwater, 12, 18, 100, 107
  lithium isotopes, 93
growth, 29–33
  batch culture, 30–33
  biomass, 66
  continuous culture, 33
  curve, 30–33, 81
  exponential phase, 30
  planktonic cells, 81
  population, 30

heat flow, 65–76
  experiment challenges, 69
  experiment data, 69–71
  metal complexation, 68–69
  protonation, 67–68
helium, 251, 350
hematite, 111, 319, 323, 326, 329–330
hexamethyldisilazane (HMDS), 153, 154
homogenization
  DRIFTS, 296
  elemental analysis, 22
  FTIR spectroscopy, 295
  nucleic acid extraction, 367
  TEM, 170, 175
HPLC (high performance liquid chromatography). *See* chromatography
humic acids, 87, 373
hydration, metal complexation, 75
hydrogen, 4, 157
  bonding, 264
hydrolysis
  fluorescein diacetate, 40
  metal ion, 69
  of phospholipids, 344
  phosphatase assay, 40
  sample degradation, 7
hydroxyl groups, 73, 85, 89, 291
  phospholipids, 347
  protonation, 74
hyperfine field (B_{hf}), 317, 319, 321–322, 323
  ferrihydrite, 327
  goethite, 327
  hematite, 329
hyperfine parameters, 317–320, 325
  green rust, 331
hysteresis, 83, 84, 135

ice sublimation, 265, 266, 280
ICP-AES (inductively coupled plasma-atomic emission spectrometry). *See* atomic spectroscopic techniques
ICP-MS (inductively coupled plasma mass spectrometry). *See* mass spectrometry
imaging
  backscatter electron, 150, 155, 159, 171
  dynamic, 126, 127–130, 132
  friction, 126, 130
  secondary electron, 150, 154
  spectral, 189, 208
  static, 126–127
infrared (IR) spectroscopy, 142, 277
instrumental broadening, 322
intact phospholipid profiling (IPP), 345
interatomic distances, 239, 244–246, 256
interferences, 97, 99–100, 103, 246, 366
internal reflection element (IRE), 298, 300
ion/ion exchange chromatography. *See* chromatography
ionic strength, 84, 86, 89
  calorimetry, 66
  ion/ion exchange chromatography, 16
  PCR, 369
  potentiometric titration, 81
  XPS, 266
ionization, 216
  chamber, 251–252
  liquid chromatography/electrospray ionization/mass spectrometry (LC/ESI/MS), 46, 345
  proton, 74
iridium, 154, 159
IRMM (Institute for Reference Materials and Measurement), 99, 104
iron, 12, 96–111
isomer shift (IS), 317–318, 321–322, 323
  ferrihydrite, 327
  goethite, 327
  hematite, 329
  lepidocrocite, 330
  siderite, 331
isotopes, isotopic analysis, 93–111
  copper, 95
  iron, 96–111

lithium, 93
magnesium, 93
molybdenum, 95
zinc, 95

kinetic energy, 98, 244, 262–264, 269
kinetics, 10, 141, *See also* growth
adsorption, 67
krypton, 252

laminar flow cabinet, 101
laser, 124, 129, 138
ablation, 111
alignment, 131–132
lateral deflection, 130
lattice distortion, 232
lattice planes, 216, 218
least-squares error, 85–87
lift/non-contact mode, 124, 126–127, *See also* contact mode
ligands, 74–75, 110, 275, 323
electronegativity, 264
light microscopy, 33–36, 121, 167
dyes, stains, 33–35
linear combination (LC) analysis, 246, 254
linear programming optimization, 85
lipidomics, 355
lipids, 269, *See also* phospholipids
analysis, 351–355
biomarkers, 361
characterization, spectroscopic techniques, 355
culture-independent methods, 41
extraction, 341, 346
Gram-negative bacteria, 34, 270
Gram-positive bacteria, 270
lipid biomarker analysis, 341–356
protein extraction, 45
staining, TEM, 182
XPS, 270, 277, 283
liquid chromatography/electrospray ionization/mass spectrometry (LC/ESI/MS), 345
liquid chromatography-mass spectrometry (LC-MS), 344, 350
long-range atomic order, 242
Lorentzian function, 138
Lorentzian line shape, 316, 317, 322
lyophilization. *See* freeze-drying
lysis
hydrolysis of phospholipids after, 344
nucleic acid extraction, 43–44, 367
protein extraction, 45

magnetic force microscopy (MFM), 127
magnetic hyperfine splitting. *See* hyperfine field
magnetic ordering, 318, 319, 323
ferrihydrite, 327
goethite, 327
hematite, 329
lepidocrocite, 330
siderite, 331
vivianite, 331
magnetic sector, 98
MALDI (matrix-assisted laser desorption/ionization), 283, 284, 356
mass bias, 100, 103
mass discrimination effect. *See* mass bias
mass resolution, 96, 97–99, 100
mass spectrometry, 280, 290
gas chromatography (GC-MS), 344
inductively coupled plasma (ICP-MS), 17, 95
and pXRD, 229
liquid chromatography/electrospray ionization (LC/ESI/MS), 46, 345
liquid chromatography-mass spectrometry (LC-MS), 344, 350
MC-ICP-MS (multi-collector inductively coupled plasma), 95–111
ToF-SIMS (time-of-flight secondary ion), 283, 284
massively parallel sequencing. *See* next-generation sequencing (NGS)
MATLAB, 85, 139
matrix, 99, 100, 104, 196, 249
FTIR spectroscopy, 290
nucleic acid sample, 367
spectrum of, ATR FTIR, 304
MC-ICP-MS (multi-collector inductively coupled plasma mass spectrometry. *See* mass spectrometry
MCR-ALS (multivariate curve resolution – alternating least squares), 306–307
media, 26–27
antibiotic-containing, amplicon cloning, 376
biomass preparation, 66
liquid, phosphate in, 331
Luria-Bertani, 27
lysogeny broth, 81
noble agar, 175
nutrient, 27
plate count, 30
resin. *See* resins
solid agar, 27
sterilization, 27
tryptic soy broth, 81
XPS, 282
metabolism, 39–41, 276, 342, 354, *See also* lipidomics
bioassays of, 39–41
metagenomics, 46, 50–53, 361, 363, 390
metals, 79, 87
accumulation, 267
complexation, 64, 68–69, 71, 73
calorimetric data, 69
isotopic analysis, 93–111
metaproteomics, 390

metatranscriptomics, 46, 53, 390, *See also* metagenomics
microarray technology, 189, 208
microbial attachment, 63
microbial communities, 3, 354
  biomacromolecule analysis, 42
  characterization, 363–366
  culture-independent methods, 41
  FISH, 209
  iron in, 106
  lipid biomarker analysis, 341–356
  low biomass, 44
  maintaining in field sample, 7
  nucleic acid analysis, 46
    functional gene microarrays, 46
    metagenomics, 50
    metatranscriptomics, 53
    qPCR, 47
  physiological measurements, 36–41
  pXRD, 215, 217, 229
  structure, 342, 351, 365, 374–380
microbial membranes, 341
microbialites, 173
microfossils, 156, 181, 184
micrographs, 149, 154, 156, 161, 166, 167, 170, 171
  whole-mount samples, 151
microscopy. *See also entries for specific techniques*
  history of, 121–122
  microscopic count, 30
    hemocytometer, 35–36
microspectroscopy, 288, 295, *See also* FTIR spectroscopy; Raman spectroscopy
mid-infrared (mid-IR) range, 289, 301
mineral identification, 182, 288, 319, 324–326
mineral precipitation, 4, 156, 190, 267, 278, *See also* biomineral precipitation
Mineral Spectroscopy Database, 325
molar absorption coefficient, 292
molecular vibration, 289, 292
monochromatic X-ray beams, 246
monochromator, 18, 251
monounsaturated fatty acids (MUFAs), 341, 349
Mössbauer effect, 314–333, *See also* Mössbauer spectroscopy
Mössbauer spectroscopy, 11, 314–333
Mösstool, 325
mothur, 380–387, 388, 389, 390
multivariate analysis, tools, 270–272, 294, 297, 306–307, 355, 375, 390

NaI phosphor, 219
nanoindentation, 136
nanometre-scale analysis, 149, 150, 154, 155, 156, 161
nebulizer, 97
next-generation sequencing (NGS), 46, 51, 190, 374, 377–380, 384
  statistical analysis, 388–390
nitrogen, 4, 157, 251
  ATR, 299
  colorimetric assay, 10
  ferrozine method, 11
  fixation, 95
  isothermal titration calorimetry, 67
  potentiometric titration, 82
  total, 21
Nitrosomona europaea, 73
Nitrosopumilis maritimus, 71
normalization, 293, 304–305
nucleic acids, 46–53, *See also* gene probing; metagenomics; PCR
  amplification, 369–373
  biomarkers, 361
  culture-independent methods, 41
  DNA (deoxyribonucleic acid), 7, 42–45, 47, 341
  extraction, 50, 367–368, 380
  growth measurement, 30
  phylogenetic techniques, 361–391
  precipitation, 368
  purification, 44, 50, 367
  quantification, 368
  RNA (ribonucleic acid), 7, 42–45
    ribosomal (rRNA), 51, 188, 196, 202, 203–204
    small subunit (SSU), 361–363, 369–371, 373, 374, 375, 376, 379, 380, 381, 382, 383, 386, 387, 388, 391
  sample collection, preservation, 6, 7, 8, 366
  stains. *See* stains, dyes
nutrients, 4, 29, 31, 63, 79, 196

oligotrophic ecosystems, 107, 188, 201
operational taxonomic unit (OTU), 384–387, 388
OPLS-DA (orthogonal projections to latent structures discriminant analysis), 306
optical density, 29, 81
optical microscopy, 142, 279
organic acids, 73
organic matter, 108, 110, 267, 268, 300, 367
oscillations, 127–130, 243–244, 252
osmium tetroxide, 191
osmium textroxide, 181–184
oxidation
  ammonia, ARISA, 374
  ATP, 41
  chromium, 12
  during drying, 5
  elemental analysis, 23
  ferrous iron, 10
  green rust, 331
  iron, 106, 107, 322–323
  lipids, 355
  metal ions, 75
  methane, ARISA, 374
  Mössbauer spectroscopy, 320, 321, 322–323
  persulfate, 20

quadrupole splitting, 319
sample degradation, 7
solution-phase oxygen, 20
state, 242, 264, 268, 317, 319
XPS, 276
oxygen, 157
oxyhydroxides, 95, 106, 110–111, 181, 324, 329, 330

paraformaldehyde (PFA), 191
fixation, FISH, 192–194
parasitic mode, 239
particle size, analysis, 15, 227, 232–234, 249, 297, 317, 319, 322, 329, 331, 333
path length, 295, 301
pattern fitting analysis, 227, 228, 231
PCR (polymerase chain reaction), 42, 208, 366, 369–373, 374, 375, 377, 379, 380
cycling, 47, 371–372
inhibition, 50, 367, 368, 371
qPCR (quantitative PCR), 29, 30, 46, 47–50, 208, 373
RT-PCR (reverse transcription), 373
peak positions, 221, 226, 228–229, 277
Peltier cooling, 220
peptides, 6, 24, 46, 268, 270, 277, 283
petrographic samples, 156–157
PFA (paraformaldehyde). *See* paraformaldehyde
phosphates, 73, 74, 75, 291, 331
chromatography, 24
colorimetric assay, 13–15
phospholipids, 24, 45, 341, *See also* intact phospholipid profiling
fatty acids (PLFA), 341–356
phosphoryl groups, 89
photoelectric effect, 216, 262
photoelectrons, 241, 243–246, 262–264, 269
photoionization, 25, 241
photomultiplier, 18, 219
phylogenetic diversity, 204, 362
phylogenetic methods, techniques, 361–391
phylogenetic placement, 384, 387
phyloseq, 387, 389
phytoplankton, 106
Picante, 389
piezoelectric materials, 123, 124, 129, 135
piezo creep, 135
pKa, 73–74, 75, 82, 83, 85, 87–88
planar orientation, 173
plate count, 29
PLFA (phospholipid fatty acids). *See* phospholipids
point of zero charge, 84
polished blocks, 156, 158–161, 162
polyimide film, 249, 250
pore water, 100, 107–111
powder diffract file, 228
principal component analysis (PCA), 254, 306, 355
proteins, 42, 45–46, 268, 269
biomarkers, 361
culture-independent methods, 41
denaturation, FISH, 191
dyes, 34
EXAFS, 239
FTIR spectroscopy, 291
ion/ion exchange chromatography, 15
sample collection, 6, 7
SEM, 151
TEM, 168
ProtoFit, 85
proton-active sites, 83
protonation, 67–68, 71, 264
amines, 74, 266
calorimetric data, 69
carboxylic acids, 73
enthalpy, 73–74
functional groups, 64, 275
hydroxyl groups, 74, 291
models, 83, 86, 87–88, 89
Nitrosomona europaea, 73
phosphates, 74
thiols, 74
pXRD (powder X-ray diffraction). *See* X-ray diffraction

Q factor, 131, 138
QIIME (Quantitative Insights Into Microbial Ecology), 380–387, 388, 389, 390
qPCR (quantitative PCR)/real-time PCR. *See* PCR
quadratic polynomial, 254
quadrupole shift, 319
ferrihydrite, 327
hematite, 329
quadrupole splitting (ΔE_Q), 317, 318, 319, 321–322, 323
goethite, 327
lepidocrocite, 330
siderite, 331
quantum mechanical wave, 244
quantum mechanics, 289
quantum tunnelling, 123

radial distribution, 245
radiation damage, 253, 256
Raman spectroscopy, 142, 189, 208, 289, 290
microspectroscopy, 290
Rayleigh criterion, 302
recoil-free fraction (*f*-fraction), 315, 329
redox
cycling, 9, 283
reactions, 93, 242, 243, 246, 250
-sensitive elements, 95, 106, 107, 253, 267, 323
state, 243
reduction
iron, 106
non-reducing reagent, 10

reflection mode
  FTIR microspectroscopy, 300
  pXRD, 221
reflections, 296–300
  attenuated total reflectance (ATR), 296, 298–300
    FTIR microspectroscopy, 300
  diffuse reflectance infrared Fourier transform spectroscopy (DRIFTS), 296–297, 299
  grazing angle reflectance, 296, 297–298
    FTIR microspectroscopy, 300
refractive index, 25, 298, 300
resins, 181
  anion ion chromatography, 15
  anion-exchange, 100, 107
  cation-exchange, 100
  embedding, 177–178
  Epon epoxy, 159
  epoxy, 157
  London Resin (LR) White, 159, 176
  Procure 812, 176–178
  ultra-thin sections, 173
resonant emission, absorption, 314–317
Reynolds number, 138
rhizosphere, 3
ribonuclease (RNase), 6
  decontamination, 42
  RNA degradation, 42, 43
  sample degradation, 7
Ribosomal Database Project Tools (RDPTools), 384
RNA (ribonucleic acid). *See* nucleic acids

SAED (selected area electron diffraction), 173
sample collection, 3–8
  FISH, 190–191
  phylogenetic methods, 366–367
Sanger sequencing, 376–377, 380, 381
scanning probe microscopy (SPM), 122–143
scanning tunnelling microscopy (STM), 34, 122, 123
scattering, 29, 216, 217
  atomic force microscopy, 136
  back-, EXAFS, 244
  background, pXRD, 227
  DRIFTS, 296
  EXAFS, 243–246
  FTIR (micro)spectroscopy, 303
  multiple-, EXAFS, 244, 246
  pXRD, 221, 222, 223, 233
  solid-state sample, 227
  X-ray spectroscopy, 241, 252
scintillation
  counter, liquid, 38
  detectors, 219
  vials, 152, 176, 177
sediments, 89, 100, 289, 366
  DRIFTS, 297
  extraction of biomacromolecules, 43
  FISH, 202
    CARD-FISH, 202
  inoculation, 27
  iron isotopes, 107, 109
  lithium isotopes, 93
  MC-ICP-MS, 107–108
  molybdenum isotopes, 95
  Mössbauer spectroscopy, 322
  nanogoethite, 327
  nucleic acid extraction, 367, 373
  permafrost, 6
self-absorption, 252
SEM (scanning electron microscopy). *See* electron microscopy
sequencing, nucleic acids, 361
sequential extraction, 109
setpoint, 123, 132
sextets, 319
  goethite, 327
  hematite, 329
  magnetite, 328
signature lipid biomarker (SLB), 342
soils, 89, 107, 366
  atomic absorption spectrometry, 18
  ATR, 300
  cell counting, 36
  chromium contamination, 12
  contaminant transformation, 239
  DRIFTS, 297
  extraction of biomacromolecules, 43
  FTIR (micro)spectroscopy, 289, 290, 300
  inoculation, 27
  ion/ion exchange chromatography, 17
  metals in, 18
  Mössbauer spectroscopy, 322
  nucleic acid extraction, 367
  omics approaches, 391
  plate count, 30
  respiration measurement, 39
  sample collection, 3, 4
solid-phase extraction (SPE), 43, 346, 355
solid-state materials, 215, 234, *See also* pXRD
  detectors, 220
solvent extraction, 355
sonication, 157, 367
  ultra-, 43, 45
space-charge effect, 100, 103
spatial resolution, 121, 239, 247, 300
spectral comparison, 291–294
spectral subtraction, 304
spectroscopic techniques. *See entries for specific techniques*
SSU rRNA (small subunit ribosomal RNA molecules). *See* nucleic acids
stability constant, 64, 65
stains, dyes, 33, 84

biomass, 66
DNA/RNA, fluorometry, 368
Gram. *See* Gram stain
heavy metals, 181–184
nucleic acids, 34, 36, 368
oligonucleotide FISH, 197, 200
osmium textroxide, 181–184
uranyl acetate, 181–184
standard solutions
chromatography, 24
ferrozine method, 12
MC-ICP-MS, 99, 104, 105
total carbon, inorganic carbon, 21
sterilization. *See also* alcohol
inoculation, 27
isolation, 28
media, 27
PCR assays, 42
sample collection, preservation, 4, 5–6
streak, 28
storage ring, 239
streak, 26, 27–28
structural characterization, 149, 151, 161, 182, 184, 239, 288
sulfhydryl groups, 89
sulphides, 95, 157, 176, 267, 279
surface charge, 83, 86, 266, 275–277
surface complexes, 63–76
surface complexation models, 63–76, 80, 87–89
surface potential, 266
suspension
bacterial, 66, 68
biomass, 65, 69, 79
cell, 28, 85
contamination, 67
culture, 29
environmental sample, 40
fast-freezing, 267
microbial, 68
purging, 82
soil, 30
specimen, microscopy, 33, 35
synchrotron, 238–256, 264, 284, *See also* FTIR spectroscopy; X-ray absorption spectroscopy; XRF microscopy; XPS
Synechococcus, 81, 83, 87

TEM (transmission electron microscopy). *See* electron microscopy
temperature
blocking (T_B), 319, 325, 327, 328
control, atomic force microscopy, 135
Curie, 329
DNA/RNA storage, 366
enrichment, 27
fixation, 151
high-temperature environments, 93, 95
low-temperature environments, 93, 95, 215, 217
Morin, 329
Mössbauer spectroscopy, 315, 321
PCR, 371–372
room, XPS, 284
sample collection, preservation, 4, 5, 7, 8
-sensitive media, 27
synchrotron experiments, 253
terminal restriction fragment length polymorphism (TRFLP), 46, 375
thermal drift, 135
thermodynamics
equilibrium, 86
speciation models, 250
surface complexation, 63
XPS, 278
thin sections, 168, 295
petrographic, polished, 156–157
ultra-, 161, 162, 168, 170, 173–181
thiols, 73, 74, 75, 76, 293
thumbscrews, 132
TIMS (thermal ionization mass spectrometry), 95–96
tip convolution, 133
titration
isothermal titration calorimetry. *See* calorimetry
potentiometric, 63, 64, 67, 74, 79–89
sample preparation, 80–81
titrants
calorimetric data, 69
isothermal titration calorimetry, 65, 66, 67, 68, 69, 71
potentiometric titration, 83
purging, 82
ToF-SIMS (time-of-flight secondary ion mass spectrometry). *See* mass spectrometry
topography, 123, 126, 154
torsional displacement, 126
TRAMPR, 375
transmission, 98, 100, 103
transmission mode, 249
EXAFS, 249
FTIR (micro)spectroscopy, 294–295, 301
Mössbauer spectroscopy, 320, 321
pXRD, 221
transmittance, 8, 289
T-REX, 375
TRFLP (terminal restriction fragment length polymorphism analysis), 374
triple layer model, 87
tunnelling current, 122, 124
two-pass technique, 127

U L_{III} shell electrons, 241
ultramicrotome, 162, 173, 175, 177–179
unit cell determination, parameters, 215, 230–232
uranyl acetate, 181–184
USEARCH, 380–387

UV (ultraviolet)
  cleaner, 130
  exposure, sample degradation, 7
  light sterilization, 42
  light vs. glow discharge system, 170

vacuum, 218, 241, 246, 280
  gap, 122
  high, 122, 149, 166
  low, 149
  ultra-high, 264, 266, 282, 284
  XPS, 282
valence, 17, 242–243, 252, 254, 264
van der Waals interactions, 125, 126, 140
Verwey transition, 328
vibrational modes, 291

wavelength, 17, 29, 216, 217, 218, 299, 300, 302
  coherent, 216
  incoherent, 216
wavenumbers, 289
wet sample analysis, 8, 221, 264
  ATR, 299
  ATR FTIR spectroscopy, 304
Woese tree, 361

XANES (X-ray absorption near-edge structure) spectroscopy. *See* X-ray spectroscopy
XPS (X-ray photoelectron spectroscopy). *See* X-ray photoelectron spectroscopy
X-ray absorption coefficient, spectra. *See* absorbance/absorption
X-ray absorption spectroscopy, 11, 17, 73, 238–256, 290
X-ray degradation, 281, 284
X-ray diffraction (XRD), 216, 324, 330
  powder (pXRD), 215, 217–234
    preferred orientation, 225
    random orientation, 224
    samples
      random orientation, 222
    slurry mount, 223, 225, 228
    top-loaded mount, 223, 225, 228
X-ray fluorescence (XRF)
  microscopy, synchrotron, 247
  photon, 246–247
  spectroscopy, 17
X-ray photoelectron spectroscopy (XPS), 262–284
  cryogenic, 264–266, 271, 272, 275, 277, 279–284
  synchrotron, 264
X-ray photons, 216, 219
X-ray spectroscopy
  extended X-ray absorption fine structure (EXAFS), 238–240, 241, 243–246, 249, 250, 252, 253, 254–256, 277, 281
  X-ray absorption near-edge structure (XANES), 238–240, 241, 242–243, 249, 252, 253, 254–256

zeta potential, 83, 275